Lecture Notes in Geoinformation and Cartography

Series Editors: William Cartwright, Georg Gartner, Liqiu Meng,
Michael P. Peterson

Christopher Pettit · William Cartwright ·
Ian Bishop · Kim Lowell · David Pullar ·
David Duncan (Eds.)

Landscape Analysis and Visualisation

Spatial Models for Natural Resource Management and Planning

Editors

Dr. Christopher Pettit
Department of Primary Industries
P.O. Box 4166, Parkville
Victoria 3052
Australia
Christopher.Pettit@dpi.vic.gov.au

Prof. William Cartwright
RMIT University
School of Mathematical
and Geospatial Sciences
GPO Box 2476V
Melbourne VIC 3001
Australia
william.cartwright@rmit.edu.au

Prof. Ian Bishop
Department of Geomatics
University of Melbourne
Grattan St, Melbourne
Victoria 3010
Australia
i.bishop@unimelb.edu.au

Prof. Kim Lowell
Department of Primary Industries
P.O. Box 4166
Parkville
Victoria 3052
Australia
kim.lowell@dpi.vic.gov.au

Dr. David Pullar
The University of Queensland
St Lucia
Brisbane 4067
Australia
d.pullar@uq.edu.au

Dr. David Duncan
Department of Sustainability
and Environment
123 Brown St
Heidelberg
Victoria 3054
Australia
david.duncan@dse.vic.gov.au

ISBN 978-3-540-69167-9   ISBN 978-3-540-69168-6 (eBook)

Lecture Notes in Geoinformation and Cartography ISSN: 1863-2246

Library of Congress Control Number: 2008928504

© 2008 Springer-Verlag Berlin Heidelberg

This work is subject to copyright. All rights are reserved, whether the whole or part of the material is concerned, specifically the rights of translation, reprinting, reuse of illustrations, recitation, broadcasting, reproduction on microfilm or in any other way, and storage in data banks. Duplication of this publication or parts thereof is permitted only under the provisions of the German Copyright Law of September 9, 1965, in its current version, and permission for use must always be obtained from Springer. Violations are liable to prosecution under the German Copyright Law.

The use of general descriptive names, registered names, trademarks, etc. in this publication does not imply, even in the absence of a specific statement, that such names are exempt from the relevant protective laws and regulations and therefore free for general use.

Typesetting: Camera-ready by the editors

Cover design: deblik, Berlin

Printed on acid-free paper

9 8 7 6 5 4 3 2 1

springer.com

Preface

Michael Batty

Centre for Advanced Spatial Analysis, University College London

Landscapes, like cities, cut across disciplines and professions. This makes it especially difficult to provide an overall sense of how landscapes should be studied and researched. Ecology, aesthetics, economy and sociology combine with physiognomy and deep physical structure to confuse our understanding and the way we should react to the problems and potentials of landscapes.

Nowhere are these dilemmas and paradoxes so clearly highlighted as in Australia — where landscapes dominate and their relationship to cities is so fragile, yet so important to the sustainability of an entire nation, if not planet. This book presents a unique collection and synthesis of many of these perspectives — perhaps it could only be produced in a land urbanised in the tiniest of pockets, and yet so daunting with respect to the way non-populated landscapes dwarf its cities. Many travel to Australia to its cities and never see the landscapes — but it is these that give the country its power and imagery. It is the landscapes that so impress on us the need to consider how our intervention, through activities ranging from resource exploitation and settled agriculture to climate change, poses one of the greatest crises facing the modern world. In this sense, Australia and its landscape provide a mirror through which we can glimpse the extent to which our intervention in the world threatens its very existence.

The team of editors have assembled an intriguing and far-reaching set of contributions which largely emanate from the study of landscape as it affects south-eastern, western and coastal Australia. They display this on a canvas that is at least as large as any which has been used hitherto to attempt a synthesis of landscape problems and the way we might approach them. Most of the chapters deal with large scale landscapes, although this does not mean that the editors or the authors are unaware of scale. In fact, the manner in which scale weaves its way through the various contributions, and the way different approaches are affected by scale is something that an astute reader can easily recognise.

Although these contributions were originally presented under the badge of the 'Place and Purpose' conference (which focused on spatial models for natural resource management and planning), as the title of this book implies, its key focus is on the analysis of landscapes and their visualisation. The contributions follow a classic organisation into sections on natural resources, land cover, ecology, and social infrastructure linked through questions of planning and management. Visualisation weaves its way through the manner in which landscapes are represented and analysed, the final section dealing explicitly with new ways of characterising and disseminating what we know about landscape through computers and the World Wide Web.

In fact representation and management are twin themes that dominate all the chapters presented here, with systems analysis and decision analysis linking the diverse contributions together. Although modelling which is accepted as being computational and statistical, is central to many of the papers, the presentation of formal models is rather low key in that the focus is much more on digital representation and ways of using associated software, typically GIS (geographic information systems), to generate landscape analysis in a media that is both visual and communicable using the latest information technologies. Modelling per se is very much subsumed here in terms of representation on the one hand and decision analysis on the other, while the management of landscape features strongly in many contributions, consistent with our increasing concern for sustainability of resources in the face of climate change and capital exploitation.

There is both an implicit and explicit sense in this book that landscapes need to be captured digitally and represented in some neutral manner prior to analysis and management. There is recognition too that science is not the only vehicle that will allow best management, but that the diversity of communities who have vested interests and views about landscapes must be included in the processes whereby such landscapes might be best managed. This of course is public participation by any other name but it is wider than this. The confluence of physical analysis concerning the sustainability of landscapes with social infrastructure, which in turn sustains their exploitation and management, must be set against the increasing challenges which link rural to urban and first world to third, thus complicating this nexus in ways that we are only beginning to realise. The contributions here present a marvellous array of this complexity, while the organisation of the book and the dedication of the contributing scholars and practitioners show just how important it is to provide a synthetic sense of what sustainable landscapes are all about. The range of models, from catchment hydrology through to decision analysis, indicate just how difficult a real synthesis of the problems of Australian landscapes in particular, and landscapes in general, really are. Themes that are captured and drive the contributions include: the evidence base, questions of uncertainty, adaptive management, the role of community in the process of landscape change, and the management and looming problem of climate change. Readers of this book, however, will need to reflect on these themes for they are manifested in a diversity of contributions which means that the arguments made here need to be understood on many levels. The pages that follow thus offer continued insights.

Digital representation and visualisation is central to these ideas and provide a useful synthesis of representation with dissemination. Since the rise of the World Wide Web (the visual interface to the Internet), desktop representation which became visual with the advent of the microcomputer, has moved quickly into a form that is widely available to anyone who has access to the Internet. These technologies are currently being refashioned in an even more interactive manner so that users can now generate their own interpretations and manipulate digital content directly through the World Wide Web by interacting with others. These technologies are increasingly referred to as Web 2.0, and in the last part of this book, exciting developments in the representation and communication of ideas about landscape are presented that inform us just how far these technologies have developed. Digital globes and games are beginning to dominate the way we can visualise spatial representations, providing us with new ways of interacting with one another through these media. Important contributions are presented by some of the key people making advances in this area, and it is fitting that the cutting edge of landscape digital representations are focussed on research and practice in Australia rather than in other parts of the world. This book provides a real sense of what is being contributed to the study of landscape from this continent using these new forms of synthesis.

A preface to a book should in one sense point the way and inform the reader how to navigate what is inevitably difficult but exciting terrain. To do this, I will select some striking themes that I urge readers to grapple with in the chapters that follow. The editors do, of course, guide readers in a more focussed way at the beginning of the book, but it is worth noting important contributions that follow.

In the first section on resource management, the focus is very much on physical representation but also on key issues of generating sustainable landscapes through the use of tools that help to bridge the gap between expert professionals and those charged with decision making. Adaptation, the use of evidence-based policy and the diversity of views about landscapes by different communities of interest are linked together through the use of tools such as catchment analysis, multi-criteria analysis and common representational infrastructures, thus impressing on the reader how landscapes always need to be analysed in ways that link science to policy.

The second part of the book deals with analytical techniques and models, mainly focusing on land cover and land use but with a strong emphasis on using models to generate alternative futures. Techniques such as cellular development of landscapes, and the way uncertainty is captured in these models are introduced, while the notion of generating an array of scenarios defining the decision or policy space in which landscape can develop is broached directly — again impressing the theme that good representation and analysis must be linked to policy through modelling and simulation.

The third part of the book then deals with ecological modelling, in particular animal habitats and vegetation, again in the context of development scenarios for the design and management of different futures. The chapter on genetic markers and the evolution of landscapes, for example, is typical of how ideas in this book make us think about the correctness of the scales that we are dealing with and the way in which different scales imply different aspects of the complexities that make up landscapes.

Social and economic conditions that interact with the physical and visual form of the landscape are presented in the fourth part. These too show another aspect of this complexity that needs to be handled where explicit techniques involving the community and experts can be used in understanding how the physical and the social interact.

All of this culminates in the last section which is about how landscapes can be represented digitally and communicated using new computer and communications technologies. This is a fitting conclusion to the book for it impresses on the reader just how extensive the study of landscape is and just how important it is to bring different interests together in communicating how they can be managed. In this, globes, games and virtual worlds have a key part to play.

There is much excitement in the pages that follow and the book can be read on many levels. The editors and the contributors have done us all a great service in providing as integrated a set of ideas as one might find in the study and management of landscape. This is a fitting contribution to a generation of research and practice fashioned in a context that provides many exemplars for others to follow.

Contents

Contributors		XXI
Abbreviations		XXXI

Introduction   **1**

1   Understanding Landscapes through Knowledge Management
Frameworks, Spatial Models, Decision Support Tools and
Visualisation   3
   1.1   Introduction   3
   1.2   Part 1: Natural Resource Knowledge
Management Frameworks and Tools   5
   1.3   Part 2: Integrating the Ecology of Landscapes
into Landscape Analysis and Visualisation   7
   1.4   Part 3: Socioeconomic Dimensions to Landscapes   9
   1.5   Part 4: Land Use Change and Scenario Modelling   11
   1.6   Part 5: Landscape Visualisation   13
   1.7   Future Challenges   15

**Part 1:   Natural Resource Knowledge Management
Frameworks and Tools**   **17**

2   Reading between the Lines:
Knowledge for Natural Resource Management   19
   2.1   Introduction   19
   2.2   Knowledge Hierarchy   20
   2.3   Timelag between Question and Answer   23
   2.4   Organising the Questions   24
   2.5   Integrating Disciplines   26
   2.6   Conclusion   27

| X | Landscape Analysis and Visualisation |

3	Improving the Use of Science in Evidence-based Policy:	
	Some Victorian Experiences in Natural Resource Management	29
	3.1	Context	29
		3.1.1	Historical Perspective	30
		3.1.2	The Policy Process: Towards Evidence-based Policy	31
		3.1.3	Use of Science as Evidence in Policy	32
	3.2	Some Victorian Experiences in	
		Natural Resource Management	35
		3.2.1	Survey of Policy Analysts	37
		3.2.2	Market Research	38
		3.2.3	Improving the Utility of Project Outputs	40
		3.2.4	Observation of How Policy Decisions Are Made	40
	3.3	Case Studies of Successful Science–Policy Influence	41
		3.3.1	Sawlogs for Salinity	42
		3.3.2	Salinity Investment Framework 3	42
		3.3.3	Soil Health	43
		3.3.4	Greenhouse in Agriculture	43
	3.4	Discussion	44
		3.4.1	Toward Better Use of Science in	
			Evidence-based Policy	44
	3.5	Conclusion	46

4	The *Catchment Analysis Tool*: Demonstrating the	
	Benefits of Interconnected Biophysical Models	49
	4.1	Introduction	50
	4.2	*Catchment Analysis Tool*: Background and Description	51
		4.2.1	The *CAT* Interface	54
		4.2.2	*CAT* Input Data	56
		4.2.3	The *CAT* Model Components	59
		4.2.4	Model Calibration and Conceptualisation	61
	4.3	Case Study	61
		4.3.1	Hypothetical Case Study	61
		4.3.2	Results and analysis	66
	4.4	Validation and Model Improvement	68
	4.5	Conclusion	69

5	The Application of a Simple Spatial Multi-Criteria Analysis	
	Shell to Natural Resource Management Decision Making	73
	5.1	Introduction	74
	5.2	Multi-criteria Analysis	74
		5.2.1	Spatial Applications	75
		5.2.2	The Decision-making Process	77

5.3	The *MCAS-S* Approach		79
	5.3.1	Design Principles	79
	5.3.2	Key Functions	80
5.4	Applications		82
	5.4.1	Prioritising Revegetation Investment	82
	5.4.2	Assessing the Sustainability of Extensive Grazing	85
5.5	Future Trends		89
5.6	Conclusion		90
5.7	Future Research Directions		91

6 *Platform for Environmental Modelling Support*:
a Grid Cell Data Infrastructure for Modellers — 97

6.1	Introduction		98
6.2	Background		100
6.3	Methodology		102
6.4	Progress and Discussions		103
6.5	The *PEMS* Demonstrator Project		105
	6.5.1	National Seasonal Crop Monitoring and Forecasting	105
	6.5.2	Develop and Demonstrate a Market-based Approach to Environmental Policy on Private Land	108
	6.5.3	Wildfire Planning: Consequence of Loss Modelling	109
	6.5.4	Land Use Data, Modelling and Reporting	111
6.6	Conclusion		115

Part 2:   Integrating the Ecology of Landscapes into
Landscape Analysis and Visualisation — **119**

7 Looking at Landscapes for Biodiversity: Whose View Will Do? — 121

7.1	Introduction		122
7.2	To be Human is to Err		122
7.3	What's Good for the Goose?		124
7.4	Consider the Lilies		127
7.5	Best is Bunkum		128
7.6	Varied Perspectives		129
	7.6.1	Mapping and Modelling Terrain, Hydrological, Pedological and Geological Features and Climate	129
	7.6.2	Vegetation Mapping Using Remotely Sensed Data, Including Vegetation Condition and Temporal Variability	130
	7.6.3	Mapping and Modelling Movement	131
	7.6.4	Integrating Multiple Perspectives	133
7.7	Conclusion		135

8		Native Vegetation Condition: Site to Regional Assessments	139
	8.1	Introduction	140
	8.2	Measuring Vegetation Condition at Sites	141
	8.3	Measuring Vegetation Condition across Regions	142
	8.4	Case Study: Vegetation Condition in the Murray Catchment, New South Wales	143
		8.4.1 Study Area	143
		8.4.2 Site Data Collection	144
		8.4.3 Modelling from the Site to the Region	146
	8.5	Results and Discussion for the Murray Catchment Case Study	149
	8.6	Conclusion	152
	8.7	Future Research Directions	153
9		Towards Adaptive Management of Native Vegetation in Regional Landscapes	159
	9.1	Introduction	159
	9.2	What Adaptive Management is and is not	161
		9.2.1 Step i: Statement of Objectives, Constraints and Performance Measures	163
		9.2.2 Step ii: Specification of Management Options	164
		9.2.3 Step iii: System Modelling and Model Credibility	165
		9.2.4 Step iv: Allocation, implementation and Monitoring — Closing the Loop	165
	9.3	Managing and Monitoring Native Vegetation	167
		9.3.1 An Example of a Formal Approach to Adaptive Management of Vegetation Condition	169
	9.4	Research	175
	9.5	Conclusion	176
	9.6	Future Directions	177
		Appendix	181
10		Revegetation and the Significance of Timelags in Provision of Habitat Resources for Birds	183
	10.1	Introduction	184
	10.2	Methodology	186
		10.2.1 Model Description	186
	10.3	Case Study	191
		10.3.1 Results	192
		10.3.2 Discussion	197
	10.4	Caveats and Extensions	199
		Appendices	204

11	The Application of Genetic Markers to Landscape Management		211
	11.1	Introduction	212
		11.1.1 The Need for Information on How Biota Occupies and Moves through Landscapes	212
		11.1.2 A Spectrum of 'Genetics' in Landscape Management and Planning	213
		11.1.3 Molecular Population Biology Supplies Information Essential for Landscape Planning and Management	213
	11.2	Background	215
		11.2.1 Three Levels of Analysis Assess Three Levels in Time and Space	215
		11.2.2 Main Molecular Tools in Landscape Molecular Population Biology	217
	11.3	Case Studies	220
		11.3.1 Impacts of Habitat Fragmentation on Cunningham's Skinks	220
		11.3.2 Dispersal and Gene Flow of Greater Gliders through Forest Fragmented by Pine Plantation	221
		11.3.3 Catchments Catch All: Congruent Patterns in Diverse Invertebrate Fauna in Decaying Wood at a Landscape Scale	222
	11.4	Future Trends	223
	11.5	Conclusion	225
	11.6	Future Research Directions	225
		Appendix	231
12	Scenario Analysis with Performance Indicators: a Case Study for Forest Linkage Restoration		235
	12.1	Introduction	236
	12.2	Background	237
	12.3	Linkage restoration	239
		12.3.1 Indicator Rule 1: Site Recovery Capacity	240
		12.3.2 Indicator Rule 2: Site Biodiversity Value	241
		12.3.3 Indicator Rule 3: Landscape Linkage Qualities	242
		12.3.4 Indicator Rule 4: Landscape Connectivity	242
	12.4	Atherton Tablelands Case Study	243
		12.4.1 Restoration scenarios	245
		12.4.2 Scenario Evaluation	246
	12.5	Conclusion	247

XIV   Landscape Analysis and Visualisation

Part 3:   Socioeconomic Dimensions to Landscapes   251

13 Strategic Spatial Governance: Deriving Social–Ecological
    Frameworks for Managing Landscapes and Regions   253
    13.1 Introduction   254
    13.2 A Potted History of Catchments for Resource Governance   254
    13.3 Defining Regions for Resource Governance   256
          13.3.1 Principle 1   256
          13.3.2 Principle 2   257
          13.3.3 Principle 3   259
    13.4 Application of Principles to Spatial Analysis   259
          13.4.1 Delineating Civic Regions from a Social Surface   260
          13.4.2 Deriving a Hierarchy of Civic Regions   262
          13.4.3 Deriving Ecoregions   264
          13.4.4 Integrating Ecoregions and Civic Regions
                  through Boundary Optimisation   265
          13.4.5 Comparing the Performance of Regions   266
    13.5 Conclusion: Past, Present and Future Resource Governance   269
    13.6 Future Directions   270

14 Placing People at the Centre of Landscape Assessment   277
    14.1 Introduction   277
    14.2 Background   278
    14.3 Methodology   279
          14.3.1 Pressure–State–Response Model   279
          14.3.2 Driving Forces–Pressure–State–Impact–Response
                  Model   281
          14.3.3 Millennium Ecosystem Assessment Framework   281
          14.3.4 Indicator Selection   282
    14.4 A Landscape Approach for Victoria   283
          14.4.1 Definitions of Five Victoria Landscapes   284
          14.4.2 The Role of Indicators   285
    14.5 Case Study 1: Semi-arid Landscape   285
          14.5.1 Overview   286
          14.5.2 Employment Indicator   288
          14.5.3 Index of Stream Condition Indicator   290
          14.5.4 Land Use Diversity Indicator   291
          14.5.5 Management Response   293
    14.6 Case Study 2: Coastal Landscape   293
          14.6.1 Overview   294
          14.6.2 Visitors to Parks and Reserves Indicator   295
          14.6.3 Ratio of Land Value to Production Value Indicator   296

	14.6.4	Land Use Diversity Indicator	297
	14.6.5	Policy Response	298
14.7		Overview of Results	299
14.8		Conclusion	299
14.9		Future Research Directions	300

15		**The Social Landscapes of Rural Victoria**	**305**
	15.1	Introduction	305
	15.2	A Narrative of Rural Transformation in Australia	306
		15.2.1 International Agricultural Competition	306
		15.2.2 Agricultural Restructuring	307
		15.2.3 Amenity Values in the Rural Land Market	307
		15.2.4 Indicators Derived from the Narrative	308
	15.3	From Indicators to Social Landscapes	310
		15.3.1 Factor Analysis Using the Principal Components Method	310
		15.3.2 Creating a Geography of Amenity and Intensification	314
	15.4	Five Social Landscapes	315
		15.4.1 The Production Landscape	316
		15.4.2 The Transitional Landscape	317
		15.4.3 The Amenity Farming Landscape	318
		15.4.4 The High Amenity Landscape	319
		15.4.5 The Intensive Agriculture Landscape	319
	15.5	Conclusion	322
	15.6	Future Research Directions	323

16		**A Decision Aiding System for Predicting People's Scenario Preferences**	**327**
	16.1	Introduction	327
	16.2	Background	328
	16.3	An Extra Step for the SDSS Discipline	329
	16.4	Description of the *Preference Prediction* Software	331
		16.4.1 Finding a Larger Set of Criteria	331
		16.4.2 Finding Relationships between Criterion Scores and Overall Scenario Merit	331
		16.4.3 The Underlying Assumption	333
	16.5	An Urban Planning Case Study Application of the *Preference Prediction* Software	334
		16.5.1 Assigning Criteria Scores to the Scenarios	335
		16.5.2 Predicting Scenario Ratings for Overall Merit	336

XVI   Landscape Analysis and Visualisation

	16.5.3	Checking the Personal Characteristics of the Advisors	338
	16.5.4	Predicting Scenario Merit Ratings on Behalf of Past Workshops	338
	16.5.5	Exploring How Scenario Ratings Were Derived	339
	16.5.6	Searching for Reasons behind Each Scenario Merit Rating	342
	16.5.7	Predicting All Groups' Preferences Simultaneously	345
16.6	Future Trends		347
16.7	Conclusion		347
16.8	Future Research Directions		348

Part 4:   Land Use Change and Scenario Modelling   351

17	Mapping and Modelling Land Use Change: an Application of the *SLEUTH* Model		353
	17.1	Introduction	353
	17.2	Methodology	355
	17.3	Results and Discussion	358
	17.4	Conclusion	364

18	Uncertainty in Landscape Models: Sources, Impacts and Decision Making		367
	18.1	Introduction	368
	18.2	Models, Variability and Sources of Uncertainty	369
		18.2.1 Model Structure	370
		18.2.2 Natural Variability, Temporal Resolution and Spatial Resolution	371
		18.2.3 Taxonomic Scale and Data Collection	375
		18.2.4 Summary on Models and Sources of Uncertainty	377
	18.3	Model Uncertainty and Decision Making	377
	18.4	Conclusion	381

19	Assessing Water Quality Impacts of Community Defined Land Use Change Scenarios for the Douglas Shire, Far North Queensland		383
	19.1	Context and Case Study Location	384
	19.2	Dialogue over Sustainable Future Landscapes and Seascapes	386
	19.3	Methodology of an Application of a Social–Ecological Framework for Sustainable Landscape Planning	387
		19.3.1 Stage I: Community Perceptions and Visions	387
		19.3.2 Stage II: Community-driven Landscape Scenarios	389

Contents   XVII

19.3.3 Stage III: Modelling of Landscape Scenarios
and Assessing Water Quality   389
19.4 Results and Discussion   391
19.4.1 Visions for the Douglas Shire Coastal Landscape 391
19.4.2 Spatially Explicit Land Use Change Scenarios   392
19.4.3 Inputs into *SedNet* for Water Quality Analysis
and Model Results   399
19.5 Conclusion   401

20 Analysing Landscape Futures for Dryland Agricultural Areas:
a Case Study in the Lower Murray Region of Southern Australia   407
20.1 Introduction   408
20.2 Futures Thinking and Scenario Analysis   409
20.3 The Lower Murray Landscape Futures study   411
20.3.1 Collaborative and Participatory Approach   412
20.3.2 Defining Targets, Scenarios and Policy Options   413
20.3.3 Landscape Futures Analysis   419
20.4 Results   425
20.5 Risk, Preference and Strategic Policy Adoption   429
20.6 Further Research   430
20.7 Application in Other Regions and Contexts   431
20.8 Conclusion   431

21 Applying the *What If?* Planning Support System for Better
Understanding Urban Fringe Growth   435
21.1 Introduction   435
21.2 The *What If?* Planning Support System   436
21.2.1 Suitability Module   438
21.2.2 Demand Module   438
21.2.3 Allocation Module   438
21.3 Mitchell Shire Application of *What If?*   439
21.3.1 Input Data Layers   442
21.3.2 Land Suitability Analysis   445
21.3.3 Demographic Projections and Land Use Demand   447
21.3.4 Future Land Use Allocation Scenarios 2031   449
21.4 Future Work   451
21.5 Conclusion   451

XVIII   Landscape Analysis and Visualisation

Part 5:   Landscape Visualisation   **455**

22   Understanding Place and Agreeing Purpose:
     the Role of Virtual Worlds   457
     22.1 Introduction   457
     22.2 Established Options for Understanding Place   459
     22.3 Emerging Options   460
     22.4 Development Methodology   461
           22.4.1 *SIEVE*   461
           22.4.2 Links to Decision Support Systems   463
           22.4.3 Virtual Decision Environment   463
     22.5 Conclusion   464

23   Geographic Landscape Visualisation in Planning Adaptation
     to Climate Change in Victoria, Australia   469
     23.1 Introduction   470
     23.2 Context of Visualisation and 'Sense of Place'   471
     23.3 Climate Change Predictions and Impacts in
           South-eastern Australia   472
           23.3.1 Climate Change and the Need for
                   Ecological Connectivity   473
           23.3.2 Biolink Zones in South-eastern Australia   474
           23.3.3 Visualisation Tools for Explaining the
                   Context of Biolinks   474
           23.3.4 Visualisation of Environmental Change
                   at a Site over Time   475
     23.4 Realism behind Visualisation Technology   479
     23.5 Realism at the Front End   480
     23.6 Future Directions   483
     23.7 Conclusion   484

24   Visualising Alternative Futures   489
     24.1 Introduction   490
     24.2 The Barwon Heads Peri-urban Development
           Visualisation Tool   491
     24.3 The Central Business District of Melbourne
           What the City Might Be? Prototype   495
           24.3.1 Marvellous Melbourne   495
           24.3.2 Melbourne and the Removal of
                   Significant Buildings   497
           24.3.3 Prototype World   498
           24.3.4 Initial Impressions   502

24.4	Visualising Proposed Landscapes: Sydney Rd, Brunswick	503
24.5	Usefulness of the Prototypes	505
24.6	Conclusion	505

25 Virtual Globes: the Next GIS? — 509
25.1	Introduction	510
25.2	Methodology	511
25.3	Results	515
	25.3.1 Hardware	515
	25.3.2 Background Data	516
	25.3.3 GIS Data Import	517
	25.3.4 Display and Data Manipulation	519
	25.3.5 Data Sharing	522
	25.3.6 Openness and Customisation	522
	25.3.7 Performance	522
25.4	Discussion	523
	25.4.1 Applications	524
25.5	Conclusion	529

26 A Virtual Knowledge World for Natural Resource Management — 533
26.1	Introduction	534
26.2	Virtual Worlds	535
26.3	*NRM Virtual Knowledge World*	536
26.4	*Bet Bet Virtual Landscape*	537
26.5	*Victorian Virtual NRM Knowledge Arcade*	544
26.6	Future Work	547
26.7	Conclusion	548

27 Computer Games for Interacting with a Rural Landscape — 551
27.1	Introduction	552
27.2	Cognitive Science	554
27.3	Conversation Theory	555
27.4	Visualisations	555
	27.4.1 Viewing Simulations	556
	27.4.2 Mobile and Computer Games	557
27.5	Game Development	558
	27.5.1 *Trainz*	559
	27.5.2 *Farcry*	561
	27.5.3 *Unreal Tournament 2004*	562
	27.5.4 *Second Life*	563
27.6	The *Bushfire Rescue Game*	565
27.7	Conclusion	568

28 Automated Generation of Enhanced Virtual Environments for Collaborative Decision Making Via a Live Link to GIS — 571

28.1 Introduction — 572
28.2 Background — 574
28.3 Methodology — 576
28.4 Case Study and Discussion — 580
28.5 Conclusion and Outlook — 587

29 Land Use Decision Making in a Virtual Environment — 591

29.1 Introduction — 592
29.2 Rational Decision Making — 593
    29.2.1 Values, Attitudes and Behaviours — 593
29.3 Methodology — 594
    29.3.1 Social: Agent-based Modelling — 595
    29.3.2 Environmental: Three-dimensional Visualisation — 595
    29.3.3 Economic: Experimental Economics — 596
    29.3.4 Experiment Design — 597
29.4 Environmental and Economic Efficiency: Results and Discussion — 600
    29.4.1 Complexity — 601
    29.4.2 Visualisation — 602
    29.4.3 Social context (ABM) — 604
    29.4.4 Value Priorities — 605
    29.4.5 Experience — 606
29.5 Conclusion — 606

Index — 609

Contributors

Anderson, Rod, Mr
  Department of Sustainability and Environment
  8 Nicholson Street, East Melbourne, Victoria 3002
  Australia
  rod.anderson@dse.vic.gov.au

Aurambout, Jean-Philippe, Dr
  Department of Primary Industries
  P.O. Box 4166, Parkville, Victoria 3052
  Australia
  Jeanphilippe.Aurambout@dpi.vic.gov.au

Austin, Mike, Dr
  CSIRO Sustainable Ecosystems
  GPO Box 284, Canberra, ACT 2601
  Australia
  Mike.Austin@csiro.au

Barr, Neil, Dr
  Department of Primary Industries
  Box 3100, Bendigo Mail Centre, Victoria 3554
  Australia
  neil.barr@dpi.vic.gov.au

Bartley, Rebecca, Dr
  CSIRO Land and Water
  Meiers Road 120, Indooroopilly, Brisbane 4068
  Australia
  Rebecca.Bartley@csiro.au

Batty, Michael, Prof.
  Centre for Advanced Spatial Analysis, University College London
  1–19 Torrington Place, London WC1E 7HB
  United Kingdom
  m.batty@ucl.ac.uk

Beverly, Craig, Dr
Department of Sustainability and Environment
PO Box 500, East Melbourne, Victoria 3002
Australia
Craig.R.Beverly@dse.vic.gov.au

Bishop, Ian, Prof.
Department of Geomatics, University of Melbourne
Grattan St, Melbourne, Victoria 3010
Australia
i.bishop@unimelb.edu.au

Bluml, Martin, Mr
Department of Primary Industries
Box 3100, Bendigo Delivery Centre, Victoria 3554
Australia
martin.bluml@dpi.vic.gov.au

Bock, Karl, Mr
Institute for Rural Futures, University of New England
Armidale, New South Wales 2351
Australia
karl.bock@une.edu.au

Bohnet, Iris, Dr
CSIRO Sustainable Ecosystems
PO Box 780, Atherton, Queensland 4883
Australia
Iris.Bohnet@csiro.au

Brodie, Jon, Mr
Australian Centre for Tropical Freshwater Research
James Cook University,Townsville, Queensland 4811
Australia
Jon.Brodie@jcu.edu.au

Brunckhorst, David, Prof.
Institute for Rural Futures, University of New England
Armidale, New South Wales 2351
Australia
dbrunckh@une.edu.au

Bryan, Brett, Dr
CSIRO
Waite Rd, Urrbrae, South Australia 5064
Australia
brett.bryan@csiro.au

Cartwright, William, Prof.
RMIT University
School of Mathematical and Geospatial Sciences
GPO Box 2476V
Melbourne VIC 3001
Australia
william.cartwright@rmit.edu.au

Chan, Tai, Dr
Department of Sustainability and Environment
PO Box 500, East Melbourne, Victoria 3002,
Australia
Tai.chan@dse.vic.gov.au

Chen, Tao, Mr
Department of Geomatics, University of Melbourne
Grattan St, Melbourne, Victoria 3010
Australia
t.chen10@pgrad.unimelb.edu.au

Cherry, Donald, Mr
Department of Primary Industries
Box 3100, Bendigo Mail Centre, Victoria 3554
Australia
Don.Cherry@dpi.vic.gov.au

Christy, Brendan, Mr
Department of Primary Industries
RMB 1145 Chiltern Valley Road, Rutherglen, Victoria 3685
Australia
brendan.christy@dpi.vic.gov.au

Clarke, Keith C, Dr
Department of Geography, University of California
Santa Barbara, California 93106-4060
United States of America
kclarke@geog.ucsb.edu

Cresswell, Hamish, Dr
CSIRO Land and Water
GPO Box 1666, Canberra City, ACT 2601
Australia
Hamish.Cresswell@csiro.au

Crossman, Neville, Dr
CSIRO
Waite Rd, Urrbrae, South Australia 5064
Australia
neville.crossman@csiro.au

Dawson, Steve, Mr
Australian Government Bureau of Rural Sciences
PO Box 858, Canberra City, ACT 2602
Australia
Steve.Dawson@brs.gov.au

Dripps, Kimberley, Ms
Department of Sustainability and Environment
PO Box 500, East Melbourne, Victoria 3002
Australia
kimberley.j.dripps@dse.vic.gov.au

Duncan, David, Dr
Department of Sustainability and Environment
123 Brown St, Heidelberg, Victoria 3054
Australia
david.duncan@dse.vic.gov.au

Ebert, Sam, Mr
Department of Sustainability and Environment
PO Box 500, East Melbourne, Victoria 3002
Australia
Sam.ebert@dse.vic.gov.au

Fitzsimons, Patricia, Ms
Department of Primary Industries
P.O. Box 4166, Parkville, Victoria 3052
Australia
Patricia.Fitzsimons@dpi.vic.gov.au

Garnett, Nichola, Ms
Department of Justice
GPO Box 4356QQ, Melbourne, Victoria 3001
Australia
Nichola.Garnett@justice.vic.gov.au

Gibbons, Philip, Dr
Fenner School of Environment and Society
The Australian National University, Canberra, ACT 0200
Australia
Philip.Gibbons@anu.edu.au

Hill, Michael, Prof.
University of North Dakota
Grand Forks, North Dakota 58202
United States of America
hillmj@aero.und.edu

Hill, Patricia, Dr
CSIRO Sustainable Ecosystems
GPO Box 284, Canberra City, ACT 2601
Australia
Patricia.Hill@csiro.au

Horrocks, Gregory, Mr
Australian Centre for Biodiversity, School of Biological
Sciences, Monash University
Clayton 3800
Australia
greg.horrocks@sci.monash.edu.au

Kennedy, Lucy, Ms
Department of Geomatics, University of Melbourne
Grattan St, Melbourne, Victoria 3010
Australia
lucykennedy@live.com.au

Keysers, Jessica, Ms
Earth Tech
71 Queens Road, Melbourne 3004
Australia
jessica.keysers@earthtec.com.au

King, Darran, Mr
CSIRO
Waite Rd, Urrbrae, South Australia 5064
Australia
darran.king@csiro.au

Klosterman, Richard, Prof.
What If?, Inc.
78 Hickory Lane, Hudson, Ohio 44236
United States of America
klosterman@whatifinc.biz

Lamb, David, Dr
The University of Queensland
St Lucia, Brisbane 4067
Australia
d.lamb@uq.edu.au

Lau, Alex, Mr
Department of Sustainability and Environment
8 Nicholson Street, East Melbourne, Victoria 3002
Australia
Alex.lau@dse.vic.gov.au

Lesslie, Robert, Dr
Australian Government Bureau of Rural Sciences
GPO Box 858, Canberra, ACT 2601
Australia
Rob.Lesslie@brs.gov.au

Lewis, Adam, Dr
Geoscience Australia
GPO Box 378, Canberra, ACT 2601
Australia
Adam.Lewis@ga.gov.au

Lewis, Hayden, Mr
Department of Primary Industries
255 Ferguson Road Tatura, Victoria 3616
Australia
Hayden.Lewis@dpi.vic.gov.au

Lowell, Kim, Prof.
Department of Primary Industries
P.O. Box 4166, Parkville, Victoria 3052
Australia
kim.lowell@dpi.vic.gov.au
Cooperative Research Centre for Spatial Information
Ground floor, 723 Swanston St, Carlton, Victoria 3052
Australia
klowell@crcsi.com.au

MacEwan, Richard, Mr
Department of Primary Industries
Box 3100, Bendigo Delivery Centre, Victoria 3554
Australia
richard.macewan@dpi.vic.gov.au

Mac Nally Ralph, Prof.
Australian Centre for Biodiversity, School of Biological
Sciences, Monash University, Clayton, Victoria 3800
Australia
Ralph.MacNally@sci.monash.edu.au

Mansergh, Ian, Dr
Department of Sustainability and Environment
8 Nicholson Street, East Melbourne, Victoria 3002
Australia
Ian.mansergh@dse.vic.gov.au

Morley, Philip, Mr
Institute for Rural Futures, University of New England
Armidale, New South Wales 2351
Australia
pmorley@une.edu.au

Pettit, Christopher, Dr
Department of Primary Industries
PO Box 4166, Parkville, Victoria 3052
Australia
Christopher.Pettit@dpi.vic.gov.au

Pullar, David, Dr
  The University of Queensland
  St Lucia, Brisbane 4067
  Australia
  d.pullar@uq.edu.au

Quinn, Brian, Mr
  RMIT University, RMIT City Campus
  Building 12, Level 11 Swanston Street, Melbourne, Victoria 3000
  Australia
  pb.quinn@bigpond.com

Reeve, Ian, Dr
  Institute for Rural Futures, University of New England
  Armidale, New South Wales 2351
  Australia
  ireeve@une.edu.au

Ryan, Paul, Mr
  CSIRO Sustainable Ecosystems
  GPO Box 284, Canberra, ACT 2601
  Australia
  Paul.Ryan@csiro.au

Seddon, Julian, Mr
  Department of Environment and Climate Change (NSW)
  GPO Box 284, Canberra, ACT 2601
  Australia
  Julian.Seddon@csiro.au

Smith, Patrick, Dr
  CSIRO Sustainable Ecosystems
  Private Bag 5, PO Wembley, Western Australia 6913
  Australia
  Patrick.Smith@csiro.au

Stock, Christian, Dr
  Cooperative Research Centre for Spatial Information
  723 Swanston St, Carlton, Victoria 3052
  Australia
  c.stock@unimelb.edu.au

Sunnucks, Paul, Dr
Monash University, School of Biological Sciences
Clayton Campus, Melbourne, Victoria 3800
Australia
paul.sunnocks@sci.monash.edu.au

Taylor, Andrea, Dr
Monash University, School of Biological Sciences
Clayton Campus, Melbourne, Victoria 3800
Australia
andrea.taylor@sci.monash.edu.au

Thankappan, Medhavy, Mr
Geoscience Australia
GPO Box 378, Canberra, ACT 2601
Australia
Medhavy.Thankappan@ga.gov.au

Thomson, James, Dr
Australian Centre for Biodiversity, School of Biological Sciences
Monash University, Clayton, Victoria 3800
Australia
jim.thomson@sci.monash.edu.au

Vesk, Peter, Dr
School of Botany, The University of Melbourne
Parkville 3010
Australia
pvesk@unimelb.edu.au

Warren, Garth, Mr
CSIRO Sustainable Ecosystems
GPO Box 284, Canberra, ACT 2601
Australia
Garth.Warren@csiro.au

Weeks, Anna, Mrs
Department of Primary Industries
RMB 1145 Chiltern Valley Road, Rutherglen, Victoria 3685
Australia
anna.weeks@dpi.vic.gov.au

Williams, Stephen Mr
Department of Primary Industries
Box 3100, Bendigo Delivery Centre, Victoria 3554
Australia
Steve.Williams@dpi.vic.gov.au

Wintle, Brendan, Dr
Commonwealth Environmental Research Facility (AEDA),
School of Botany
The University of Melbourne, Parkville, Victoria 3010
Australia
brendanw@unimelb.edu.au

Wu, Yingxin, Ms
Department of Primary Industries
PO Box 4166, Parkville, Victoria 3052
Australia
yingxin.wu@dpi.vic.gov.au

Wyatt, Ray, Assoc. Prof.
University of Melbourne
221 Bouverie Street, Carlton, Victoria 3010
Australia
ray.wyatt@unimelb.edu.au

Zerger, Andre, Dr
CSIRO Sustainable Ecosystems
GPO Box 284, Canberra, ACT 2601
Australia
Andre.Zerger@csiro.au

Abbreviations

ABS	Australian Bureau of Statistics
AHP	Analytical Hierarchy Process
CAT	Catchment Analysis Tool
CMA	Catchment Management Authority
CMF	Catchment Management Framework
CRC	Cooperative Research Centre
CRCSI	Cooperative Research Centre for Spatial Information
CSDL	Corporate Spatial Data Library
CSIRO	Commonwealth Scientific and Industrial Research Organisation
DAS	Decision Aiding System
DEM	Digital Elevation Model
DPI	Department of Primary Industries, Victoria
DSE	Department of Sustainability and Environment, Victoria
DSS	Decision Support System
ESRI	Environmental Systems Research Institute
GFS	Groundwater Flow Systems
GIS	Geographical Information System
GPS	Geographical Positioning System
GUI	Graphical User Interface
IPCC	Intergovernmental Panel on Climate Change
KML	Keyhole Markup Language
KMZ	Keyhole Markup Zipped
LiDAR	Light Detection and Ranging
LGA	Local Government Authority
LUIM	Land Use Impact Model
MCA	Multi-Criteria Analysis
MCAS-S	Multi-Criteria Analysis Shell for Spatial Decision Support

NRM	Natural Resource Management
OECD	Organisation for Economic Co-operation and Development
ORL	Our Rural Landscape
PSS	Planning Support System
SDI	Spatial Data Infrastructure
SDSS	Spatial Decision Support Systems
SIEVE	Spatial Information Exploration and Visualisation Environment model
SLA	Statistical Local Area
SLEUTH	Slope, Land use, Exclusion, Urban extent over time, Transportation, Hill-shaded model
VRML	Virtual Reality Modelling Language
VRO	Victoria Resources Online
XML	Extensible Markup Language

INTRODUCTION

1 Understanding Landscapes through Knowledge Management Frameworks, Spatial Models, Decision Support Tools and Visualisation

Christopher Pettit[1], William Cartwright[2], Ian Bishop[3], Kim Lowell[1, 4], David Pullar[5] and David Duncan[6]

[1] Department of Primary Industries, Parkville Centre, Victoria, Australia
[2] RMIT University, Melbourne, Victoria, Australia
[3] Department of Geomatics, University of Melbourne, Victoria, Australia
[4] Cooperative Research Centre for Spatial Information, Carlton, Victoria, Australia
[5] Geography, Planning and Architecture, The University of Queensland, Australia
[6] Arthur Rylah Institute for Environmental Research, Department of Sustainability and Environment, Heidelberg, Victoria, Australia

1.1 Introduction

This book is about *landscape* analysis and visualisation. But what do we mean by landscape? A landscape can simply be defined as the features which comprise an area of land. When analysed, a landscape can be interpreted as a dynamic system of living and non-living objects. There are many dimensions to a landscape, which can be categorised by biophysical (fauna and flora), geomorphological, social (anthropogenic) and economic (natural resource) considerations. Understanding and managing landscapes in all their complexity, is an ongoing challenge that we must get right if we are to fulfil our obligation to future generations through the public policy goal of sustainability. Both expert and local knowledge is essential in order to achieve this goal, and requires management and effective communication to decision makers, planners, resource managers, landholders and communities.

There are a number of sophisticated spatial models and tools available for addressing the range of critical land use issues facing society today. In compiling this volume we have selected a collection of papers from the 'Place and Purpose – Spatial Models for Natural Resource Management and Planning' conference held on the 30–31 May 2007 in Bendigo, Victoria, Australia. The research provides a comprehensive exposé of cutting-edge spatial tools and approaches for analysing, simulating and visualising natural, agricultural and urban landscapes. The research in this volume is typically focused on frameworks, tools and models applied in Australia, though there are a select number of international applications. We believe all research put forth in this volume provides examples of international 'best practice' in their respective fields, and as such presents a useful reference point for the applied researcher and practitioner interested in landscape analysis and visualisation

The five parts of this volume deal with:

- Part 1: Natural Resource Knowledge Management Frameworks and Tools
- Part 2: Integrating the Ecology of Landscapes into Landscape Analysis and Visualisation
- Part 3: Socioeconomic Dimensions to Landscapes
- Part 4: Land Use Change and Scenario Modelling
- Part 5: Landscape Visualisation.

While the contributions within each theme sit together naturally, there is also considerable complementarity between the chapters presented in each part. In many cases such overlap indicates pathways for new multi-disciplinary approaches and learnings. In compiling this volume one of the challenges has been working with a number of scientists across a range of disparate disciplines including: ecology, hydrology, pedology, geology, systems modelling, geospatial sciences, economics and social sciences. This mix of disciplines has made this project exciting. Understanding and defining a common language to exchange ideas across disciplines often proves challenging in multi-disciplinary environments, but it is critical if we are to *sustainably manage* our landscapes for future generations. If we are to achieve this goal it will take the collective energy and expertise of the science community to work together in asking and framing the right questions. To tackle these critical questions we need to develop and apply the best available frameworks, models and tools — if these widgets do not sufficiently address these questions, then we need to adapt and try new methods and techniques. Furthermore a real challenge is the ability to expediently and efficiently communicate research findings simply and clearly so that users can understand and apply such critical learnings in both policy and community settings.

We offer a volume to facilitate collaboration across policy, science and community. A range of frameworks, models, decision support and communication (visualisation) tools that span across the strategic to the applied continuum of research are presented. Some tools are quite mature, and can offer 'just-in-time' policy support through an evidenced-based approach, as discussed in the chapter by Dripps and Bluml. Whilst other tools like Pullar's scenario analysis and Wyatt's *Preference Predictor* software system are relatively new and are yet to be put through their paces and tested in the policy arena. Landscape visualisation tools such as computer game engines like *SIEVE*, as discussed by Bishop, are becoming more prevalent in understanding place and space and provide a powerful interface to the array of sophisticated land use change impact models such as those discussed in the ecological and land use change modelling sections of this volume.

1.2 Part 1: Natural Resource Knowledge Management Frameworks and Tools

2 Reading between the Lines: Knowledge for Natural Resource Management
Richard MacEwan
3 Improving the Use of Science in Evidence-based Policy: Some Victorian Experiences in Natural Resource Management
Kimberley Dripps and Martin Bluml
4 The *Catchment Analysis Tool*: Demonstrating the Benefits of Interconnected Biophysical Models
Anna Weeks, Brendan Christy, Kim Lowell and Craig Beverly
5 The Application of a Simple Spatial Multi-criteria Analysis Shell to Natural Resource Management Decision Making
Robert Lesslie, Michael Hill, Patricia Hill, Hamish Cresswell and Steve Dawson
6 *Platform for Environmental Modelling Support*: a Grid Cell Data Infrastructure for Modellers
Tai Chan, Craig Beverly, Sam Ebert, Nichola Garnett, Adam Lewis, Chris Pettit, Medhavy Thankappan and Stephen Williams

Natural resources are increasingly being managed holistically. This reflects a real-world trend in decision making by policy makers and resource managers of considering multiple impacts of management decisions. This is a marked philosophical contrast to former approaches to natural resources management in which decision makers focussed primarily or solely on a single 'commodity' or 'service' produced by landscapes such as critical wildlife habitat, economically productive soil, or high quality water. Consideration of multiple landscape services complicates decision making because resource managers are put in a position of having to make trade-offs based on the perceived and measured social, economic and environmental worth of, for example, a hectare of endangered habitat or a ton of topsoil.

The increased real-world complexity of natural resource management and decision making has increased interest in systems-based tools underpinned by sound discipline-specific scientific knowledge. Real-world resource managers and policy makers want tools that provide information about the potential impacts of management actions on a number of 'landscape services' and that provide such information in a form that will facilitate efficient decision making. For scientists, therefore, the challenge is two-fold. Firstly, scientists must have the capacity to combine accumulated knowledge from multiple disciplines in a logical, realistic and transparent manner. Secondly, the tools created by scientists must produce useful information that can be presented in ways that can be understood and absorbed by non-scientists.

The chapters in Part 1 address these two challenges. The first chapter by MacEwan sets the scene by outlining the relationship between planning and decision making, and the value of spatial modelling. MacEwan discusses this in the context of the data to knowledge hierarchy and introduces the conceptual framework for organising critical landscape questions. The second chapter by Dripps and Bluml discusses a new generation of science-based tools for addressing policy questions. It describes the need for such tools from the point of view of policy makers, provides information on the type of information desired, and describes examples where information produced by systems-based tools may facilitate natural resource management. The chapter by Chan et al. addresses the difficulty of creating systems-based models from the perspective of data provision. Not only is knowledge of natural systems dispersed in different scientific domains, but the data necessary to create and use systems-based models are geographically and administratively dispersed, and available in non-standard formats. Chan et al. describe a grid-based spatial data system being developed that should resolve many of the modelling problems related to data. In their chapter, Weeks et al. emphasise the importance of tools that provide information about the results of possible land management actions on multiple landscape services. This is done by presenting and describing one such tool, whose focus is the linkage of knowledge from diverse scientific disciplines. Its use is then demonstrated using a case study approach. Finally, Lesslie et al. demonstrate how better information about multiple landscape services can be organised, presented and analysed so that improved decisions can be made. Hence the focus of this chapter is on facilitating the uptake of knowledge by non-scientists rather than improved methods of generating such knowledge.

For those working in the area of natural resource management, it is encouraging to see this diversity of efforts being made to improve the management of natural resources. Dripps and Bluml demonstrate clearly that high level decision makers are not only open to, but actually desirous of, the complex information that can be produced by systems-based tools — provided the decision makers understand the information, and that it is presented in a way that meets their needs. Chan et al. also recognise the value of the information desired by Dripps and Bluml, but note that the development of tools to produce it requires data custodians and managers to collaborate within and across organisations, in order to compile information and provide it in a format that will help developers and users of computer-based natural resource management tools. Weeks et al. are an example of scientists who will benefit from such efforts to standardise data and compile them in a common repository; their work also is indicative of the level of sophistication that systems-based tools can achieve through the combination of relevant scientific knowledge and associated data. Lesslie et al. closes the loop from decision maker to data provider to tool developer to information producer back to decision maker. Their work recognises that just being able to produce systems-based information is not the ultimate goal of those working with natural resource knowledge management frameworks and tools. Instead, it is the uptake of such information and the development of better natural resources management decisions that should be the focus of all.

1.3 Part 2: Integrating the Ecology of Landscapes into Landscape Analysis and Visualisation

7   Looking at Landscapes for Biodiversity: Who's View Will Do?
   F Patrick Smith

8   Native Vegetation Condition: Site to Regional Assessments
Andre Zerger, Philip Gibbons, Julian Seddon, Garth Warren, Mike Austin and Paul Ryan
9   Towards Adaptive Management of Native Vegetation in Regional Landscapes
David Duncan and Brendan Wintle
10 Revegetation and the Significance of Timelags in Provision of Habitat Resources for Birds
Peter Vesk, Ralph Mac Nally, James Thomson and Gregory Horrocks
11 The Application of Genetic Markers to Landscape Management
Paul Sunnucks and Andrea Taylor
12 Scenario Analysis with Performance Indicators: a Case Study for Forest Linkage Restoration
David Pullar and David Lamb

Natural ecosystems are comprised of an infinite number of complex interactions between plants, animals, fungi and bacteria that play out over variable spatial and temporal scales. Ecological sustainable development (ESD) raised the protection of these interactions and the biological diversity that supports it to an equal footing with social and economic criteria, and the phrase 'ecosystem services' is now widely applied to summarise the benefits that human civilisations derive from natural ecosystems. However, these interactions, and the impact of human activities on other species' persistence are as poorly understood at landscape planning scales ('landscape ecology') as they are richly understood at the organismal level and small spatial scales. The chapters in this section explore challenges to integrating biodiversity and ecological processes into landscape analysis and visualisation. To begin with, Smith's purposefully accessible chapter opens the minds of readers from all backgrounds of the frailties of our assumptions about how other species respond to altered landscapes, and mixtures of natural and artificial landscape elements, with examples from the wheat and sheep farming areas of Western Australia.

In practice, because we know so little about the requirements of the vast majority of our biodiversity, objective setting in natural resource management for biodiversity and ecosystem services in Australia is framed around the amount and quality of native vegetation. We assume that native vegetation quality is a surrogate for habitat quality, and therefore that by maintaining amounts of native vegetation of sufficient quality, we may provide for the ongoing coexistence of plant and animals, and continue to benefit from the services they provide. In their chapters, Zerger et al. explore how models of native vegetation condition (quality) might be produced at spatial scales that will be of greatest utility to landscape planning and management, while Duncan and Wintle look at what progress has been made applying an adaptive management framework to the management of native vegetation at regional scales. The following chapter from Vesk et al. continues Smith's theme of noting how glib the representation of the ecology of landscapes can be at broad decision-making scales. It begins with the simple observation that restoration aimed at providing habitat for species such as birds will not provide that habitat for many years to come, centuries in some cases. They combine an aspatial model of habitat development with freely available data on bird species biology to illustrate the implication of delays in habitat maturity for the desired objective of bird species' persistence. On a different tack, in their chapter Sunnucks and Taylor ask why the potential of molecular genetics are so under-utilised in analyses of the ecological function of landscapes, given that the persistence of sexual species essentially comes down to demographic parameters of births, deaths, finding a partner (migration) and mating. Their review emphasises that combined with the right questions, the information value, cost and time efficiency of genetic methods can detect signatures of landscape function and dysfunction at surprisingly short time scales. Part 2 concludes with a case study of improving forest connectivity at the scale of a local landscape by Pullar and Lamb. This example demonstrates how simple 'rules of thumb' based on landscape ecological principles may be combined, analysed and represented in a GIS environment, which may allow stakeholders to grasp some of the spatial and opportunity cost trade-offs in landscape restoration. This chapter demonstrates one way in which more detailed ecological understandings of the preceding chapters may be brought into the kind of application environment more familiar to landscape planners.

1.4 Part 3: Socioeconomic Dimensions to Landscapes

13 Strategic Spatial Governance: Deriving Social–Ecological Frameworks for Managing Landscapes and Regions
   David Brunckhorst, Ian Reeve, Phil Morley and Karl Bock
14 Placing People at the Centre of Landscape Assessment
   Patricia Fitzsimons and Don Cherry
15 The Social Landscapes of Rural Victoria
   Neil Barr
16 A Decision Aiding System for Predicting People's Scenario Preferences
   Ray Wyatt

The socioeconomic dimensions of natural resource management are becoming increasingly important as decision makers, planners, land managers, landholders and communities come to grips with how we can best manage our landscapes collectively. Natural resource management and planning has traditionally taken a biophysical-centric approach for understanding land use change and impact. However, there is a growing acknowledgement that social and behavioural sciences need to play a more important role in understanding landscape.

Part 3 begins with a chapter from Brunckhorst et al. which questions the philosophy of spatial definition of resource governance regions. The chapter does so by reviewing existing governance arrangement within New South Wales. In this research the authors propose a new set of land management boundaries based upon a social–ecological framework, also known as 'eco–civic' regions. A comparative analysis with existing governance boundaries is conducted. Importantly, this chapter presents three principles which underpin the development of regionalisation for government administration through a combined top-down and bottom-up process of spatial–social contextualisation.

There exist a number of frameworks which can be used to interpret landscape change and impact. Conceptual frameworks for understanding landscapes and their intrinsic functions and processes have been around for years. In Part 1 MacEwan discusses the frameworks of McHarg (1969) and Steinitz (1990). More recent are the development and application of frameworks such as Hajkowicz et al. (2003) which acknowledge the importance of community and social values in understanding landscape change. The chapter by Fitzsimons and Cherry reviews three additional frameworks: Pressure–State–Response (PSR), Driving forces–Pressure–State–Impact–Response (DPSIR), and the Millennium Ecosystem Assessment (MA). The authors describe the first Australian application of the MA framework, downscaling five landscape types from a global to a local level of interpretation for the State of Victoria. Two specific landscapes are then described in more details and interpreted through a set of social, economic and environmental indicators, which link to State of the Environment reporting.

Building upon the concept of socioeconomic indicators for understanding rural landscapes, Barr constructs five social landscapes, each with a divergent trajectory of rural restructuring in Victoria. This research applies a traditional principal component analysis (PCA) approach which incorporates nine statistical indicators and results in two orthogonal factors used to create the five social landscapes. Barr combines extensive local knowledge of rural Victoria with a robust statistical analysis to provide insights into the restructuring of Victoria's farm industries. In interpreting trends across the landscape — for example a move towards farm aggregation in some areas and a move towards rural amenity in others — provides valuable insight to policy makers, allowing them to better understand the drivers of change in Victorian rural landscapes.

Wyatt provides the final chapter in Part 3, and presents a decision aiding system (DAS) which has its roots in Choice Theory (Glasser 1998), and Theory of Planned Behaviour (Ajzen 1991). The system uses people's previous responses to a range of questions in order to put forward likely preferences for various options or scenarios. In this instance the chapter discusses an urban planning application in Santa Barbara, California, to demonstrate its applicability in providing people's preferences across land use decision making and taking. This chapter offers a good transition into the Part 4, as it provides a technique for evaluating the scenarios generated from land use change and spatial decision support systems.

1.5 Part 4: Land Use Change and Scenario Modelling

17 Mapping and Modelling Land Use Change: an Application of the *SLEUTH* Model
Keith Clarke
18 Uncertainty in Landscape Models: Sources, Impacts and Decision Making
Kim Lowell
19 Assessing Water Quality Impacts of Community Defined Land Use Change Scenarios for the Douglas Shire, Far North Queensland
Iris Bohnet, Jon Brodie and Rebecca Bartley
20 Analysing Landscape Futures for Dryland Agricultural Areas: a Case Study in the Lower Murray Region of Southern Australia
Brett Bryan, Neville Crossman and Darran King
21 Applying the *What If?* Planning Support System for Better Understanding Urban Fringe Growth
Christopher Pettit, Jessica Keysers, Ian Bishop and Richard Klosterman

Land use change modelling operates at the interface between physical and social changes in our world. In understanding the causes and consequences of land use dynamics it is necessary to consider the open and complex nature of systems and their interactions (Veldkamp and Verburg 2004; Lambin et al. 2003). Scenarios are a way to explore pathways for the most significant causes of change and impacts. Collectively the chapters in Part 4 contribute new insights and tools for land use change modelling.

They cover diverse topics to answer key questions concerning cross-scale interactions on theoretical issues and practical applications. All the authors are experienced researchers from academia or industry. They describe relatively advanced application areas, whilst providing ample explanation and background knowledge so the concepts are easily understood by researchers and practitioners alike.

Part 4 starts with two technically oriented chapters. Clarke presents a land use change model, *SLEUTH*, which uses cellular automata and is based on advanced concepts of structural and transitional processes causing urban land use change at a micro-spatial scale. *SLEUTH* is the most popular tool of this kind in use today. This chapter explains the interaction between two key processes for urban growth and change in land use. As computational models develop more sophistication, it is important to appreciate the level of complexity required to represent natural processes and at the same time to provide useful information to decision makers. The next chapter, by Lowell, describes fundamental sources of uncertainty in landscape representations — namely natural variability, temporal scale, spatial scale, taxonomic scale and data collection — that affect model development and usage. Lowell gives intuitive insights and stimulates the reader to think about forms of uncertainty associated with models and the confidence users have in model outputs.

The next three chapters present application oriented topics. Land use scenarios are described for landscapes from the north to the south of Australia, ranging from coastal to catchment environments. Bohnet outlines a framework that links landscape analysis with community developed scenarios. While these scenarios apply sophisticated hydrological models that give robust predictions, they hide the computational complexity of process models to be able to engage in meaningful discussions with the community on their vision of landscape futures in an agricultural watershed. Underlying the scenario modelling is a tension between protecting high valued natural assets, like the Great Barrier Reef, and addressing livelihood concerns for people who live in reef catchments. The Bryan et al. chapter also applies futures thinking and scenario analysis to the management of one of Australia's most significant waterway systems — the Murray River. Land use models at this scale need to incorporate linkages between landscapes and regions. The scale of the model is impressive in terms of its geographical extent and the integration of social, economic and environmental elements of natural resource management. The their chapter, Pettit et al. describe a planning support tool, *What If?*, used for developing and analysing the impact of land use scenarios. This chapter presents the benefits of planning support to bridge the gap between modelling experts and the planning practitioner.

These chapters touch on key concepts important to land use modelling:

- level of analysis
- cross-scale dynamics
- driving forces
- spatial interaction and neighbourhood effects
- temporal dynamics
- level of integration (Veldkamp and Verburg 2004).

They also draw attention to future research and relevant problems. In particular, Clarke suggests that models need to deal with temporal aspects of land use change and the sequencing and path dependence of land use transitions. This readily extends to themes of uncertainty and utility of information for making a series of decisions based upon model outputs. One future direction will be the development of theories, methods and tools to further explore land use change and the natural resource management problems described in Part 4.

1.6 Part 5: Landscape Visualisation

22 Understanding Place and Agreeing Purpose: the Role of Virtual Worlds
   Ian Bishop
23 Geographic Landscape Visualisation in Planning Adaptation to Climate Change in Victoria, Australia
   Ian Mansergh, Alex Lau and Rod Anderson
24 Visualising Alternative Futures
   William Cartwright
25 Virtual Globes: the Next GIS?
   Jean-Philippe Aurambout, Christopher Pettit and Hayden Lewis
26 A Virtual Knowledge World for Natural Resource Management
   Christopher Pettit and Yingxin Wu
27 Computer Games for Interacting with a Rural Landscape
   Brian Quinn and William Cartwright
28 Automated Generation of Enhanced Virtual Environments for Collaborative Decision Making via a Live Link to GIS
   Tao Chen, Christian Stock, Ian Bishop and Christopher Pettit
29 Land Use Decision Making in a Virtual Environment
   Lucy Kennedy and Ian Bishop

Sound decisions about landscape futures depend on good science, good communications and good decision making. The science and, to a degree, the decision making have been discussed elsewhere in this volume. The further contribution of landscape visualisation to decision making through effective communication is considered in Part 5.

Traditionally, landscapes have been visually depicted by artists and cartographers. In more recent times the use of digital tools such as graphics packages, geographical information system (GIS) software and computer game engines are being used to enable visual analyses of past, present and future landscapes. In Part 5 we look at a range of technologies at the cutting-edge of computer graphics and review the ways in which these can assist people in many aspects of landscape management. The first chapter begins with Bishop's review of the literature of place, the criteria for effective representation of a sense of place within virtual environments, and an example of the use of a geographic information system and a linked game engine to communicate not only the physical nature of a place but also, ideally, the culture and human aspects of that environment. Bishop argues that an ability to communicate place, in addition to space, is imperative to achieve agreement on purpose and hence good decision making. Mansergh et al. follow this theme in the specific context of understanding and responding to climate change. They illustrate the way in which very large area planning can be visualised, and statewide initiatives understood, through effective visualisation. The remaining chapters delve into various technologies used to support visualisation and present case studies of quite different applications.

Many software packages and protocols exist for three-dimensional representation of everything from the atom to the universe. Landscape sits within this continuum; however it has special requirements because it is so visually familiar to everyone. We have natural interpretation capabilities but we can be extremely critical because of this familiarity. In addition a new generation of end users will not want to simply look at simulated or virtual landscapes, but will want to interact with them, query the data on which they are based, meet people within them, play-out scenarios and learn through the process. The different ways in which a variety of tools can add value to the visualisation process are explored in Part 5. Aurambout et al. review virtual globes (such as *Google Earth*). Web-based mark-up languages such as VRML are the basis of developments outlined in the chapters by Pettit and Wu, and Cartwright. Quinn and Cartwright discuss how computer games can support a range of virtual activities. Chen et al. also base their development on computer game platforms, but link their development directly to a geographic information system, while Kennedy and Bishop extend the scope of their visualisation through agent-based modelling.

The applications are similarly diverse: rural landscape scenarios (Aurambout et al.), natural resource data access (Pettit and Wu), the past and the future of the urban and peri-urban fabric (Cartwright), emergency response training (Quinn and Cartwright), revegetation of saline landscapes (Chen et al.) and understanding individual land use decision making (Kennedy and Bishop). The potential of visualisation to support basic research, personal development, local decision making and regional policy making is clearly revealed.

1.7 Future Challenges

There are a growing number of frameworks, models and decision support tools now available to assist in managing our landscape. However, it is not always clear which tool is best for the job and which one can best provide results in a timely manner that are digestible to the end user. This is one of the major challenges — bridging the gaps between theory and practice, and between scientist and the end user — which spans from policy makers to communities.

In this volume we have endeavoured to assemble a resource which will arm both academic and practitioner alike, with a better understanding of the breadth of available natural resource knowledge management frameworks, spatial models and communication (visualisation) tools. The level of maturity across research products presented is varied. However, we believe such tools can and should be developed further in order to better inform policy and planning processes concerned with critical land use issues including: urban sprawl, climate change, salinity, biodiversity, and changing landscapes, both socioeconomic and biophysical.

References

Ajzen I (1991) The theory of planned behavior. Organizational Behavior and Human Decision Processes 50(2):179–211
Glasser W (1998) Choice theory: a new psychology of personal freedom. Harper Collins, New York
Hajkowicz S, Hatton T, Meyer W, Young M (2003) Exploring future landscapes: a conceptual framework for planned change. Land and Water Australia, Canberra
Lambin EF, Geist HJ, Lepers E (2003) Dynamics of land use and land cover change in tropical regions. Annual Review of Environment and Resources 28:205–41
McHarg IL (1969) Design with nature. Natural History Press, New York

Steinitz C (1990) A framework for the theory applicable to the education of landscape architects. Landscape Journal 9:136–143

Veldkamp A, Verburg PH (2004) Modelling land use change and environmental impact. Journal of Environmental Management 72(1–2):1–3

PART 1
NATURAL RESOURCE KNOWLEDGE MANAGEMENT FRAMEWORKS AND TOOLS

2 Reading between the Lines: Knowledge for Natural Resource Management

Richard MacEwan

Department of Primary Industries, Epsom Centre, Victoria, Australia

Abstract: There are a number of spatial models available to support natural resource management decision making, however which one is best suited to answering the question at hand? Specific tools and approaches have various data, information and knowledge requirements. In this chapter the concept of the knowledge hierarchy (Ackoff 1989) is used to explain the dimensions of modelling and planning. This hierarchy is important as it provides a sound theoretical basis from which scientists and modellers can better relate to the planners and policy makers. To facilitate a better relationship between science and decision making, Steinitz's (1960) landscape decision framework provides a basis for linking six levels of questions and models. In conclusion, natural resource managers and planners need to better utilise spatial models to simplify and better understand complex phenomena, and utilise communication tools such as landscape visualisation to improve discipline integration. There is an urgent need to bring together expertise in the fields of natural resource management so that scenarios for change can be explored and critical decisions concerning our natural resources can be made based on collective wisdom.

2.1 Introduction

In May 2007 the Victorian Department of Primary Industries hosted the 'Place and Purpose' conference. This event was a significant showcase for, and forum to debate, the latest tools, directions and research questions on the theme 'Spatial Models for Natural Resource Management and Planning'. The contributions in this book were developed as a consequence of the Place and Purpose conference (DPI 2007). The accounts of models and their visual outputs that are reported in this volume are a credit to the collective efforts of modellers to make their (modelled) advices and options more accessible to the next user or decision maker. However, there are assumptions underlying these endeavours and they are embedded in our tools and approaches to decision making. I offer this short dissertation on the topic of knowledge for natural resource management (NRM) to provide a perspective on this area. I want to take a step away from any particular applications, consider some general theory, provide some lessons from a science fiction allegory, and to advocate for simplicity out of complexity.

2.2 Knowledge Hierarchy

Ackoff (1989) proposed a knowledge hierarchy relating data, information, knowledge and wisdom to an increasing degree of connectedness and understanding. I have added the dimensions of modelling and planning to Ackoff's system to illustrate how these components of knowledge supply the modelling and planning needs of NRM in a general way (Fig. 2.1).

Modelling requires knowledge, that is, understanding of a system's pattern of relationships. A model may be conceptual, qualitative or numeric. Modelling in landscape analysis typically uses real data, their relationships and process behaviour in systems. Modelling may be used to generate a better representation of landscape, such as, using digital elevation data and airborne geophysics to model and delineate land units. Modelling is also used to explore scenarios by altering model inputs or environmental variables, for example, a change in land use and its resulting hydrological impact. Planning requires decisions. Planning puts things in order for the future. Successful planning depends on foresight and wisdom. Proficient planning may use the results of models but decisions may not be dictated by the results of models, in fact the models used in planning may simply be conceptual. In Fig. 2.1 it is therefore suggested that modelling and planning do not directly overlap. This is a good illustration of the different domains occupied by scientist or modeller, and by those concerned with policy and planning decisions. This is not meant to imply that modellers are without wisdom, nor that planners necessarily exercise it!

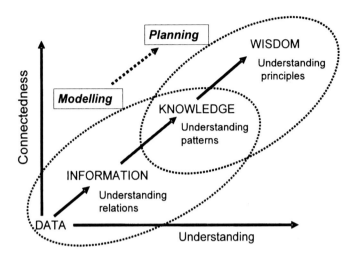

Fig. 2.1. From data to wisdom and modelling to planning; their relation to degree of connectedness and understanding

Decisions need to be made — where to do this or that, how much of this land to use, which ecological remnants to preserve, how much water to allocate to different uses. These decisions are in the domain of NRM and planning. They are both private and public decisions as they affect economic success as well as community wellbeing. Such decisions have wide reaching and lasting impact, many of them will take us, and the resources we use, down a one-way street or a road from which there will be no turning back. They are ephemeral decisions with terminal effects — we cannot readily reclaim agricultural productivity from bitumen and concrete, and we cannot effectively reconstruct lost ecosystems. The lines we draw on maps through natural systems can have irreversible consequences (Fig. 2.2). Some degree of reading between the lines is required to ensure harmonious interaction or sustained protection of use across artificial boundaries. McHarg, in his seminal work *Design with Nature* (McHarg 1969), pioneered the way and set the foundation principles now commonly applied in GIS and spatial models. Meine (1997) in his essay 'Inherit the grid' draws attention to the issues raised by disjunction between a north–south grid and the earth's curvature and the ecological disruption generated by land tenure.

Sound decisions require reliable methods to present and evaluate options; those who make the final decisions need effective means to engage others in the decision-making process. In the last two decades the decision processes for management of natural resources in Australia have become increasingly complex. Different levels and divisions of government, non government organisations serving the interests of primary industries or conservation, and individuals with a stake in the land are involved in seemingly endless iterative planning cycles. Strategies, action plans and consultative processes with open and transparent logics are all mandated for any aspect of government funded NRM investment.

Fig. 2.2. Lines that define land use artificially divide the natural world (Photograph: Peter Hyett, *The Bendigo Advertiser*, June 2003)

This, on one hand, is very positive as it is easy to become involved in providing specialist advice into these planning activities. In fact, as specialists, we are frequently asked to comment on strategies, provide scoping studies, model processes in the landscape, and propose research that can assist with answering NRM questions. On the other hand, it is difficult, or often feels difficult, to provide the quality of advice that is truly useful to decision makers. We need to find ways to close this gap and to improve communication and understanding from both sides. There is a general need to improve the timeliness and the quality of support for NRM. There are many issues that arise in consideration of improvement. From the research side, requirements for better data, increased understanding of processes and ability to validate model outputs are all seen as essential. From the decision maker's side, requirements are for simple indicators of NRM performance and clear options for moving forward.

2.3 Timelag between Question and Answer

Campbell (2006) espouses the need for a well-designed NRM knowledge system to focus on the interfaces between scientists and decision makers. He acknowledges that scientists are often frustrated that policy is often poorly informed by the best science, whilst policy makers are frustrated by the lack of timeliness in scientists' answers which usually carry requests for more research funding. Results of research are consequently delivered long after the political imperative has defined the question.

This issue of timelag between question and answer is wonderfully parodied in the *Hitchhiker's Guide to the Galaxy* (Adams 1979). Inhabitants of the planet Magrathea had become so frustrated by constant philosophical bickering that they built a super computer, 'Deep Thought', to provide the answer to the ultimate question concerning 'life, the universe and everything'. After 7.5 million years Deep Thought gave an annoyingly simple answer: 'forty-two'. This seemed incomprehensible and irrelevant because the ultimate question (terms of reference for the consultancy?) had been insufficiently defined. A more powerful computer (the Earth) was then designed by Deep Thought to run a ten million-year program to follow up the initial research and actually define the ultimate question. In Adams' fictional universe the follow-up research was unfortunately not completed as, five minutes before the ten million-year program was completed, the Earth was demolished by Vogons to make way for a hyperspatial express route. The demolition notice had been posted somewhat inaccessibly to Earth in Alpha Centauri and the planning approval signed by the galactic president, Zaphod Beeblebrox, 'with love', because he thought someone was asking for his autograph. There is a quaint bureaucratic allegory here too which could be elaborated, but which, at the very least, illustrates major failures in communication.

2.4 Organising the Questions

Adams' story provides an amusing allegory for many principles that govern our own work:

- research takes time
- there is always a lag between question and answer
- the answer needs to satisfy the question
- research should be driven by clear questions
- the answer needs to be simple, but not too simple
- usually the answer generates a new question
- answering the next question is usually more costly than the first
- general trend in science understanding and in models is towards greater complexity
- completion of research can depend on external factors that have nothing to do with the research itself.

The issues and processes are complex. The challenge, particularly for decision making, is to make these issues and complexes simple, or, perhaps more correctly, the challenge is to make them appear simple. Perhaps we need the equivalent of Douglas Adams' Babel Fish which is a (fictional) kind of universal knowledge broker and translator?

There are many approaches that can assist in engagement and closing the gap between scientists modelling processes, planners making decisions and the communities affected by the decisions. Visualisation methods and their associated models described in other chapters in this volume are all important examples of how this can be done in different or specific circumstances. However, to return to the mission stated in the opening paragraph, to step away from particular instances and consider some general overarching theory, it is worth considering an approach by Steinitz (1990). Simple principles are often the best, they can be returned to time and again and applied to different situations.

Carl Steinitz, an educator and Geographical Information Systems (GIS) specialist, sought an integrative approach to examine: the questions we ask, what we know about what we do, and what we teach. He was convinced that, despite individual differences and some collective professional differences in emphasis, there is an overwhelming and necessary structural similarity among the *questions* asked by, and of, landscape architects and other environmental design professionals. This led him to the development of an overarching framework for organising the questions associated with altering the landscape. This framework can be used: to organise applicable knowledge or models directed towards landscape change, to identify areas where contributions of theory are needed, and to assist in the decision-making process.

The landscape decision framework proposed by Steinitz (1990) links six levels of questions and models that are deemed essential to landscape design and landscape planning and is illustrated in Fig. 2.3. The framework is simple to explain and its logic is rapidly grasped.

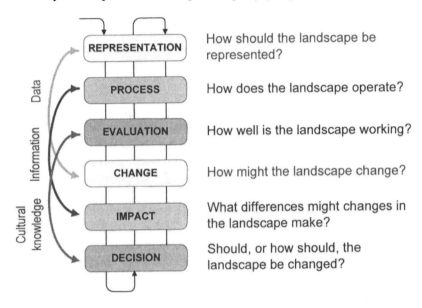

Fig. 2.3. Landscape decision framework (after Steinitz 1990)

This framework complements Ackoff's framework (Ackoff 1989) that is represented in Fig. 2.1. The requirements for data, information or knowledge can be related to each question. Representation of landscapes, present or future, simply require data, in this case these are attributes of land at a spatial scale sufficient to the question. Questions concerning processes in the landscape and potential impacts of change have information requirements, such as, relationships between land attributes and activities in the landscape. Evaluations concerning landscape (e.g. indicators of economic, social or environmental health) and preferred options for landscape change or preservation of the status quo sit firmly in the realm of (cultural) knowledge.

The questions are to a large degree interdependent but all should be driven by the decision question. The scope of the decision determines the spatial scale for landscape representation and the attributes that need to be included. It will also dictate the processes that need to be understood or modelled — hydrological, social, economic, climate, for example. In turn, process models have certain requirements for input data parameters and scale, and so on. As well as providing structure for a decision-making process, the framework thus provides a means of auditing the adequacy of data, information and knowledge required to make a sound decision. As a test of the simplicity and overarching applicability of this framework it is worth the reader's time, when reading the other contributions in this volume, to envisage how each of the areas of work presented are accommodated by, and contribute to, the schema presented in Fig. 2.3.

2.5 Integrating Disciplines

No single person in the NRM business can possibly have full command of each of these necessary building blocks for decision making. Multidisciplinary projects are therefore common in NRM. As the complexities of issues become recognised these projects also tend to be large. There is an underlying, sometimes explicit, assumption that these large integrated projects are the best way to reach the solution. Is this true, what can be learned from past efforts? What are the requirements for a successful large integrated multi-disciplinary NRM project?

Tress et al. (2007) investigated the issues surrounding these types of project. They used an online survey but also contacted some organisations and groups directly. Participants in the study represented thirty countries. Responses were in agreement about a number of fundamental issues. High on the list was the need for an *integration* plan that would define how the different disciplines would actually contribute to a single solution. There was also universal recognition that extra time was needed to enable communication between disciplines and to overcome the language and conceptual differences of the participating specialists. They emphasised the need to have regular meetings and events to develop a common language and to gain familiarity, trust and common understandings in the team. Contrary to popular practice they also concluded that it was better to plan smaller rather than larger projects. They also highlighted problems which can exist in any project but which become amplified in the larger projects. Their advice to plan realistic outputs that can be delivered on time, and to avoid setting expectations too high in order to please funding agencies and stakeholders should ring true for all of us.

2.6 Conclusion

So, we acknowledge that the job is not easy, would seem to need more time, certainly needs more clarity, but also needs to be done faster and smarter. A framework such as that of Steinitz can at least simplify project planning by partitioning the expertise and designing a process for effective communication between each of the six questions.

Returning to Adams' allegory, the super computer, Earth, was destroyed, but Slartibartfast and a team of terra-forming engineers were able to build 'Earth mark two' based on Deep Thought's original blueprint. We do not have that option — there is no backup for the system that we are trying to manage. It is therefore of the utmost importance that NRM is served by the best that can be provided in terms of decision support in the ascendancy from data to wisdom. Major challenges are to make what is complex, simple, and to provide expedient advice regarding scenarios for change. These challenges are in urgent need of our service.

Acknowledgement

The Victorian Government's Department of Primary Industries is acknowledged for provision of funding for the Our Rural Landscape project.

References

Ackoff RL (1989) From data to wisdom. Journal of Applied Systems Analysis 16:3–9

Adams DN (1979) The hitchhikers guide to the galaxy. Pan MacMillan, London

Campbell A (2006) The Australian natural resource management knowledge system. Land and Water Australia. Canberra, ACT

DPI (2007) Place and purpose – spatial models for natural resource management and planning'. Department of Primary Industries conference, 30–31 May 2007, Bendigo. Available at http://www.dpi.vic.gov.au/placeandpurpose

McHarg IL (1969). Design with nature. Natural History Press. New York

Meine C (1997). Inherit the grid. In: Nassauer JI (ed) Placing nature: culture and landscape ecology. Island Press, Washington, DC, pp 45–62

Steinitz C (1990) A framework for the theory applicable to the education of landscape architects. Landscape Journal 9:136–143

Tress G, Tress B, Fry G (2007) Analysis of barriers to integration in landscape research projects. Land Use Policy 24:374–385

3 Improving the Use of Science in Evidence-based Policy: Some Victorian Experiences in Natural Resource Management

Kimberley J Dripps[1] and Martin Bluml[2]

[1] Department of Sustainability and Environment, East Melbourne, Victoria, Australia
[2] Department of Primary Industries, Epsom Centre, Victoria, Australia

Abstract: This chapter describes the characteristics of effective science and policy linkages using several current examples relating to natural resource management, landscape analysis and visualisation. It ties these findings to the notion of effective evaluation and the pursuit of evidence-based public policy. Key features of successful case studies and empirical evidence from policy 'next users' reinforce the importance of research design tailored to address a particular policy need when policy makers are expected to utilise the findings. In a more general sense, research projects seeking to communicate more effectively with policy are most successful when project leaders pursue personal networks and exposure through mechanisms such as policy advisory committees. Research projects are also more likely to have policy influence when they provide a web presence that is easily located using commonly used search engines such as Google.

3.1 Context

The issue of science informing policy has perplexed researchers, research investors and policy makers for many years (e.g. Davies et al. 2000; Owens 2005; Lawton 2007; Sarewitz 2000). Most agree that science informing policy is a good thing, and that there should be more of it, but opinion is divided about how best to pursue this objective. What is clear is that improved communication is insufficient, and that scientists benefit from a greater understanding of the policy process, both to enable appropriate interventions into the policy cycle and to manage scientists' expectations of impact (Sarewitz 2000).

This chapter examines evidence-based policy from the perspective of a research investor, and offers insights into how scientists can increase the relevance and impact of the work they pursue. The chapter is based on observation, case study analysis and literature review. The chapter includes analysis of survey and market research data collected by the Integrating Farming Systems into Landscapes research and development investment program (IFSL) in the Victorian Department of Primary Industries (DPI). It also explores some interventions that were made by the program to increase researchers' understanding of potential 'next users' of their work. Several successful case studies of research informing policy are also included.

3.1.1 Historical Perspective

Mulgan (2006) provides a detailed analysis of the history of government across the history of civilisation. The context is briefly summarised here, to inform understanding of the purpose of government. The concept of government appeared with the emergence of surpluses of food. Pre-agrarian civilisations needed little in the way of government, as they moved across the landscape, harvesting resources and conducting battles where necessary to displace others. With the emergence of agrarian civilisation, humans began to produce surpluses of food and needed to be able to ward off intruders who sought to over-run their communities. Increased investment in houses and infrastructure meant that societies were much more reluctant to 'move on' when intruders threatened (Mulgan 2006).

Governance arrangements began to be set up to enable communities to distribute surpluses, wage wars and manage disputes. Over time governments have evolved to manage more complex affairs, embracing accountabilities for the wellbeing of communities, managing health and welfare, as well as security. As populations increased governments also embraced roles relating to 'tragedies of the commons' (Hardin 1968) managing access and use of common areas to avoid their over-exploitation by single individuals (Mulgan 2006). In policy terms these roles are referred to as forms of public value and there is much argument within advanced democracies about the balance of investment across the range of different forms of value that could be invested in.

Ultimately, in advanced democracies, governments wish to make wise decisions. Governments aim to optimise progress and minimise outrage, which means that decisions must be sufficiently popular with the electorate to ensure that they are not overturned. Mulgan (2006) argues that in pursuing these aims 'an advanced democracy cannot function without the constant production and dissemination of data and knowledge'. However he also notes that '…whose facts are reliable becomes all important, and a critical battleground between journalists and politicians, civil servants, non governmental organisations and scientists.'

3.1.2 The Policy Process: Towards Evidence-based Policy

To better understand the policy process it is worth considering the roles of some of the key players, and what policy is. Policy makers are those in public service who advise Ministers and other elected officials, who actually make the decisions on the appropriate course to choose. Many others have an active role in shaping policy, including activists, industry organisations, lobbyists and practitioners. Policy defines the rules of society, as agreed through democratic processes. Policy is a mixture of art and science, and tends to emerge rather than being consciously chosen. Policy making is a complex process and it is difficult to identify the beginning, middle or end of the process (Nutley and Webb 2000).

In order to make wise decisions politicians and policy makers often seek evidence-based policy. Evidence may relate to the science that informs the policy choice, or to evidence of program efficacy. According to Lubchenco (1998) the role of science in informing decisions is becoming a crucial and unmet challenge for civilisation. Evidence-based policy builds on the tradition of evidence-based medicine, and aims to build public policy interventions based on a solid body of knowledge and evidence (DEFRA 2007). From an evaluation perspective 'evidence-based policy and practice' refers to areas where evaluative methodologies are built into program design. Evaluation results are used throughout a project to refocus and improve actions to optimise outcomes. Knowledge is built through the use of short learning loops. At the end of the program, evaluation data are used to demonstrate outcomes and to facilitate learning of 'what worked best, for whom, why and when?' (Davies et al. 2000; Funnell and Rogers 2006).

Adaptive management, or adaptive resource management (Lee 1999), has a similar construct. Adaptive management uses a repeated decision-making process, often underpinned by modelling and reflective practice, with regular monitoring, to reduce uncertainty and optimise natural resource management (NRM) outcomes. Evidence-based policy and adaptive management are sufficiently similar for the approaches outlined in this chapter to be applied in both endeavours.

Victoria's quest for evidence-based policy has followed the lead of countries like the United Kingdom, where the Blair Government's focus on 'what works?' is thought to have increased interest in underpinning government interventions with evidence of effectiveness (CM4310 1999; DEFRA 2003; Solesbury 2001). Other advanced democracies, such as the United States, have resources such as the Coalition for Evidence-based Policy (CEBP 2007) to assist policy makers in understanding how best to intervene in complex and complicated program areas. New Zealand and Canada have also adopted the evidence-based approach in health care, biotechnology, and other contentious areas (e.g. HRCNZ 2003; De Broucker and Sweetman 2001; Marsh 2006). Recent thought highlights the need for outcomes focused planning, evaluation, and the development of a body of evidence to support interventions made by government (Pawson 2006).

3.1.3 Use of Science as Evidence in Policy

There is a body of work that explores why the science–policy interface does not work well, and what might be considered to address this shortfall. Lawton (2007) provides an excellent summary of the literature in this area, and offers 11 possible reasons for this shortfall. These are summarised as follows:

- We (scientists) are to blame, as we do not transfer messages clearly enough.
- There is too much science, and policy makers do not know where to look for appropriate evidence.
- The science is ambiguous.
- There is insufficient public support for what needs to be done.
- Policy has to consider many things other than the science.
- Scientists and policy makers work in very different time scales.
- Policy makers are impacted by powerful interest groups with other agendas.
- There is a failure in governance and institutions which means the information does not get to the right people at the right time.
- The solutions require international agreement, which is difficult to achieve.
- Scientific advice is inconsistent with political wisdom or public values.
- Some politicians are corrupt and will make decisions for money.

Owens (2005) offers three different cognitive constructs that might be seen at play in the science–policy interface. 'Technical rationality' suggests that the interface is linear, with outputs of science being inputs into policy processes. While this notion has powerful intuitive appeal and may sometimes be true, it is also true that policy is rarely informed by research in a straightforward manner. 'Strategic use of knowledge' exposes the use of knowledge by policy makers during the period of decision making. In this context power comes into play and counter evidence may be presented by different sides of an argument. The context of 'cognitive perspectives' understanding is more complex, but (Owens argues) more promising. It suggests that information is a critical part of the decision-making process and that although questions may be cross-disciplinary and problems unstructured, different kinds of knowledge are critical in reaching conclusions. Clearly specific research has a role in this system, and the key is to ensure that relevant knowledge may be easily found by policy makers.

Sarewitz (2000) purports that scientists are aiming at the wrong goal if they seek to assist in the resolution of political controversies. He indicates that scientific influence on policy will be more satisfactory if researchers focus on areas where they can alert society to potential future challenges and problems, and where they can be part of guiding action after political consensus has been achieved. Sarewitz (2000) refers to the case of the management of acid rain where $600 million in research and ten years work generated much excellent science, but no progress in resolving the political differences among the states needing to cooperate to resolve the issue. He contrasts this with the progress made by science supporting the economic policy instrument (a cap and trade system to control acid rain) after political agreement had been reached.

Nutley and Webb (2000) remind us that the construct of the science–policy interface is in science informing policies that are being considered, and in assisting in the administration of them. Science will never be the sole determinant of a policy decision-making process. This construct has been in existence at least since the Fabian Society was formed in 1883 (Nutley and Webb 2000). Approaches have evolved and the emergence of new public management in the 1980s and 1990s resulted in a focus on:

- the practice of evaluation
- other forms of social research
- using evaluative findings as evidence to inform future policy.

Lubchenco (1998) notes that science does not provide the solutions: but has a critical role in understanding the consequences of different policy options. She offers a new social contract for scientists suggesting that scientists have three responsibilities:

- to address the most urgent needs of civilisation

- to communicate knowledge broadly to inform decision making at multiple scales
- to exercise good judgment and humility in their approach.

In this context it is important to recognise the dominance of humans on the functioning of the planet and to pursue truth with vigour. Lubchenco (1998) notes that:

> 'scientific understanding can help frame the questions to be posed, provide assessments about current conditions, evaluate the likely consequences of different policy or management options, provide knowledge about the world and develop new technologies'.

Lubchenco (1998) also highlights the need for better mechanisms to allow the investigation of complex multi-disciplinary issues spanning multiple temporal and spatial scales. Current challenges in natural resource management include salinity, catchment management and climate change. She also suggests that there is value in specifically training scientists to work at the science–policy and science–management interfaces, and in enhancing science communication using scientific assessments from large groups of credible experts, such as is done by the International Panel on Climate Change (IPCC). This is a challenge as most science uses a reductionist approach which may conflict with a systems solution. Nutley and Webb (2000) chart the emergence of the Blair Government and the focus on 'what counts is what works'. They discuss in detail models for six different ways in which evidence informs policy, including: knowledge driven model, problem solving model, interactive model, political model, tactical model, and the enlightenment model.

Nutley and Webb (2000) note that research is more likely to inform policy where:

- implications are non-controversial
- changes required are within a program's scope or easy to implement
- the program environment is relatively stable, or
- when a program is in crisis and nobody knows what to do.

They also note the different cultural constructs between the USA and Europe and note that to a degree these influence the uptake of research findings. The implications of these for Australia are unexplored.

Nutley et al. (2007) refine this thinking by raising four mechanisms by which research informs policy:

- knowledge impelling action
- percolation

- knowledge grabbing for problem solving or tactical considerations
- interaction resulting in co-production of knowledge.

It is clear that policy can be directly informed by research if the research and development is designed with policy input and with policy users in close engagement. It is also clear that research that is *not* designed with a direct policy need in mind can nonetheless be communicated more effectively.

3.2 Some Victorian Experiences in Natural Resource Management

The Victorian Department of Primary Industries (DPI) is a government department that incorporates policy, research, development and extension within a single organisational structure. A potential strength of this department therefore, is in the ability of research to directly inform policy. The remainder of this chapter examines how strengthened science–policy interfaces have been pursued within one of the research and development investment programs within DPI, and provides insight into what might work well in an integrated government department. The findings may also have more general applicability for all researchers seeking to influence public policy.

The study on which this chapter is based has five components:

- a semi-structured telephone survey of DPI policy staff
- market research into next user requirements of the Integrating Farming Systems into Landscapes investment program
- a project commissioned to improve the utility of project outputs
- analysis of eight years of experience (by the lead author) in working in the public policy environment in the Victorian government
- analysis of case studies where the connection between research and policy has been mutually informing.

The five components were conducted simultaneously as part of the management of the Integrating Farming Systems into Landscapes investment program (IFSL). The program was one of six parts of the Agriculture Development Division's investment portfolio, which was directed at delivering the DPI's aspirations in the development of agricultural industries.

The program was developed on the basis that increasing our understanding of the impacts of agriculture on the environment and surrounding ecosystem will allow agricultural industries (including private forestry) to proactively manage the inherent risks and consequences of farming activities, including issues around the license to operate within an authorising environment. This is expected to lead to improved catchment and landscape level biophysical outcomes in the longer term, through a behavioural change at the landholder level. Behavioural change in land managers may be encouraged through voluntary or involuntary policy instruments, thus the interest of the program in evidence-based policy.

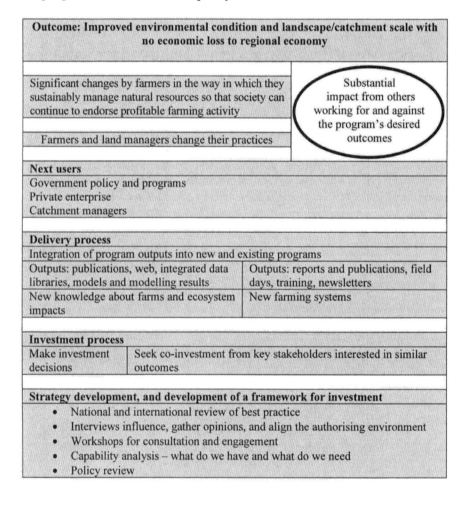

Fig. 3.1. Integrating Farming Systems into Landscapes outcome hierarchy

The IFSL program has an underpinning 'theory of action' (Fig. 3.1) that informs understanding of how the projects contribute to outcomes. The work described in this chapter was part of focused interventions to enhance performance in areas where evidence of program effectiveness was weak. This was in the area of transfer of outputs to next users. Following the collation of material from the five components the aim was to interpret and understand the meaning behind the data, so that practice is informed as a result. This approach is known as a hermeneutic phenomenological approach (Laverty 2003), and contrasts with a more traditional reductionist problem-solving methodology. For optimum readability each component is presented sequentially and then insights from synthesis across the five areas are presented.

3.2.1 Survey of Policy Analysts

The first component of the study was a survey of policy analysts. Survey results confirmed the lead author's observation and experience of the work of policy analysts. In February 2007, six policy analysts from DPI were surveyed using a semi-structured interview approach to determine how policy analysts seek scientific information and knowledge pertaining to their policy areas. Respondents were also asked in what form information based on research projects would be most useful and accessible. Respondents reported that personal networks play the greatest role in sourcing information, and that these networks generally extend across DPI and in some instances into other departments, statutory authorities and rural industry research organisations. Linkages to networks were generally established through attendance of policy officers at project working groups, technical support groups, seminars or forums. Other sources of information that were noted were Google searches, historical policy documents and the assistance of DPI librarians. It was noted that DPI has a strong culture of oral history and that quite often formal records did not provide the rich picture and responsiveness that policy officers require in sourcing information. Interestingly, tools built specifically to extend information from DPI research, such as the Victorian Resources Online website <http://www.dpi.vic.gov.au/dpi/vro/vrosite.nsf/pages/vrohome>, fact sheets and workshops were not cited as information sources. The business management tool — Contract Management Interface — which is the repository of all research, development and extension projects undertaken was also an unused resource for policy staff. It should be noted that the system had only recently been introduced.

The time and effort involved in sourcing information is critical for policy staff, and most policy officers feel that sourcing key messages and 'learnings' from research projects involves significantly more work than is warranted by what is learned. In general, access to key facts for succinct briefings was stated to be more important than access to a depth of information. Policy officers reported that they were willing to be more informed by research projects provided that material was not difficult to access and contained a succinct statement of project findings and key messages. There is a tension between the nature of scientific thought with its detail, and the pragmatic need of policy to provide direction in succinct and time-bound circumstances.

3.2.2 Market Research

The second component of the study was the 2006–2007 market research conducted to better understand the needs of anticipated 'next users' of research project outputs. Next users are the people who will use the outputs of a project, either during the project life, or after it has been completed. For example, if a researcher discovers a new way of saving water on a farm, the next user is most likely to be a farm management consultant or extension officer who works with farmers to facilitate changed farmer behaviour as part of an integrated farm management package.

The Lucas Group (2007) undertook the market research, conducting a telephone survey of 91 people using a scripted interview process. Respondents included policy staff from DPI, policy staff from DSE (Victorian Department of Sustainability and Environment), staff from Catchment Management Authorities (CMAs), representatives of Landcare groups and representatives of agribusiness service providers. These groups were selected as they are the anticipated (or expected) next users of the IFSL program's outputs. Most participants had become aware of the program and its projects through their work, although few could elucidate its purpose. Most people did not recognise the name of the program, but had some awareness of its activities.

The majority of participants were aware of Victorian Resources Online and most were aware of 'Greenhouse in Agriculture' and the Environmental Management System 'Connecting Farms to Catchments' projects. The satisfaction with interactions with the program and projects was generally high, with the exception of DSE policy staff, where a significant number rated the program poorly. Anecdotal evidence collected independently of this market research, suggests that DSE policy staff have a perception that DPI projects are sometimes not well managed and fail in delivery. The preferred methods for policy officers, CMAs, Landcare groups and agribusiness firms to receive communication were mail, email plus website and one-to-one meetings, in that order (see Fig. 3.2).

Improving the Use of Science in Evidence-based Policy   39

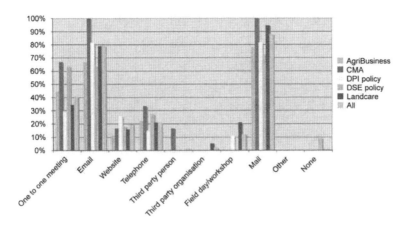

Fig. 3.2. Lucas Group survey responses to the question: 'How do you prefer to receive information?'

When looking for information on science-based questions, the majority reported that they would search the Internet (see Fig. 3.3). When viewed together website and Internet are overwhelmingly the preferred source of information. Although a coarse instrument, the Lucas Group survey confirmed the need for information to be easily sourced as a key feature in research informing policy development.

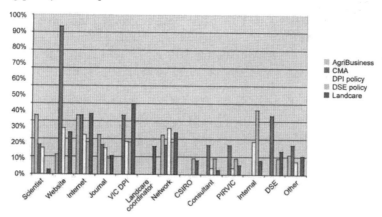

Fig. 3.3. Lucas Group survey responses to the question: 'Where do you go for information on a science-based question?'

3.2.3 Improving the Utility of Project Outputs

The third component of the study was a series of workshops to improve the utility of project outputs. The Theory of Social Marketing (Dripps 2001; Kotler and Roberto 1989) was exposed to researchers to underpin an increased understanding of the 'product' that is being developed and the requirements of the users of those products (Linehan and Wright 2007). One of the challenges that emerged from the workshops was the ability of researchers to consider the needs of the end users of their work at an early stage in planning. The analogy used by the presenter was that 'people do not go to the hardware store because they want to buy drill bits — they go because they want holes'. This is helpful, as it simply shows how far apart the minds of the creators and users of products can be. Research producers are no different. The ability to focus on outcomes, and whether the outcomes would be of use to the eventual end users, was a significant achievement of the project.

3.2.4 Observation of How Policy Decisions Are Made

Observational studies can offer insight into science–policy linkages. The fourth component of the study was analysis of personal policy experience. During eight years of experience in the Victorian public sector policy environment, the manner in which policy decisions are made in a number of departments (Treasury and Finance, Premier and Cabinet, Natural Resources and Environment, Sustainability and Environment, Industry Innovation and Regional Development, and Primary Industries) was observed by the lead author. This experience has been mapped and the roles of various types of analysis interpreted.

Across the Victorian Government departments examined, four 'overlays' are evident in policy decision making. These are economic impact, environmental impact, social impact and political implications. These may be informed at various stages by research and development. It is important for both researchers aspiring to inform policy development, and policy makers aiming to provide a greater evidence base for policy decisions, to be aware of these overlays. In addition, it is important for both parties to be aware of the differences in work environment, which will impact the communications process.

Scientists and policy makers work in extremely different environments and assumptions of both parties about how and where to seek data and knowledge prevent beneficial interactions. It is instructive to consider the work environment of each party briefly.

In a policy business the key clients are the Minister and the Secretary, followed by senior management of the department. The units of work prepared by policy analysts include briefing papers, responses to correspondence and parliamentary questions, and are often quite small. While the work of an analyst will be focussed on a particular issue and there is usually a strategic direction in which the issue is being managed, there is a high degree of responsiveness (or reactiveness) in day-to-day work. This means that policy analysts may not be able to keep appointments that have been made or to travel to extended workshop format events. The results of the interviews and market research support this observation. Policy analysts, overwhelmingly, gather information from internal or external networks and from the Internet. When they use the Internet they generally use a common search engine, such as Google.

In a research business, work is carefully considered and planned on a project basis. The key clients are the investors, who have had to be satisfied about a level of detail of the proposed work before a funding decision is made. Units of work include literature reviews, experiments, analysis and scientific papers, and are often quite large. Researchers have time and space to explore the detail of an area of work and develop a high level of technical expertise in their discipline. Researchers can plan their work schedules to capitalise on opportunities for interactions. However, given the organised nature of projects, it can be difficult to reschedule appointments or cancelled workshops. Researchers gather evidence comprehensively based on the discipline of the literature review and are reluctant to make assertions that cannot be confirmed by a body of evidence. Mulgan (2006) however suggests that 'most intellectual disciplines have a far stronger sense of service to themselves than of service to the people, and most assume that the public are incapable of interpreting their own lives'. This might be an important point for researchers contemplating greater policy influence.

The different work environments of policy and research are like a cog and a flywheel — one running extremely rapidly and responding to significant changes in pressure from the external environment, and the other operating at a comparatively more measured and considered pace. This analogy graphically illustrates the differences between the businesses. It is suggested then that a gear-like mechanism might be helpful in mediating the interface.

3.3 Case Studies of Successful Science–Policy Influence

One of the challenges for researchers is to understand how and when to interact with policy thinking to ensure the greatest policy impact from the scientific work undertaken. The fifth part of the study examines a number of case studies that exemplify the science–policy interface. The case studies included the following projects: Sawlogs for Salinity, Salinity Investment Framework 3, Soil Health, and Greenhouse in Agriculture.

3.3.1 Sawlogs for Salinity

The Sawlogs for Salinity project used hydrological catchment modelling to determine the externalities imposed, and benefits generated, by the establishment of plantation forests for sawlog production. Hydrological modellers interacted with policy makers and economists to design and trial an auction-based system to allocate funds available for sawlog plantation development.

Evaluation of the program and personal communication with those involved in the program exposed issues including the following:

- At one point the science was challenged, creating a public controversy for the project that needed to be managed.
- Policy needed to make decisions beyond the scope of what researchers could do with certainty and this created tensions in the project team.
- The scope of the project changed as time passed and policy issues were discussed and debated in other forums.

Researchers reported frustration at the planning process and the level of certainty in the project plan. Nevertheless, the extraordinary focus of the research leader on communication and his willingness to take on and find a home for tasks (beyond the scope of a normal research role) were instrumental in the success of the project.

3.3.2 Salinity Investment Framework 3

The Salinity Investment Framework 3 project proposes a model for considering which policy mechanisms are most likely to be beneficial in delivering salinity outcomes at a catchment scale. While the processes and issues relating to the conduct of the project have not been examined in depth, it is clear that the lead researchers show the following characteristics:

- One of the researchers runs a regular web-based discussion list, providing 'brief pieces on issues and ideas in economics, science, the environment, natural resource management, politics, agriculture and whatever else'.

- The material provided is topical, readable, and links to further work.
- The researcher also responds extremely promptly to email enquiries about his discussion list posts, even when he/she does not know the person seeking information and discussion.
- The researchers show a willingness to be involved in national specialist reference groups and to engage in debate and discussion about ideas with policy makers and investors.
- The researchers are also keen to travel to brief key policy makers and policy forums, and show an acute understanding of the importance of ideas being available and digestible in a policy friendly form.

3.3.3 Soil Health

The Soil Health — Leaving a Legacy for South Eastern Australia project provides an example where a public policy process of parliamentary enquiry has exposed an issue where the engagement of researchers in helping policy respond resulted in a research program. In this case the foresight of science managers in making researchers available on a short-term basis to undertake policy tasks has been rewarded with funding for a new program of work. It remains to be seen whether the close interaction between policy and science that developed in this case will continue in the absence of a future driver. It also remains to be seen if the research undertaken meets policy makers' needs on an ongoing basis.

3.3.4 Greenhouse in Agriculture

The Greenhouse in Agriculture work is an example where public policy has evolved with very close connections to research problems, and projects have been specifically designed to inform policy gaps. The Victorian Greenhouse Office and the Australian Greenhouse Office were both new entities servicing areas of emerging policy in the late 1990s. As part of their agenda setting they identified gaps in knowledge that were of policy relevance and commissioned research and development to fill those gaps. In agriculture early work focused on understanding and measuring emissions from production systems. In Victoria the work was undertaken in partnership with New Zealand and Canada in a ground breaking effort that resulted in the refinement of Australia's greenhouse gas inventory.

Key features of the success of this science–policy interface were:

- The energy and enthusiasm of the project leader.

- The international standing of the research leader and the team assembled.
- The presence of a cooperative research centre that focused effort, provided funding, and coordinated strategic planning.
- A dedicated investment manager with a strong commitment to ensuring smooth relationship management.

More recently policy differences between the State and Commonwealth have put strain on the research–policy interface in this area.

3.4 Discussion

These case studies show that the two areas where the science–policy interfaces work well are; where policy mechanisms are evolving and scientific evidence is required to inform specific gaps in understanding, or where research is required to deliver on an innovative policy mechanism. These areas are characterised by policy problems that are difficult and multidisciplinary. Choosing an appropriate and topical policy area to research is insufficient to ensure success at the science–policy interface. Researchers involved in well functioning science–policy interfaces appear to have a distinct set of interpersonal and communication skills that enable appropriate and timely communication. These researchers used these skills to participate in influencing activities, including national reference and advisory groups, well beyond the scope of the standard role of a researcher.

3.4.1 Toward Better Use of Science in Evidence-based Policy

It is clear from the preceding analysis that policy users want precise, rapid, easily accessible information provided in summary form for their day-to-day briefings. Ideally they also desire a forum where the interpretation of the data and information can be debated. It is ideal for knowledge to be easily available using common search engines, and the conduct of email lists may be considered if the researcher is sufficiently confident and committed to that form of communication.

A challenge for researchers is how to appropriately engage policy makers in the design stage of research projects where a policy impact is desired. Researchers may identify appropriate policy personnel for engagement in the early conceptual planning of work, with a key challenge being the analysis of policy networks to determine whether policy analysts, policy managers, central agencies or other influencing departments are appropriate targets.

This will vary on a case-by-case basis. Policy analysts can be targeted for individual briefings, invited onto design groups or steering committees, or provided with succinct briefings focusing on implications and uses, rather than descriptions of the state of knowledge.

The policy context ought to also be considered. There will be situations where new knowledge can inform policy makers about issues to be managed, or the consequences of particular policy decisions. There will also be situations where new knowledge and new technologies can be harnessed in the development of policy mechanisms, such as market based mechanisms. However, it needs to be remembered that in a democracy, decisions are taken by politicians, and that many sources of information and advice are considered. In this arena researchers may consider efforts to engage and inform politicians. These are often best pursued in an organisational context to ensure that potential audiences are not flooded with information.

One important element of improving the science–policy interface is the management of expectations of researchers. Not every research area will have policy relevance, or require such relevance, and not every eminently sensible idea will be incorporated into policy decision making. Also, well-planned policy interactions may not work due to personalities involved. Owens (2005) also observes the phenomenon of research undertaken many years previously having emerging policy relevance. This challenged her, as her work had moved beyond the research that was gaining policy traction.

It is clear that there is a degree of serendipity about some policy–science interactions. This suggests that networks have value, even without specific intent. Researchers might therefore consider cultivating policy linkages with no particular intent apart from the sharing of ideas. In time some of these ideas may gain traction or the relationship may be transferred among policy analysts.

One of the notions also worth exploring further is the role of scientists and other lobby groups, in advocating for an outcome that is well beyond the decision that government is able to make. This can be extremely helpful in public policy debates as governments are usually unwilling to move beyond the court of public opinion. Sarewitz (2000) points out that the role of science in directly informing policy decisions is small, but that science plays a critical role in alerting society to potential challenges that lie ahead and in guiding action after political consensus is achieved. He states that the role in actual decision making is minor because opposing sides of an argument can always find science to back up their positions. He also notes that conducting another study is often used in the policy process to delay a decision, without appearing to be obstructionist. Despite this, research can be useful in moving public opinion on an issue, in some cases.

The availability of information, counter-information and mis-information are changing the nature of public policy development. The emergence of network governance is changing the hierarchical nature of policy development. Decisions are informed by more sources and therefore are likely to be more confusing in the process, but of better quality in the long-term. In the future, data and information intensity will increase, which will challenge researchers and policy analysts — both of whom are knowledge workers. It will be important that the interpretation of data, information and knowledge is based on a range of mental paradigms, including both research and policy.

The management of data and knowledge will be an increasing challenge and the potential use of modelling outputs will increase and become more contested as models become more complex. An equal danger is the increasing pressure to use models that are either poorly conceptualised or have not yet been properly validated. As citizens become more able to bypass experts the possibilities of mis-interpretation and mis-information will increase. However this environment also provides opportunities for innovative science communicators. It also is worth considering the future role of intellectual property protection in the context of network governance. It may become more appropriate for work to be freely available, rather than protected. This will clearly be a priority decision for research organisations moving forward.

3.5 Conclusion

Based on research on the policy–science interface, positive interaction and incorporation of scientific findings into policy is most effective when:

- mechanisms for engaging policy in the design stage are pursued, with these mechanisms designed to suit the working style and needs of policy analysts
- researchers adopt a proactive approach to the policy–science interface, based on networks and the translation of findings into policy friendly language
- findings are placed so that they can be accessed easily using common internet search engines
- contextual considerations are taken into account
- the expectations of researchers regarding 'visible evidence of impact' of research on policy decisions are well managed.

In conclusion, Lubchenco's advice (1998) on the three responsibilities of scientists is worth bearing in mind: first to address the most urgent needs of civilisation, second to communicate knowledge broadly to inform decision making at multiple scales, and third to exercise good judgment and humility in the approach.

Acknowledgements

We are grateful to Chloe Munro for stimulating thought in this area, Michonne van Rees, Bruce Kefford, Catherine Hollywell, Kevin Love and all our policy mentors and colleagues. Also to DPI staff working on Integrating Farming Systems into Landscapes projects, and the Program Investment Managers. In particular, Kimberley would like to thank Anna Ridley, David Halliwell and Bron McDonald for their fearless editorial advice, and Roger Armstrong for providing the Lawton (2007) paper at a critical time.

References

CEBP (2007) Centre for Evidence-based Policy. Retrieved 16 July 2007, http://www.excelgov.org/admin/FormManager/filesuploading/Coalition_purp ose_agenda_3_06pdf and http://www.evidencebasedpolicy.org/defaultasp?sURL=ombI

CM4310 (1999) Modernising Government White Paper. The Stationary Office, London

Davies HTO, Nutley SM, Smith PC (2000) What works? Evidence-based policy and practice in public services. The Policy Press, University of Bristol

De Broucker P, Sweetman A (2001) Towards evidence-based policy for Canadian Education. McGill-Queen's University Press, Queen's University, Kingston, Ontario

DEFRA (2003) Delivering the evidence: DEFRA's science and innovation strategy, 2003–2006. Department for the Environment, Food and Rural Affairs, London

DEFRA (2007) Evidence-based policy making. Retrieved 8 August 2007, http://www.defra.gov.uk/science/how/evidence.htm

Dripps K (2001) Will marketing science make a greater contribution to government organisations which contribute to natural resource management in this millennium than it did in the last? MBA assignment, RMIT University, Australia

Funnell S, Rogers P (2006) Evaluating complex and complicated programs: issues, approaches, implementation and implications — lessons from the evaluation of the stronger families and communities strategy 2000–2004. Australasian Evaluation Society Conference, 4–7 December 2006, Darwin, Australia

Hardin G (1968) The tragedy of the commons. Science 162:1243–1248

HRCNZ (2003) Partnerships for evidence-based policy and practice. Health Research Council New Zealand. Retrieved 27 August 2007, http://www.hrc.govt.nz/assets/pdfs/publications/EvidencedBasedPolicy.pdf

Kotler P, Roberto EL (1989), Social marketing. The Free Press, Macmillan Inc, New York

Lawton JH (2007) Ecology, politics and policy. Journal of Applied Ecology 44:465–474

Laverty SM (2003) Hermeneutic phenomenology and phenomenology: A comparison of historical and methodological considerations. International Journal of Qualitative Methods 2(3) Article 3. Retrieved 8 August 2007, http://www.ualberta.ca/~iiqm/backissues/2_3final/html/laverty.html

Lee KN (1999) Appraising adaptive management. Conservation Ecology 3(2):3

Linehan C, Wright V (2007) What is your product? Department of Primary Industries workshop presentation delivered 7 February and 22 March 2007, Melbourne, Australia

Lubchenco J (1998) Entering the century of the environment: a new social contract for science. Science 279:491–497

Lucas Group (2007) Market research report for DPI Victoria, integrating farming systems into the landscape program. Unpublished report for the Department of Primary Industries, Victoria

Marsh D (2006) Evidence-based policy: framework, results and analysis from the New Zealand biotechnology sector. International Journal of Biotechnology 8:206–224

Mulgan G (2006) Good and bad power: the ideals and betrayals of government. Allen Lane, London, pp 257–262

Nutley S, Webb J (2000) Evidence and the policy process In: Davis HTO, Nutley SM, Smith PC (eds) What works? Evidence-based policy and practice in public services. The Policy Press, University of Bristol

Nutley SM, Walter I, Davies HTO (2007) Using evidence: how research can inform public services. The Policy Press, University of Bristol

Owens S (2005) Making a difference? Some perspectives on environmental research and policy. Transactions of the Institute of British Geographers, 30:287–292

Pawson R (2006) Evidence-based policy: a realist perspective. Sage Publications, London

Sarewitz D (2000) Science and environmental policy: an excess of objectivity. Retrieved 17 July 2007, http://www.cspo.org/products/articles/excess.objectivity.html

Solesbury W (2001) Evidence-based policy: whence it came and where it's going. Economic and Social Research Council UK, Centre for Evidence-based Policy and Practice, University of London, retrieved 7 May 2007, http://evidencenetwork.org/Documents/wp1.pdf

4 The *Catchment Analysis Tool*: Demonstrating the Benefits of Interconnected Biophysical Models

Anna Weeks[1], Brendan Christy[1], Kim Lowell[2] and Craig Beverly[3]

[1] Department of Primary Industries, Rutherglen Centre, Victoria, Australia
[2] Department of Primary Industries, Parkville Centre, Victoria, Australia
[3] Department of Sustainability and the Environment, East Melbourne, Victoria, Australia

Abstract: The Catchment Analysis Tool (CAT) *is a hydrological model that helps to define the surface and subsurface movement of water and nutrients in a catchment, and evaluate the impact of different farming systems and land management strategies on vegetative growth and productivity, stream quality, streamflows and groundwater. It was created in response to a growing desire by decision makers to be able to examine modified landscapes and look at the trade-offs in biophysical responses such as, crop yield, soil erosion, salinity and water dynamics.*

The CAT *captures the hydrological dynamics of the whole landscape by combining a suite of mature models that individually describe a variety of landscape processes such as crop growth, forest growth, grazing systems, water balance and groundwater models. The basic spatial and temporal input data required to run the* CAT *are generally commensurate with available data. These include land use, elevation, soil characteristics and climate data. Detailed analysis of groundwater systems and watertable mapping requires a full geophysical conceptualisation of the underlying groundwater systems and an extensive calibration process based on statistical optimisation, expert consultation, and modification of underlying assumptions about the groundwater systems being modelled.*

At the catchment scale there are very few points in the landscape where relevant data are measured and captured, and as such, validation of the CAT *remains challenging. Where possible modelled results are validated against data captured from stream gauges, bore data and point studies. This chapter describes the* CAT*, discusses briefly the individual components and their linkages, and provides a case study example of how the* CAT *can be used to inform a policy decision. A discussion of the weaknesses and difficulties inherent in the creation of such integrative modelling frameworks is also presented.*

4.1 Introduction

Policy makers function in an environment in which they are constantly called upon to make decisions that require balancing the costs and benefits of different outcomes relative to the social, economic, and environmental values of different assets. In an agricultural and natural resource context, favouring economics alone will lead to inevitable environmental degradation, whereas focussing solely on environmental health will eliminate the capacity of a landscape to provide the financial prosperity on which societies depend. This difficulty is exacerbated in natural systems because of the interconnectedness among different landscape elements. One cannot, for example, increase irrigation without decreasing the amount of water available to maintain environmental flows.

The preceding paragraph argues the need in landscape management for models that:

- reflect the interconnectedness of landscape systems
- can accommodate a variety of land management options (e.g. irrigation, forest establishment, change of crop type)
- furnish information on a number of 'environmental services' (e.g. clean water, agricultural productivity) provided by a landscape.

Unfortunately, most biophysical models focus on specific processes rather than considering the holistic integrated environment. As such the philosophy underlying them addresses a limited number of outputs. Such models address a relatively limited set of the dynamics that drive landscape processes.

This is not to say that individual models do not have a high level of sophistication. On the contrary, since the early 1970s when complex, computer-based models became relatively commonplace, models have become increasingly sophisticated. As such, they have expanded in functionality and their ability to use complex mathematical constructs to describe dynamic natural processes. However, individual modellers have tended to focus on a single domain in which they are experts such as forest growth or soil erosion.

One of the biggest current and future modelling challenges is the linking of mature models so that landscape processes can be described holistically. By creating such linkages, questions will not be limited to a single domain, but instead, policy makers will have the capacity to ask 'Will changing the forest management regime to *regime X* increase timber supply, maintain soil erosion at current levels, and provide sufficient streamflow for aquatic populations?' Linking models that reflect the real-world interconnectedness of landscape processes will also allow an inversion of this question, 'Are there any forest management regimes that will increase timber supply while maintaining current levels of soil erosion and not cause a substantive decrease in streamflow?' And if such models are spatially based, decision makers can also ask questions about location, such as, 'Where should one change the timber management regime so as to minimise the impact on soil erosion and streamflow?'

The *Catchment Analysis Tool* has adopted this holistic approach linking selected models in a way that reflects interactions among real-world processes and that uses data that are widely available.

4.2 *Catchment Analysis Tool*: Background and Description

The *CAT* was originally developed in the early 2000s by the Department of Primary Industries (DPI) Victoria in collaboration with the Cooperative Research Centre (CRC) for Plant-based Management of Dryland Salinity (CRC for Salinity). The underlying philosophy of incorporating models from a variety of disciplines has made it almost inevitable that the *CAT* would continue to evolve. Since its creation, the *CAT* has grown to include modules developed in conjunction with groups including the CRC for Catchment Hydrology, the University of Melbourne, Charles Sturt University, the Department of Agriculture (NSW), and the Department of Sustainability and Environment (Victoria).

The *CAT* evaluates the impact of land use and climate variability from paddock- to catchment-scale on a variety of 'landscape services[1]'. Depending on the landscape service of interest, modelled temporal response to changing landscapes can be estimated on a daily, monthly, yearly or mean

---

[1] 'Landscape services' are any commodity provided by landscapes that has social, economic or environmental value even if it cannot be explicitly quantified. Examples of landscape services are biodiversity, clean water and agricultural production.

annual basis. The fundamental question of scale, both spatial and temporal, plays a major role in defining the extent and detail of input data and model complexity.

The *CAT* allows for the analysis of multiple land use and management scenarios, generating different spatial and temporal system responses under alternative scenarios. Typical example questions include:

- What is the impact of different land use and land management strategies on surface recharge, lateral flow, streamflow, water yield, salt and nutrient loads, groundwater recharge, watertable level, evapotranspiration, carbon sequestration, erosion and biomass yield?
- What land management strategies can be implemented to assure protection of natural assets?
- What is the impact of production forestry and tree growth over time on carbon sequestration and water use?
- How much afforestation would be required to reduce end-of-catchment salt loads by a specified amount?
- Where in the catchment would afforestation make the greatest impact?
- Can rotations of perennial and annual pastures assist in decreasing depth to watertable at a farm scale?
- What is the impact of predicted climate change on end-of-catchment stream export?
- Where farmers can apply for government funding to revegetate their land, can applications be prioritised according to their environmental merits?

To answer these questions, the *CAT* links paddock-scale land use, soils, topography and climate data to catchment-scale groundwater systems and streamflows on a daily time-scale. A suite of crop growth and farming systems models evaluate the impact of different types of vegetation, farming systems and land management strategies on surface hydrology and system productivity. The description of surface hydrology dynamics is derived from a process-based model that partitions rainfall at all points on the landscape into components of evaporation, transpiration (from vegetation), surface runoff, subsurface lateral flows and recharge to groundwater systems (Fig. 4.1). The surface hydrology is modelled using a one-dimensional water balance model based on *PERFECT* (Littleboy et al. 1992). To improve the prediction of recharge derived from this purely vertical analysis, a two-dimensional hill slope model (Rassam and Littleboy 2003) is used to partition excess water into lateral and vertical pathways. Point-based knowledge of water partitioning across a landscape provides an opportunity to include models linked to specific pathways of water such as nutrient movement, salt transport, erosion, crop or forest yield, livestock and carbon sequestration models.

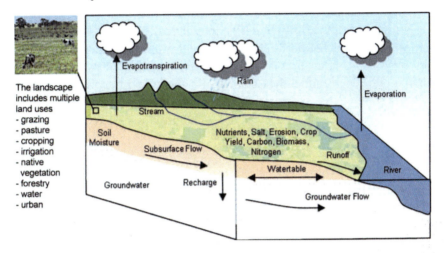

Fig. 4.1. Generalised hydrological processes considered in the *CAT*

There have been significant recent investigations of the impact of different landscapes, climate, topography and soils on the catchment-scale water balance. These range from simplistic approaches such as that described by Zhang et al. (2001) that map mean annual evapotranspiration to vegetation type using readily available parameters, to the more complex fully distributed models such as *SHE* (Abbot et al. 1986) and *MODFLOW* (McDonald and Harbaugh 1988) that require comprehensive calibration and parameterisation. Each approach has its strengths and weaknesses. Whilst the simplistic models generate rapid assessments, they often do not account adequately for soil and/or vegetation effects, land management and land practice impacts, and often operate only at broad spatial and temporal scales. Such models have been shown to poorly represent within-catchment responses (Beverly et al. 2005) but are valuable tools for community engagement and describing relative impacts among different catchments and/or scenarios. Conversely, some of the more complex models require such extensive calibration and have run times that are so long that they become prohibitive to use as management tools.

The *CAT* is one of a number of broad-scale frameworks that exploit the extensive range of 'process-based' models, generally based on fundamental equations that define the movement of water and nutrients. These models rely on readily available information such as soil type, vertical hydraulic conductivity and soil saturation point, such that model parameterisation is kept to a minimum. Some of the currently available frameworks used to describe catchment dynamics include *E2* (Argent et al. 2005), *SWAT* (Neitsch et al. 2005), and *CLASS* (Tuteja et al. 2004). E2 is a nodal-based framework and describes streamflow, dams and stream extractions well, but is limited in its' ability to define the internal dynamics of a catchment and provide a linkage between land use and water movement. *CLASS* differs from the *CAT* in that it employs a fully distributed two-dimensional surface hydrology model that accounts for the lateral transfer of water between adjacent cells. This is beneficial in describing nutrient and sediment transfer, however adds significant complexity to the data requirements and processing, resulting in prohibitive run times. In the context of groundwater *CLASS* only considers groundwater discharge to streams. *SWAT* takes a similar approach to the *CAT* in terms of surface hydrology; however *SWAT* uses a spatially lumped groundwater model that does not account for lateral distribution of groundwater. The strengths and weaknesses of a number of broad-scale frameworks are discussed in length in MacEwan et al. (2006).

Key strengths of the *CAT* are described as follows:

- Whilst stand-alone fully distributed groundwater models exist, such as *MODFLOW* (McDonald and Harbaugh 1988) and *FEFLOW* (Diersch 1998), none are explicitly linked to the landscape. The *CAT* fills this gap by seamlessly providing landscape-based recharge estimates to the fully distributed groundwater model.
- The *CAT* has the ability to link landscape-based surface hydrology to a suite of groundwater models that range in complexity from a simple model such as *BC2C* (Evans et al. 2004) that measures the length of time for a groundwater system to reach a state of equilibrium, to a fully distributed groundwater model that maps the depth to watertable whilst accounting for time varying recharge, groundwater pumping and stream extractions.
- The *CAT* has the ability to address relatively complex landscape scenarios such as multiple rotations of different cropping systems and forest growth over time.
- Data requirements for the *CAT* are commensurate with available data.

4.2.1 The *CAT* Interface

Figure 4.2 presents the conceptual structure of the *CAT* showing the links among the software interface, input data and models. The *CAT* interface provides a framework to explicitly link data inputs to modelled outputs, while storing any information pertinent to the modelled outcome in a structure described as a 'scenario'. A number of embedded analytical tools, visualisation techniques and reporting methods facilitate flexible analysis and presentation of results.

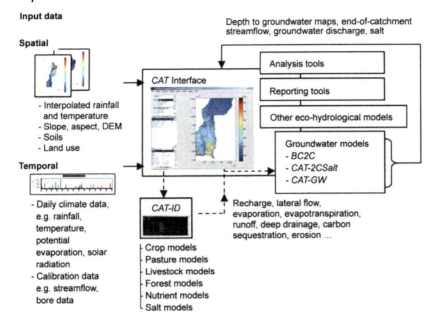

Fig. 4.2. The conceptual structure of the *Catchment Analysis Tool*

This conceptual framework enables end users to visually interrogate spatial and temporal data and map modelled outputs to inputs. This framework also has the advantage of enabling all information used in producing a particular modelling outcome to be recorded, thereby enabling results to be easily replicated or expanded as improved input data become available. Some of the key functionality of the *CAT* interface includes:

- tools to manipulate spatial data, for example, data re-sampling, gap filling for missing data, data smoothing and interpolation functions and data banding techniques
- display tools, for example, user defined colour maps, three-dimensional display based on a digital elevation model (DEM), query functions to detail input layer conditions and modelled temporal response for outputs
- the capability to filter and modify input layers
- various summary and analytical tools, for example, descriptive statistics, correlation coefficients, layer arithmetic operations, and temporal analysis
- multiple reporting tools, for example, tabular output such as weighted mean annual values per subcatchment, and maps for presentation that include scale bar, north arrow and exports.

4.2.2 *CAT* Input Data

The data required to run the *CAT* vary depending on the type of question asked and the temporal and spatial scales of a given problem. Typically the *CAT* works with data that are commensurate with publicly available data. However in some cases, specific data must be adapted to suit the particular needs of the model. Table 4.1 lists some of the standard layers used in the *CAT* with each of the different categories of data subsequently described in more detail. Where possible, the source of data on which the authors generally rely is indicated, although the *CAT* is not linked to any associated proprietary third party data format(s).

Table 4.1. Typical data utilised in the *CAT*

Input data	Category	Description	Source[a]
Climate station location	Climate	Spatial layer	DNRMQ
Daily climate data	Climate	Temporal file for each climate station	'Patched point' data (Jeffrey et al. 2001)
Mean annual rainfall	Climate	Spatial layer	*ANUClim* (Hutchinson 2001)
Mean annual temperature	Climate	Spatial layer	*ANUClim* (Hutchinson 2001)
Digital elevation model (DEM)	Topographical	Spatial layer	SII
Slope	Topographical	Spatial layer	Derived from DEM
Aspect	Topographical	Spatial layer	Derived from DEM
Soil layer	Soil	Spatial layer	CLPR (Smith 2000)
Current land use	Land use	Spatial layer	CSDL
Groundwater flow systems (GFS)	Groundwater	Spatial layer	Coram et al. (2000)
Groundwater conceptualisation	Groundwater	Spatial layers	Kevin (1993), Day (1985)
Gauged streamflow, salt and nutrients	Calibration	Temporal file for each stream gauge in catchment	Victorian Water Resources Data Warehouse (VWD 2003)
Bore data	Calibration	Temporal file measuring depth to watertable	DPI and DSE groundwater monitoring network

[a]Source data varies by catchment. Sources shown are for the Bet Bet catchment. *DNRMQ*: Department of Natural Resources and Mines, Queensland. *SII*: Victorian Governments Spatial Information Infrastructure Group. *CLPR*: Centre for Land Protection Research. *CSDL*: DPI Victoria Corporate Spatial Data Library.

Climate Data

The point location of each climate recording station is used to create a Voronoi diagram that maps the nearest climate station for each point in a catchment. Daily climate data for each climate station are sourced from Queensland Department of Natural Resources and Environment's *SILO* service. The daily data are called 'patched point data' and contain rainfall, minimum and maximum temperatures, radiation, evaporation and vapour pressure. They combine original Bureau of Meteorology measurements for a given climate station with a process for infilling any gaps in the record using interpolation methods discussed in Jeffery et al. (2001). As climate stations are located sparsely, daily rainfall and temperature data are scaled according to interpolated mean annual spatial layers. Interpolated surfaces are created using the *ANUClim* software (Hutchinson 2001) that combines a DEM and temporal climatic data to generate a smoothed climate surface.

Topographical Data

A DEM represents the elevation of a surface relative to mean sea level (MSL). The slope and aspect layers are derived from the DEM using a triangulated irregular network (TIN). For each triangular tile, slope is reported as degrees from horizontal and is important in determining the partitioning of excess water into vertical and lateral components. The aspect layer reports the azimuth of a tile. These data are used to estimate the daily solar radiation received by each tile.

Soils Data

Soil properties are estimated using a simple two-layer model of the soil consisting of an A and B horizon. The following properties are estimated for each point on the landscape:

- number of soil profile layers
- depth of soil profile layers (mm)
- volumetric water contents of soil profile layers (%) at air dry
- volumetric water contents of soil profile layers (%) at wilting point
- volumetric water contents of soil profile layers (%) at field capacity
- volumetric water contents of soil profile layers (%) and saturation
- saturated hydraulic conductivity of soil profile layers (mm/hr)
- stage II soil evaporation parameter (CONA) for top two soil profile layers
- upper limit of stage I (U) soil evaporation for top two soil profile layers (mm)

- runoff curve number for average antecedent moisture conditions and bare soil
- reduction in curve number at 100% cover
- maximum reduction in curve number due to surface roughness
- cumulative rainfall required to remove surface roughness (mm)
- modified USLE (Universal Soil Loss Equation) soil erodibility factor, K
- field slope (%)
- slope length or contour bank spacing (m)
- revised USLE rill/interill ratio factor
- bulk density of the top profile layer (g/cm^3)
- soil cracking option (yes/no)
- maximum infiltration into soil cracks (mm).

Soil properties are estimated for the A and B horizons. To obtain this information, the authors generally rely on the soil profile dataset held by CSIRO Land and Water that contains descriptions for over 7000 profiles across Australia. The estimates of thickness, texture, bulk density and pedality are used in the *CAT* to estimate parameters that describe the soil water retention curve that in turn provides for the calculation of soil plant available water holding capacity.

Land Use Data

Land use is classified using the Australia Land Use Mapping (ALUM) classification Version 5 (BRS 2001). For the *CAT*, the ALUM taxa are loosely grouped into two categories, land cover and land use, the latter of which require management strategies to be specified. There are a large number of land use options, including production of a variety of different annual and perennial crops such as salt bush, lucerne and native grasses. There are also land uses that represent urban landscapes, water bodies and pastures for grazing. The land use options currently available in the *CAT* are reasonably extensive and can be further augmented by detailing the values of a number of land use parameters. The *CAT* also provides an option to implement different land management strategies such as crop rotations, for example, a lucerne crop in year 1, followed by wheat in years 2 and 3, and then back to lucerne in year 4.

Groundwater Data

To take full advantage of the *CAT* capability, additional spatial layers that address groundwater are generally required. These provide information on sub-surface saturated groundwater flow systems in the area of interest.

Some of the simpler groundwater models utilise groundwater flow systems (GFS) maps containing attributes related to lateral hydraulic conductivity, depth to bedrock, depth to watertable, specific yield and groundwater salinity. The more complex fully distributed groundwater models require a full conceptualisation of the groundwater systems based on bore hole data, drilling transects and soil structure. The creation of a full groundwater conceptualisation is a major undertaking and must be completed before the modelling of groundwater through the area of interest can begin.

4.2.3 The *CAT* Model Components

A number of models have been incorporated into the *CAT*. Many sit within the interface itself whilst others such as the more complex surface (*CAT-1D*) and *MODFLOW* groundwater models have been explicitly linked to the *CAT* structure. For more information on the source of specific models and model capability refer to the *CAT* Technical Manual (Beverly 2007) and the *CAT* User Manual (Christy and Beverly 2004). Table 4.2 provides a brief overview of some of the key models associated with the *CAT*, and Fig. 4.2 provides an indication of how they are structured and linked.

Table 4.2. Overview of some of the key models associated with the *CAT*

Model name	Description	Outputs
CAT-1D[a] Crop models Pasture models Livestock models Forest models Nutrient models Salt models	Surface hydrology model *CAT-1D* combines a suite of crop growth and farming systems models to evaluate the impact of different types of vegetation, farming systems and land management strategies on surface hydrology and system productivity	Recharge, lateral flow, evaporation, deep drainage, runoff, carbon sequestration, erosion, nitrogen, salt, biomass yield, sediment yield, soil moisture Temporal scale: daily Spatial scale: paddock
BC2C[b]	Groundwater model Describes the groundwater discharge over time in response to a change in recharge	Groundwater discharge Temporal scale: annual Spatial scale: catchment
Zhang[c]	Groundwater model Maps mean annual evapotranspiration to vegetation type to give estimates of recharge, then uses simple relationship to calculate groundwater discharge	Groundwater discharge Temporal scale: annual Spatial scale: catchment

Table 4.2. (cont.)

2CSalt[d]	Groundwater/salt export model Quantify surface and groundwater contribution of salt export to catchments and to predict the impacts of land use change on salt losses from catchments	End-of-catchment stream-flow, groundwater discharge and salt export Temporal scale: monthly Spatial scale: lumped hydrologically contained units
MODFLOW[e]	Groundwater/ nutrients model Fully distributed spatio-temporal groundwater model	Depth to groundwater, streamflow, groundwater discharge Temporal scale: daily/monthly Spatial scale: paddock
Landscape Preferencing[f]	Biodiversity model Used to assess the biodiversity value of native vegetation within a catchment. Based on the general paradigm that round patches of native vegetation, connected to other remnants provide better habitat opportunities than isolated remnants	Landscape rating based on biodiversity value Spatial scale: pixel
Base Flow Separation[g]	Streamflow separation model Uses a digital filter technique to separate measured streamflow into surface and groundwater components	Quick flow and ground-water discharge Temporal scale: daily Spatial scale: point
Terrain Analysis[h]	Terrain model A number of models such as genera-tion of flow paths based on slope and disaggregation of catchment into hill slope and alluvial zones based on a surface flatness factor.	Flow paths, mapping of hill slope and alluvial zones Spatial scale: pixel
Climate Change[i]	Climate change model Climate sequence projections from 2000–2070	Climate sequence Temporal scale: daily Spatial scale: point

[a]Beverly (2007), [b]Evans et al. (2004), [c]Zhang et al. (2001), [d]Christy et al. (2006), [e]McDonald and Harbaugh (1988), [f]Ferwerda (2003), [g]Arnold and Allen (1999), [h]Christy and Beverly (2004), [i]Anwar et al. (2007).

4.2.4 Model Calibration and Conceptualisation

The level of calibration depends largely on the type of model run. Calibration is not required for parametric models such as *BC2C* and *Zhang* where the modelled response is based on empirical relations derived from field observations and fundamental concepts. Additionally there are the biophysically parameterised models such as the *CAT-1D* that require little calibration and can indeed be validated against measured stream gauge data. At the other end of the scale there are the spatially distributed groundwater models such as *2CSalt* and *MODFLOW* that rely heavily on groundwater conceptualisations that are often sparsely populated and have a large associated uncertainty. These models are highly parameterised and require a high level of calibration.

Groundwater systems are extremely complex, important components of which can only be inferred since they are below ground. Much of the calibration effort involves developing an appropriate conceptualisation of the sub-surface flow of water including factors such as underlying geology, depth to bedrock, soil texture and topography. Once this is developed, long-term information from boreholes, weather stations and weirs are obtained and used to calibrate the groundwater model. The initial step in this calibration involves the use of a least squares/maximum likelihood fitting procedure. Finalising a calibration then involves an iterative process that includes consultation with experts, modification of the groundwater conceptualisation, observation data matching (including base flow, groundwater discharges and groundwater observation bores) and statistical optimisation.

4.3 Case Study

To provide an indication of the utility of the *CAT* in landscape management, a hypothetical case study has been developed. Though fictitious, this case study is related to actual situations in which the *CAT* has been employed, and also a geographical area for which the *CAT* has been calibrated.

4.3.1 Hypothetical Case Study

To meet statewide salt targets, the North Central Catchment Management Authority (CMA) wishes to introduce a forest revegetation scheme to reduce the total salt export from the catchment whilst minimising the impact on the end-of-catchment streamflow. The North Central CMA and private landowners held a general meeting, the outcome of which recognised five locations across the catchment that could potentially be revegetated given adequate funding for fencing, infrastructure and compensation for loss of farming land. The North Central CMA has the necessary resources to fund the revegetation of one area.

> *Key Question*: Of the five locations, where would revegetation have the largest impact on the end-of-catchment stream salinity?

Description of Catchment Area

The Bet Bet catchment is located in north central Victoria, within the Loddon catchment of the North Central CMA region (Fig. 4.3). The catchment covers an area of 64,342 ha and currently exports approximately 20,000 tonnes of salt per year. Within this catchment the annual rainfall ranges between 490–815 mm and grazing is the dominant land use (approximately 70% of the catchment).

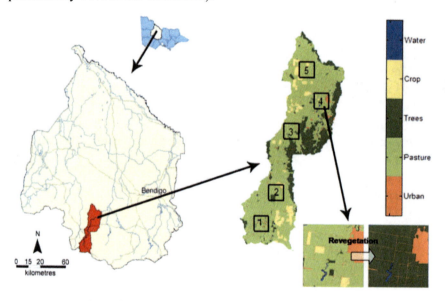

Fig. 4.3. Location of Bet Bet catchment within the North Central CMA region, Victoria. Map of CMA region showing current practice land use with hypothetical revegetation sites and example of revegetation on Site 4

Land Use Scenarios

The five 1600 ha sites nominated for revegetation are outlined in Fig. 4.3. For this case study, a simple revegetation scheme was applied to each site

The *Catchment Analysis Tool*   63

— any land currently utilised for pasture or cropping was replaced by trees. Table 4.3 lists the percentage allocation of land under current practice and the total area revegetated under the modified land use scenario.

Table 4.3. Percentage land use under current practice and the total area revegetated for each site in hypothetical case study

	% Crop	% Pasture	%Tree	%Urban	%Water	% Revegetated (area in ha)
Site 1	0	86	3	10	1	86 (1376)
Site 2	4	79	14	3	0	83 (1328)
Site 3	0	83	12	5	0	83 (1328)
Site 4	4	77	2	16	1	81 (1296)
Site 5	7	89	1	3	0	96 (1536)

Spatial input data

A list of the spatial input layers required to model the surface hydrology is provided in Table 4.4. General statistics including mean, minimum and maximum values provide an indication of the range and mean value of the climatic and topographical layers.

Table 4.4. Summary of spatial inputs used to estimate surface hydrology (cell resolution of all inputs is 20 m)

Input	Mean (min. – max.)
Rainfall [mm]	555.7 (490–815)
Temperature [°C]	13.2 (11–13.5)
Slope [°]	2.7 (0–13.1)
Aspect	0: flat
	1: north facing
	2: south facing
Climate stations	79000 Amphitheatre 81000 Avoca (Post Office)
	81002 Bealiba 81038 Natte Yallock
	81085 Dunolly 88038 Lexton
	88043 Maryborough 88056 Talbot (Post Office)
Soil[a]	Red duplex soils (Dr2.22, Dr2.23, Dr2.41, Dr2.42 , Dr2.43)
	Yellow & yellow-grey duplex soils (Dy3.22, Dy3.41, Dy3.42)
	Massive and structured earths (Gn3.11, Gn3.22, Gn4.11)
	Cracking clays (Ug5.20, Ug5.40)
	Loams (Um5.21)
Land use	Urban Pasture
	Tree Crop
	Water

[a] Soils classified using Northcote factual key (Northcote 1979).

Modelling

Using the input layers described in Table 4.4, the *CAT-1D* surface hydrology model was run for the current and modified land uses. This procedure generated monthly outputs of recharge, runoff, subsurface lateral flow and potential evaporation. The estimated decrease in the mean annual recharge after the five candidate areas have been revegetated is shown in Fig. 4.4. This figure suggests that the decrease in recharge will be greater in the upland areas of the catchment (b), a logical result given that higher rainfall occurs at higher elevations thereby providing greater water availability for plant use.

Change in recharge is just one aspect of revegetation on the five candidate sites — this case study also requires the prediction of end-of-catchment streamflow and salt export. When considering catchment salinity dynamics, it becomes particularly important to consider groundwater processes. For example, in the Bet Bet catchment, although most (around 80%) of the streamflow generally comes from surface runoff and subsurface lateral flows, geological and hydrological studies of the area (Kevin 1993; Clark 2005) suggest that a large portion of the salt export can be attributed to groundwater discharge.

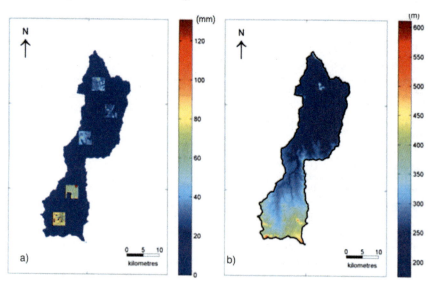

Fig. 4.4. (a) Decrease in recharge (mm) after revegetation of five candidate sites. (b) DEM for Bet Bet catchment (m)

To investigate salinity processes the *CAT-2CSalt* model was applied to estimate groundwater discharge and salt export via streams using the monthly recharge and other surface components from the *CAT-1D* surface hydrology model. This model relies on existing groundwater flow system (GFS) attribution to generate a simple lumped model of groundwater processes within the catchment. Figure 4.5 illustrates the GFS and attribution for the Bet Bet catchment.

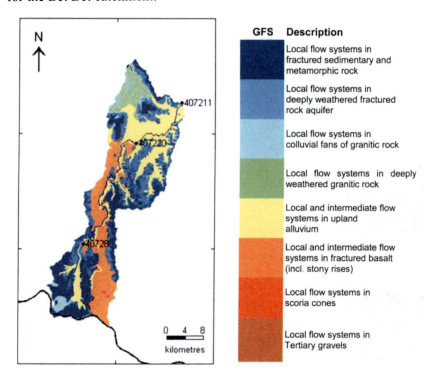

Fig. 4.5. Groundwater flow systems of Bet Bet catchment

Calibration

The model was calibrated against measured end-of-catchment stream gauge information using an automated, constrained non-linear optimiser (The MathWorks Inc. 2007). An example of the end-of-catchment calibration for the Bet Bet catchment is illustrated in Fig. 4.6. Statistics associated with calibration show that while the model captures the temporal nature of the stream data reasonably well, the magnitude of large streamflow events tends to be underestimated.

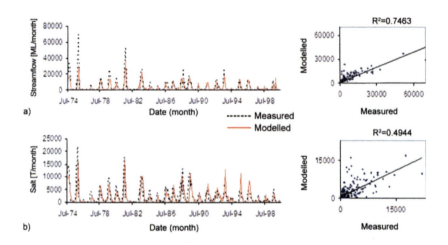

Fig. 4.6. Comparison of measured and modelled end-of-catchment (**a**) streamflow (ML/month) and (**b**) salt export (t/month) (Weeks et al. 2005)

4.3.2 Results and analysis

The model was run for the current practice scenario and then for each of the five revegetation scenarios. As expected, for each of the five sites revegetation led to an overall decrease in the catchment streamflow and salt export.

Figure 4.7 presents the percentage decrease in mean annual streamflow and salt export, with the total decrease separated into surface and groundwater components. This figure provides a general sense of the response to land use change of surface and groundwater responses and highlights the relationship between salt export and hydrological pathways. In sites 1, 2 and 3 it is the reduction of groundwater flow that has led to the greatest reductions in salt export.

Of interest is the fact that the model has predicted no change in groundwater discharge for sites 4 and 5. This could be due to a couple of reasons. Firstly, sites 4 and 5 are situated on lower lying terrain with low rainfall patterns, resulting in a smaller change in recharge under the (re)vegetated land use. Secondly, the *CAT-2CSalt* model, designed specifically for upland areas defined by local groundwater systems, approaches the limits of its reliability at elevations as low as sites 4 and 5 where the underlying groundwater system tends towards being an intermediate–regional system. Although the *CAT-2CSalt* model provides a useful indication of impacts of local land use change, comprehensive analysis of regional areas might be better undertaken using the more complex fully distributed groundwater models.

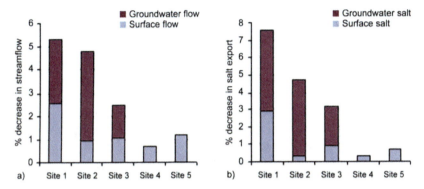

Fig. 4.7. Percentage decrease in end-of-catchment mean (**a**) annual streamflow and (**b**) salt export separated into surface and groundwater components

To address the key question of this case study, the effective end-of-catchment mean annual salinity (mg/L) has been plotted for current practice and each of the revegetation scenarios (Fig. 4.8). The *CAT* results suggest that revegetating site 1 will result in the lowest end-of-catchment stream salinity. In fact, this result is consistent with the mapped spatial discharge regions of the Bet Bet catchment (Clark 2005). Interestingly, the model predicts that the revegetation of sites 2, 4 and 5 would result in higher stream salinity. That is, revegetating any of these sites would reduce streamflow more than salt export.

Therefore, to address the original question: The revegetation of site 1 would have the largest impact on the end-of-catchment stream salinity.

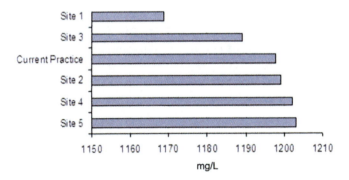

Fig. 4.8. End-of-catchment effective mean annual stream salinity for current practice and each of the five revegetation scenarios

4.4 Validation and Model Improvement

To this point, little has been said about model validation. This is because the *CAT* model, like virtually all landscape-based biophysical models, suffers from a lack of data relative to its complexity. At the catchment scale there are very few points in the landscape where relevant data is measured and captured, and as such, validation of the *CAT* remains challenging. There is very little measured information available that describes within-catchment hydrological dynamics, such as surface and groundwater flow and soil moisture, however it is these processes that are fundamental to modelling plant growth and nutrient runoff. Where possible modelled results are validated against data captured from stream gauges, bore data and point studies. However, in the case of groundwater modelling such validation often becomes problematic because the models themselves are calibrated on this data.

As stated at the beginning of this chapter, the *CAT* captures the hydrological dynamics of the whole landscape by combining a suite of mature models that individually describe a variety of landscape processes. However, interactions among different landscape processes are not always well understood. This means that the algorithmic linkage of sub-models exposes gaps in knowledge of interactions among real-world processes. For example, the *CAT* provides a capability to estimate the impact on streamflow of forestation by different species, tree densities, and timber management regimes. The reality, however, is the uptake of water by different tree species planted at different densities under a variety of ecophysiographic conditions is poorly understood. The value of models such as the *CAT* is that they challenge researchers in specific disciplines to develop and test biophysical hypotheses, design appropriate field trials to produce datasets that are appropriate and useful for model validation, and provide insight into the dynamics of dominant interactions and processes that operate across landscapes.

This difficulty in validating the *CAT* for any area to which it is applied highlights the need for ongoing model improvement and improved data collection. Indeed, one of the benefits of growing the *CAT* usage in real-world land management problems is the identification of the most urgent needs for *CAT* improvement. Future work on the *CAT* will involve continued validation and improvements of existing sub-models as well as the addition of models that specifically look at nutrient transfer to determine the impacts of changing landscapes on water quality. Additionally, climate change projections will be incorporated into the *CAT* to reflect the impact of a changing climate on the landscape.

4.5 Conclusion

The *CAT* is an example of the positive trend for models to address complex systems in a holistic and interconnected manner. Its ongoing development aims to better inform the growing demand for landscape managers to consider multiple outputs in the management of landscapes. The *CAT* captures the hydrological dynamics of the whole landscape by combining a suite of mature models that individually describe a variety of landscape processes. The creation of this interconnected biophysical model has continued to expose gaps in our knowledge about landscape connection to surface and sub-surface water movement and the controlling processes. Validation of the *CAT* remains a challenging issue, and leads us to pursue new and integrative ways of studying landscape dynamics.

References

Abbot MB, Bathurst JC, Cunge JA, O'Connell PE, Rasmussen J (1986) An introduction to the European hydrological system SHE, 1 History and philosophy of a physically-based distributed modelling system. Journal of Hydrology 87:45–59

Anwar MR, O'Leary G, McNeil D, Hossain H, Nelson R (2007) Climate change impact on rainfed wheat in south-eastern Australia. Fields Crops Research 104:139–147

Arnold JG, Allen PM (1999) Automated methods for estimating base flow and ground water recharge from streamflow records. Journal of the American Water Resources Association 35:165–179

Argent RM, Murray N, Podger G, Perraud J, Newham L (2005) E2 Catchment modelling software component models. Cooperative Research Centre for Catchment Hydrology, Canberra

Beverly, C (2007) Technical manual models of the catchment analysis tool (CAT1D Version 22). Department of Sustainability and Environment, Victoria

Beverly C, Bar M, Christy B, Hocking M, Smettem K (2005) Salinity impacts from land use change: comparisons between a rapid assessment approach and a detailed modelling framework. Australian Journal of Experimental Agriculture 45(2):1453–1469

BRS (2001) Land use mapping at catchment scale: principles, procedures and definitions, 1st edn. Department of Agriculture Fisheries and Forestry, Bureau of Rural Science, Canberra

Clark R (2005) Mapped discharge sites North Central catchment. Corporate Spatial Data Library, Department of Primary Industries, Victoria

Coram JE, Dyson PR, Houlder PA, Evans WR (2000) Australian groundwater flow systems contributing to dryland salinity. Bureau of Rural Sciences report for the National Land and Water Research Audit, Dryland Salinity Theme, Canberra

Christy B, Beverly C (2004) User manual catchment analysis tools: user manual version 6. Department of Primary Industries, Victoria

Christy B, Weeks A, Beverly C (2006) Application of the 2CSalt model to the Bet-Bet, Wild Duck, Gardiner and Sugarloaf catchments in Victoria. Department of Primary Industries, Victoria

Day C (1985) A study of the geomorphic, soil and geo-hydrological conditions of the Timor West/Black Ranges area. Land Protection Service, Conservation Forests and Lands, Victoria

Diersch HG (1998) FeFLOW: Finite element subsurface FLOW system. Institute for Water Resources Planning and Systems Research Ltd, Berlin

Evans R, Gilfedder M, Austin J (2004) Application of the Biophysical Capacity to Change (BC2C) model to the Little River (NSW). CSIRO Land and Water Technical Report No 16/04, March 2004

Ferwerda F (2003) Assessing the importance of remnant vegetation for maintaining biodiversity in rural landscapes using geospatial analysis. M.Sc. thesis, RMIT University, Melbourne

Hutchinson MF (2001) ANUClim Version 51. Retrieved 29 July 2004, http://cres.anu.edu.au/outputs/anuclim.php

Jeffrey S, Carter J, Moodie K, Beswick A (2001) Using spatial interpolation to construct a comprehensive archive of Australian climate data. Environmental Modelling and Software 16:309–330

Kevin P (1993) Groundwater and salinity processes in the uplands of the Loddon River catchment. Technical Report No. 5, Centre for Land Protection Research, Victoria

Littleboy M, Silburn DM, Freebairn DM, Woodruff DR, Hammer GL, Leslie JK (1992) Impact of soil erosion on production in cropping systems. I. Development and validation of a simulation model. Australian Journal of Soil Research 30:757–774

MacEwan R, Pettit C, Dorrough J, Suter H, Hossain H, Cherry D, Beverly C, Cheng X, Sposito V, Melland A, Huang Z, McNeill J, Hood A, Rab A, Sheffield K, Duncan D (2006) Landscape analysis models and frameworks – a review. Our Rural Landscape 11 – New Dimensions for Agricultural Landscapes. Department of Primary Industries, Victoria

McDonald MC, Harbaugh AW (1988) MODFLOW, a modular three-dimensional finite difference groundwater flow model. US Geological Survey, Washington DC

Neitsch SL, Arnold JG, Kiniry JR, Williams JR (2005) Soil and water assessment tool theoretical documentation version 2005. Grassland, Soil and Water Research Laboratory, Agricultural Research Service, Texas

Northcote K (1979) A factual key for the recognition of Australian Soils. Rellim Technical Publications Pty Ltd, South Australia

Rassam D, Littleboy M (2003) Identifying vertical and lateral components of drainage flux in hillslopes. In: Post, DA (ed) MODSIM 2003 International

Congress on Modelling and Simulation, July 2003. Modelling and Simulation Society of Australia and New Zealand, pp 183–188

Smith C (2002) 1:250 000 Statewide soil attribute coverage documentation. Centre for Land Protection Research, Department of Primary Industries, Victoria

The MathWorks Inc (2007) Optimisation Toolbox 3 User's Guide. Available at http://www.mathworks.com

Tuteja NK, Vaze B, Murphy G, Beale G (2004) CLASS: Catchment scale multiple-land use atmosphere soil water and solute transport model. Technical Report 04/12, Department of Infrastructure, Planning and Natural Resources, and Cooperative Research Centre for Catchment Hydrology, Canberra

VWD (2003) Victorian Water Database. Available at http://www.vicwaterdata.net

Weeks A, Beverly C, Christy B, McLean T (2005) Biophysical approach to predict salt and water loads to upland REALM nodes of Victorian catchments. In: Zerger A, Argent RM (eds) MODSIM 2005 International Congress on Modelling and Simulation Modelling and Simulation Society of Australia and New Zealand, 12–15 December 2005, pp 2776–2782

Zhang L, Dawes W, Walker G (2001) Response of mean annual evapotranspiration to vegetation changes at a catchment scale. Water Resources Research 37:701–708

5 The Application of a Simple Spatial Multi-Criteria Analysis Shell to Natural Resource Management Decision Making

Robert G Lesslie[1], Michael J Hill[2], Patricia Hill[3], Hamish P Cresswell[4] and Steve Dawson[1]

[1] Australian Government Bureau of Rural Sciences, ACT, Australia
[2] Department of Earth System Science and Policy, University of North Dakota, USA
[3] CSIRO Sustainable Ecosystems, ACT, Australia
[4] CSIRO Land and Water, ACT, Australia

Abstract: Natural resource management decision making generally requires the analysis of a variety of environmental, social and economic information, incorporating value judgement and policy and management goals. Justifiable decisions depend on the logical and transparent combination and analysis of information. This chapter describes the application of spatial multi-criteria analysis to natural resource assessment and priority setting at regional and national scales using a newly developed spatial multi-criteria analysis tool — the Multi-Criteria Analysis Shell for Spatial Decision Support (MCAS-S). *MCAS-S is designed for use in participatory processes and workshop situations where a clear understanding of different approaches to spatial data management and information arrangement is necessary. The* MCAS-S *work environment provides for multiple map display, combination and manipulation, live update of changes, and development of spider/radar plots important in ecosystem service assessments. These and other capabilities promote clear visualisation of the relationships among the decision, the science, other constraints and the spatial data. The regional scale example illustrates the analysis of biodiversity and salinity mitigation trade-offs in revegetation in a participatory process. The national scale application illustrates reporting to policy clients on the tensions between resources use and conservation in Australian rangelands — essentially an expert analysis.*

5.1 Introduction

Government policy makers, local authorities and land managers with responsibility for natural resource management decision making are often required to analyse large amounts of environmental, social and economic information. The transparent and logical treatment of this information including the incorporation of community opinion, public policy and management goals can be achieved using a spatial multi-criteria analysis (MCA) approach (Kiker et al. 2005; Hill et al. 2005a; Malczewski 2006).

Spatial MCA capability is available in a range of commercial geographic information systems (GIS) and this may be customised for particular purposes in spatially explicit decision support tools. However, simple, flexible spatial multi-criteria shells with easy adaptability to any problem are not readily available, particularly without a programming requirement. For participatory processes and workshop situations, a spatial multi-criteria analysis shell requires a flexible interface and transparency among spatially explicit assessments, the input data, and the classification and combination rules used to generate them.

This chapter describes the application of spatially explicit multi-criteria analysis to natural resource management decision making using a flexible, easy-to-use, spatial multi-criteria analysis shell — the Multi-Criteria Analysis Shell for Spatial Decision Support (*MCAS-S*). The utility of the software, which provides visual cognitive links between the MCA participatory process and spatially explicit data, is illustrated using two case studies. The first is a regional level land use planning process that maps priority locations for revegetation in the West Hume region of south-eastern New South Wales (NSW). The second is a national level assessment of factors affecting the sustainability of extensive livestock grazing in the Australian rangelands.

5.2 Multi-criteria Analysis

MCA is a technique that allows for the measurement and aggregation of the performance of alternatives or options, involving a variety of both qualitative and quantitative dimensions. As a means of considering the links among biophysical, economic and social data with human imperatives, it is therefore particularly useful for approaching complex interactions and effects in the context of land use and land management.

There are many variants of the general MCA method that can be applied in a wide variety of contexts. Many approaches are based on the pair-wise comparison method of the Analytical Hierarchy Process (Saaty 2000; Ramanathan 2001). Well-developed MCA methods usually share a number of characteristics. Generally, they are flexible, enable the capture of quantitative and qualitative data and issues, are relatively simple for clients and stakeholders to use, permit the development of many alternative scenarios, allow the exploration of trade-offs, and enable stakeholders to factor results into decision making.

The MCA process is a tool to assist decision makers in reaching outcomes — it does not do the decision making, or produce a solution. Attention must be given to how information quality and uncertainty is factored in and integrated with stakeholder viewpoints and biases, political and structural realities, and achievability versus optimality. It is important that each stage of the MCA process is carried out rigorously, in parallel with stakeholder engagement. Matching the spatial and temporal scale of the input information and analysis to the issues and processes under consideration is also critical.

5.2.1 Spatial Applications

There is a long history of the use of MCA in operations research, but these applications are essentially non spatial although there are recent crossover developments with traditional methods such as *DEFINITE* (DEcisions on a FINITE set of alternatives; Janssen and Van Herwijin 2006). The development of spatial applications of MCA has accelerated as GIS software has improved, and as computer operating systems have become more suited to implementation of GUI (Graphical User Interface) approaches. There are three major groupings:

- GIS-based applications utilising either established MCA modules (e.g. *IDRISI*) or involving the use of GIS modelling and object-based programming linkages (e.g. *ArcInfo* and *ArcGIS*)
- hybrid approaches that combine GIS, MCA, models and other capability via programming
- stand-alone software specifically designed for an application.

Generic MCA capability within existing GIS products is tied to specific software, although in some cases standalone executables can be spawned. Table 5.1 provides a selected bibliography of the development of spatial MCA applications. Increased publication of these approaches in recent years indicates a heightened interest in MCA in combining methods and in applications that aid decision making in the coupled human–natural system.

Table 5.1. Multi-criteria analysis applications and software development: a selected list of GIS-based and standalone software-based applications

Software/analysis	Application	Reference
1. GIS-based applications		
IDRISI (®Clark University) GIS-based MCA	Earthquake hazards; crop suitability; soil erosion in Ethiopia	Ceballos-Silva and Lopez-Blanco (2003); Dragan et al. (2003)
ASSESS (A System for SElecting Suitable Sites) written in *ArcInfo* AML (®ESRI)	Radioactive waste repository; soil conditions; catchment condition	Veitch and Bowyer (1996); Bui, (1999); Walker et al. (2002)
ArcView (®ESRI) GIS-based MCA	Planning tool; urban land use	Pettit and Pullar (1999); Dai et al. (2001)
Model-GIS (*ArcInfo*) coupling with MCA	Non point source farm pollution	Morari et al. (2004)
ILWIS GIS	Nature conservation value of agricultural land	Geneletti (2007)
MapInfo (®) GIS-based DSS	Urban transport policies	Arampatzis et al. (2004)
Other GIS-based MCA	Watershed management; biodiversity conservation	Horst and Gimona (2005)
2. Hybrid applications		
PROMETHEE integrated with *ArcGIS*	Land parcels ranked for housing suitability	Brans et al. (1984); Marinoni (2005)
SIMLAND – cellular automata, MCA and GIS written in C and using *ArcInfo* GIS	Land use change	Wu (1998).
HERO (Heuristic multi-objective optimisation) combined with GIS, AHP and Bayesian analysis	Forest planning; habitat suitability	Kangas et al. (2000); Store and Kangas (2001); Store and Jokimaki (2003)
3. Stand-alone software		
LMAS – Land Management Advice System	Spatial expert system	Cuddy et al. (1990)
MULINO-DSS (MULti-sectoral, INtegrated and Operational DSS) combines simulation models, mapping and MCA	Water resources	Giupponi et al. (2004)
GIWIN (Geographic Information Workshop for WINdows) LRA (Land Resource Allocation)	Land resource allocation for rice paddy fields	Ren (1997)
MEACROSS – Multi-criteria Analysis of Alternative Cropping Systems	Cropping systems	Mazzetto and Bonera (2003)
IWM – decision support system for Management of Industrial Wastes	Industrial waste	Manniezzo et al. (1998)
GSA (Global Sensitivity Analysis) in *SimLab* (Software for Uncertainty and Sensitivity Analysis)	Hazardous waste disposal	Gomez-Delgado and Tarantola (2006)

5.2.2 The Decision-making Process

Policy and program development and decision making about natural resources are often driven by questions that are deceptively simple. Questions such as: 'Where in the landscape is agricultural production approaching the limits of sustainable resource use?', or 'Where can we most effectively invest in revegetation?' raise complex issues of equity, economic performance, and biophysical impact. Revegetation, for example, may have benefits for biodiversity, water quality and amenity, and costs associated with reduced water supply and agricultural production. Usually there is no 'right' answer — complex trade-offs may be involved, with cost and benefit considerations resting on value judgment or opinion. The generalised spatial multi-criteria analysis and the priority-setting process are illustrated in Fig. 5.1, and its stages are described below.

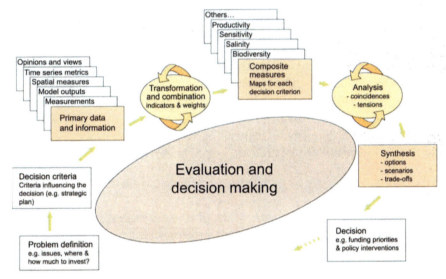

Fig. 5.1. The spatial multi-criteria analysis and the priority-setting process

Problem Definition

The more specific the problem, the easier it will be to detail decision criteria. For example, an investment decision might be restricted to a particular set of outcomes such as water quality or improved farming practices, or to a particular set of mechanisms such as market-based instruments.

Decision Criteria

The criteria that are to be taken into account in making investment decisions. These criteria will depend on the nature and scope of the problem. They will usually be derived from a number of sources including legislation, agreements, policy initiatives and programs.

Primary Data and Information

Primary data and information is the store of available primary information upon which decision making ultimately depends. In a natural resource decision-making context this includes land, soil, water, biodiversity, social and economic information. It takes the form of measurements, model outputs, time series information, and expert knowledge.

Transformation and Combination

This is a participatory process that involves using available *primary data and information* (above) to obtain an assessment against each *decision criterion*. Outputs are usually represented as maps of *composite measures* (below), or indexes. For example, the creation of a map of 'species conservation value' may involve the combination of species distribution data, with measures of diversity, endemism, rarity and endangerment along with landscape information. The process is usually iterative.

Composite Measures

Composite measures, or indices, which provide an assessment against each decision criterion. These could include measures such as species conservation value, biodiversity threat, agricultural potential, agricultural value, water availability and salinity risk. More than one composite measure may be needed to represent a decision criterion.

Analysis

Analysis involves the exploration of coincidences and tensions among *composite measures*.

Synthesis

Synthesis is the outcome of the *analysis* process. Consensus positions may not be reached. The synthesis process usually involves the development of options and scenarios that represent particular stakeholder perspectives,

The Application of a Simple Spatial Multi-Criteria Analysis Shell   79

and alternative pathways forward. These options and scenarios are the choices for decision makers.

Decision Making

Decision making involves the resolution of inconsistencies, contradictions and competing claims.

5.3 The *MCAS-S* Approach

As MCA is useful for understanding coupled human–environment systems, it is a process that should be routinely available to natural resource policy makers and practitioners. With this in mind, the design goal for *MCAS-S* was to create a portable and easy-to-use exploratory multi-criteria analysis shell with generic functionality, independence from data and project type, and freedom from GIS-based barriers to flexibility and interactive use (Hill et al. 2005b). *MCAS-S* functionality promotes participatory processes that require the understanding of relationships between decision-making requirements and the available data. Of particular value in this regard are software features that enable interactive cognitive and 'live update' mapping of alternative views.

5.3.1 Design Principles

In designing *MCAS-S*, the decision was made to focus functionality on assisting the decision-making process by promoting improved visualisation and transparency of relations among spatial data, value judgments, decisions and results. The tool thus requires the use of readily available GIS for spatial data management and analysis, and does not seek to provide these functions.

The first key design principle of *MCAS-S* is coupling the human cognitive process with the display and interrogation of spatial data (Bisdorff 1999) so that the origins of complex summary indices and queries are transparent. In *MCAS-S*, direction arrows connect data elements showing their contributions to combined layers. The second key principle is the automatic update of available functions according to the type of user-selected data layers. These two features make the interface highly intuitive and easy to use.

The third key principle is the maintenance of a live link between the primary data and all subsequent indicator, composite and analysis layers.

This allows the user to make adjustments to the treatment of primary data and immediately see the consequences as derived layers.

A final major construct behind *MCAS-S* is enabling simple visual comparison of multiple themes in pairs (using a two-way comparison matrix) and in groups (using multi-way radar or spider plots).

5.3.2 Key Functions

MCAS-S requires the user to prepare raw spatial data for a project in a raster format of consistent geographical extent, pixel resolution and projection. Primary data can then be selected from the menu and dragged into a display workspace whereupon the spatial data layer can be classified using a variety of simple classification methods and tools for tailoring input data. With the creation of individual class or rule layers, the user can apply them in weighted combinations to construct composite layers contributing to themes of interest. Key functions for combining and comparing spatial layers are illustrated below in two natural resources management applications, and represented schematically in Fig. 5.2.

A *composite* layer interface will appear into the workspace when selected from the menu button. The weighted contribution of any selected layer to any composite can be set by using this interface. The composite map dynamically updates as the user changes the weightings on input layers. The user can see in the workspace a hierarchical, cognitive 'map' showing the development of individual layers culminating in a final summary layer representing a theme. Relationships among themes, specific views and particular indicators may be examined using several methods.

Two-way comparison enables the user to create a two-way comparison map, explore the association among input classes, and define a colour ramp and value scale to highlight association of high or low values, or feature a particular geographical region.

Multi-way comparison is used when the spatial association of two or more data layers is required. The multi-way analysis uses the radar plot as the basis for visualisation. The user can create radar plots with spokes representing layers or themes. The multi-way function provides for the creation of a binary layer that spatially delimits the area either inside or outside the limits of the radar plot, that is, the area that meets or does not meet certain user-defined criteria levels. The multi-way comparison may also be displayed as a grey-scale surface showing 'distance' from selected criteria values.

A range of ancillary facilities includes a viewer that presents — for a selected grid cell — dynamic feedback of values for all related layers, including two-way and multi-way comparison. Masking and overlay functions, a reporting utility, image and process logging, and data export are also available.

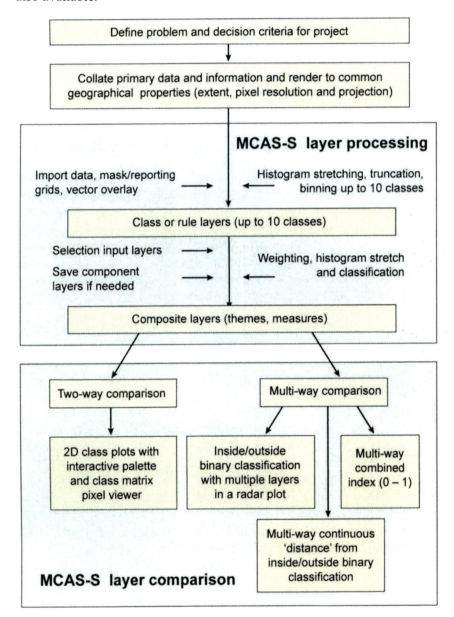

Fig. 5.2. Flow chart for *MCAS-S* showing stages and functionality

5.4 Applications

The remainder of this chapter demonstrates the use of *MCAS-S* functionality in two natural resources management applications:

- land use planning process that involved mapping priorities for revegetation in the West Hume region of southern NSW
- national assessment of factors affecting the sustainability of extensive livestock grazing in the Australian rangelands.

5.4.1 Prioritising Revegetation Investment in West Hume NSW

A practical catchment investment planning process needs to focus on indicative zones for investment in landscape change, consistent with maximising multiple environmental outcomes. In a regional context an effective prioritisation process provides for the best possible use of existing datasets and the technical expertise of participants, integrating knowledge and balancing landscape options in a transparent and objective way, and at a level that allows prioritisation of on-ground works and incremental improvement over time, bringing together new information to help decision making.

In 2006, CSIRO Land and Water and the Murray Catchment Management Authority (CMA) collaborated in identifying priority areas for revegetation — to decide 'what needs to be planted where' — in the West Hume area of southern NSW (Hill et al. 2006a). The West Hume area is approximately 86,000 ha of predominantly mixed cropping agricultural land, in the southern part of NSW within a region managed by the Murray CMA (Fig. 5.3). This area was previously identified by the CMA as being within a salinity and biodiversity management priority area (Murray Catchment Management Board 2001). The Murray CMA co-invests with land owners in local-scale environmental activities such as revegetation with native plant species and establishment of perennial pasture. Targeting such investment in locations where environmental outcomes are likely to be maximised, and having a transparent and defensible prioritisation mechanism, is a primary CMA objective.

Fig. 5.3. Location of the West Hume area of southern New South Wales

Priority mapping was developed using *MCAS-S* from a series of rules or guidelines and spatial analysis procedures for the definition and creation of suitability ranked input data layers for biodiversity and salinity. Spatial data used for the creation of these input data layers included:

- historical rainfall and temperature
- elevation (and derivatives, e.g. slope)
- soil landscapes and profile classes
- soil water holding capacity
- current (or recent) land use
- pre-settlement vegetation
- threatened species (point) data
- groundwater pressure and quality data
- groundwater flow systems mapping
- drainage networks, stream flow and water quality
- location of important assets or infrastructure such as towns and roads.

The relative importance of rules or guidelines was assessed using a pair-wise comparison procedure by a CSIRO and Murray CMA focus group. The pair-wise comparison used the method of the Analytical Hierarchy Process (AHP) (Saaty 2000; Ramanathan 2001). In this method, each guideline rule was compared with every other guideline rule and ranked as more or less important by each member of the focus group. The results were combined in a cross-tabulation and converted to a series of relative importance weightings and rankings for the salinity and biodiversity guidelines (Table 5.2). The focus group also assessed the relative importance of the biodiversity and salinity themes.

Table 5.2. Weighting and ranking for biodiversity and salinity guidelines

Guidelines		Weighting (rank: 1=most important)
Biodiversity guidelines (rules)		
1	Revegetate for geographical dispersal	0.0974 (6)
2	Revegetate biophysically heterogeneous areas	0.0541 (9)
3	Revegetate rare broad vegetation types	0.1038 (5)
4	Revegetate areas with rare species	0.0935 (7)
5	Revegetate in areas with dense patch distribution	0.1502 (1)
6	Revegetate close to large patches	0.1302 (4)
7	Revegetate close to streams	0.1492 (2)
8	Revegetate enclosed areas	0.07312 (8)
9	Revegetate to form corridors	0.1305 (3)
10	Revegetate land with low production potential	0.0179 (10)
Total		1.000
Salinity guidelines (rules)		
11	Revegetate areas: responsive groundwater flow systems	0.1678 (1)
12	Protect high value water resources	0.0939 (6)
13	Protect high value biodiversity assets	0.1039 (5)
14	Protect high value built assets	0.0730 (7)
15	Revegetate soils with high salt stores	0.1262 (3)
16	Revegetate high recharge potential areas	0.1551 (2)
17	Revegetate in areas with high rainfall	0.0683 (8)
18	Revegetate away from saline discharge zones	0.0641 (9)
19	Revegetate low value agricultural land	0.1180 (4)
20	Revegetate areas with high forest production potential	0.0299 (10)
Total		1.000

Each of the 20 spatial suitability surfaces for each of the guidelines was imported into an MCAS-S project. Three new composite layers were added to the project: *Salinity*, *Biodiversity* and *Combined*. The 10 biodiversity guidelines were used to create the *Biodiversity* composite, using the weightings developed in the AHP pair-wise comparison. The same process was repeated for the *Salinity* composite using the 10 salinity guidelines and their associated weightings. The *Salinity* and *Biodiversity* composites were then combined to create a *Combined* composite, with the biodiversity layer receiving a 0.4 weighting and the salinity layer receiving 0.6 weighting (Fig. 5.4).

The spatial multiple-criteria analysis with *MCAS-S* used within a participatory process defined locations appropriate for revegetation based on the number of desirable criteria (guidelines) that were met, and the relative importance of these criteria (weighting). The analysis provided a straightforward and logical means for allocating investment in revegetation according to relative suitability in a catchment context, using simple first-principle criteria with available spatial data. Further decisions on sites for revegetation work are made in the context of local farm plans and farmer land management objectives.

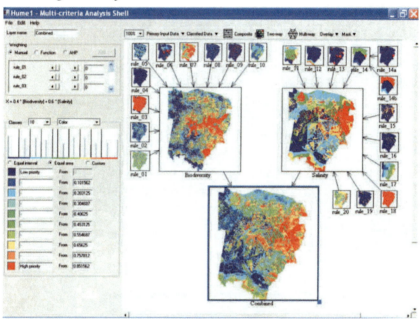

Fig. 5.4. A *MCAS-S* workspace linking spatial rule layers to prioritise revegetation in West Hume

5.4.2 Assessing the Sustainability of Extensive Livestock Grazing in the Australian Rangelands

The Australian rangelands occupy approximately three-quarters of the Australian continent. Land use in the rangelands is dominated by extensive sheep and cattle grazing on native pastures, with rainfall generally being too low or too variable for dryland cropping or grazing on improved pastures. Understanding the implications of alternative land management options is important in the rangelands because of the large area of Australia they occupy and the reliance of rangeland communities and industries on sustainable management of natural resources. Spatial multi-criteria analysis approaches have been used to support policy analysis on aspects of sustainability in the Australia rangelands (Stafford-Smith et al. 2000; Hill et al. 2006b).

From a pastoral perspective, livestock grazing in the rangelands can be characterised as sustainable where economic resilience and stability can be achieved in conjunction with the regional maintenance of native species and other ecosystem services. Informed public policy development requires an understanding of where in the landscape these ecological and economic influences are operating. Sustainability may be at risk in locations where the resource base has limited resilience to livestock grazing and pastoral land use is viable in terms of potential productivity (Fig. 5.5) (Stafford-Smith et al. 2000).

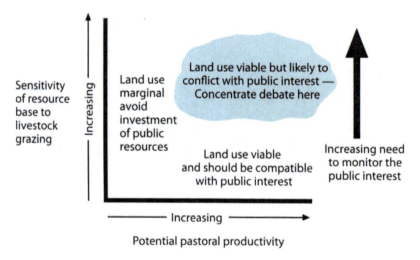

Fig. 5.5. Scope for public intervention in rangeland management (adapted from Stafford-Smith et al. 2000)

Lesslie et al. (2006b) illustrate how appropriate spatial data layers and expert opinion can be combined using the *MCAS-S* tool to allow spatial exploration of these relationships. First, relevant input data were combined using the *MCAS-S* composite function to produce composite indexes representing the attributes on each axis shown in Fig. 5.5. Primary data input layers were selected and their role in contributing to composite measures examined in a workshop process utilising expert advice. The potential pastoral productivity index was, for instance, developed by the weighted combination of input layers using the logic that potential productivity will generally be greater where:

- landscapes have higher productive potential (forage potential)
- rainfall is relatively consistent within and between seasons
- there is better access to markets, supplies and labour.

The result of composite development within *MCAS-S* for potential pastoral productivity is shown in Fig. 5.6.

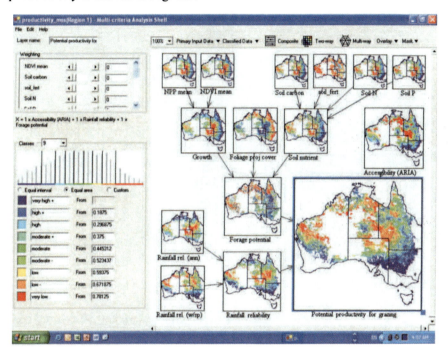

Fig. 5.6. A representation of potential productivity for livestock grazing in the Australian rangelands

Second, the *MCAS-S* two-way function was used to enable exploration of the spatial relationship among derived composite layers of potential pastoral productivity and the sensitivity of the resource base to livestock grazing. The two-way layer on the right hand side of the *MCAS-S* desktop in Fig. 5.7 highlights locations where conditions in the upper right quadrant of the matrix shown in Fig. 5.5 apply on the basis of derived index values (i.e. where there is a coincidence of high potential productivity and high resource sensitivity index values). Locations where other relationships apply (e.g. high potential productivity and low resource sensitivity; low potential productivity and high resource sensitivity; low potential productivity and low resource sensitivity) can be interactively explored in *MCAS-S* by simply clicking on the appropriate cell in the matrix at bottom left. In this way the two-way function facilitates the spatially explicit landscape condition assessment and policy analysis.

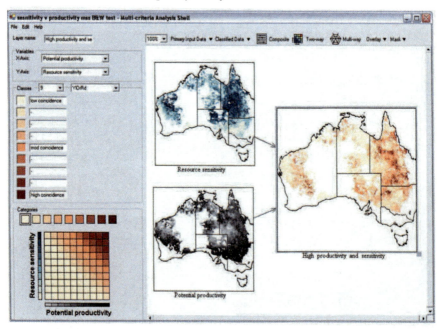

Fig. 5.7. A two-way analysis of potential productivity and resource sensitivity

The spatial relations among more than two layers can be interactively explored using the *MCAS-S* multi-way comparison facility. Figure 5.8 shows the creation of a binary layer (at bottom right) that identifies locations that meet specified conditions for each of three composite layers developed for the Australian rangelands study: potential productivity, sensitivity of the resource base to livestock grazing, and total grazing pressure. The spatial association and visualisation process involves the use of the radar plot, with spokes that represent the scale of each respective input layer. Class limits on these spokes are interactively manipulated to delimit locations that satisfy user defined criteria levels. The multi-way binary layer shown in Fig. 5.8 identifies locations where there is a coincidence of relatively low potential productivity, relatively high sensitivity of the resource base to livestock grazing, and relatively high total grazing pressure. The multi-way layer may also be displayed as a grey-scale surface showing 'distance' from selected criteria values.

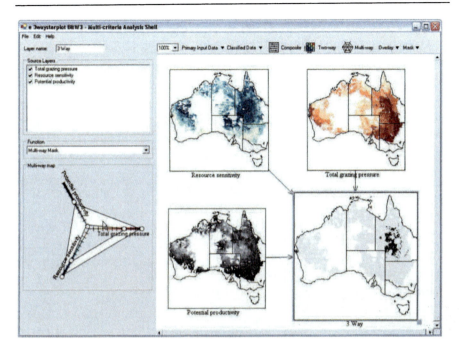

Fig. 5.8. A multi-way analysis of potential productivity, resource sensitivity and total grazing pressure

5.5 Future Trends

As a versatile spatial multi-criteria analysis shell, *MCAS-S* has many practical applications, including the assessment of ecosystem services and the visualisation of complex analyses of land use management and land management practices. The *MCAS-S* composite and multi-way analysis functions enable the development and comparison of multiple spatial data themes (e.g. carbon, water, aesthetics, production and ecosystem function) as described by Foley et al. (2005). Similarly, *MCAS-S* multi-criteria visualisation procedures can be applied with land use and land management practices information (Lesslie et al. 2006a) to assist evaluation of complex land management change and trade-off options.

Additional functionality will be added to *MCAS-S*. In the West Hume study a pair-wise comparison of guidelines was completed in order to rank and weight them for analysis. This was conducted outside the *MCAS-S* software with weightings applied in the creation of the biodiversity and salinity composite layers. Further development of the *MCAS-S* software will add a pair-wise comparison module to *MCAS-S* so that the rankings, ranking matrix table, and weightings can all be created within the software and associated with each layer. A map calculator function for composite development is also under development.

Description of time and space-based complexity will enhance the functionality of *MCAS-S*. Time series describe many important and spatially explicit processes in the landscape. Full information extraction from time series may require application of a suite of approaches including classical time series decomposition (Roderick et al. 1999), calculation of curve metrics (Reed et al. 1996), piece-wise logistic functions to capture curve trajectories (Zhang et al. 2003), time-based classifications (Moody and Johnson 2001) and Fourier and wavelet analysis (Li and Kafatos 2000). Recent research has highlighted the importance of global sensitivity analysis (Gomez-Delgado and Tarantola 2006) and multi-criteria weight sensitivity analysis (Feick and Hall 2004) to define the response space and impact of input layers on MCA outcomes.

The implementation of a Bayesian Belief Network within *MCAS-S* is also being explored. There are recent examples of the application of Bayesian network approaches and multi-objective model land system and environmental assessments (Dorner et al. 2007; Pollino et al. 2007). Consideration could be given to inclusion of soft systems approaches that enable users to create partial weight tables to capture soft attributes such as belief, commonality and plausibility (Beynon 2005).

5.6 Conclusion

Natural resource management decision making by communities, rural industries and governments is usually undertaken in a context of uncertainty where, ideally, the decision-making process engenders trust and engagement among stakeholders. Informed debate about competing demands on natural resources and the desirability of alternative futures can benefit from the use of versatile spatially explicit decision support tools such as *MCAS-S*.

The particular advantages of the *MCAS-S* tool are its flexibility, simplicity, the capability to visualise spatial data and the linkages between spatial data, and its suitability for use in group and participatory processes. The West Hume land use planning application combines simple first principle criteria with readily available spatial data to prioritise locations appropriate for revegetation based on the number of desirable criteria that were met, and the relative importance of these criteria. The process demonstrates how a regional investment strategy can be translated into mapped priorities by focussing on identifying where multiple environmental and production outcomes can be achieved at minimal cost. The utility of the tool in a broader context of policy analysis and development is demonstrated by the Australian rangelands application. The wider adoption of this type of planning process in Australia could substantially improve the effectiveness of current investment in natural resource management.

5.7 Future Research Directions

There is a growing expectation that spatial science and information can be successfully applied to decision making on complex issues affecting coupled human–environment systems. This expectation is promoted by the increasing sophistication of spatial multi-criteria analysis tools and more readily accessible economic, biophysical and social spatial data. While tools such as *MCAS-S* can help integrate factual information with value judgements and policy goals in a transparent and flexible way, there is still considerable scope for improvement in spatial data handling technologies and analytical methods. A particular need is for technologies that are capable of delivering information at many levels of scale and complexity with a cognitive context, through visualisation, that aligns more naturally with human thinking and decision making.

Key research challenges include:

- the analysis of time series data, particularly the discrimination of patterns in trend, periodicity and magnitude
- the linkage of time series and signal processing methods to spatial analysis
- resolving scale, particularly in relation to economic and social data and issues arising from changes in scale
- managing uncertainty, especially in the linkage of scientific and social domains.

Improvement in visualisation is important so that scenario outputs from spatially explicit modelling and decision processes can be made more meaningful to non-experts.

Acknowledgements

The *MCAS-S* system was developed for the Natural Resource Management Division of the Australian Government Department of Agriculture, Fisheries and Forestry with funding from the Natural Heritage Trust. We acknowledge funding support for the West Hume study from the Murray Catchment Management Authority and the CSIRO Water for a Healthy Country Flagship program. Jean Chesson is thanked for her advice on the multi-criteria analysis and priority-setting process. The chapter also benefited from the comments of two anonymous referees.

References

Arampatzis G, Kiranoudis CT, Scaloubacas P, Assimacopoulos D (2004) A GIS-based decision support system for planning urban transportation policies. European Journal of Operational Research 15:465–475

Beynon MJ (2005). A method of aggregation in DS/AHP for group decision-making with the non-equivalent importance of individuals in the group. Computers and Operations Research 32:1881–1896

Bisdorff R (1999) Cognitive methods for multi-criteria expert decision making. European Journal of Operational Research 119:379–387

Brans JP, Mareschal B, Vincke PH (1984) PROMETHEE: A family of outranking methods in multicriteria analysis. Operational Research 22:477–490

Bui E (ed) (1999) A soil information strategy for the Murray-Darling Basin (MDBSIS). Report to Murray Darling Basin Commission, Project D5038, available at http://www.brs.gov.au/mdbsis/publications.html

Ceballos-Silva A, Lopez-Blanco J (2003) Evaluating biophysical variables to identify suitable areas for oat in Central Mexico: a multi-criteria and GIS approach. Agriculture, Ecosystems and Environment 95:371–377

Cuddy SM, Laut P, Davis JR, Whigham PA, Goodspeed J, Duell T (1990) Modelling the environmental effects of training on a major Australian army base. Mathematics and Computers in Simulation 32:83–88

Dai FC, Lee CF, Zhang XH (2001) GIS-based geo-environmental evaluation for urban land-use planning: a case study. Engineering Geology 61:257–271

Dorner S, Shi J, Swayne D (2007) Multi-objective modelling and decision support using a Bayesian network approximation to a non-point source pollution model. Environmental Modelling and Software 22:211–222

Dragan M, Feoli E, Fernetti M, Zerihun W (2003) Application of a spatial decision support system (SDSS) to reduce soil erosion in northern Ethiopia. Environmental Modelling and Software 18:861–868

Feick RD, Hall GB (2004) A method for examining the spatial dimension of multi-criteria weight sensitivity. International Journal of Geographical Information Science 18:815–840

Foley JA, DeFries R, Asner GP, Barford C, Bonan G, Carpenter SR, Chapin FS, Coe MT, Daily GC, Gibbs HK, Helkowski JH, Holloway T, Howard EA, Kucharik CJ, Monfreda C, Patz JA, Prentice IC, Ramankutty N, Snyder PK (2005) Global consequences of land use. Science 309:570–574

Geneletti D (2007) An approach based on spatial multi-criteria analysis to map the nature conservation value of agricultural land. Journal of Environmental Management 83:228–235

Giupponi C, Mysiak J, Fassio A, Cogan V (2004) MULINO-DSS: a computer tool for sustainable use of water resources at the catchment scale. Mathematics and Computers in Simulation 64:13–24

Gomez-Delgado M, Tarantola S (2006) GLOBAL sensitivity analysis, GIS and multi-criteria evaluation for a sustainable planning of hazardous waste disposal site in Spain. International Journal of Geographical Information Science 20:449–466

Hill MJ, Braaten R, Lees B, Veitch SM, Sharma S (2005a) Multi-criteria decision analysis in spatial decision support: the ASSESS analytic hierarchy process and the role of quantitative methods and spatially explicit analysis. Environmental Modelling and Software 20:955–976

Hill MJ, Lesslie R, Barry A, Barry S (2005b) A simple, portable, spatial multi-criteria analysis shell – MCAS-S. In: Zerger A, Argent RM (eds) MODSIM 2005 International Congress on Modelling and Simulation. Modelling and Simulation Society of Australia and New Zealand, 12–15 December 2005, pp 1532–1538, available at http://www.mssanz.org.au/modsim05/papers/hill.pdf

Hill P, Cresswell H, Hubbard L (2006a) Spatial prioritisation of NRM investment in the West Hume area (Murray CMA region). Technical Report, CSIRO, Water for a Healthy Country National Research Flagship: Canberra, pp 1–28

Hill MJ, Lesslie RG, Donohue R, Houlder P, Holloway J, Smith J (2006b) Multi-criteria assessment of tensions in resource use at continental scale: A proof of concept with Australian rangelands. Environmental Management 37:712–731

Horst D, Gimona A (2005) Where new farm woodlands support biodiversity action plans: a spatial multi-criteria analysis. Biological Conservation 123:421–432

Janssen R, Van Herwijnen M (2006) A toolbox for multi-criteria decision-making. International Journal of Environmental Technology and Management 6:20–39

Kangas J, Store R, Leskinen P, Mehtätalo L (2000) Improving the quality of landscape ecological forest planning by utilising advanced decision-support tools. Forest Ecology and Management 132:157–171

Kiker GA, Bridges TS, Varghese A, Seager TP, Linkov I (2005) Application of multicriteria decision analysis in environmental decision making. Integrated Environmental Assessment and Management 1:95–108

Lesslie R, Barson M, Smith J (2006a) Land use information for integrated natural resources management – a coordinated national mapping program for Australia. Journal of Land Use Science 1:45–62

Lesslie R, Hill MJ, Woldendorp G, Dawson S, Smith J (2006b) Towards Sustainability for Australia's Rangelands: Analysing the Options. Australian Government Bureau of Rural Sciences: Canberra, pp 1–12

Li Z, Kafatos M (2000) Interannual variability of vegetation in the United States and its relation to El Nino/Southern Oscillation. Remote Sensing of Environment 71:239–247

Malczewski J (2006) GIS-based multi-criteria decision analysis: a survey of the literature. International Journal of Geographical Information Science 20:703–726

Maniezzo V, Mendes I, Paruccini M (1998) Decision support for siting problems. Decision Support Systems 23:273–284

Marinoni O ΄(2005) A stochastic spatial decision support system based on PROMETHEE. International Journal of Geographical Information Science 19:51–68

Mazzetto F, Bonera R (2003) MEACROS: a tool for multi-criteria evaluation of alternative cropping systems. European Journal of Agronomy 18:379–387

Moody A, Johnson DM (2001) Land-surface phenologies from AVHRR using the discrete Fourier transform. Remote Sensing of Environment 75:305–323

Morari F, Lugato E, Borin M (2004) An integrated non-point source model-GIS system for selecting criteria of best management practices in the Po Valley, North Italy. Agriculture, Ecosystems and Environment 102:247–262

Murray Catchment Management Board (2001) Murray catchment management plan; a blueprint for action. Dated 17 October 2001 (draft)

Pettit C, Pullar D (1999) An integrated planning tool based upon multiple criteria evaluation of spatial information. Computers, Environment and Urban Systems 23:339–357

Pollino CA, Woodberry O, Nicholson A, Korb K, Hart BT (2007) Parameterisation and evaluation of a Bayesian network for use in an ecological risk assessment. Environmental Modelling and Software 22:1140–1152

Ramanathan R (2001) A note on the use of the analytic hierarchy process for environmental impact assessment. Journal of Environmental Management 63:27–35

Reed BC, Loveland TR, Tieszen LL (1996) An approach for using AVHRR data to monitor US Great Plains grasslands. Geocarto International 11:1–10

Ren F (1997) A training model for GIS application in land resource allocation. ISPRS Journal of Photogrammetry & Remote Sensing 52:261–265

Roderick ML, Noble IR, Cridland SW (1999) Estimating woody and herbaceous vegetation cover from time series satellite observations. Global Ecology and Biogeography 8:501–508

Saaty TL (2000) Fundamentals of Decision Making and Priority Theory with AHP Analytic Heirarchy Process. RWS Publications, Pittsburg

Stafford-Smith DM, Morton SR, Ash JA (2000) Towards sustainable pastoralism in Australia's Rangelands. Australian Journal of Environmental Management 7:190–203

Store R, Jokimaki J (2003) A GIS-based multi-scale approach to habitat suitability modelling. Ecological Modelling 169:1–15

Store R, Kangas J (2001) Integrating spatial multi-criteria evaluation and expert knowledge for GIS-based habitat suitability modelling. Landscape and Urban Planning 55:79–93

Veitch SM, Bowyer JK (1996) ASSESS: A System for Selecting Suitable Sites. In: Morain S, Lopez Baros S (eds) Raster Imagery in Geographic Information Systems. OnWord Press, Santa Fe, p 495

Walker J, Veitch S, Dowling T, Braaten R, Guppy L, Herron N (2002) Assessment of catchment condition. CSIRO Land and Water, Canberra

Wu F (1998) SimLand: a prototype to simulate land conversion through the integrated GIS and CA with AHP-derived transition rules. International Journal of Geographical Information Science 12:63–82

Zhang X, Friedl MA, Schaaf CB, Strahler AH, Hodges JCF, Gao F, Reed BC, Huete A (2003) Monitoring vegetation phenology using MODIS. Remote Sensing of Environment 84:471–475

6 *Platform for Environmental Modelling Support*: a Grid Cell Data Infrastructure for Modellers

Tai Chan[1], Craig Beverly[1], Sam Ebert[1], Nichola Garnett[2], Adam Lewis[3], Christopher Pettit[4], Medhavy Thankappan[3] and Stephen Williams[5]

[1] Department of Sustainability and Environment, East Melbourne, Victoria, Australia
[2] Department of Justice, Melbourne, Victoria, Australia
[3] Geoscience Australia, ACT, Australia
[4] Department of Primary Industries, Parkville Centre, Victoria, Australia
[5] Department of Primary Industries, Epsom Centre, Victoria, Australia

Abstract: Australian Spatial Data Infrastructure (SDI) is in need of a grid cell-based data component that supports the needs of landscape or environmental process modellers and other GIS users. Spatial data infrastructure and innovation diffusion concepts based on the Organisational Innovation Process are used to establish the innovative nature of this proposed component of the SDI and the process for its acceptance by key stakeholders in Australia.

The experience of development of a Cooperative Research Centre for Spatial Information (CRCSI) Demonstrator Project, namely, Platform for Environmental Modelling Support (PEMS) *is described to illustrate a collaborative model used by a group of public and private organisations to promote the diffusion of grid cell data infrastructure at the state and national level in Australia. The experience of the project suggests that in addition to the established collaborative environment offered by the CRCSI, investment in robust and structured project management is also important to facilitate future adoption of the grid cell data infrastructure by the participating organisations.*

6.1 Introduction

Many governments worldwide are developing Spatial Data Infrastructures (see Nebert 2004 for a detailed definition) to provide the standards, technical, institutional and policy means to facilitate access to spatial information (Onsrud 1999). The main advantage of a spatial data infrastructure lies in the typical cost structure of developing GIS applications. According to a study conducted by the Federal Geographic Data Committee of the USA (FGDC 2000) about 75% of costs are expended in locating and assembling data while only about 18% and 7% are expended in application development and actual productive operations respectively. Therefore, it is imperative for governments to facilitate the documentation, discovery and access to existing spatial information assets within their jurisdictions so that these assets can be used effectively and efficiently to address the increasing complex social, economic and environmental challenges the world is facing. In Australia, the effort is called the Australian Spatial Data Infrastructure (ASDI).

Typically SDIs in Australia have a strong vector and imagery focus but lack support for grid cell data. According to the Cooperative Research Centre for Spatial Information (CRCSI 2007), a grid cell data structure is a matrix where the coordinate of a cell (also called a pixel) can be calculated if the origin point is known, and the size of the cells is known. Such a matrix-based structure lends itself to manipulation via two-dimensional arrays in computer encoding and as such allows for considerable efficiencies in the development of analytical (modelling) processes. Most grid cell-based GIS software requires that each grid cell contains only a single discrete value, thus generating single attribute maps. This is in contrast to most conventional vector data models that maintain data as multiple attribute maps, for example, forest inventory polygons linked to a database table containing all attributes as columns. This basic distinction of grid cell data storage is that it provides the foundation for quantitative analysis techniques based on sophisticated mathematical modelling processes. This makes cellular (tessellated) data structures the paradigm of choice for the vast majority of spatial modelling processes, particularly those involving continuous data, such as elevation. On the other hand, vector data systems are better suited to linear data analysis (e.g. shortest path) and to high quality cartographical requirements. Therefore it is important for the user to pick the best fit-for-purpose data structure for application development.

Owing to the above characteristics of grid cell data, scientists and government departments across Australia are increasingly relying on grid cell data in their planning, analyses and modelling activities to provide scientific evidence for environmental management policies and decisions. Unfortunately grid cell data often are only available locally in work groups or business units and are not published with proper metadata, making it difficult for potential users to discover the data needed and assess their currency, relevance and value. Even where metadata are available, modellers often have to go through another process of data extraction and conversion (and potential re-sampling) to ensure that the grid cell data from various sources can be used together. Governments at all levels in Australia are being challenged by serious issues such as climate change, water shortage and environmental sustainability that affect a wide range of sectors such as health, education and security. The ability to utilise grid cell data as a fundamental input into developing solutions to these issues is poorly supported in the ASDI.

Strategic benefits of a grid cell component in the ASDI include access to consolidated[1], authoritative and up-to-date spatial information, and a greatly improved capability to integrate a range of social, economic and environmental spatial information. Both provide direct improvement to routine business queries and modelling for policy development and decision making. Short-term tactical benefits include significant efficiencies in undertaking queries, analyses and modelling to higher standards than in the past, and the ability to perform these activities in a transparent and repeatable manner. To realise these benefits the grid cell component of the ASDI must be actively managed to ensure the currency of the grid data and their associated metadata is maintained.

To address this gap in ASDI, the Victorian Government Spatial Information Infrastructure Branch, in concert with its counterparts in other sectors, formulated the idea of a grid cell data component for the ASDI almost two years ago. After following an approach based on SDI and innovation diffusion theories that incorporated a rigorous process of stakeholder engagement, the idea is being realised in the form of a Demonstrator Project called Platform for Environmental Modelling Support (PEMS). This project is sponsored by the Cooperative Research Centre for Spatial Information (CRCSI). The organisations involved include Geoscience Australia, the Department of Sustainability and Environment, the Department of Primary Industries, the Office of Emergency Services Commissioner of Victoria, and Spatial Vision of the private sector.

---

[1] Derived from multiple sources including local, state/territorial and national agencies.

6.2 Background

The concept of this data infrastructure was built on the process behind the Broadhectare Study first published for south-east Queensland by the Queensland Department of Housing, Local Government and Planning (QDHLGP 1992). The fundamental design concept underpinning *PEMS* is a managed system that will format, prepare and deliver spatial information to a standardised grid-based topological paradigm. Proposed key aspects of the grid cell-based approach in *PEMS* are:

- A broad range of spatial data will be available to users, keyed on a consistent national (or jurisdictional) grid. These datasets will be maintained and progressively updated allowing time series analysis and modelling.
- Each grid cell, a standard geographic unit identified by a unique key and spatially located by grid coordinates, will have a list of attributes such as land use, slope, annual precipitation, prevailing wind direction, vegetation type, valuation, planning zone, salinity (i.e. a broad range of social, environmental and economic data). One or more grid cells with their associated attributes can be seen as a 'data cube'.
- Stored in this manner, the data will allow fast and effective sieve mapping[2] and other simple forms of data manipulation, whilst retaining their spatial attributes.
- Processed data may be queried and the required results displayed spatially, or formatted for input to an external application or more specialised modelling tools.

The target user groups include modellers, GIS experts and analysts, business and professional users and data custodians.

While each of the above aspects of the PEMS approach is not new to many organisations across Australia, the combination of these aspects is innovative requiring new investments. For the PEMS approach to be adopted as the mainstream component of the ASDI, adoption by many key organisations across Australia is a prerequisite. Promoting this adoption is a complex and challenging process. By using Rogers' (1995) diffusion paradigm, Chan and Williamson (1998) identified a way forward by applying Rogers' theory of Organisational Innovation Process (see Fig. 6.1) to GIS diffusion in government.

---

[2] Sieve mapping is a GIS technique used in land planning. In this technique all of the land is initially considered eligible. Then certain areas are eliminated because of one reason or another until all that is left is the target area.

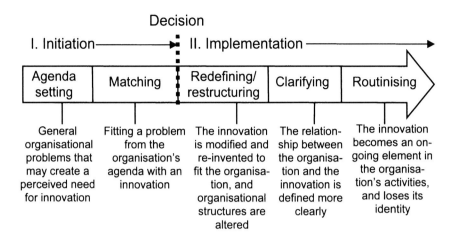

Fig. 6.1. The Organisational Innovation Process (adapted from Rogers 1995, p. 392)

Chan and Williamson (1998) argue that the diffusion of a complex and strategic collection of technologies such as a corporate GIS does not go 'cleanly' through the Organisational Innovation Process as an integral entity. Instead its diffusion can be 'messy' and viewed as the adoption and implementation of a number of separate but related technologies called infrastructure modules and business process modules through the Organisational Innovation Process. One or a collection of these modules may go through the process at different points in time and experience different levels of success. Based on Fig. 6.1 the modules that have their potential values recognised (i.e. that have passed the agenda setting stage), possess functions matched to one or more specific problems (matching), and contain components that are clearly defined, piloted, adjusted and tested to suit the operational environment of the organisation (redefining/restructuring), are most likely to complete the diffusion earlier. Other modules not matched to any specific problems will remain in the initiation phase and their adoption by an organisation will be problematic.

At the current early stage, PEMS is very much like a corporate GIS as it has a broad conceptual design that promises strong strategic benefits and efficiency gains for a wide range of stakeholders. The breadth of stakeholder interest helps make it a candidate to be a potential component of the ASDI. As PEMS is a key SDI component, naturally each stakeholder sees it as an opportunity, but views this opportunity through a lens formed from its own business perceptions, needs and application requirements. There are a number of needs for the PEMS innovation to be adopted by key organisations and become part of the ASDI. They include:

- Assembling a small group of like-minded organisations to articulate and develop the concept of *PEMS* that is representative of the interest of all major groups of target users.
- Identifying and applying a means to allow the participants to match the concept of *PEMS* to their organisational agenda and have it tested within their operational environment as much as possible.
- Identifying an appropriate model of collaboration (McDougall et al. 2005) and the associated funding model (Giff and Coleman 2005).
- Identifying a means to promote and extend the services enjoyed by existing participants to new organisations, and generally to promote wider adoption across Australia (de Bree and Rajabifard 2005).

6.3 Methodology

Spatial Information Infrastructure of the Victorian Government has been an active core participant of the CRCSI since its formation in 2003, and sees the CRCSI as a vehicle for overcoming the challenges of diffusion identified above. The CRCSI provides management resources, formal governance and a legal framework to guide and control collaborative project development, cash and in-kind contributions, project implementation, and intellectual property management and utilisation — all of which lead to commercialisation and government adoption.

In particular, apart from simply funding the usual research projects the CRCSI maintains a Demonstrator Program that allows existing technologies to be packaged into innovative products and services that are then demonstrated, promoted and brought to the market to meet perceived demands. As part of its Demonstrator and Commercialisation programs, the CRCSI has the mechanism to work with interested users in both the public and private sectors to further develop existing Demonstrator prototypes to meet specific requirements in the organisations. The entire process is well known to all the participants that are active in the spatial industry and include major government departments, universities and private sector companies across Australia. As a result the CRCSI can be the vehicle to overcome the challenges mentioned above by offering a:

- ready pool of like-minded organisation to participate in *PEMS*, and in the future to promote *PEMS*
- Demonstrator Program to allow participants to match *PEMS* functions to real-world problems

- formal operational environment that offers a legal, financial and institutional framework of collaboration that is well known to potential participants
- Demonstrator Program and Commercialisation Program designed to promote wider adoption.

Having identified the broad approach to realise the design concept of *PEMS*, the following steps were taken to initially gain the support of the Demonstrator Program of the CRCSI, and then to implement the *PEMS* Demonstrator Project according to the theory of Organisational Innovation Process. These include to:

- appoint a full-time project manager to work with the project sponsor (Director of Spatial Information Infrastructure) to document the *PEMS* concept in a discussion paper
- develop a 'proof-of-concept' to supplement the discussion paper when discussing with CRCSI and like-minded business managers in member organisations to gain in-principle support and identify potential participants
- engage actively with the CRCSI and potential participants to develop a more detailed project proposal that meets the expectation of key stakeholders and gains the in-principle support of the respective national peak bodies of Australia New Zealand Land Information Council and Australian Spatial Information Business Association in the public and private sectors
- secure funding approval of the Governing Board of the CRCSI
- establish an internal governance arrangement for *PEMS* to complement that of the CRCSI to ensure that by adopting project management best practices, the project: meets the needs of the CRCSI and project participants; promotes communication among project participants; engages with parent organisations, state and national bodies and the wider user communities and; responds to strategic opportunities such as the Victorian eResearch Strategic Initiative and the National Collaborative Research Infrastructure Strategy.

6.4 Progress and Discussions

PEMS started formally in February 2007 and the CRCSI Demonstrator Project agreement that formalises the cash and in-kind funding of over $2.25M from the CRCSI and participating organisations was executed in May 2007. A Project Control Group (PCG) that comprises a representative from each of the five participating organisations and chaired by the project sponsor has been meeting monthly to put in place policies, procedures and guidance to implement the project and to ensure proper management of user and participant expectations.

With strong inputs from members of the Technical Reference Group the PCG oversaw the production of three consensus documents that guided the development of the *PEMS* Demonstrator Project. The first document was the Specification Report, which articulated a common vision for the *PEMS* concept — including the design of a system of national nested grids — a prototype spatial framework for Australia led by Geoscience Australia. The second document was the Demonstrator Implementation Report that specified the work to be included in the Demonstrator Project, which is necessarily a subset of the full *PEMS* concept. The third document was the Evaluation Report that scoped the assessment work needed to document the value of *PEMS*. System testing and functional evaluation started in February 2008.

The process took three months longer than expected despite active involvement of the participants. Instead of an unnecessary delay, the PCG regarded the process as a positive step to address the identified risk of not managing participants' expectations resulting in loss of support or even bad publicity. Through pragmatic debates over what should be included in the final Demonstrator Project, the participants were able to reassess the original design of their use cases, which were effectively prototypes or pilots of the business processes to be demonstrated to their constituencies. They were able to make the difficult but informed decisions needed to redefine the use cases and match their organisational needs to the resources available.

This experience suggests that *PEMS* means different things to different participating organisations. The delay in getting a consensus on the Specification Report and the Demonstrator Implementation Report indicates that even like-minded participating organisations were grappling with the concept of *PEMS* and how it fits in with their respective organisational requirements. Through the discipline imposed by the project management process, they were forced to go through the 'matching' and 'redefining' phases in Fig. 6.1. In the process they came to a better understanding of what the *PEMS* Demonstrator could do for them and how it could fit in with their organisations.

This improved understanding of *PEMS* and its relation with their requirements allows the participating organisations to articulate the benefits and the associated means of measurement in the *PEMS* Evaluation Report — the preparation of which has been delayed due to the challenges faced when preparing the Specification Report and Implementation Report. They are now encouraged to follow a robust and structured process to identify, measure and evaluate the benefits to the immediate stakeholders in their respective organisations as demonstrated by *PEMS*. This is seen as an important step to help the immediate stakeholders involved to clarify the value of *PEMS* to their organisation, allowing them to plan for wider adoption later on.

6.5 The *PEMS* Demonstrator Project

The *PEMS* Demonstrator Project is designed to support the participant sponsored use cases as listed below (and described in the sections that follow):

- national seasonal crop monitoring and forecasting
- develop and demonstrate a market-based approach to environmental policy on private land
- wildfire planning: consequence of loss modelling
- land use data, modelling and reporting.

As a demonstrator project and with the limited budget available, the project aims not at automating all functions specified, but rather putting together a collection of automated and mock-up functions that help the participants demonstrate how the *PEMS* grid cell data capabilities could work in, or for, their constituencies. The documents produced and the functions developed have laid much of the ground work necessary to raise the awareness of the importance of grid cell data infrastructure as a component of the national spatial data infrastructure. It will, in the least, take most interested organisations through the *agenda setting* stage of the Organisational Innovation Process in Fig. 6.1. For the *PEMS* participants it is expected that through their respective use cases in the project, they are able to take their constituencies to more advanced stages such as *matching, redefining/restructuring* and *clarifying*.

6.5.1 National Seasonal Crop Monitoring and Forecasting

The main objective of this use case is to enable the monitoring and modelling of spatio–temporal changes in a pilot study area covering the Wimmera–Mallee region in Victoria using time series remote sensing data, within the *PEMS* environment. This use case also establishes the:

- value of *PEMS* for monitoring temporal changes in landscape at regional scales

- capability for *PEMS* to provide agricultural crop condition information on a periodic basis
- standards for a hierarchical, national grid that enables multi-scale data analysis.

The use case demonstrates scalability of *PEMS* to support similar applications at a national level and is expected to trigger wider discussion within Australian Government of the *PEMS* approach to spatially enabling government and effectively informing national policy.

Time series satellite data from the *MODIS* Terra satellite sensor with a spatial resolution of 250 m and *Landsat Thematic Mapper (TM)* with a spatial resolution of 30 m, was used to derive Vegetation Index (VI) information. Monitoring temporal changes in VI provides information about changes in the density and vigour of vegetation over time. Temporal profiles of VI can also help identify key phenological stages and types of vegetation where field data are available. In the case of agricultural crops, time series VI allows comparison of the spatial extent and severity of negative VI departures from the long-term mean caused by heat stress, droughts or frosts. *PEMS* provides a hierarchical grid-based environment for comparing the temporal and spatial behaviour of VI at target locations and interpreting the statistical patterns with respect to the phenology of vegetation including crops. The difference in the spatial resolution of *MODIS* and *Landsat TM* data enables multi-resolution analysis within *PEMS*, leading to the development of consistent time series datasets representing vegetation dynamics over a given region. Crop type, area, and condition information based on analysis of spatially complete VI data within *PEMS* could enhance the quality of commodity forecasts based purely on sampling techniques.

For the pilot study area chosen for this use case, two forms of VI, Normalised Difference Vegetation Index (NDVI) and Enhanced Vegetation Index (EVI), served as inputs for modelling and monitoring spatio–temporal changes within *PEMS*. Time series NDVI and EVI images derived from the satellite data and conforming to multiple resolutions of the hierarchical grid were input into *PEMS*. The *PEMS* environment then enabled the querying of these time series VI datasets based on user-specified geographical extent or point location in order to extract information about vegetation condition. The onset of greenness, peak greenness and senescence, which are useful indicators of vegetation condition or drought, can be determined by examining the temporal profiles of VI for any location within the *PEMS* grid (Fig. 6.2).

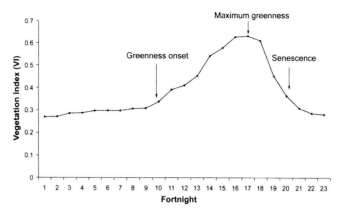

Fig. 6.2. Vegetation Index profile showing key phenological stages

Comparison of the current crop season's VI with a reference (normal crop season) or the long-term average VI could determine if a crop season is expected to be below normal, above normal or normal for one or more grid cells or pre-defined administrative units (e.g. Statistical Local Area or Statistical Division) within *PEMS*. Significant shifts in the VI trends over time or inter-annual changes in land use or land cover can be identified through analysis within *PEMS*.

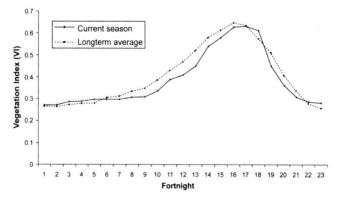

Fig. 6.3. Vegetation Index profile for current season crop compared to the long-term average

Basic VI statistics by location or land cover type queried using *PEMS* (e.g. mean, standard deviation or histogram of VI for a user selected area) can be output as a chart or simple text. *PEMS* also enables queries over multiple grids to compare VI statistics for datasets with different spatial resolutions. Trend analysis of the time series VI data within *PEMS* produces charts showing departure from long-term average for the selected location or cover type, similar to the chart shown in Fig. 6.3. The ability to export query results as GIS data also enables further analysis of the results within a GIS environment.

6.5.2 Develop and Demonstrate a Market-based Approach to Environmental Policy on Private Land

The Market-Based Solutions to Redress Landscape Decline (MBS) is a Victorian Government project to develop and trial one comprehensive set of policy mechanisms/instruments. The mechanisms aim to improve governments' ability to procure environmental services from private landholders. The approaches are based in part on a pilot project called EcoTender.

According to Eigenraam et al. (in press), EcoTender uses an auction mechanism to reveal the private cost of procuring multiple environmental outcomes. The mechanism is fully integrated from desk to field with a non-point production biophysical simulation model called the *Catchment Management Framework* (*CMF*) that links paddock-scale land use and management to environmental outcomes at both the paddock and catchment scale. The *CMF* builds upon the *Catchment Analysis Toolkit* (*CAT*) that accounts for biophysical processes for soil erosion, water, carbon and saline land to estimate environmental outcomes (Beverly 2007; Beverly et al. 2005; Christy et al. 2006; Weeks et al. 2005). Biodiversity algorithms evaluate the current location of native vegetation and biodiversity landscape preference, and assess the future spatial needs of key mobile fauna species. As a result market-based instruments, such as EcoTender, have much greater demands for high quality data and information. They use these to score or measure environmental impacts in the form of metrics. A lot of these metrics are informed by spatial and biophysical modelling of grid cell-based biophysical data.

In this context the objective of the use case in the *PEMS* Demonstrator is to assess the effectiveness of a generic grid cell-based data management environment offered by *PEMS* in supporting the use of respective spatial and biophysical models (such as *CMF* and the underpinning *CAT* model) as tools in the MBS Project. Inherent in these predictive models are a variety of input biophysical raster data layers including soil, climate (average annual temperature and rainfall), slope, land use (current and proposed), aspect, tree cover (ecological vegetation classes), climate station proximity and a digital elevation model (DEM). Categorical data (soils, land use, tree cover) traditionally have been sourced from the government's Corporate Spatial Data Library (CSDL), whereas the continuous data (DEM, slope, aspect, climate) were derived from GIS and splining software (*ANUClim©*) based on the DSE Forests 20 m DEM. All data were generated on a 20 m resolution for each of the 10 Victorian Catchment Management Authorities (CMAs). This resulted in the development of a seamless statewide coverage. Depending on the scale of a study area, data were re-sampled from 20 m (subcatchment scale) to 100 m (CMA or regional scale).

New data layers derived from the MBS project include spatial groundwater conceptualisation and attribution such as thickness and extent of discrete aquifers, aquifer lateral conductivities/transmissivities, aquifer storages and aquifer vertical conductivities. Model output from *CAT* includes mapped recharge, discharge, erosion, carbon sequestration, nutrient transport, salt transfer and time varying water movement (surface and groundwater) and depth to watertable maps. These derived data will have immense value to all land managers ranging from farmers to policy makers and the MBS project would like to identify a means to publish the data, through the CSDL for example, to make them available to these people.

Given the nature of the *CMF* model requiring teams of field staff running on-farm assessments across all Victorian CMAs, there is a need for a web-based raster data dissemination tool to allow field staff access to the basic eight input layers required by *CMF* and *CAT* models. At the same time there is also a need to publish the data derived from these models.

PEMS provides a standard data environment to manage the above grid cell data in a consistent format, and integrates with an existing spatial data ordering application called *Spatial Datamart* in *Spatial Information Infrastructure* to provide the web-based request and delivery of the above data in a standardised format. Re-sampling of data to a different resolution is also possible. The *PEMS* standard data environment also allows the easy publication of outputs from models such as *CMF* and *CAT* so that new data can be viewed together with existing data to better inform the decision-making process at the farm-scale enterprise level through to statewide initiatives.

6.5.3 Wildfire Planning: Consequence of Loss Modelling

Planning for emergencies in Victoria currently suffers from limited access to consistent, relevant and reliable risk-based information, which is essential for facilitating effective risk mitigation and emergency response. The Victorian Office of Emergency Services Commissioner (OESC) Wildfire Project is an initiative of Fire Safety Victoria (FSV) — a strategic framework for delivering continuous improvements in fire safety to the Victorian community. The long-term goal of FSV is a Victorian community that is safer from the social, economic and environmental effects of fire. The strategy recognises the need for coordinated knowledge management across fire agencies and the community.

The first phase of the Wildfire Project has been completed and has established an evidence-based methodology that is designed to inform and support integrated wildfire planning and decision making. The project has developed approaches, principles and tools to support a consistent and shared understanding of the consequence of wildfire on assets across Victoria. The Wildfire Project products include a methodology, derived spatial datasets, and maps to inform integrated strategic wildfire management planning for DSE, CFA and Local Government Authorities across public and private land. The spatial products provide a resource for integrated fire management planning initiatives such as the Integrated Fire Management Planning (IFMP) Project that can be used to support the response and recovery phases of wildfire management and are sufficiently flexible to apply to hazards other than wildfire.

The Wildfire Project data and methodology, however, are not currently delivered on an accessible system that is available for multiple users with limited knowledge or access to specialist spatial data and tools. The *PEMS* platform is seen as a mechanism by which these data and tools can be delivered to fire management planning officers in the local planning forums. The Wildfire Project identifies over 200 source datasets that are aligned to 173 asset categories representing assets present in the landscape across Victoria. Through stakeholder consultation, these 173 asset categories have been grouped into eight asset classes and aligned with triple bottom line reporting levels (Table 6.1).

Table 6.1. Three-tiered asset classification, land and air asset classes were considered but not implemented in the Wildfire Project methodology

Triple bottom line reporting level	Asset classes reporting level	Examples of asset categories[a]	Number of asset categories[a]
Environmental	Biodiversity	Old growth forest, parks, threatened flora and fauna	29
	Water	Streams, wetlands and water bodies	2
Social	Cultural heritage	Aboriginal and post-European contact significant sites	12
	Social infrastructure	Schools, hospitals, recreational sites and community facilities	40
	Human life	Residential, daytime and peak seasonal populations	10
Economic	Economic production	Agriculture, forestry, mining, commerce and industry	24
	Infrastructure	Transport, utilities, buildings and equipment	37
	Property	Residential, rural, commercial and industrial	19

[a]Source datasets were analysed and collated into asset categories.

The Total Consequence of Loss (Tot CoL) of an asset is assigned to each asset category from a statewide perspective and is summarised into the eight asset classes. These values are represented in 1 km^2 reporting units with a nested 500 m^2 reporting unit representing urban centres and areas of significant interest. The Tot CoL of an asset is defined as:

$$Consequence = damage + disruption \qquad (6.1)$$

where *damage* is the total loss of value (replacement or intrinsic), and *disruption* is the impact from the loss of an asset (based on service and/or function it performs).

The statewide perspective provides a valuable starting point from which to assess the Tot CoL of an asset. However, the relative values of an asset may differ at state, regional and local levels, particularly when planning for fire management at these different levels. A requirement exists to interact with the assigned Tot CoL values to capture local knowledge and provide a local representation of the loss of an asset. The Wildfire Project methodology provides for this interaction by allowing users to apply weightings to asset categories or, where there is no damage but a disruptive effect to an asset, a disruption value can be applied.

It is anticipated that *PEMS* will provide an environment where the Wildfire Project spatial products are accessible and available in a standardised, interactive, visually rich format to contribute to integrated wildfire planning and decision making.

The *PEMS* Demonstrator Project is seen to be developing a foundation for an application that has the potential to make a considerable contribution to community awareness by increasing understanding, developing commitment and building community capacity via practical interaction in wildfire planning. The ability to seamlessly incorporate local knowledge into an integrated wildfire planning process will be seen by communities as a significant improvement in community engagement in public processes of significant local relevance, importance and priority.

Figure 6.4 provides an example of the anticipated interaction that a fire management planner would have with the Wildfire Planning–Consequence of Loss methodology and spatial information in the *PEMS* environment.

6.5.4 Land Use Data, Modelling and Reporting

In an era of rapid urbanisation where the global population now exceeds 6.6 billion people, there is an increasing pressure and duty of care on land custodians to manage our natural resources in a more sustainable manner. This use case spans the data, modelling and reporting continuum (Fig. 6.5) where we are interested in developing spatial data infrastructure to support better understanding of the impacts of current and future land use practices in Victoria. This includes improving perspectives on both above-ground and below-ground landscape phenomena including hydrological modelling, land use transition and impact modelling.

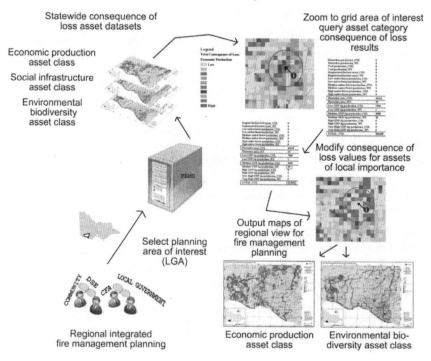

Fig. 6.4. Typical wildfire planning process using *PEMS*

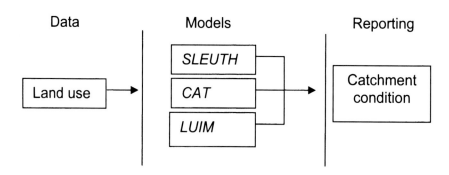

Fig. 6.5. Land use data, modelling and reporting methodology

Data

There are a number of topographical datasets required to support land use modelling activities to better understand landscape process and predictive behaviour of landscape change. Critical to these modelling efforts are the underpinning datasets. Catchment-scale land use data are a fundamental dataset used as an input into a number of land use models applied in Victoria. The catchment-scale land use data have been created in accordance with the Australian Land Use and Management (ALUM) classification published by the Australian Bureau of Rural Sciences (BRS 2006). This national classification is a three-tiered schema comprising 130 classifications at the third level. The creation and compilation of this dataset is costly and requires considerable effort and expertise. The current statewide dataset is the combined result of regionally based projects delivered over a number of years that were co-funded by national, state and regional government bodies. It is likely that further improvements to the dataset to maintain its currency will again be funded and delivered on a regional basis or targeted to areas of greatest land use change. This approach has and will continue to deliver a dataset whose currency varies spatially. Each region is almost certain to have a different time series that is associated with collection events or instances of the data.

In the overall scheme, two core views into the data need to be supported. Firstly, a consolidated view of the most current known land use (or best estimate) for any area is required to support analysis and modelling to address current issues and impacts. Secondly, an extraction of all data instances for nominated areas is required to support retrospective analysis and predictive modelling of land use change and associated future impact scenario modelling. Both of these operations or queries are facilitated by a grid cell-based infrastructure. It is envisaged *PEMS* will solve or advance a processing methodology to accurately convert the current vector land use dataset to a grid-based form and demonstrate support for the aforementioned queries. To accommodate these requirements *PEMS* provides a data-management environment supporting data versioning and associated metadata. As an adjunct or variant to the first data view, users with a focus on a particular industry or issue will use the *PEMS* query interface to specify and select individual land use classes or groupings for further analysis, processing or download.

Modelling

It is estimated that up to 80% of the time required for delivery of a landscape-based modelling project is in data preparation. Therefore, an environmental modelling platform that can provide modellers easy access to a number of fundamental datasets and minimises further data manipulation and transformation is an exciting prospect. There are a number of landscape-scale models applied in Victoria to inform a range of landscape questions such as:

- What are the likely impacts of a change in land use on water resources and catchment processes?
- What areas are sensitive to soil or wind erosion?
- How might urbanisation impact on agricultural land in peri-urban landscapes?

Landscape models such as the Catchment Analysis Toolkit (CAT), the *Land Use Impact Model* (*LUIM*), and the *Slope Land use Elevation Urbanisation Transportation and Hillshading* (*SLEUTH*) model are being applied in Victoria to address the aforementioned questions respectively (see Clarke Chapter 17). These models currently require or can use raster input data to operate. Providing modellers access to a standardised raster data repository with a toolset or suite of functions to support data selection, manipulation and extraction will enable modellers to spend more time modelling rather than undertaking data preparation.

This component of the use case focuses on evaluating the efficiency of *PEMS* for providing access to fundamental data and in supporting elements of workflows associated with modelling activities. To enable this, *PEMS* will be required to support model metadata ranging from what data a model requires for execution, through to recording details of model runs associated with data stored in *PEMS*. Ultimately in the future, beyond the Demonstrator Project, is the ability of the *PEMS* to support an eScience style implementation for model execution, chaining, coupling and delivery.

Reporting

It is important to not only understand the impacts on the environment but also the social and economic dimensions of the landscape. This is known as a triple bottom line (TBL) approach to environmental reporting. Underpinning TBL reporting are a number of spatial and aspatial functions supporting the formulation and assembly of indicator information or views from core datasets and modelling data products stored in *PEMS*. Since 1997, Victorian catchment condition reporting occurs on a five-yearly basis. *PEMS* will be evaluated for its ability to accurately convert and store an instance of TBL indicators and to support 'just in time' policy and requests for information from decision makers. By storing the indicators in a common repository

PEMS will enhance the ability to perform time series analysis so that a better understanding of performance trends will be possible.

6.6 Conclusion

The *Platform for Environmental Modelling Support* (*PEMS*) is a grid cell-based data infrastructure that has the potential to become an integral component of the Australian Spatial Data Infrastructure. It offers significant strategic and short-term benefits to governments and the wider community in Australia, by supporting spatial queries, analyses, reporting, and in particular, modelling. However, being an innovative infrastructure based on SDI and innovation diffusion theories creates two major challenges for the development of *PEMS*. Firstly, it requires a mechanism for potential like-minded organisations to collaborate to demonstrate the value of the infrastructure through real business applications. Secondly, it requires a formal process that takes the participants through the steps of agenda setting, matching, redefining/restructuring and clarifying the value of *PEMS* (Fig. 6.1) to facilitate the adoption of *PEMS* by themselves and stakeholders in their respective organisations.

This Demonstrator Program of the CRCSI was used as the mechanism to provide the legal, financial and institutional framework to enable four state and national government departments and one private company to collaborate to develop the grid cell data infrastructure and promote its adoption across Australia. To complement the CRCSI model, *PEMS* decided to invest in project management to ensure a structured and robust implementation of the original project plan. The experience of *PEMS* confirms that robust project management provides the formal process to allow participants to match, redefine and clarify the value of *PEMS*. The initial benefit is illustrated by the updating of the use cases, including the objectives and expectations as described in this chapter.

Acknowledgements

This development work is being conducted within the CRCSI which was established and supported under the Australian Government's Cooperative Research Centres Program. The views expressed in this chapter are those of the authors and do not represent the views of their organisations.

References

BRS (2006) Guidelines for land use mapping in Australia: principles, procedures and definitions, 3rd edn. Bureau of Rural Sciences, Canberra

Chan TO, Williamson IP (1998) The different identities of GIS and GIS diffusion. International Journal of Geographical Information Science 13(3):267–281

Beverly C (2007) Technical manual models of the Catchment Analysis Tool (CAT1D Version 22). Department of Sustainability and Environment, Victoria

Beverly C, Bari M, Christy B, Hocking M, Smettem K (2005) Salinity impacts from land use change: comparisons between a rapid assessment approach and a detailed modelling framework. Australian Journal of Experimental Agriculture 45(2):1453–1469

Christy B, Weeks A, Beverly C (2006) Application of the 2CSalt Model to the Bet-Bet, Wild Duck, Gardiner and Sugarloaf catchments in Victoria. Department of Primary Industries, Victoria

CRCSI (2007) Use of raster data — summary. Cooperative Research Centre for Spatial Information, Victoria. Retrieved 30 December 2007, http://www.spatialvision.com.au/html/pemsproject.htm

de Bree F, Rajabifard A (2005) Involving users in the process of using and sharing geo-information within the context of SDI initiatives. Paper presented at FIG Working Week 2005 and GSDI-8 Cairo, Egypt, April 16–21. Retrieved 28 April 2007, http://www.gsdi.org/gsdiconfproceedings/gsdi-9.asp

Eigenraam M, Strappazzon L, Lansdell N, Beverly C, Stoneham G (in press) Designing frameworks to deliver unknown information to support market based instruments. In: Otsuka K, Kalirajan K (eds) Contributions of agricultural economics to critical policy issues. Blackwell Malden, Massachusetts

FGDC (2000) Financing the NSDI: National Spatial Data Infrastructure – aligning federal and non-federal investments in spatial data, decision support and information resources, revision 20 for public comments. Federal Geographic Data Committee. Retrieved 30 December 2007, http://www.fgdc.gov/library/whitepapers-reports

Giff G, Coleman D (2005) Using simulation to evaluate funding models for SDI implementation. Paper presented at FIG Working Week 2005 and GSDI-8 Cairo, Egypt, April 16–21. Retrieved 28 April 2007, http://www.gsdi.org/gsdiconfproceedings/gsdi-9.asp

McDougall K, Rajabifard A, Williamson I (2005) Understanding the motivations and capacity for SDI development from the local level. Paper presented at FIG Working Week 2005 and GSDI-8 Cairo, Egypt, April 16–21. Retrieved 28 April 2007, http://www.gsdi.org/gsdiconfproceedings/gsdi-9.asp

Nebert D (ed) (2004) Developing Spatial Data Infrastructures: The SDI cookbook version 20. Global Spatial Data Infrastructure Association. Retrieved 28 April 2007, http://www.gsdi.org/gsdicookbookindex.asp

Onsrud H (1999) Status of NSDI around the World. University of Maine

QPHLGP (1992) Broadhectare study south-east Queensland 1992. Queensland Department of Housing, Local Government and Planning, Brisbane

Rogers EM (1995) Diffusion of innovations. The Free Press, New York

Weeks A, Beverly C, Christy B, McLean T (2005) Biophysical approach to predict salt and water loads to upland REALM nodes of Victorian catchments. In: Zerger A, Argent RM (eds) MODSIM 2005 International Congress on Modelling and Simulation, 12–15 December 2005, Melbourne, Australia. Modelling and Simulation Society of Australia and New Zealand, pp 2776–2782

PART 2
INTEGRATING THE ECOLOGY OF LANDSCAPES INTO LANDSCAPE ANALYSIS AND VISUALISATION

7 Looking at Landscapes for Biodiversity: Whose View Will Do?

F Patrick Smith

CSIRO Sustainable Ecosystems, Wembley, Western Australia

Abstract: Biodiversity conservation is increasingly one of the multiple outcomes for which landscapes are managed, alas, this may not be as straight-forward as we would like. Despite our intentions, we tend to look at landscapes from a human perspective, but when we are aiming to conserve biodiversity this perspective can be decidedly unhelpful. Other biota do not perceive landscapes in the same way as people do, nor is one species' 'view' necessarily like another.

Biodiversity research within the highly modified landscapes of Australia reinforces our understanding that different taxa w— even from within the same family of organisms — can have strikingly different needs for persistence in a landscape. Different organisms can respond in completely different ways to the same landscape change, some for better, some for worse. The notion that a landscape can have — or at some time in the past has had — a single state that is 'best' is fanciful. One must always ask 'best for what?' for as soon as a change is made to improve the lot of one group or taxon, another may suffer.

Landscape managers have many tools at their disposal to help take in these varied perspectives. Firstly mapping and modelling of climate, terrain, hydrological, pedological and geological features are an important start, especially in helping to inform our understanding of the temporal and spatial distribution of both biota and biologically important processes. Secondly the mapping and modelling of vegetation distribution, type and condition is an important output in its own right and is also an important input to higher biodiversity related analyses. Finally we can map and model the movement of biota or biological entities (such as pollen or genes) throughout landscapes, both explicitly and implicitly. A present challenge is for us to learn to integrate these tools in order that we avoid unintended bias in our analyses.

Whenever we seek to map, model, 'design' or in any way understand landscapes we need to ask ourselves 'whose view will do?' If our objective is a landscape that is 'best' for humans then our task is (relatively) easy. However, if our objective is a landscape that is 'best' for non-human biodiversity then typically there is no 'right' answer, there is no 'best' outcome. We simply must be clear about our landscape objective and then do our best to 'view' the landscape with the appropriate — likely multiple, and hopefully well integrated — perspectives.

7.1 Introduction

'The real voyage of discovery consists not in seeing new landscapes, but in having new eyes.' (Marcel Proust 1871–1922)

It has been said that an accent is the one thing that we all have without knowing it. I would like to suggest that perspective can be much the same. By perspective I mean 'a way of looking at things', or perhaps more precisely 'the way we perceive and understand our surroundings'. Most of us realise that we each have a particular perspective on the world and that it can differ from that of other people. In human relations we are well served by realising this and by understanding and managing those differing perspectives, but how often do we stop to think about the importance of perspective in our non-human relations? How often have we stopped to think about how a frog's perception of its surroundings might differ from ours? Or the perception of a bird or spider? Does it matter if we don't? This volume discusses landscape analysis and visualisation and contains a showcase of some of the finest contemporary examples of mapping, modelling and understanding landscapes. The purpose of this chapter then is to remind the reader that when we look at landscapes we can take any of a number of different views — each with their own inherent biases — and in each different situation we need to ask ourselves 'Whose view will do?'

7.2 To be Human is to Err

Humans tend to look at landscapes from a human perspective. This statement is neither surprising nor, arguably, inappropriate if we are looking at the landscape with a view to some human objective or outcome. However, the human perspective can be decidedly unhelpful if we are meant to be looking at a landscape for the benefit of other biota, because other biota don't see landscapes — literally or figuratively — in the same way we do.

There are two problems with human perspective. Firstly — like all biota — humans are adapted to perceive landscapes with a bias towards human needs. Both our physical senses and our mental capacity are evolved towards extracting *our* needs from any environment we encounter. Take human sight for example. Our two large forward facing eyes give us a keen sense of sight within the (ahem) 'visible' spectrum, with excellent depth perception to boot. These are highly desirable traits for an omnivorous tree dwelling primate. We are, however, quite insensitive to the UV-visible markings of various nectar bearing flowers that are a key resource for the many invertebrates and birds whose sight *is* adapted to perceive UV. The same sorts of arguments can be made about our physical stature and upright posture, the range of our hearing, our sensitivity to touch, heat and cold, and our senses of smell and taste. All are adapted — and often finely tuned — to enable us to resource our basic human needs from the environment.

Secondly, and here we are unlike other biota (as far as we know), humans also endow landscapes with what I will call 'cultural' values (incorporating aspects variously described as 'spiritual', 'emotional' and 'philosophical' values). These are not directly related to our basic physical needs but rather are linked to them through the lens of human culture. Thus we feel an attraction to, and affinity with, certain landscapes (e.g. green grassy woodlands with clear running water) through our cultural links and associations, rather than because we directly and presently derive food, shelter or mates from them. The Prospect–Refuge Theory of Appleton (1975, 1995) has played a key role in our developing understanding of some of the links between landscapes and human psychology.

In combination, these two phenomena *left unchecked* steer us towards a view of landscapes that may not be in the best interest of other biota. When we map a landscape — even as ecologists — we tend to include features that align with the human perspective (which includes our incomplete knowledge and understanding) and often leave out features that can be quite significant to other biota (Fig. 7.1). For example when considering water bodies we may include natural lakes but not consider farm dams and roadside ditches, even though the latter may be highly relevant to amphibians. Or when mapping vegetation we may include defined blocks of native vegetation but not scattered paddock trees or thin strips of roadside vegetation, even though they may be very important to invertebrates and often higher organisms as well.

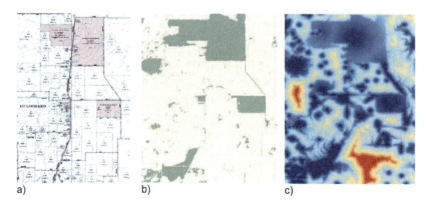

Fig. 7.1. Contrasting views of the same agricultural landscape in the wheatbelt of Western Australia. The general human perspective (**a**) highlights human infrastructure and land use, while a natural resource manager's perspective (**b**) may rather focus on native vegetation cover derived from remotely sensed data. An understanding of an animal's habitat requirements may allow us to go further and map measures of habitat suitability (**c**), as in this hypothetical example of landscape conductivity for a reptile: where the landscape ranges from highly amenable (dark blue) to highly hostile (red)

7.3 What's Good for the Goose?

If a human perspective won't do, then how do we arrive at a perspective that *is* good for the goose, or the gecko, or the grasshopper? One way is to consider what each of these organisms require to survive and persist. We can break the requirements of animals down into three fundamentals: sustenance (food, energy and water), refuge (from threats such as 'the elements' or predators), and (in most but not all cases) mates. Thus to animals a 'landscape' may be viewed as merely a spatial arrangement of those resources. A landscape may support a population of a given organism if these resources are provided in sufficient quantity and in a suitable spatial arrangement (including at a scale that makes them accessible).

The challenge for humans is both to know what provides these resources for animals, and to be able to recognise this provision at a spatial arrangement and scale relevant to the animals in question. This is no small task. For very few taxa do we have a good understanding of these resource requirements or the way they are provided in different landscapes (e.g. Dennis et al. 2003). Such knowledge only comes through intense research effort and is available for a very small minority of the animal taxa within Australian landscapes.

The usual 'way around' this widely acknowledged lack of critical information for animal taxa is to assume that native vegetation is habitat (i.e. provides these fundamental resources) for native animals (Parkes et al. 2003; Miller 2000). This is not, of itself, a bad assumption since native vegetation certainly did provide these resources prior to the widespread transformation of Australian landscapes after European settlement. However there are two important cautions that should be noted about making this assumption in the highly modified landscapes of Australia. Firstly much native vegetation is highly degraded and its capacity to provide essential resources for animals may be highly compromised (Bennet and Ford 1997; Cunningham 2000; Doherty et al. 2000). This is well understood and is usually taken into account by grading the vegetation condition and hence 'habitat quality' of native vegetation. However, because we don't know the specific resource requirements of taxa we cannot necessarily tell how 'degraded' vegetation may or may not supply those needs. Using relatively 'intact' vegetation as a surrogate for habitat is problematic. Using degraded vegetation as a surrogate for degraded habitat is possibly an order of magnitude more so.

The second caution is that in highly modified landscapes native vegetation is now not the only source of resources for animals. Water is now provided by water troughs, dams and other water management earthworks such as banks and drains. Some water birds and frogs have benefited from these disturbances (Braithwaite and Stewart 1975; Hazell 2003; Hazell et al. 2004; James et al. 1999). Human structures including buildings, transport infrastructure and even rubbish dumps provide 'unnatural resources' for other taxa (Dufty 1994; Shine and Fitzgerald 1996). The exotic vegetation and animals within farming systems and invasive species in more natural areas also provide resources (Frith et al. 1974; Law 2001; Miller 1979) (Fig. 7.2).

So a map of critical resources for any given taxon may look quite different to a map of native vegetation in any given landscape. Take for example the western pygmy possum (*Cercartetus concinnus*) (Fig. 7.3). Recent research from the wheatbelt of Western Australia shows that this tiny marsupial (barely larger than a house mouse) is frequently found foraging in planted perennial vegetation within the agricultural matrix (author's unpublished data). The possums have been found to nest in the hollows of large paddock or roadside trees and forage within the plantations of oil mallees (*Eucalyptus* spp.) or other mixed native species typically found as belts traversing paddocks or along drainage lines. Interestingly, they are two to three times more likely to be found in these paddock/plantation situations than they are in tracts of intact native vegetation on farms or in public reserves.

Fig. 7.2. Carcass of a feral piglet (*Sus scrofa*) high in the eyrie of a wedge-tailed eagle (*Aquila audax*), one example of an 'unnatural resource' now being provided for native wildlife within highly modified landscapes. Photographer: Simon Cherriman

Fig. 7.3. The tiny western pygmy possum (*Cercartetus concinnus)* has been found to favour foraging in plantations of oil mallees (*Eucalyptus* spp.) on farms in the wheatbelt of Western Australia. The possums nest in old isolated trees within paddocks and along roadsides, foraging at night within the plantations. Neither the isolated trees nor the plantations are resources that feature on most maps. Photographer: Georg Wiehl, CSIRO

The point is that neither the isolated trees nor the plantations appear in the typical map of an agricultural landscape — even one focussed on native biota — which means that any attempt to map or model the habitat of this species based on a 'typical' view of the landscape will not be starting from the best premise.

7.4 Consider the Lilies

Plants may also be considered to have a perspective on landscapes since they too have fundamental resource requirements that are met within a spatial context. Plants' needs include sunlight, water and a favourable substrate (e.g. soil type) all of which can be context sensitive. Furthermore plants, like animals, can depend utterly on the presence and activity of other organisms in order to survive and reproduce. Some plants are symbiotic and survive only in association with other organisms (e.g. legume symbioses with *Rhizobia* and *Bradyrhizobia*, parasitic and hemi-parasitic plants, plants of nutrient poor soils that form symbioses with mycorrhiza). Others rely on animals for pollination services or for the dispersal of their propagules (Kearns et al. 1998; Murphy et al. 2005; Reid 1989; Williams et al. 2001). Some plants are also dependent upon environmental phenomena such as fire or flood (Pickup et al. 2003) and the nature of these may also be modified by land use (Bren 2005; English and Blythe 2000; Jenkins et al. 2005; Lunt 1998a).

All plant needs are provided by the physical and biotic environments within a spatial context which can be quite dynamic. A map of vegetation is a static representation of the distribution of plants at one point in time, but it is known that the distribution of plants changes in response to the dynamics of both disturbance and other organisms (mutualists, competitors, herbivores, pathogens etc.). This is particularly well documented in response to herbivory (Fensham et al. 1999) and fire (Bradstock et al. 1997; Lunt 1998b). Other organisms and processes may also modify species distributions, such as the presence/absence of beneficial mycorrhiza (Bougher and Tommerup 1996) or soil disturbing small mammals (Murphy et al. 2005). These examples are given to emphasise the reliance of plants both on other biota and on dynamic processes within the landscape. Without an understanding of these it is not possible to properly map, model and understand plants in the landscape.

7.5 Best is Bunkum

Once we recognise that plants and animals can have very different perspectives on landscapes to humans, we then have to recognise that they also can have very different perspectives to one another. Different organisms can respond in completely different ways to the same landscape change, some for 'better', some for 'worse' (Bradstock et al. 1997; Lunt 1998a, 1998b; Fensham et al. 1999; Frith et al. 1974; Hodgson et al. 2007 Jansen and Robertson 2005; McIntyre and Lavorel 2007; Prober et al. 2007). With every change there will be both winners and losers. The notion that a landscape can have — or at some time in the past has had — a state that is 'best' is fanciful, and highly anthropocentric. One must always ask 'best for what?' for as soon as a change is made to improve the lot of one group or taxon, another will suffer.

Whenever we seek to map, model, design or in anyway understand landscapes we need to ask ourselves 'whose view will do?'. If our objective is a landscape that is 'best' for humans then our task is (relatively) easy. We already think like humans, perceive the landscape like humans and value its components like humans. So, designing and managing a landscape that is best for humans (or at least one that appears to be from a short-term human point of view) comes quite naturally. If our objective is a landscape that is 'best' for non-human biodiversity then our task becomes much more difficult. Which bits of biodiversity do we mean — the plants? (If so, which ones — the natives or the exotics, the annuals or the perennials, the inbreeders or the outbreeders?) What about the reptiles? (Would that be the arboreal geckos, or the perhaps the burrowing skinks?) Figure 7.4 illustrates how simple differences in the preferences or needs of different animals can result in quite different maps of hypothetical resource distributions.

If we have difficulty answering the question: 'Which bits of biodiversity do we mean?', then it's a clear sign we need to think more carefully about our landscape objectives. We may be tempted to answer: 'all the native biodiversity', or 'everything that is supposed to be there naturally', or 'everything that was there before'. However, these statements are laden with the problems of human perspective. What is native? What is natural? Before when, and why then? The point is that there are no 'right' answers, there is no 'best' outcome. We simply must be clear about our landscape objective and then do our best to 'view' the landscape with the appropriate perspective/s.

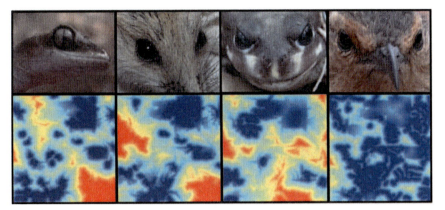

Fig. 7.4. Whose view will do? Different organisms can have vastly different perspectives on the same landscape. Each may utilise different resources and face different threats. Attempts to model and analyse landscapes may need to take account of these differences, as indicated by these four contrasting hypothetical habitat suitability resource maps for the same landscape

7.6 Varied Perspectives

Acknowledging the need for different perspectives is an important first step, but it must necessarily be followed by analyses which allow us to generate, compare and synthesise these differing perspectives. Technically there are many different ways to view a landscape, to map it, model it and understand it. Listed below are some of the many perspectives that are often considered in landscape visualisation and analysis, and a selection of different technical approaches and their applications. The list is by no means exhaustive but is intended to reflect the complexity of the issue and emphasise the need to target and select tools and approaches appropriate to one's landscape objectives.

7.6.1 Mapping and Modelling Terrain, Hydrological, Pedological and Geological Features and Climate

While often overlooked, the foundation of all landscapes is, quite literally, the earth beneath our feet, so the mapping and analysis of this substrate is possibly an appropriate place to start. Far more than just informing us about the physical form and appearance of a landscape — although important from a human visualisation perspective (see Bishop Chapter 22) —

analysis of terrain, hydrology, geology, pedology and climate can be crucial to our understanding of the distribution and ecology of organisms and processes within a landscape. Terrain, geology and climate are important determinants of many environmental features at both macro and micro scales. Characteristics such as elevation, slope and aspect influence environmental variables such as temperature, solar radiation, precipitation patterns (rain, snow, fog, cloud-water), soil formation and erosion processes, and surface and groundwater flows. In combination with climate and geology these variables determine the suitability of different parts of a landscape for all biota. Terrain and climate modelling and analysis can make a major contribution to the prediction of the distributions of both plants and animals (Austin 2002; Guisan and Zimmermann 2000).

In addition to these, geology, pedology and hydrology may also inform our understanding of the spatial and temporal distribution of the biota and important processes. For example, plant species' distributions are often closely linked to soil type and landform so an analysis of the pedogeology of a landscape may inform our understanding of species distributions in the past (e.g. pre-clearing vegetation, Fensham and Fairfax 1997), present, or future (e.g. climate change planning, Hilbert and Ostendorf 2001). Remote sensing of mineralogical features using surface or air-borne equipment (electromagnetic and radiometric) is enhancing our capacity to map and understand the pedogeology of landscapes and can contribute to the analysis of important processes such as salinisation (McKenzie and Ryan 1999; Williams and Baker 1982) and erosion/deposition (Pickup and Marks 2000). The modelling of hydrological processes such as surface and groundwater flows can also inform our ecological understanding through modelling phenomena such as flooding, inundation and salinisation (Overton 2005).

7.6.2 Vegetation Mapping Using Remotely Sensed Data, Including Vegetation Condition and Temporal Variability

Vegetation maps are a typical starting point for mapping and modelling landscapes, whether it is the vegetation itself that is of interest, or its role as habitat for other native biota. Traditional maps based on on-ground surveys are increasingly being augmented and/or replaced by remotely sensed and modelled data, to give vegetation cover, composition and condition (Moisen and Frescino 2002; Newell et al. 2006). However derived, a map of vegetative cover opens up the possibility of a vast array of landscape analyses based on the spatial characteristics of the patches of vegetation. Such analyses are particularly used in studies of 'fragmented landscapes' where the spatial arrangement of patches of habitat within a matrix of non-habitat is a key consideration (Henle et al. 2004). Observed or modelled vegetation type (composition) and condition can add further value to considerations of spatial context, especially when modelling the likely movement or persistence of biota (Gibson et al. 2004; McAlpine et al. 2006). In addition to the determination of spatial variation in vegetation, one major benefit of the use of remotely sensed data in constructing vegetation maps is that it allows the map to capture temporal variation in cover and, increasingly, condition and composition (Wallace et al. 2004, Zerger et al. Chapter 8) opening up the prospect of cost-effective broad-scale monitoring.

7.6.3 Mapping and Modelling Movement

The movement of biota within a landscape is of fundamental interest to many areas of landscape science, especially with respect to conservation planning and management. The capacity of biota to move about within a landscape is a key to successful resource acquisition, dispersal and migration, and for many organisms — especially the vertebrates — movement at the landscape scale is fundamental to survival and persistence. There are many ways that movement can be incorporated into the modelling and analysis of landscapes, ranging from behaviour-based models of the movement of individuals of a given species through to very general models of implied movement of diverse taxa based on landscape characteristics. A selection of methodologies is presented below.

Explicit Behaviour-based Movement Modelling

Some approaches to movement modelling attempt to replicate the known behaviour of specific animals or plants within a spatial modelling framework. This approach requires detailed knowledge of the organism's behaviour and the way it interacts with elements within its environment. Situations where such detailed knowledge is available are limited, but the value of this approach can be great. Take, for example, work in the wet tropics in Australia where detailed knowledge of bird movement behaviour has led to improved prediction of the seed shadows of bird-dispersed plants (Westcott et al. 2005). This understanding has in turn been used to better predict and map the likely spread of new bird-dispersed weeds within the Daintree World Heritage area, greatly improving the efficiency of weed control efforts (Buckley et al. 2006; Westcott et al. in press).

Implicit Behaviour-based Movement Modelling

Where detailed animal behaviour is not known or where explicit movement details are not required, more general behaviour-based models can be developed. These typically use simple movement rules (e.g. threshold movement or dispersal distances) in combination with landscape features (e.g. habitat patch separation distances) to model the likelihood of animal or plant movement between points within the landscape (e.g. dispersal success between two habitat patches). The applications of this approach are many, particularly within the field of meta-population studies, where the connectivity within habitat networks is paramount. An example of this approach is the use of observed bird patch occupancy to predict likely patch occupancy in other landscapes (Brooker 2002; Freudenberger and Brooker 2004).

One popular way to map and model landscapes for biodiversity conservation is to take a graph theoretic approach (Urban and Keitt 2001) that represents the landscape as a network of nodes and links. Using graph theory, landscape habitat networks can be analysed for their size and connectivity and proposed modifications to the landscape can be tested for their impact on various network metrics (Fig. 7.5). Habitat patch size and patch quality are important landscape components that can be included in the analysis as modifiers of dispersal success between patches.

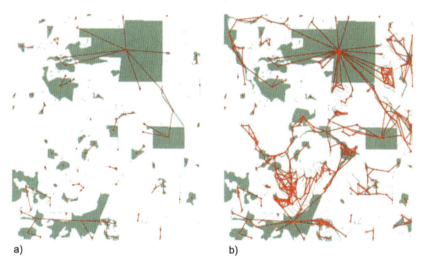

Fig. 7.5. Graph analysis confirms the reconnection of this landscape by farmers planting trees and shrubs. Before the revegetation activity (**a**) the graph had 45 components (connected sub-units) and 51 edges (connections). After revegetation (**b**) the components have been more than halved (21) and the connections have increased by an order of magnitude (491)

Implicit Landscape-based Movement Modelling

Of course, it is not always possible or desirable to specify behavioural rules in movement models. For example, if one is concerned with conserving all biota within a landscape, then it is not possible to specify a single meaningful dispersal threshold or minimum patch size or habitat condition metric since — as discussed in the previous sections, different organisms will have very different perspectives on what these values should be. In such cases it is possibly more appropriate to model movement implicitly based on general landscape characteristics rather than specific behavioural rules. Landscape friction analysis is one means of achieving this since it represents the amenity of a landscape to movement in a general way, taking account of the spatial relationships among landscape components, but not specifying thresholds of distance or size. This style of analysis allocates a friction or 'resistance' factor to all points within a landscape (e.g. each cell in a mapped raster or grid) based on the known or perceived conduciveness of the cover, land use or habitat in that cell to wildlife movement. GIS-based algorithms can then be used to model resistance to movement along various pathways within the landscape. By allocating different resistance profiles to different types of cover/land use/habitat, the same landscape can be modelled in a way deemed appropriate for a range of different assemblages or communities, based on a known habitat preference but without any knowledge of movement behaviour (Fig. 7.1). The landscape friction analysis can be extended to specify least cost paths (or alternatively less prescriptive 'least cost corridors') between specific points in the landscape (e.g. nearby habitat patches) to identify the most readily gained improvements in landscape connectivity (Fig. 7.6). These methods have been incorporated into conservation planning tools (e.g. Drielsma et al. 2007).

7.6.4 Integrating Multiple Perspectives

Just as a human-centric view can be unhelpful when attempting to understand a landscape from the perspective of non-human biota, we can, ironically, fall into just the same trap if we adopt a non-human perspective in too narrow a way. For example, if one was to work mainly with birds it would be possible to replace one's natural human-centric perspective with a bird-centric perspective. Of course this would be fine if it was only birds that needed to be considered, but it may not be helpful in considerations of the needs of mammals, or reptiles or plants. The challenge is for us to generate and then integrate multiple perspectives so that we avoid unintended bias. It does not appear that we are meeting this challenge at present.

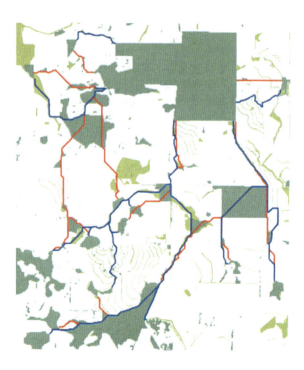

Fig. 7.6. From the perspective of many animals the addition of many patches and corridors of planted native vegetation (light green) within this landscape has improved its amenity for movement and dispersal. For species that require vegetative cover (e.g. small birds) movement between patches of native habitat (dark green) was previously often through risky and hostile agricultural fields (red paths), but the planting of connecting vegetation has reduced the "cost" of moving through this landscape by about 50% (blue paths)

All of the methodologies for looking at landscapes presented in the previous section are limited to subsets of the components of the landscape or subsets of the biota present. While some of the more generalised approaches do attempt to take into account broad subsections (e.g. the friction approach to movement modelling) they still generate a perspective more relevant to one subset (e.g. animals that move) than others (e.g. plants). The next real challenge for those that develop and utilise these tools is to begin to develop frameworks and approaches that allow these many different — and complimentary — perspectives to be integrated in meaningful ways.

7.7 Conclusion

Humans need to be aware of the inherent biases in the way they perceive landscapes if they are to properly map, model and understand them for biodiversity conservation. The non-human biota have vastly different resource requirements to humans, and one organism can have vastly different resource requirements to another, even from within the same taxonomic group or family. The challenge for humans is to understand these various requirements and map and model landscapes in a way that is cognisant of them. A diverse range of technical approaches are available to this end, all with differing emphases, levels of complexity and degrees of specificity. The development of approaches to integrate these various techniques for more comprehensive and multi-perspective outcomes would be a welcome advance. We do well to remember that with landscapes there is no 'right' answer, there is no 'best' outcome. We simply must be clear about our landscape objective and then do our best to 'view' the landscape with the appropriate perspective/s.

References

Appleton J (1975) Experience of landscape. John Wiley, London

Appleton J (1995) Experience of landscape, revised. John Wiley, Chichester

Austin MP (2002) Spatial prediction of species distribution: an interface between ecological theory and statistical modelling. Ecological Modelling 157(2–3):101–118

Bennett AF, Ford LA (1997) Land use, habitat change and the conservation of birds in fragmented rural environments: a landscape perspective from the northern plains, Victoria. Pacific Conservation Biology 3:244–261

Bougher NL, Tommerup IC (1996) Conservation significance of ectomycorrhizal fungi in Western Australia: their co-evolution with indigenous vascular plants and mammals. In: Hopper SD, Chappill J, Harvey M, George R (eds) Gondwanan heritage: past, present and future of the Western Australian biota. Surrey Beatty and Sons, Chipping Norton

Bradstock RA, Tozer MG, Keith DA (1997) Effects of high frequency fire on floristic composition and abundance in a fire-prone heathland near Sydney. Australian Journal of Botany 45:641–655

Braithwaite LW, Stewart DA (1975) Dynamics of water bird populations on the Alice Springs Sewage Farm, Northern Territory. Wildlife Research 2:85–90

Bren LJ (2005) The changing hydrology of the Barmah-Millewa forests and its effects on vegetation Proceedings of the Royal Society of Victoria 117:61–76

Brooker L (2002) The application of focal species knowledge to landscape design in agricultural lands using the ecological neighbourhood as a template. Landscape and Urban Planning, 60:185–210

Buckley YM, Anderson S, Catterall CP, Corlett RT, Engel T, Gosper CR, Nathan R, Richardson DM, Setter M, Spiegel O, Vivian-Smith G, Voigt FA, Weir JES, Westcott DA (2006) Management of plant invasions mediated by frugivore interactions. Journal of Applied Ecology 43 (5):848–857

Cunningham SA (2000) Effects of habitat fragmentation on the reproductive ecology of four plant species in mallee woodland. Conservation Biology 14:758–768

Dennis RLH, Shreeve TG, Van Dyck H (2003) Towards a functional resource-based concept for habitat: a butterfly biology viewpoint. Oikos 102:417–426

Doherty M, Kearns A, Barnett G, Sarre A, Hochuli D, Gibb H, Dickman C (2000) The interaction between habitat conditions, ecosystem processes and terrestrial biodiversity: a review. Environment Australia, Canberra

Drielsma M, Manion G, Ferrier S (2007) The spatial links tool: automated mapping of habitat linkages in variegated landscapes. Ecological Modelling 200:403–411

Dufty AC (1994) Habitat and spatial requirements of the eastern barred bandicoot (*Perameles gunnii*) at Hamilton, Victoria. Wildlife Research 21:459–71

English V, Blyth J (2000) Interim recovery plan: shrubland and woodlands on Muchea limestone. Interim recovery plan 57, 2000–2003, Department of Conservation and Land Management, Wanneroo, Western Australia

Fensham RJ, Fairfax RJ (1997) The use of the land survey record to reconstruct pre-European vegetation patterns in the Darling Downs, Queensland, Australia. Journal of Biogeography 24:827–836

Fensham RJ, Holman JE, Cox MJ (1999) Plant species responses along a grazing disturbance gradient in Australian grassland. Journal of Vegetation Science 10(1):77–86

Freudenberger D, Brooker L (2004) Development of the focal species approach for biodiversity conservation in the temperate agricultural zones of Australia. Biodiversity and Conservation 13:253–274

Frith HJ, Brown BK, Barker RD (1974) Food of the crested and common bronzewing pigeons in inland New South Wales Australian. Wildlife Research 1:129–144

Gibson LA, Wilson BA, Cahill DM, Hill J (2004) Modelling habitat suitability of the swamp antechinus (*Antechinus minimus maritimus*) in the coastal heathlands of southern Victoria. Australia Biological Conservation 117 (2):143–150

Guisan A, Zimmermann NE (2000) Predictive habitat distribution models in ecology. Ecological Modelling 135(2–3):147–186

Hazell D (2003) Frog ecology in modified Australian landscapes: a review. Wildlife Research 30:193–205

Hazell D, Hero J-M, Lindenmayer D, Cunningham R (2004) A comparison of constructed and natural habitat for frog conservation in an Australian agricultural landscape. Biological Conservation 119(1):61–71

Henle K, Lindenmayer DB, Margules CR, Saunders DA, Wissel C (2004) Species survival in fragmented landscapes: where are we now? Biodiversity and Conservation 13 (1):1–8

Hilbert DW, Ostendorf B (2001) The utility of artificial neural networks for modelling the distribution of vegetation in past, present and future climates. Ecological Modelling 146(1–3):311–327

Hodgson P, French K, Major RE (2007) Avian movement across abrupt ecological edges: differential responses to housing density in an urban matrix. Landscape and Urban Planning 79:266–272

James CD, Landsberg J, Morton SR (1999) Provision of watering points in the Australian arid zone: a review of effects on biota. Journal of Arid Environments 41(1):87–121

Jansen A, Robertson AI (2005) Grazing, ecological condition and biodiversity in riparian river red gum forests in south-eastern Australia. Proceedings of the Royal Society of Victoria 117:85–95

Jenkins KM, Boulton AJ, Ryder DS (2005) A common parched future? Research and management of Australian arid-zone floodplain wetlands. Hydrobiologia 552:57–73

Kearns CA, Inouye DW, Waser NM (1998) Endangered mutualisms: the conservation of plant-pollinator interactions. Annual Review of Ecology and Systematics 29:83–112

Law BS (2001) The diet of the common blossom bat (*Syconycteris australis*) in upland tropical rainforest and the importance of riparian areas. Wildlife Research 28:619–626

Lunt ID (1998a) Two hundred years of land use and vegetation change in a remnant coastal woodland in southern Australia. Australian Journal of Botany 46:629–647

Lunt ID (1998b) Allocasuarina (*Casuarinaceae*) invasion of an unburnt coastal woodland at Ocean Grove, Victoria: structural changes 1971–1996. Australian Journal of Botany 46:649–656

McAlpine CA, Rhodesa JR, Callaghan JG, Bowen ME, Lunney D, Mitchell DL, Pullar DV, Possingham HP (2006) The importance of forest area and configuration relative to local habitat factors for conserving forest mammals: A case study of koalas in Queensland, Australia. Biological Conservation 132(2):153–165

McIntyre S, Lavorel S (2007) A conceptual model of land use effects on the structure and function of herbaceous vegetation. Agriculture, Ecosystems and Environment 119:11–21

McKenzie NJ, Ryan PJ (1999) Spatial prediction of soil properties using environmental correlation. Geoderma 89(1–2):67–94

Miller B (1979) Ecology of the little black cormorant, *Phalacrocorax sulcirostris*, and little pied cormorant, *P. melanoleucos*, in inland New South Wales: food and feeding habits. Wildlife Research 6:79–95

Miller C (2000) Vegetation and habitat are not synonyms. Ecological Management and Restoration 2:102–104

Moisen GG, Frescino TS (2002) Comparing five modelling techniques for predicting forest characteristics. Ecological Modelling 157:209–225

Murphy MT, Garkaklis MJ, Hardy GEStJ (2005) Seed caching by woylies *Bettongia penicillata* can increase sandalwood *Santalum spicatum* regeneration in Western Australia. Austral Ecology 30(7):747–755

Newell GR, White MD, Griffioen P, Conroy M (2006) Vegetation condition mapping at a landscape-scale across Victoria. Ecological Management and Restoration 7:S65–S68

Overton I C (2005) Modelling floodplain inundation on a regulated river: integrating GIS, remote sensing and hydrological models. River Research and Applications 21(9):991–1001

Parkes D, Newell G, Cheal D (2003) Assessing the quality of native vegetation: the 'habitat hectares' approach. Ecological Management and Restoration 4:S29–S38

Pickup G, Marks A (2000) Identifying large-scale erosion and deposition processes from airborne gamma radiometrics and digital elevation models in a weathered landscape. Earth Surface Processes and Landforms 25:535–557

Pickup MA, McDougall KL, Whelan RJ (2003) Fire and flood: soil-stored seed bank and germination ecology in the endangered Carrington Falls grevillea (*Grevillea rivularis*, Proteaceae). Austral Ecology 28:128–136

Prober SM, Thiele KR, Lunt ID (2007) Fire frequency regulates tussock grass composition, structure and resilience in endangered temperate woodlands. Austral Ecology 32(7):808–824

Reid N (1989) Dispersal of mistletoes by honeyeaters and flowerpeckers: components of seed dispersal quality. Ecology 70(1):137–145

Shine R, Fitzgerald M (1996) Large snakes in a mosaic rural landscape: the ecology of carpet pythons *Morelia spilota* (serpentes: pythonidae) in coastal eastern Australia. Biological Conservation 76(2):113–122

Urban D, Keitt T (2001) Landscape connectivity: a graph-theoretic perspective. Ecology 82:1205–1218

Wallace JF, Caccetta PA, Kiiveri HT (2004) Recent developments in analysis of spatial and temporal data for landscape qualities and monitoring. Austral Ecology 29(1):100–107

Westcott DA, Bentrupperbäumer J, Bradford MG, McKeown A (2005) Incorporating patterns of disperser behaviour into models of seed dispersal and its effects on estimated dispersal curves. Oecologia 146(1):57–67

Westcott DA, Setter M, Bradford MG, Setter S (in press) Cassowary dispersal of the invasive pond apple in a tropical rainforest: the contribution of secondary dispersal modes in invasion, diversity and distribution

Williams BG, Baker GC (1982) An electromagnetic induction technique for reconnaissance surveys of soil salinity hazards. Australian Journal of Soil Research 20(2):107–118

Williams GA, Adam P, Mound LA (2001) Thrips (*Thysanoptera*) pollination in Australian subtropical rainforests, with particular reference to pollination of *Wilkiea huegeliana* (Monimiaceae). Journal of Natural History 35(1):1–21

8 Native Vegetation Condition: Site to Regional Assessments

Andre Zerger[1], Philip Gibbons[2], Julian Seddon[3], Garth Warren[1], Mike Austin[1] and Paul Ryan[1]

[1] CSIRO Sustainable Ecosystems, ACT, Australia
[2] Fenner School of Environment and Society, The Australian National University, ACT, Australia
[3] Department of Environment and Climate Change (NSW), ACT, Australia

Abstract: There is an increasing emphasis in Australia on the use of vegetation condition information for regional conservation planning. The use of site-based vegetation condition assessments is a relatively mature management application with methods developed for different landscapes (e.g. rangelands and riparian) and to support a variety of natural resource management requirements (property vegetation planning or market-based instruments). On the other hand, the creation of regional-scale maps of native vegetation condition is still a developing methodology. This chapter argues that regional-scale maps of native vegetation condition are an important tool to complement site-based assessments. When combined they can provide a powerful integrated tool for regional conservation planning. Through a case study we describe a methodology for extending site-based data to maps of two vegetation condition attributes based on the BioMetric site assessment method. The case study, in the Murray Catchment in New South Wales, Australia, illustrates how an understanding of faunal response to native vegetation condition can be combined with modelled data to develop regional conservation planning maps. A spatial data aggregation approach is applied to model outcomes to address concerns about uncertainty and data confidentiality.

8.1 Introduction

Vegetation condition is assessed routinely as part of natural resource management in Australia. Information on vegetation condition is used for landholder education, development applications, distributing incentive funds, guiding restoration and identifying areas of high conservation value. In recent years several vegetation condition indices have become established tools for natural resource management in Australia (Gibbons et al. 2005; Parkes et al. 2003; Tongway and Hindley 2004). Although there is no standard definition for native vegetation condition, it is generally a measure of modification relative to a reference state defined according to:

- the requirements of one or more species
- a desired functional state
- sites relatively unmodified by humans
- notional pre-European conditions.

Vegetation condition is usually assessed using biophysical attributes intended as surrogates for biodiversity or ecosystem function (e.g. tree hollows, native ground cover, fallen timber, soil stability) rather than primary data such as species inventories, because it is typically used in the context of simple, rapid assessments.

Vegetation condition is commonly measured on the ground at the scale of the site using a plethora of techniques ranging from simple check lists (Goldney and Wakefield 1997) to dedicated software (Gibbons et al. 2005) that have been developed for measuring it at this scale. However, there has been an increasing emphasis on regional-scale decision making in natural resource management in Australia. Decisions made in a regional context are likely to give rise to a more efficient allocation of resources for conservation (Pressey and Nicholls 1989) and are more effective for conserving biota and ecosystem processes that operate at broad scales. Regional bodies such as catchment management authorities require information on vegetation condition to:

- set regional vegetation management targets
- undertake on-ground actions in the context of these targets
- monitor and report achievement towards these targets.

These objectives cannot be met by measuring vegetation condition only at the scale of individual sites.

In this chapter we review methods used to assess vegetation condition at scales ranging from individual sites to entire regions (covered by two 1:100,000 scale map sheets). Using the Murray Catchment area of New

South Wales, Australia, we demonstrate the development of vegetation condition information at multiple scales and apply the outputs to a simple case study of fauna management. The case study examines issues of site stratification, GIS and remote sensing database development and modelling vegetation condition, and includes a novel approach to presenting model results which addresses issues of model accuracy, precision and confidentiality.

8.2 Measuring Vegetation Condition at Sites

Vegetation condition assessments at the site scale are generally based on attributes that are surrogates for biodiversity, are assumed to indicate the likelihood of persistence of biodiversity and are informative for managing a site. Many attributes of native vegetation that are relatively simple to measure have been statistically associated with the occurrence or abundance of biota or ecosystem function. These include direct measures of biota (e.g. native plant species richness), surrogates for the occurrence or abundance of one or more species (e.g. numbers of hollow-bearing trees), or surrogates for ecosystem function. Oliver et al. (2007) provide a list of variables identified by experts that are surrogates for biodiversity (including ecosystem function), and McElhinny et al. (2006) provide a list of surrogates for the occurrence and abundance of fauna in Australia. There are also many attributes that can be measured in native vegetation at individual sites that are indicative of threats. Information on threats can be important for predicting the future trajectory of a site and are instructive for setting management objectives. Threats include exotic species, erosion, salinity and the nature of management such as grazing regime or fertilizer application. The occurrence of threats are not generally sufficient to assess vegetation condition at sites because ecosystems have varying resilience to modification and it is often the nature or regime of the threat over time (e.g. intensity and duration of grazing), or a threat occurring distal to the site that are critical in terms of their impacts on native ecosystems. Such threats cannot always be assessed at a single visit in time or at the scale of the site.

Measures from multiple attributes of the type discussed above are typically combined into an index. Indices that combine multiple attributes are conceptually easier to understand and compare than multiple attributes measured on different scales. Several approaches are employed to combine raw vegetation and habitat attributes measured on sites into an index (Goldney and Wakefield 1997; Newsome and Catling 1979; Parkes et al. 2003; Gibbons et al. 2005 reviewed by McElhinny et al. 2005). Though often based on a simple metric, several indices have been statistically associated with the occurrence, abundance or richness of selected biota (Catling and Burt 1995; Watson et al. 2001) or ecosystem function (Tongway and Hindley 2004).

If constructed carefully these indices are transparent and repeatable and therefore lead to consistent and equitable decision making. As a result of being undertaken on the ground they are accurate to a fine scale and are instructive for land management. Although the aggregation of site attributes into a final index is common, this study focuses on spatially modelling individual attributes of an index. Examining and modelling individual attributes provides greater flexibility as these can be aggregated in different ways to address specific natural resource management needs.

8.3 Measuring Vegetation Condition across Regions

Methods for measuring vegetation condition across regions can be generally partitioned into those that rely on remote sensing and those that utilise spatially explicit modelling with GIS surrogates. In the later, remotely sensed data may also be one of the explanatory variables. In terms of remote sensing, compositional, structural and functional surrogates of vegetation condition have been mapped using aerial photographic interpretation (Woodgate et al. 1994), videography (Coops and Catling 1997; Dare et al. 2002), airborne laser scanning (Hinsley et al. 2002), hyperspectral imagery (Ludwig et al. 2002), and multispectral imagery (Dare et al. 2002; Levesque and King 1999; Lucas et al. 2001; Scarth and Phinn 2000; Wallace and Furby 1994). Wallace et al. (2006) describe an approach to vegetation condition assessment which relies on time series satellite imagery to detect temporal change as an indicator of condition. Remote sensing techniques have not yet been adopted routinely for broad-scale vegetation condition mapping, partly because they rely on technologies which are too costly to be applied at regional scales (e.g. videography).

In terms of modelling approaches for vegetation condition mapping, there are relatively few examples in the literature that focus specifically on native vegetation condition (Frescino et al. 2001; Moisen and Frescino 2002; Newell et al. 2006; Overton and Lehman 2003; Zerger et al. 2006). However, if we also examine methods for mapping compositional attributes of native vegetation condition, a number of important examples can be cited. Examples include the use of statistical approaches such as generalised linear models and generalised additive models (Cawsey et al. 2002; Guisan and Zimmermann 2000; Pearce and Ferrier 2000), genetic algorithms (Anderson et al. 2003), classification and regression trees (Franklin et al. 2000) and expert-based approaches such as multi-criteria decision making (Clevenger et al. 2002). For a full synopsis of the use of modelling for mapping vegetation characteristics, see Ferrier et al. (2002). These methods all take advantage of the array of GIS databases that have recently become available for natural resource management. These include high-resolution digital elevation models and their derived variables (e.g. slope, aspect and topographic position), soil and geology layers, climatic variables, land use and land tenure. Owing to the scale of commonly available GIS data, modelling is typically conducted at regional or national scales, rather than at local scales. Through a case study evaluation in the Murray Catchment of New South Wales, Australia, we describe an approach for moving from site-scale vegetation condition data to regional-scale maps of condition through the use of GIS data and complimented with satellite imagery.

8.4 Case Study: Vegetation Condition in the Murray Catchment, New South Wales

The case study presents methods for developing maps of native vegetation condition at regional scales for the Murray Catchment of New South Wales. After describing the study area, we present the *BioMetric* site assessment methodology and examine how sites were spatially allocated in the landscape. Site allocation is an important challenge of the study as the aim, in the context of vegetation condition mapping, is to place sites across all disturbance gradients. This contrasts with approaches typically adopted for mapping vegetation composition which attempt to capture environmental gradients (Austin 2002). To explore practical applications of the methodology a scenario is presented for the Murray Catchment that integrates current science knowledge regarding faunal responses to native vegetation condition attributes and modelled maps of native vegetation condition.

8.4.1 Study Area

The New South Wales section of the Murray Catchment spans approximately 35,362 km^2 extending from east of Khancoban to 50 km west of Swan Hill. The catchment is considered one of the most modified regions in Australia owing to a history of agricultural production resulting in extensive clearing of native vegetation. It is estimated that 22% of the catchments' woody native vegetation remains, with half of this reserved in several major national parks (Miles 2001). Consequently, much remnant native vegetation occurs on private land, roadside vegetation and in travelling stock reserves. The agricultural landscape is highly fragmented with many small isolated patches of remnant vegetation not linked to any major conservation easements.

The project study area is situated across two 1:100,000 scale map sheets (561,316 ha). This area was selected for a number of operational and strategic reasons. Firstly, existing GIS data were already available for this region, including vegetation mapping. Secondly, there was a significant change in terrain relief moving from the alpine regions in the east towards the west of the catchment. Incorporating altitudinal variation was critical as earlier research (Zerger et al. 2006) had highlighted the importance of terrain as a surrogate for native vegetation condition. This also ensures that climatic and biotic variation is incorporated into the mapping. Finally, selecting two 1:100,000 scale map sheets allowed the project to adopt sampling intensities similar to other state vegetation mapping programs.

8.4.2 Site Data Collection

Site data were collected using the *BioMetric* methodology (Gibbons et al. 2005) that underpins biodiversity assessments in the NSW Property Vegetation Planning (PVP) process. As the study needed to have practical relevance it was important to adopt a site assessment methodology that is widely used. In *BioMetric*, the primary sampling unit is the plot (20 m x 50 m) and each plot contains one 50 m transect and a nested 20 m by 20 m plot (Fig. 8.1) A plot occurs in a uniform vegetation community, and where practical it should lie along a 'contour' rather than traversing multiple landforms. Primary structural vegetation attributes collected along the transect included native over-storey cover, native mid-storey cover, native ground cover (shrubs), native ground cover (grasses), native ground cover (other), exotic plant cover and organic litter. Indigenous plant species richness was recorded in the 20 m x 20 m plot. The following *BioMetric* variables were collected in the 50 m x 20 m plot: regeneration, number of trees with hollows, total volume of fallen logs, and number of stems in specified DBH (diameter at breast height) classes.

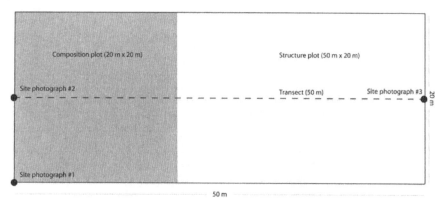

Fig. 8.1. Field sampling areas (plots and transects) using the *BioMetric* methodology

Field Work Stratification — Sampling Across Disturbance Gradients

The objective of the site stratification was to ensure that *BioMetric* plots sampled disturbance gradients. In other words, field sampling should sample both high quality and low quality vegetation across the two 1:100,000 map sheets. Zerger et al. (2006) described how site sampling biased towards better quality vegetation can impact upon the efficacy of the modelling. To overcome this in the Murray Catchment, the field sampling was stratified across disturbance gradients using regional-scale GIS data. Once environmental stratification polygons were generated, a random subset of these was made and provided to the mapping teams as candidate sites.

To generate environmental stratification units, a two-stage sampling scheme using a combination of GIS data was used, including: tenure (private and public), patch size, topographic position and elevation (see Table 8.1). Final categorical layers were combined by addition using raster analysis functions in *ArcGIS* to develop a theoretical maximum of 384 unique environmental stratification units (ESUs). A 'non-weighted by area' allocation of stratification units was used as some 'weighted by area' allocations could bias sampling towards the larger or smaller polygons, and towards theoretically higher or lower quality native vegetation. Forty-seven stratification units were found to occur in the study area (12% of possible ESUs). Given only 47 stratification units occurred in the study area, and taking into account the field budget for the project, the study could sample approximately 125 plots in each map sheet. Consequently a random selection of three polygons (non-weighted by area) was conducted per ESU. This generated approximately 141 candidate polygons in each map sheet. To minimise spatial autocorrelation of combinations, separate random polygon allocations were conducted for the east and west map sheets.

Table 8.1. GIS variables used to derive environmental stratification units (ESU)

Variable	Derivation	Levels (categories)
Tenure	Derived from Bureau of Rural Sciences land cover mapping	Private or public
Elevation	Derived from NSW statewide 25 m digital elevation model (DEM)	0–300 m, 300–600 m, >600 m
Patch size	Derived from CSIRO *SPOT5* mapping of woody vegetation	0–10000 m^2, 10000–100000 m^2, >100000 m^2
Topographic position	Derived using algorithm of Gallant and Wilson (2000) and 25 m statewide DEM	Drainage lines, flats, slopes, ridges

The stratification was challenging in that public land was over-sampled relative to private land owing to the presence of a greater number of unique ESUs on public land (i.e. greater variation in topographic position, elevation and patch size). Consequently we partitioned the random selection of ESU polygons into those selected on public land and those selected on private land. For those that occurred on private land we selected two polygons from each ESU. For those ESUs which occurred on public land, we selected one polygon per ESU to compensate for this variation. Between December 2006 and February 2007, 249 *BioMetric* plots were sampled using a stratification tailored to capture disturbance gradients (Fig. 8.2). Figure 8.3 shows some example sites with their associated *BioMetric* scores.

8.4.3 Modelling from the Site to the Region

The predictive spatial modelling methodology uses the *GRASP* framework running under 'R' (Version 2.5.0) for building generalised additive models (GAM), (Lehmann et al. 2003). The methodology uses site data to calibrate a statistical model that captures the relationship between field observations and statistically significant GIS and remote sensing variables. As the GIS and remote sensing data are available at regional scales, model inferences can be made at these scales, albeit with inherent predictive uncertainties. Explanatory variables used in this study include remotely sensed predictors derived from both *SPOT5* multispectral and *Landsat 5 TM* (Thematic Mapper) satellite imagery such as the Normalised Difference Vegetation Index (NDVI) and the Perpendicular Vegetation Index.

Native Vegetation Condition: Site to Regional Assessments 147

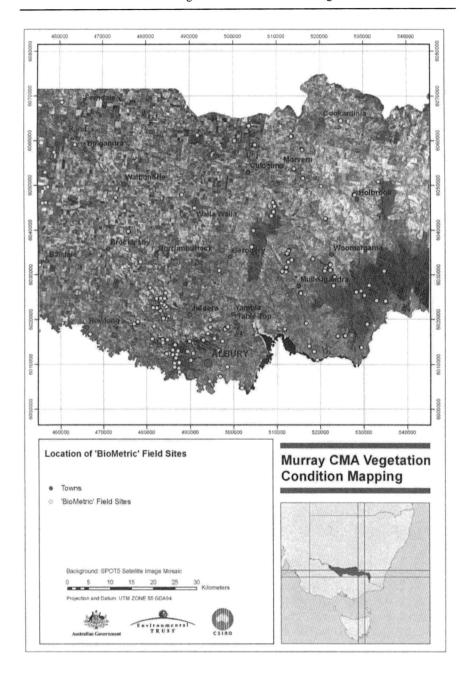

Fig. 8.2. Satellite image showing study site and location of *BioMetric* plots in the Murray Catchment of New South Wales

Fig. 8.3. Example sites (and *BioMetric* condition scores) for different vegetation types in the Murray Catchment (Ayers et al. 2007). (**a**) Southern Tableland Grassy Woodland (79); (**b**) Box Ironbark Forest (18); (**c**) River Red Gum Forest (61); (**d**) White Box Yellow Box Red Gum Woodland (30); (**e**) Tablelands and Slopes Box-Gum Woodland (76); (**f**) Tablelands and Slopes Box-Gum Woodland (79)

Remote sensing is an important element of such modelling because it has the potential to be used as a temporal monitoring tool, as it measures direct attributes of native vegetation. On the other hand, GIS data acts as surrogates for disturbance thereby limiting their potential ongoing use for monitoring. For the purposes of this chapter, reference is made only to *Landsat* imagery as ongoing research has found that *Landsat* imagery provides important operational advantages over *SPOT5* (Zerger et al. 2007). GIS-based predictors included topographic roughness, land use, topographic position and landscape connectivity metrics (Table 8.2).

Table 8.2. Candidate vegetation condition predictor variables. Correlated variables ($r^2 > 0.8$) were removed for GAM modelling and are not shown in this table

GIS and remote sensing predictor variables	Abbreviation	Description
Landsat 5 TM PVI	pvi_ls5tm	Perpendicular Vegetation Index (plot centroid value)
Landsat 5 TM NDVI focal mean	ndvi_ls_fmr25	Normalised Difference Vegetation Index (plot mean)
Landsat 5 TM NDVI focal standard deviation	ndvi_ls_fst25	Normalised Difference Vegetation Index (plot variability)
Landsat 5 TM vegetation cover (4 x 4 window)	lvcover4	Percentage vegetation cover (4 cells by 4 cells)
Landsat 5 TM vegetation cover (10 x 10 window)	lvcover10	Percentage vegetation cover (10 cells by 10 cells)
Landsat 5 TM landscape connectivity	ls_cba	Landscape connectivity (Drielsma et al. 2007)
Landsat 5 TM vegetation patch area	ls_woody	Area of each vegetation patch (m^2)
Elevation (25 m DEM)	dem25	25 m NSW state DEM
Topographic position (continuous)	tpos150	25 m DEM using algorithms of Gallant and Wilson (2000)
Topographic roughness (2 x 3 window)	dem25_fstd	25 m DEM using focal standard deviation function to assess terrain variability
Land use	land use	NSW state land use mapping program

8.5 Results and Discussion for the Murray Catchment Case Study

Model performance for only two of a possible ten *BioMetric* attributes are shown in Table 8.3 and are reported in terms of r^2 using cross-validation methods used by Lehman et al. (2003). Cross-validation is an appropriate approach to validation, as using independent validation data risks comparing different sampling strategies rather than actual model performance (Lehman et al. 2003). Results show that the accuracy of the final prediction varies in terms of model r^2 values from 0.27 for the Volume of Fallen Logs to 0.47 for Native Ground Cover – Grasses. Relationships between individual predictor variables and the vegetation condition attribute are logical. For example, predictions for 'Native Ground Cover – Grasses' were associated with low woody vegetation cover (i.e. tree cover), and gradually increased as woody cover increased until the canopy density reached some maximum threshold above which the cover of native grasses commences to decrease, presumably due to a shading effect.

Table 8.3. Final model performance (r^2) for two *BioMetric* attributes derived from cross-validation (249 samples and 10 groups)

BioMetric attribute	Selected explanatory variables in GAM modelling	r^2 (cross validation: 10 groups, n=249)
Native Ground Cover – Grasses	*Landsat TM* derived vegetation cover, elevation from 25 m DEM, *Landsat TM* derived mean plot NDVI value	0.47
Volume of Fallen Logs	Topographic position, *Landsat TM* derived PVI at plot centroid, *Landsat TM* derived mean plot NDVI value	0.27

Figure 8.4 shows the relative contribution of each selected variable to the final GAM model. The results show that for both models, remotely sensed variables make a relatively important contribution to the final model compared to GIS-derived surrogates (dem25 and topographic position). This has implications for the ongoing monitoring of native vegetation condition, as regional-scale monitoring can be best achieved through the use of remote sensing to directly sense changes in vegetation attributes. Static GIS surrogates such as topographic position and terrain roughness will not support ongoing monitoring and are best used for stratification purposes, and to establish baseline vegetation condition maps that capture historical disturbance patterns.

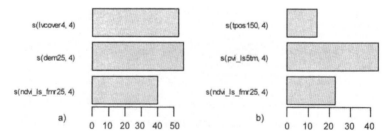

Fig. 8.4. Model contributions of each explanatory variable in the GRASP modelling for Native Ground Cover – Grasses (**a**) and Volume of Fallen Logs (**b**). Model contributions on the x-axis are calculated by the range between the maximum and minimum contribution of each variable in the linear predictor scale before transformation by the link function (Lehman et al. 2003). Model contribution provides an indication of the contribution of that variable to the final predictive model

Recent research by Cunningham et al. (2008) illustrates the potential use of predicted vegetation condition data beyond native vegetation management to explicitly address fauna conservation planning. In their study examining the value of farms for birds, Cunningham et al. (2008) developed a simple site-based additive vegetation condition index that explained approximately 50% of the variation in native bird species richness. Farms with larger numbers of paddock trees, more fallen timber, and greater cover of native pasture plus larger remnants of native vegetation supported more native bird species. Using *GRASP* modelled spatial layers of *BioMetric* attributes, including fallen timber and native grass cover, and *Landsat TM* derived vegetation cover, a comparable regional-scale index can be calculated using the following equation:

$$BH = TC + FT + NGC \tag{8.1}$$

Thus the bird habitat (*BH*) depends on tree cover including paddock trees (*TC*), fallen timber (*FT*) and native grass cover (*NGC*).

Prior to spatially modelling this habitat relationship across the Murray study, it is important to raise and address a concern regarding the implied spatial precisions in spatial model outputs. For example, in the Murray Catchment study a grid cell resolution of 25 m has been used owing to the resolution of the digital elevation model and *Landsat TM* data. The concern with such spatial data is that it implies that the final modelled map contains levels of spatial accuracy and precision that are not warranted. For example, such data should not be used for property-scale planning, given the relatively small sampling intensities of vegetation condition data across the study area (249 *BioMetric* plots) and the uncertainty inherent in the modelling (r^2 values). By presenting sub-paddock-scale data in map form at a raster grid cell resolution of 25 m, modelled results may also introduce confidentiality issues.

To overcome potential misuse we propose a data aggregation approach which takes the disaggregated final model outcomes (GIS rasters) and expresses them as a regular tessellation at a coarser resolution than the source data. For this study we adopt methodologies developed by the US Environmental Protection Agency (USEPA 2007) which applies a hexagonal tessellation of 648 km^2 for the conterminous United States. A 100 ha grid has been used for the Murray Catchment and the sum of the attributes in Eq. 8.1 were calculated for each hexagon. For a more detailed justification for the use of such sampling schemes see White et al. (1992). Figure 8.5 shows the hexagonal sampling scheme for the three explanatory vegetation condition variables described in Eq. 8.1. This map could be used for several purposes including:

- to identify farms that are likely to support a greater numbers of native bird species
- to develop networks that act as links for native birds between conservation reserves
- to prioritise where restoration would be beneficial for native birds (e.g. areas adjacent to farms that are good for native bird conservation).

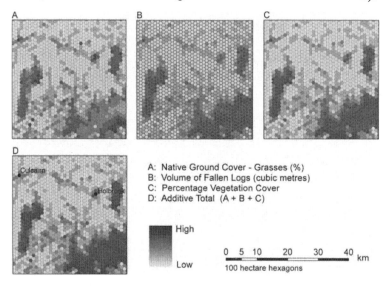

Fig. 8.5. Final aggregated predictions derived from *GRASP* modelling for native ground cover grasses, volume of fallen logs, and actual percentage vegetation cover derived from *Landsat TM* mapping. Results have been aggregated for a section of the Murray Catchment to a 100 ha hexagonal grid where the sum of each attribute has been calculated for each hexagon. The additive total of these attributes is shown

8.6 Conclusion

As the modelling of native vegetation condition using methods presented in this chapter relies on both direct measures (remote sensing) and spatial surrogates of vegetation disturbance (topographic position, elevation, vegetation cover), users of such data need to be aware that the models are regional summaries of vegetation condition, with inherent temporal and spatial accuracy and precision limitations. For example, maximum r^2 values of only 0.47 were attained for *BioMetric* vegetation condition attributes. If developing maps of native vegetation condition is the primary objective of collecting field data, final map accuracies may have major limitations at finer decision-making scales such as the scale of the individual site or property. However, the acquisition of site data to underpin such regional mapping may be justified if it serves multiple purposes. For example, appropriately stratified *BioMetric*-type plot data could serve the needs of a site-based vegetation monitoring and evaluation system, while also acting as a primary input to develop regional-scale maps of native vegetation condition for strategic natural resource management planning.

8.7 Future Research Directions

The key research challenge facing such studies is to develop more accurate predictive models for *BioMetric*-type vegetation condition attributes, and to develop a capability to regionally monitor vegetation condition. At present, the model inaccuracies and uncertainties make regional-scale monitoring impossible and present advances can best be seen as developing a baseline of native vegetation condition from which to plan. These two objectives are not mutually exclusive and improvements in remote sensing image acquisition, calibration and analysis could address both needs. The recent availability of very high spatial resolution systems such as *LiDAR* (Light Detection and Ranging) could provide a rich platform for detecting structural elements of native vegetation. However owing to costs there would be significant operational limitations to using such data at regional scales for monitoring vegetation condition. To develop operational solutions for landscape management, research should examine improved approaches for correcting, calibrating and acquiring and analysing commonly available satellite data (e.g. *SPOT5* and *Landsat 5 TM*).

Acknowledgements

We are grateful to Annemarie Watt and Peter Lyon (Australian Government, Department of Environment and Water Resources) and the NSW Environmental Trust for funding this research. Thanks are extended to Jack Chubb, Alexandra Knight and Emmo Willinck (Murray Catchment Management Authority, NSW) for their ongoing support of the project. The study would not have been possible without the contribution of Damian Wall of Red-Gum Consulting. We are grateful to the landholders in the Murray Catchment who generously provided access to their properties.

References

Anderson R P, Lew D, Peterson AT (2003) Evaluating predictive models of species' distributions: criteria for selecting optimal models. Ecological Modelling 162:211–232

Austin MP (2002) Spatial prediction of species distribution: an interface between ecological theory and statistical modelling. Ecological Modelling 157:101–118

Ayers D, Seddon J, Briggs S, Doyle S, Gibbons P (2007) Interim vegetation types of NSW developed for the BioMetric tool. Department of Environment and Conservation (NSW) unpublished, p 7

Catling PC, Burt RJ (1995) Studies of the ground dwelling mammals of Eucalypt forests in south-eastern New South Wales: the effect of habitat variables on distribution and abundance. Wildlife Research 22:271–88

Cawsey EM, Austin MP, Baker BL (2002) Regional vegetation mapping in Australia: a case study in the practical use of statistical modelling. Biodiversity and Conservation 11(12):2239–2274

Clevenger AP, Wierzchowski J, Chruszcz B (2002) GIS generated, expert-based models for identifying wildlife habitat linkages and planning mitigation passages. Conservation Biology 16(2):503–514

Cunningham RB, Lindenmayer DB, Crane M, Michael D, MacGregor C, Montague-Drake R (2008) What factors influence bird biota on farms? Putting restored vegetation into context. Conservation Biology (in press)

Dare PM, Zerger A, Pickett-Heaps CA (2002) Investigation of space borne and airborne remote sensing technologies for mapping manna gum (*Eucalyptus viminalis*) health in south-eastern Australia. Asian Journal of Geoinformatics 3(1):63–69

Drielsma M, Manion G, Ferrier S (2007) The spatial links tool: automated mapping of habitat linkages in variegated landscapes. Ecological Modelling 200(3–4):403–411

Ferrier S, Watson G, Pearce J, Drielsma M (2002) Extended statistical approaches to modelling spatial pattern in biodiversity in north-east New South Wales I Species-level modelling. Biodiversity and Conservation 11(12):2275–2307

Franklin J, McCullough P, Gray C (2000) Terrain variables used for predictive mapping of vegetation communities in Southern California. In: Wilson JP, Gallant JC (eds) Terrain analysis: principles and applications. John Wiley and Sons, New York

Frescino TS, Edwards Jr TC, Moisen GG (2001) Modelling spatially explicit forest structural attributes using generalised additive models. Journal of Vegetation Science 12:15–26

Gallant JC, Wilson JP (2000) Primary topographic attributes. In: Wilson JP, Gallant JC (eds), Terrain analysis: principles and applications. John Wiley and Sons, New York

Gibbons P, Ayers D, Seddon J, Doyle S, Briggs S (2005) BioMetric Version 18 a terrestrial biodiversity assessment tool for the NSW property vegetation plan, developer operational manual. Department of Environment and Conservation (NSW), unpublished

Goldney D, Wakefield S (1997) Assessing farm bushland. In: Charles Sturt University and Orange Agricultural College (ed) Save the bush toolkit. Charles Sturt University, Bathurst, pp 1–12

Guisan A, Zimmermann NE (2000) Predictive habitat distribution models in ecology. Ecological Modelling 135(2–3):147–186

Hinsley SA, Hill RA, Gaveau DLA, Bellamy PE (2002) Quantifying woodland structure and habitat quality for birds using airborne laser scanning. Functional Ecology 16:851–857

Lehmann A, Overton JMcC, Leathwick JR (2003) GRASP: generalised regression analysis and spatial prediction. Ecological Modelling 160:165–183

Levesque J, King DJ (1999) Airborne digital camera image semivariance for evaluation of forest structural damage at an acid mine site. Remote Sensing of the Environment 68:112–124

Lucas RM, Tickle PK, Witte C, Milne AK (2001) Development of multistage procedures for quantifying the biomass, structure and community composition of Australian woodlands using polarimetric radar and optical data. In: Proceedings of the International Geoscience and Remote Sensing Symposium, 9–13 July 2001, Sydney

Ludwig JA, Eager GN, Bastin VH, Chewings VH, Liedloff AC (2002) A leakiness index for assessing landscape function using remote sensing. Landscape Ecology 17:157–172

McElhinny C, Gibbons P, Brack C, Bauhus J (2005) Forest and woodland stand structural complexity: its definition and measurement. Forest Ecology and Management 218:1–24

McElhinny C, Gibbons P, Brack C, Bauhaus J (2006) Fauna-habitat relationships: a basis for identifying key stand structural attributes in Australian Eucalypt forests and woodlands. Pacific Conservation Biology 12:89–110

Miles C (2001) NSW Murray Catchment biodiversity action plan. Nature Conservation Working Group, Murray Catchment, New South Wales, p 86

Moisen GG, Frescino TS (2002) Comparing five modelling techniques for predicting forest characteristics. Ecological Modelling 157(2–3):209–225

Newell GR, White MD, Griffioen P, Conroy M (2006) Vegetation condition mapping at a landscape-scale across Victoria. Ecological Management and Restoration 7(S1):65–68

Newsome AE, Catling PC (1979) Habitat preferences of mammals inhabiting heathlands of warm temperate coastal, montaine and alpine regions of south-eastern Australia. In: Specht RL (ed) Ecosystems of the World Volume 9A: Heathlands and related shrublands of the world. Elsevier, Amsterdam

Oliver I, Jones H, Schmoldt DL (2007) Expert panel assessment of attributes for natural variability benchmarks for biodiversity. Austral Ecology 32(4):453–75

Overton J, Lehmann A (2003) Predicting vegetation condition and weed distributions for systematic conservation management: an application of GRASP in the central South Island. Science for Conservation (220), Department of Conservation, Wellington

Parkes D, Newell G, Cheal D (2003) Assessing the quality of native vegetation: the 'habitat hectares' approach. Ecological Management and Restoration 7(S1):29–38

Pearce J, Ferrier S (2000) Evaluating the predictive performance of habitat models developed using logistic regression. Ecological Modelling 133(3):225–245

Perkins I (2002) Harrington Park stage 2 ecological assessment: final report. Ian Perkins Consultancy Services and Aquila Ecological Surveys, Sydney

Pressey RL, Nicholls AO (1989) Efficiency in conservation evaluation: scoring versus iterative approaches. Biological Conservation 50:199–218

Scarth P, Phinn S, (2000) Determining forest structural attributes using an inverted geometric-optical model in mixed Eucalypt forests, southeast Queensland, Australia. Remote Sensing of the Environment 71:141–157

Tongway D, Hindley N (2004) Landscape function analysis: a system for monitoring rangeland function. African Journal of Range and Forage Science 21:109–113

USEPA (2007) Environmental Monitoring and Assessment Program. United States Environmental Protection Agency. Retrieved 11 September 2007, http://www.epa.gov/emap/

Wallace J, Furby S (1994) Assessment of change in remnant vegetation area and condition. CSIRO Mathematical and Information Sciences and Agriculture Western Australia. Available at http://www.cmis.csiro.au/rsm/research/remveg/vegassess_all.html.

Wallace J, Behn G, Furby S (2006) Vegetation condition assessment and monitoring from sequences of satellite imagery. Ecological Management and Restoration 7(S1):31–36

Watson J, Freudenberger D, Paull D (2001) An assessment of the focal-species approach for conserving birds in variegated landscapes in southeastern Australia. Conservation Biology 15:1364–1373

White DA, Kimerling JA, Overton SW (1992) Cartographic and geometric components of a global sampling design for environmental monitoring. Cartography and Geographic Information Systems 19(1):5–22

Woodgate PW, Peel WD, Ritman KT, Coram JE, Brady A, Rule AJ, Banks JCG (1994) A study of the old-growth forests of East Gippsland. Department of Conservation and Natural Resources, Victoria

Zerger A, Gibbons P, Jones S, Doyle S, Seddon J, Briggs SV, Freudenberger D (2006) Spatially modelling native vegetation condition. Ecological Management and Restoration 7(S1): 37–44

Zerger A, Warren G, Austin M, Gibbons P, Seddon J (2007) Implications of scale change on native vegetation condition mapping. In: Oxley L, Kulasiri D (eds) MODSIM 2007 International Congress of Modelling and Simulation. Modelling and Simulation Society of Australia and New Zealand, 10–13 December 2007, Canterbury, New Zealand. Available at http://www.mssanz.org.au/MODSIM07/papers/21_s46/ImplicationsOfScale_s46_Zerger_.pdf

Additional Reading

Elith J, Graham CH, Anderson RP, Dudík M, Ferrier S, Guisan A, Hijmans RJ, Huettmann F, Leathwick JR, Lehmann A, Li J, Lohmann LG, Loiselle BA, Manion G, Moritz C, Nakamura M, Nakazawa Y, Overton JMcC, Townsend Peterson A, Phillips SJ, Richardson K, Scachetti-Pereira R, Schapire RE, Soberón J, Williams S, Wisz MS, Zimmermann NE (2006) Novel methods to improve prediction of species' distributions from occurrence data. Ecography 29(2):129–151

Gibbons P, Zerger A, Jones S, Ryan P (2006) Mapping vegetation condition. Special Issue of Ecological Management and Restoration 7(1)

Hobbs R, Harris JA (2001) Restoration ecology: Repairing the earth's ecosystems in the new millennium. Restoration Ecology 9:239–246

McCarthy MA, Parris KM, Van Der Ree R, McDonald MJ, Burgman MA, Williams NSG, McLean N, Harper MJ, Meyer R, Hahs A, Coates T (2004) The habitat hectares approach to vegetation assessment: an evaluation and suggestions for improvement. Ecological Management and Restoration 5(1):24–27

Noss RF (1990) Indicators for monitoring biodiversity: a hierarchical approach. Conservation Biology 4:335–364

Parkes D, Lyon P (2006) Towards a national approach to vegetation condition assessment that meets government investors' needs: a policy perspective. Ecological Management and Restoration 7(S1):3–5

9 Towards Adaptive Management of Native Vegetation in Regional Landscapes

David H Duncan[1] and Brendan A Wintle[2]

[1] Arthur Rylah Institute for Environmental Research, Department of Sustainability and Environment, Heidelberg, Victoria, Australia
[2] School of Botany, The University of Melbourne, Victoria, Australia

Abstract: Landscape modellers are now capable of combining high resolution spatial data with process models to explore natural resource management scenarios at scales appropriate for decision making, but what of the process of decision making itself? In this chapter we review the applicability of the 'adaptive management' paradigm to natural resource management, using regional management of native vegetation by Catchment Management Authorities as an example. We find that progress has been made in the approach to defining management objectives and specifying assumptions behind vegetation change models; however, there remain significant challenges in instituting true management experiments and identifying performance indicators appropriate to support continuous learning. We argue that the ecological and institutional complexity of native vegetation management reinforces the importance of systematic decision protocols. Adaptive management is the most logical approach to decision making where there is uncertainty about the effectiveness of management options, and the opportunity exists to learn and update understanding. This iterative process offers continuous improvements to investment efficiency in native vegetation management.

9.1 Introduction

Today's natural resource managers, be they land owners or policy makers, have access to an increasing range of data, tools and models. Unprecedented computing power allows for increasingly complex system models, higher resolution spatial data and models (for example, see Clarke Chapter 17; Chan et al. Chapter 6), and more engaging methods for stakeholder interactions (see Bishop Chapter 22; Cartwright Chapter 24). These developments help us to understand our resources and assets and make complex utilisation and conservation decisions. However, one of the greatest challenges in natural resource management is the least technological — dealing with uncertainty about the effectiveness of actions when making resource management decisions. How can we monitor and evaluate programs and investments to achieve institutional learning and ensure that future management decisions are scientifically and socially defensible?

Adaptive management is 'learning by doing', a structured iterative processes of decision making with the capacity to gradually reduce uncertainty through system monitoring. It offers transparency and accountability to decision making and resource prioritisation, while providing a formal theoretical foundation for learning and improving management. Adaptive management remains widely cited as the most logical and elegant framework for continuous improvement in natural resource management (Johnson 1999). In this chapter we explore the distance still to travel between current natural resource management, and a future where our management decisions and actions are determined within a formal adaptive management framework (Holling 1978; Walters and Holling 1990).

The number of references to adaptive management in the natural resource management literature is testament to its intuitive appeal (Commonwealth of Australia 1992, 1996, 1998; Tuchmann 1998; VCMC 2007). However, to date there are few successful examples of adaptive management in practice (Stankey et al. 2003, 2005). Stakeholder risk aversion, failure to identify clear and measurable management goals, inappropriate statistical approaches to inference and learning, and failure to monitor the performance of management actions have all been identified as impediments to successful adaptive management. A reluctance to invest in long-term management experiments is also a key hurdle. Together, these factors could perhaps be summarised as 'entrenched social norms and institutional frameworks' (Allan and Curtis 2005). In natural resource management, an emphasis on best practice management tends to have been construed as championing a single best practice (Bormann et al. 1996; Meredith 1997), rather than spreading management over competing options in order to learn about them. Spreading investment over a range of competing strategies is seen as standard practice in financial investment, but has failed to take hold in natural resource management (Bormann et al. 1996).

In natural resource management, objectives are typically more complex and feedback about success of management is more difficult to obtain than for financial investment. Of particular interest for natural resource management is the way in which adaptive management formalises the relationship among management goals, performance measures, and monitoring. Monitoring is central to adaptive management because it provides the quality control and performance evaluation function that is central to continuous improvement. Monitoring the performance of investments in natural resource management is central to the government's credibility as a prudent investor (ANAO 2008). Beyond the issue of reporting and accountability, monitoring provides the basis on which to learn about the state of the system and to identify system properties that may be unstable or out of control. However, without the formal link to management goals and performance measures provided by the adaptive management framework, monitoring has a tendency to be inefficient, meaningless and wasteful (Nichols and Williams 2006). Furthermore, monitoring management outcomes only makes sense if there is a genuine commitment and plan for change in response to monitoring results.

In the following section we will review the basic principles of adaptive management and discuss the integration of traditional adaptive management methods and recent developments in environmental risk assessment. Then, later in this chapter, we explore its potential for application in the context of native vegetation management in Victoria (Australia), identifying some of the major challenges that we expect to arise when attempting to implement adaptive native vegetation management. Native vegetation at the scale of whole regions is a particularly complex application environment, and we do not prescribe a solution in this chapter. We do attempt to provide some guidance towards the development of true adaptive management strategies for vegetation management and monitoring at the catchment level.

9.2 What Adaptive Management is and is not

Formal approaches to adaptive management integrate information gained from research, monitoring and management to evaluate and improve management practices. Many managers and policy makers confuse adaptive management with ad hoc approaches to continuous improvement that may be better characterised as trial and error management. Strategies for continuous improvement might be broadly categorised into trial and error management and passive or active adaptive management. Trial and error management commonly entails persisting with the management option thought to be the best at the time until such time as it is shown to be inadequate, at which point management actions may be changed in the hope of achieving a better outcome. Trial and error management is not underpinned by a formal model (or models) for the system being managed, it does not explicitly recognise uncertainty about the system model in the allocation of management effort, does not involve a plan for learning, and is usually neither replicated nor statistically rigorous. Unfortunately, the bulk of management efforts described as 'adaptive management' fit into the category of trial and error management.

In contrast, passive adaptive management explicitly recognises uncertainty about how management actions contribute to management outcomes and usually involves concurrent application of competing management options so that learning about the system and the relative efficacy of management may be achieved as management progresses. Learning should be based on a formal evaluation of evidence, preferably supported by statistical inference. Active adaptive management (AAM) involves a more aggressive program of learning about the efficacy of management options. AAM aims to maximise long-term gains through a strategic allocation of resources to management and learning, sometimes at the expense of short-term gain. As far as we are aware, there are no practical applications of active adaptive management in natural resource management and very few theoretical contributions (refer to McCarthy and Possingham 2007; and the 'Future Directions' section below). In this chapter we focus on describing approaches to (passive) adaptive management and the numerous challenges it poses. We describe adaptive management in four steps (Fig. 9.1):

- Step i: Identification of management goals, constraints and performance measures
- Step ii: Specification of management options
- Step iii: Identification of competing system models and model weights
- Step iv: Allocation of resources, implementation of management actions and monitoring of management performance.

We chose to aggregate the implementation of management actions and monitoring to emphasise that monitoring is central to good management and not an optional extra. This entails a conscious decision about how a budget should be allocated between management and monitoring. In the following sections we describe each step in the adaptive management cycle (Fig. 9.1).

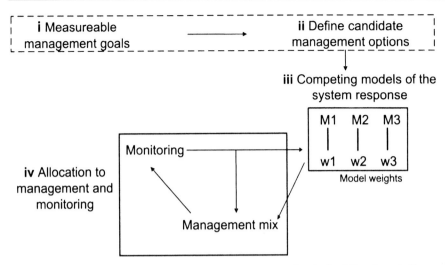

Fig. 9.1. Steps in an adaptive management strategy. The dashed-line box indicates steps that require elicitation of social preferences, while the remainder of the process is largely the domain of technical experts

9.2.1 Step i: Statement of Objectives, Constraints and Performance Measures

The first step in adaptive management is to clearly define management objectives, constraints and measures by which management performance will be assessed. While objective setting is a social process, there is an important role for scientists in helping to articulate these objectives and their management responses in ways that are amenable to measurement, open to statistical inference and comparison, and therefore appropriate for adaptive management. Appropriately constructed statements of objectives and constraints convert broad (often opaque) policy objectives such as 'reverse the long term decline in native vegetation extent and quality' (Victoria's Biodiversity Strategy, DNRE 1997) into operational and measurable goals. An example might be:

> 'within 15 years, demonstrate with at least 90% confidence that a 15% improvement in the condition score for native vegetation within the catchment has been achieved, subject to the constraint of maintaining (with at least 90% confidence) no less than 90% of the expected commercial output of local farms within the catchment'.

This statement is characterised by a clearly defined performance measure (in units of dollars and condition score) and an explicit spatial (catchment) and temporal (15 years) context. It also explicitly states acceptable levels of uncertainty (>90% confidence). The objectives are social preferences that must be elicited throughout the management planning process via community engagement. Clear statements of goals make trade-offs explicit. Here, some loss of productive potential may be tolerated for some gain in vegetation condition. Management performance can then be assessed against goals and constraints.

Performance thresholds are defined as states that must be achieved in order to satisfy objectives. They are often implicit in clearly stated management objectives. For example, in the hypothetical management objective defined above, a vegetation condition performance threshold is identifiable: the manager must ensure, with 90% confidence, that vegetation condition has improved by 15% on the current estimated condition within a pre-specified time. Setting a threshold has limited value unless there is an identified action that will be undertaken if that threshold is breached or not met, as the case may be. In our example, one such action might include a change in vegetation clearing permit conditions or a switching of revegetation strategies.

The inherent unpredictability of natural systems means that unforeseen results may (probably will) occur even when the best available knowledge is applied to future projections. This should not reflect badly on a manager. Rather, a manager should be judged by how quickly the failure of management is detected (i.e. how robust was their monitoring strategy) and the speed of implementation of remedial or alternative actions.

9.2.2 Step ii: Specification of Management Options

Specification of management options is partly a social and partly a scientific process. Management options are usually generated by opinions of stakeholders and scientists about the best means to achieve management objectives. The need for multiple management options arises from uncertainty about the outcomes of particular management options. An adaptive management strategy would explore the plausible strategies proposed by scientists and stakeholders. The extent to which investment would be distributed among options will depend on the benefits predicted to arise from each management option under assumptions about how the system will respond to management. These issues are discussed in more detail in later sections of this chapter.

9.2.3 Step iii: System Modelling and Model Credibility

There is usually substantial uncertainty about how a system will respond to management actions and the ecological processes that mediate that response. It is common for different experts to support qualitatively different models of ecological processes. Qualitatively different forest management models usually imply different views about how species and environmental processes interact with humans and natural disturbances.

When experts support qualitatively different models of the system under management, there is usually divergent opinion about the best approach for achieving desired management outcomes. When such uncertainty exists (and is acknowledged), there is value in implementing management options that will facilitate learning about the relative merits of competing models, and ultimately, the best long-term strategies for achieving management outcomes.

In some instances, data and expert opinion may favour some models over others. When this is the case, formal methods for weighting competing models may be utilised (Burnham and Anderson 1998; Wintle et al. 2003). Competing model weights may be used to assist in the allocation of effort among competing management options. If there is no substantial evidence in favour of one model over another, then uninformative (equal) model weights may be appropriate until further evidence arises that supports one model over others. The Appendix provides an example of how uncertainty about the process under management might be formulated as competing models that are iteratively evaluated with imperfect monitoring data.

9.2.4 Step iv: Allocation, implementation and Monitoring — Closing the Loop

Decision Theory provides a framework for determining the optimal allocation of resources to competing management options (Possingham 2001), though few authors are able to provide guidance and examples of how this allocation is achieved in practice (however see Johnson et al. 1997; Hauser et al. 2006; McCarthy and Possingham 2007). Having clearly defined goals and constraints, identified competing management options and characterised the predicted response of key performance measures to management (preferably as mathematical models) it is possible to apply formal optimisation methods to determine the optimal allocation of effort among management options. However, it is not critical to the adaptive management process that allocation of investment among management options be formally (mathematically) optimised. In some instances, 'rules of thumb' for allocation may work quite well in achieving management goals and facilitating learning. If particular management options are supported by particular system models (as in the Mallard duck example, see Appendix), then allocating management units in proportion to the degree of support assigned to competing models may be a defensible rule of thumb.

The role of monitoring in providing feedback on management efficacy is clearly evident from the example given in the Appendix. Without monitoring, there would be no grounds on which to discern among the four competing models, each of which support substantially different 'optimal' management strategies. Unfortunately, determining the optimal allocation of resources *to monitoring* is not a trivial undertaking. Monitoring often costs substantial amounts of money and, at some level, comes with an opportunity cost because that money could have been used for something else, such as implementing management. The question of what proportion of a conservation management budget should be spent on monitoring is not addressed in the statistical and management literature, and is an interesting area of current research (Hauser et al. 2006).

Classical approaches to hypothesis testing and statistical power calculation (Underwood 1994) do not answer the question of how much monitoring is enough, though power analysis does provide a means to determining how much effort should be invested to achieve desired type 1 and type 2 error rates (statistically, the balance between alarmism — concluding an effect when there is none — and over-confidence, concluding no effect when in fact there is one). Arguably, one of the main reasons for the lack of convincing monitoring programs is that researchers have focussed too much on the statistical minutia of monitoring design and not enough on monitoring within a coherent decision framework. While it is theoretically feasible, we were unable to find practical examples of where the allocation of effort to monitoring was optimised along with the allocation to other management activities. Gerber et al. (2005) and Hauser et al. (2006) show how such an allocation could be theoretically achieved, though there is a substantial technical overhead to utilising the approaches they present. Nonetheless, Decision Theory provides some general guidance on the characteristics of a good monitoring program:

- There should be a clear purpose to monitoring. Monitoring should be undertaken in order to maximise the probability of achieving management objectives and not simply as a stand-alone activity.
- Monitoring should be undertaken on clearly defined performance measures (e.g. population size of forest dependent species) that have a direct link to management objectives.

- The planned response to findings should be articulated; there is no point in monitoring if there is no intention to change management in light of findings.
- A sampling design that adequately reflects the precision (confidence) requirements implied by the stated management objectives.
- A spatial extent and resolution of monitoring that is congruent with the stated management objectives.
- Monitoring a small number of things well rather than many things poorly.
- The targeting of taxa for which detailed risk assessments are also undertaken (e.g. Haynes et al. 2006, Lindenmayer and McCarthy 2006).

In the context of adaptive management it is worth distinguishing between monitoring and research. Monitoring should largely be thought of as the ongoing collection of information about the degree to which competing management actions influence management performance measures. Monitoring is also important as a way of informing managers about the state of the system which is particularly important for state-based management (sensu Nichols and Williams 2006). State-based management simply implies that the optimal management decision at any given time depends on the state of the system at that time. For example, the optimal number of kangaroos to cull depends critically on the number of kangaroos thought to exist at the time harvest quotas are set. If the state of the kangaroo population is low, it would be sub-optimal to set a high quota if the management goal was to ensure a minimum number of kangaroos at all times. The role of monitoring is to ensure sensible state-based decisions are made. Monitoring has a dual role in adaptive management of facilitating learning about the effectiveness of management and informing managers about the state of the system. Research can be thought of as any targeted study that aims to improve knowledge about a particular aspect of the system (or a particular parameter in a system model). Research may improve management indirectly by improving system models and therefore assisting in better identification of optimal management strategies.

9.3 Managing and Monitoring Native Vegetation

In the previous section we reviewed adaptive management and its main components. Here we discuss the application of the framework to regional management of native vegetation in Victoria (Australia). Australian governments have committed to reversing the long-term decline in extent and quality of native vegetation (NRMMC 2002b). In Victoria, both the current extent and quality of native vegetation are assessed with reference to an estimated pre-1750 benchmark state for the vegetation type (DNRE 2002). While 'extent' is a relatively unambiguous notion, being the spatial extent of a vegetation type, 'vegetation quality' (and type) is more difficult to define. We use 'native vegetation quality' to refer to the structure and composition of vegetation relative to a long undisturbed benchmark. High quality is encountered where species diversity and vegetation structure is most similar to sites where disturbance has been absent for decades if not centuries (Parkes et al. 2003). 'Quality' is more or less synonymous with the contemporary usage of native vegetation 'condition' or 'integrity' (ESCAVI 2007).

Improving native vegetation extent and quality is an issue of enormous importance because it is widely believed to be a surrogate indicator of relevance for biodiversity, land sustainability, regional climate change, soil and water quality, and other ecosystem services (Parkes and Lyon 2006). Native vegetation extent and condition (quality) were identified as *Matters for Target* for national investment (NRMMC 2002a) in recognition of their importance for landscape sustainability. In Victoria for example, it was recently estimated that $15million of State and Commonwealth funds are being spent per annum on native vegetation protection, enhancement and revegetation (DNRE 2002). This contribution is likely to be at least matched by Landcare and community groups (Brunt and McLennan 2006; DNRE 2002; Smith forthcoming). Leaving aside the question of whether or not the level of investment has been sufficient to address the magnitude of the problem, what have we to show for the effort and investment?

There is currently great uncertainty as to the magnitude and direction of impact from vegetation protection and enhancement and revegetation activities on catchment targets (e.g. Brunt and McLennan 2006; Robinson and Mann 1998). This is not surprising; the ecological response of native vegetation condition to perturbation is the cumulative response of the various component species, which is mediated by the timing, frequency and intensity of various drivers at different temporal scales. In addition to great ecological complexity there is considerable institutional complexity related to the involvement of many stakeholders, agents and land managers. Here we examine how regional natural resource management agencies are trying to implement an adaptive management approach, and the major challenges that are yet to be overcome. At the outset we acknowledge that some of these challenges are common to both resource managers and scientists, and that much collaborative effort is required to support the adaptive management of native vegetation.

9.3.1 An Example of a Formal Approach to Adaptive Management of Vegetation Condition at the Scale of the Catchment Management Authority

Regional management of native vegetation is a complex institutional and ecological problem. In Victoria, this task falls to Catchment Management Authorities (CMAs). CMAs are statutory authorities established under the *Catchment and Land Protection Act* 1994. Each CMA must:

'develop and co-ordinate the implementation of [a] five-year Regional Catchment Strategy (RCS) in partnership with its communities' (VCMC 2007).

The activities of the CMA are outlined in the RCS document, and in the case of vegetation, there are supporting Native Vegetation Management Plans that further outline the strategic approach. CMAs have a statutory reporting responsibility for the condition of native vegetation within their jurisdiction. These typically large regions ($13,000$–$39,000$ km^2) may comprise diverse land uses and a suite of land managers including private interests and state government agencies, over which CMAs have limited power to direct management. Thus, along with their regional partners (including state agencies and non-government organisations) CMAs allocate resources to encourage private landholders to alter management over specified land parcels via a range of approaches described below. Within this discussion we have not attempted to capture the full ecological and institutional complexity of the problem of native vegetation management, rather we attempt to characterise the major challenges and look for improvements that might arise through closer adherence to formal decision protocols and adaptive management. We structure our discussion of these issues according to the key steps required under the adaptive management model (Fig. 9.1).

Statement of Objectives, Scope, Constraints and Performance Measures

Objective statements should clearly identify the scope of management, performance measures and constraints that may interfere with achieving goals. The current strategic documents indicate substantial progress in the area of defining and describing the objectives for native vegetation management (e.g. GBCMA 2003, NCCMA 2003, NECMA 2003). Most CMAs in Victoria have already identified medium- to long-range objectives (resource condition targets) that are largely consistent with adaptive management principles. They express quantifiable goals that are to be achieved by a particular point in the future. For example, the North Central

CMA aims to 'increase native vegetation coverage [from 13%] to 20% of the [CMA] region' by 2030 (NCCMA 2003). The Goulburn Broken CMA aims to 'improve the quality of 90% of existing (2003) native vegetation by 2030' (GBCMA 2003). We make no comment on the merits of attainable versus aspirational targets, suffice to say that CMAs are routinely specifying measurable goals in terms of extent of vegetation cover and identifying the appropriate scale at which progress towards goals could be assessed.

The current set of objectives and targets could be improved by development of explicit statements about acceptable limits of uncertainty, both in terms of the likely achievement of goals and the ability to precisely measure progress. That uncertainty exists is acknowledged to varying degrees in strategic and evaluation documents (GBCMA 2003; NECMA 2006b), and in some cases the significance of uncertainty for achieving targets has been explored (Brunt and McLennan 2006). However the implications of uncertainty or competing assumptions about ecosystem responses for achieving stated targets is rarely tested and the probability of achieving targets under various management and investment strategies is rarely compared and evaluated.

The scale at which objectives are set must be clearly defined. Adaptive management is a coherent framework that can be implemented at any scale, but the individual components are necessarily scale dependent. For example, the goals, performance measures and constraints imposed at a catchment level are likely to be very different to those imposed at a state (or possibly even regional) level. Moreover, it is important to acknowledge that state or national level goals and performance measures should not necessarily be imposed at the regional level. Setting of goals and identifying suitable performance measures for regional vegetation condition remains a substantial technical challenge.

Constraints on managers' abilities to achieve native vegetation targets are rarely explicitly recognised. If there were no constraints, the ability to achieve nominated revegetation targets would be greatly enhanced. Obvious constraints that inhibit managers' ability to achieve targets include limits to:

- available revegetation budgets
- the number of people who can be engaged to undertake projects
- the amount of land suitable for restoration
- the amount of land that can be converted to native vegetation while maintaining the economic viability of the farming business.

In order to attempt an optimal management strategy, these constraints need to be identified and, where possible, quantified.

Identifying appropriate performance measures is a critical component of adaptive management. A good performance measure is one that adequately reflects progress towards stated goals. This is far from a trivial task where native vegetation condition is concerned. We manage native vegetation for multiple benefits including biodiversity conservation and the provision of a range of ecosystem services from fundamental biogeochemical soil formation processes to less tangible life fulfilment and spiritual benefits. The link between native vegetation condition and these benefits is not always clear and rarely easy to quantify. Consequently, CMAs have tended to use system inputs such as the number of hectares fenced off for protection or revegetation activities as performance indicators (NECMA 2006a; WCMA 2007). These are the activities over which they have most direct control. Unfortunately, there are currently few data or process models to connect these activities with change in the native vegetation or biodiversity resources, which are the target of the activity.

Reporting on the number of hectares fenced, or the number of people participating in vegetation improvement activities would be appropriate if the goal of management is to increase the amount of fencing in the region, or to increase the number of people engaged in native vegetation management (possibly a very important goal). However, such measures provide scant indication about progress towards a stated goal such as increasing the amount or quality of native vegetation cover from 15% to 20% in ten years time. Clearly, the variable of interest here that must be measured is the total area of successful revegetation (requiring a definition of success that should be explicit in the objective statement), or the amount of vegetation condition gain expected from protective actions (see VCMC 2007, p. 45 for a simple case study from North Central Victoria).

It is important to gain a better understanding of the relationship between fencing, community engagement and vegetation restoration success so that the value of these activities may be predicted in terms of the performance criteria (change in extent and quality) and used for planning investments. Learning about the relative value of competing management options is central to adaptive management, however, this requires an objective and a performance measure that is directly linked to the management goal (Lake 2001). There remain significant challenges in identifying such a measure for goals relating to vegetation condition that we will discuss in more detail below.

Specification of Management Options

Identifying distinct management options is central to adaptive management, but not often practiced. The full range of potential management options available which address the management objectives should be identified jointly by managers and stakeholders. Initially, it is important that the process of identifying management options is not dominated by consideration of the likelihood of success — the probability that proposed actions will achieve management objectives. It is quite common for competing interests to have different beliefs about the probability that particular options will achieve a particular outcome. The process of assigning credibility to competing models (and consequently evaluating the probability of success under the various management options) happens in later stages of the adaptive management process, hopefully in the presence of data.

The principal vegetation management activities being deployed by CMAs and landholders include:

- remnant protection and enhancement
- revegetation
- pest and weed control
- identification and mapping of high conservation value areas
- creating protection overlays for local planning schemes
- reverse auctions
- community engagement and education
- doing nothing (sometimes the most efficient action).

At the beginning of a planning and resource allocation process, each of these actions might be considered as candidates. The process of evaluating actions in terms of their potential contribution toward management objectives, and allocating resources to the different candidate models, is dealt with in the following sections.

System Modelling and Model Credibility

One of the major technical challenges to the application of adaptive management in vegetation management is the development of system models that can reliably predict the efficacy of competing management options at regional scales. The major impediment to the development of such models is to identify how local level management activities mediate vegetation condition or vegetation establishment success. Victorian CMAs have access to the most advanced statewide datasets relating to the native vegetation resource in Australia (Griffin 1999). These include the modelled distribution of fine-scaled pre-European and extant Ecological Vegetation

Classes (EVCs) (Woodgate et al. 1994); and modelled fine-scale vegetation condition for the entire state (VCMC 2007), which relate the structure and composition of the vegetation to long undisturbed benchmarks for the EVC type. These models are the fundamental inventory that describes the native vegetation assets the CMAs have within their boundaries.

Anecdotally, there appear to be two broad assumptions (or conceptual system models) upon which reporting of native vegetation condition change is currently based. The first assumption is that where direct interventions such as fencing out stock or replanting occur, vegetation condition is likely to improve, albeit slowly (Brunt and McLennan 2006; DSE 2006). The second assumption is that broad landscape vegetation condition is declining across private and public land tenures, as a result of ongoing utilisation, overgrazing, and the absence of regenerative ecological disturbances. These assumptions about site and landscape trends are certainly plausible; however, they are recognised as being too generic (one assumption for all vegetation types) and uncertain to inform continual improvement in decision making.

The required system models inform managers about the expected gains in performance measures (e.g. area successfully revegetated, or area of vegetation in a minimally acceptable condition) likely to arise from the various competing management options. Further development of credible native vegetation change models for anticipating resource change at the landscape scale in response to management actions is urgently required. In the absence of scientific data at appropriately broad scales, broad rules of thumb for expected change within existing remnant wooded and non-wooded vegetation sites have been developed for policy purposes. These represent a credible starting point and similar change models have been subsequently, or independently, adopted within some regions (e.g. Brunt and McLennan 2006). However, these models were not designed to account for expected sources of variation such as EVCs, site productivity and land use history, and thus cannot be deployed within a spatially referenced modelling context with any degree of certainty. Furthermore, there is currently scant information on the impact of engagement and education programs on people's *subsequent* land management behaviour and the specific implications for vegetation condition and extent.

Without models to predict the relative gains likely to be achieved through competing management options, there is no basis for discerning among options such as spending money on public education and engagement, or providing subsidies for onground fencing or planting, when making resource allocation decisions. This is a major impediment to adaptive management of native vegetation. The uncertainty associated with the effectiveness of competing management options in achieving vegetation management goals highlights the importance of structuring management efforts to facilitate learning and adaptation to new findings.

Allocation, Implementation and Monitoring

Without empirical or process models from which to predict the expected gain in vegetation condition and extent resulting from competing management or investment options, it is difficult to transparently optimise investments. The prioritisation phase of most native vegetation restoration projects is an ad hoc procedure based on implicit conceptual models or intuition. This creates three problems. Firstly, because models are not written down and their assumptions not stated, the process of prioritisation lacks transparency. Secondly, without writing out the model and assumptions, there is a danger that institutional knowledge will be lost as experienced managers move on and there is no formal basis on which to learn about the plausibility of the model(s) that underpinned the prioritisation. Finally, intuitive models are often of insufficient precision to distinguish which of the proposed actions are most likely to bring the greatest benefits in terms of vegetation condition, extent or protection. Ongoing development of models describing the ecological benefits arising from competing management actions (e.g. Dorrough et al. 2008) will provide the foundation for optimising management benefit; however, this is currently a substantial impediment.

Monitoring the effectiveness of management actions implies that the performance measure (metric) being monitored is directly relevant to the stated objective (Lake 2001; ESCAVI 2007). Information gained from monitoring may then be fed back into system models to update knowledge and make predictions about the relative benefits of competing management options. In practice, this amounts to re-weighting competing models after each monitoring cycle (see Appendix). Monitoring change in vegetation quality is complex and as yet there seem to be no strong candidates for short- to medium-time scale monitoring of change. A national approach was recently proposed for reporting on native vegetation integrity at different scales using a broad composite vegetation condition metric, 'habitat hectares' (ESCAVI 2007). However, its use in target setting and monitoring for regional bodies was explicitly discouraged. The concept of habitat hectares was designed to address the policy need to assess and compare the quality of diverse vegetation types for investment prioritisation. The strengths of techniques such as habitat hectares are in allowing relative non-specialists to capture much of the information about a site or habitat unit that we expect to matter for biodiversity conservation. It has been extremely successful in gaining acceptance and there are now data from over

15,000 assessments in Victoria, data that have now been used to model the statewide distribution of native vegetation condition (Newell et al. 2006; VCMC 2007).

Given that vegetation quality assessments are required by state and federal governments to justify expenditure on vegetation protection or enhancement, there has been considerable expectation that these data may be useful for monitoring the biophysical objectives of native vegetation condition improvement. However, these assessments were not designed to track changes at individual sites through time. Furthermore, for monitoring purposes they are unlikely to be appropriately stratified to capture key combinations of management actions, vegetation types and land use history (e.g. Zerger et al. Chapter 8). Thus, while it is likely that components of the habitat hectares metric may contribute to an appropriate metric for monitoring long-term change at the national or state scale (DNRE 2002), it is unlikely that these data will yield information to inform CMAs about their *interim* progress towards the targets or help distinguish among the relative merits of alternative management strategies for improving vegetation condition. Clarifying appropriate condition metrics for evaluating restoration performance is an important research priority.

9.4 Research

There are a great many important research questions regarding native vegetation condition and its responses to management, environment and climate change, some of which we have outlined already. Within the adaptive management framework, research priorities should emerge from the questions:

- Which unknowns most impact on our ability to predict the outcomes of our management using our system models?
- What do we need to learn in order to make better predictions and better management decisions about native vegetation condition?

Encouraging examples of this type of approach to research priority setting have begun to emerge (e.g. Brunt and McLennan 2006). Some examples are:

- How has the quality of 'unmanaged' native vegetation changed over the past decade/s?
- How does native vegetation quality, degraded to various degrees, change when a primary threat such as stock grazing is removed with the aid of public money?

- What is the magnitude of privately funded and implemented vegetation enhancement or improvement works?
- Does involvement in a fixed-term conservation agreement influence the subsequent management of the land subject to the agreement?
- At a more fundamental level, we might ask to to what extent is native vegetation condition an effective surrogate for biodiversity conservation values, and provision of ecosystem services (Keith and Gorrod 2006)?

Research to address some of these critical questions will serve to improve the confidence in process models representing links among policy decisions, management actions and resource change. Some of these questions (including points 2 and 4 above) may be answered with adaptive management experiments, while others are more amenable to targeted ecological and social research.

9.5 Conclusion

We have outlined the key components of adaptive management to align them with the major native vegetation management challenges facing natural resource management agencies, such as CMAs. We believe that some aspects of the adaptive management cycle, such as goal setting, are already reasonably well understood and addressed, while others, such as systems modelling are under-utilised in current management and resource allocation strategies. We do not suggest that adaptive management is a silver bullet for all complex natural resource management problems. Even with a perfect adaptive management experiment in place, there would still be considerable uncertainty as to the effectiveness of some of our actions to improve complex natural resource assets such as native vegetation condition. The timelag in vegetation responses to management is an inherent difficulty in predicting and understanding vegetation change, though it also reinforces the importance of having a structured approach to decision making and learning under uncertainty. Eradicating uncertainty is not a realistic goal. However, coherent management behaviour in the face of uncertainty is highly desirable and not easy to achieve without a formal decision framework such as adaptive management.

9.6 Future Directions

If CMAs are able to encourage an appetite for adaptive management amongst their staff, stakeholders and communities, then Decision Theory offers some further advances for learning more rapidly. Active adaptive management (AAM) is similar to adaptive management in that it explicitly recognises uncertainty about the process being managed and the most effective means to achieve management goals. However, AAM differs from adaptive management in the explicit recognition that long-term benefits of learning may outweigh the benefits of maximising management outcomes in the short-term. For example, under adaptive management a manager may chose a 70:30 mix of management strategies A and B on the basis that A was found to be slightly better than B in an inconclusive trial the previous year. Under an AAM strategy, the manager may decide that a 50:50 mix of strategies A and B is a better initial management mix because it will allow faster learning about the true relative performance of A and B, leading to greater long-term benefits.

Identifying the optimal management mix under an AAM strategy requires an appraisal of the:

- expected rate of learning
- benefits that are expected if knowledge was more complete
- costs in terms of reduced management performance associated with (possibly) sub-optimal management over the period for which learning is occurring.

Approaches to formal AAM are relatively new in natural resource management (McCarthy and Possingham 2007) and represent an important area of current research.

Acknowledgements

This work was supported by the Commonwealth Environmental Research Facilities scheme through the research 'Hubs Landscape Logic' (Duncan) and 'Applied Environmental Decision Analysis' (Wintle). We thank the North Central Catchment Management Authority for hosting a thought provoking workshop on monitoring biodiversity and native vegetation in 2006. This chapter was improved by discussions with biodiversity and native vegetation staff from DSE, DPI and CMAs, and an earlier version of the text was improved by the comments and suggestions of Kim Lowell, Peter Vesk and Ted Lefroy. The opinions expressed in this chapter are those of the authors and do not represent those of the Department of Sustainability and Environment or The University of Melbourne.

References

Allan C, Curtis A (2005) Nipped in the bud: Why regional scale adaptive management is not blooming. Environmental Management 36:414–425

ANAO (2008) Regional delivery model for the Natural Heritage Trust and the National Action Plan for Salinity and Water Quality. ANAO Audit Report No. 21, 2007–08. Australian National Audit Office, Canberra

Bormann BT, Cunningham PG, Gordon JC (1996) Best management practices, adaptive management, or both? In: Proceedings of the National Society of American Foresters Convention, 28 October–1 November 1995, Portland, Maine, p 6

Brunt K, McLennan R (2006) Biodiversity monitoring action plan. Goulburn Broken Catchment Management Authority, Shepparton, Victoria

Burnham KP, Anderson DR (1998) Model selection and inference: A practical information-theoretic approach. Springer, New York

Commonwealth of Australia (1992) National forest policy statement. a new focus for Australia's forests. Commonwealth of Australia, Canberra

Commonwealth of Australia (1996) National strategy for the conservation of Australia's biodiversity. Commonwealth of Australia, Canberra

Commonwealth of Australia (1998) A framework of regional (sub-national) level criteria and indicators of sustainable forest management in Australia. Commonwealth of Australia, Canberra

DNRE (1997) Victoria's Biodiversity: Directions in Management. Department of Natural Resources and Environment, Melbourne

DNRE (2002) Victoria's native vegetation management: a framework for action. Department of Natural Resources and Environment, Melbourne

Dorrough J, Vesk PA, Moll J (2008) Incorporating ecological uncertainty and farm-scale economics when planning restoration. Journal of Applied Ecology 45:288–295

DSE (2006) Native vegetation gain approach – technical basis for calculating gains through improved native vegetation management and revegetation. Department of Sustainability and Environment, East Melbourne

ESCAVI (2007) An interim approach to the native vegetation condition indicator. Executive Steering Committee for Australian Vegetation Information, Commonwealth of Australia, Canberra

GBCMA (2003) Goulburn Broken regional catchment strategy. Goulburn Broken Catchment Management Authority, Benalla, Victoria

Gerber LR, Beger M, McCarthy MA, Possingham HP (2005) A theory for optimal monitoring of marine reserves. Ecology Letters 8:829–837

Griffin NRM (1999) Native vegetation national overview. Report for Australia and New Zealand Environment and Conservation Council. Environment Australia, Canberra

Hauser CE, Pople AR, Possingham HP (2006) Should managed populations be monitored every year? Ecological Applications 16:807–819

Haynes RW, Bormann BT, Lee DC, Martin JR (2006) Northwest Forest Plan – the first 10 years (1993–2003): synthesis of monitoring and research results. PNW-GTR-651, Forest Service, United States Department of Agriculture

Holling CS (1978) Adaptive environmental assessment and management. John Wiley, New York

Johnson BL (1999) The role of adaptive management as an operational approach for resource management agencies. Conservation Ecology 3:Article 8

Johnson FA, Moore CT, Kendall WT, Dubovsky JA, Caithamer DF, Kelley JR, Williams BK (1997) Uncertainty and the management of mallard harvests. Journal of Wildlife Management 61:202–216

Keith D, Gorrod E (2006) The meanings of vegetation condition. Ecological Management and Restoration 7:7–9

Lake PS (2001) On the maturing of restoration: Linking ecological research and restoration. Ecological Management and Restoration 2:110–115

Lindenmayer DB, McCarthy MA (2006) Evaluation of PVA models of arboreal marsupials. Biodiversity and Conservation 15:4079–4096

McCarthy MA, Possingham HP (2007) Active adaptive management for conservation. Conservation Biology 21:956–963

Meredith C (1997) Best practice in performance reporting in natural resource management. ANZECC Working Group on National Parks and Protected Area Management – Benchmarking and Best Practice Program. Department of Natural Resources and Environment, Port Melbourne, Victoria

NCCMA (2003) North Central regional catchment strategy 2003–2007. North Central Catchment Management Authority, Huntly, Victoria

NECMA (2003) North East Regional Catchment Strategy, North East Catchment Management Authority, Wodonga, Victoria

NECMA (2006a) Annual report 2005/2006. North East Catchment Management Authority, Wodonga, Victoria

NECMA (2006b) Catchment condition report 2005/2006. North East Catchment Management Authority, Wodonga, Victoria

Newell G, White M, Griffioen P, Conroy M (2006) Vegetation condition mapping at a landscape-scale across Victoria. Ecological Management and Restoration 7:S65–S68

Nichols JD, Williams BK (2006) Monitoring for conservation. Trends in Ecology and Evolution 21:668–673

NRMMC (2002a) National framework for natural resource management standards and targets. Natural Resource Management Ministerial Council, Commonwealth of Australia, Canberra

NRMMC (2002b) National natural resource management monitoring and evaluation framework. Natural Resource Management Ministerial Council, Commonwealth of Australia, Canberra

Parkes D, Lyon P (2006) Towards a national approach to vegetation condition assessment that meets government investors' needs: A policy perspective. Ecological Management and Restoration 7:S3–S5

Parkes D, Newell G, Cheal D (2003) Assessing the quality of native vegetation: The 'habitat hectares' approach. Ecological Management and Restoration, 4:S29–S38

Possingham HP (2001) The business of biodiversity: Applying decision theory principles to nature conservation. Australian Conservation Foundation, Melbourne

Robinson D, Mann S (1998) Effects of grazing, fencing and licensing on the natural values of Crown Land frontages in the Goulburn Broken catchment. Goulburn Valley Environment Group, Shepparton, Victoria

Smith FP (forthcoming) Who's planting what, where and why – and who's paying? An analysis of farmland revegetation in the central wheatbelt of Western Australia. Landscape and Urban Planning

Stankey GH, Bormann BT, Ryan C, Shindler B, Sturtevan V, Clark RN, Philpot C (2003) Adaptive management and the Northwest forest plan; rhetoric and reality. Journal of Forestry 101:40–46

Stankey GH, Clark RN, Bormann BT (2005) Adaptive management of natural resources: theory, concepts, and management institutions. US Department of Agriculture, Forest Service, Pacific Northwest Research Station, Portland

Tuchmann ET (1998) The Northwest forest plan: A Report to the President and Congress. DIANE Publishing, Washington

Underwood AJ (1994) On beyond BACI: sampling designs that might reliably detect environmental disturbances. Ecological Applications 4:3–15

USFWS (1999) Adaptive harvest management – 1999 duck-hunting season. US Fish and Wildlife Service, US Department of Interior, Washington, DC

VCMC (2007) Catchment condition report 2007. Victorian Catchment Management Council, Melbourne

Victorian Government (1994) Catchment and Land Protection Act 1994. No. 52, Parliament of Victoria

Walters C, Holling CS (1990) Large-scale management experiments and learning by doing. Ecology 71:2060–2068

WCMA (2007) Annual report 2006–2007. Wimmera Catchment Management Authority, Horsham, Victoria

Wintle BA, McCarthy MA, Volinsky CT, Kavanagh RP (2003) The use of Bayesian Model averaging to better represent the uncertainty in ecological models. Conservation Biology 17:1579–1590

Woodgate PW, Peel WD, Ritman KT, Coram JE, Brady A, Rule AJ, Banks JCG (1994) A study of the old growth forests of East Gippsland. Department of Conservation and Natural Resources, Melbourne, Victoria

Appendix

Using Bayes' Theorem to Assign Credibility to Competing Models with Monitoring Data: the Management of Mallard Ducks

Models that predict a system response to management actions are needed to optimise management decisions (Nichols and Williams 2006). Typically, multiple competing views (opinions, hypotheses) about how a system will respond to management exist and these views can be formalised as competing models. The plausibility of competing models may be assessed by comparing their predictions to data obtained from monitoring. In developing an adaptive management strategy for Mallard duck harvest, Johnson et al. (1997) describe a process of updating belief about the plausibility of competing models based on Bayes' theorem, such that the plausibility of a given model given the newly observed data (D) is:

$$\Pr(S_i \mid D) = \frac{\Pr(D \mid S_i)\Pr(S_i)}{\sum_{j=1}^{s}\Pr(D \mid S_j)\Pr(S_j)} \qquad (9.1)$$

where $\Pr(S_i|D)$ is known as the 'posterior probability' or 'weight' of model S_i (i.e. the degree of belief in S_i *after* considering the available data). $\Pr(D|S_i)$ is the likelihood that a given set of data would be observed if S_i were true, $\Pr(S_i)$ is the prior probability assigned to model S_i and the denominator represents the sum across the products of prior probabilities and likelihoods for all competing models including model S_i.

Models describing duck population responses to hunting pressure are central to the sustainable management of duck harvests. Managers of Mallard ducks use the equation above to iteratively update their belief in competing models as yearly monitoring data are collected (Johnson et al. 1997; USFWS 1999). Various scientists and stakeholders hold alternative views about how duck hunting impacts on duck population dynamics. Debate focuses on whether population growth would compensate for harvest mortality (compensatory mortality versus additive mortality) and whether reproductive success was strongly or weakly linked to habitat availability (strong versus weak density-dependence).

In developing an adaptive management system for duck hunting, competing views were summarised as four models of duck hunting population response (USFWS 1999):

- Model 1: additive mortality (am), strongly density-dependent recruitment (sdd)
- Model 2: additive mortality, weak density dependent recruitment (wdd)
- Model 3: compensatory mortality (cm), strongly density dependent recruitment
- Model 4: compensatory mortality, weakly density-dependent recruitment.

The implication of strong density dependent and compensatory hunting mortality is that higher hunting quotas may be sustainable. More conservative harvesting may be warranted if density-dependence is low and hunting mortality is not compensated by increased reproductive success and a reduction in other forms of mortality. Table 9.1 shows how model probabilities were updated with duck population monitoring data over the years 1995–1999. Note that prior to the collection of monitoring data in 1995, all models shared equal prior probability (i.e. $Pr(S_i) = 0.25$). As monitoring data were collected and compared against the predictions of the four competing models, it rapidly became apparent that the compensatory mortality hypothesis was not supported by the data. The data provided slightly more support for strong density-dependence than weak.

Table 9.1. Trends in probabilities for competing hypotheses of Mallard duck population dynamics taken from USFWS (1999)

Model[a]	1995	1996	1997	1998	1999
1: am, sdd	0.25	0.65	0.53	0.61	0.61
2: am, wdd	0.25	0.35	0.46	0.39	0.38
3: cm, sdd	0.25	0.00	0.00	0.00	0.00
4: cm, wdd	0.25	0.00	0.00	0.00	0.00

Note: model probabilities have been rounded to two decimal places. [a]am, sdd, cm, and wdd are defined in text above.

10 Revegetation and the Significance of Timelags in Provision of Habitat Resources for Birds

Peter A Vesk[1], Ralph Mac Nally[2], James R Thomson[2] and Gregory Horrocks[2]

[1] School of Botany, The University of Melbourne, Victoria, Australia
[2] Australian Centre for Biodiversity, School of Biological Sciences, Monash University, Victoria, Australia

Abstract: In many approaches to landscape visualisation and reconstruction for biodiversity management, vegetation is represented as being either present or absent. Revegetation is assumed to be possible, and new vegetation appears 'immediately' in a mature state, which is likely to drastically overestimate habitat suitability in the short-term. We constructed a simple temporal model of resource provision from revegetated agricultural land to estimate habitat suitability indices for woodland birds in south-eastern Australia. We used this model to illustrate the trajectory of change in biodiversity benefits of revegetation. As vegetation matures, its suitability for a given species changes, so a time-integrated assessment of habitat value is needed. Spatial allocation strategies, such as offsets, that may provide high value habitat in the long-term but imply shorter term population bottlenecks from a paucity of key resources (e.g. tree hollows) must be avoided. Given that vegetation may not meet both foraging and breeding requirements of a given species, populations may be limited continuously — by foraging constraints at some times, and by breeding constraints at other times. Animal species differ in their resource requirements so that optimisation involves compromises among species. Temporal processes associated with revegetation and differences in resource requirements of species complicate landscape reconstruction. Nevertheless, our analyses suggest that the time-course of vegetation development must be incorporated in models for optimising landscape reconstruction and for calculating revegetation offsets.

10.1 Introduction

Land use decisions rest upon models of system responses. These models may be implicit or explicit and their focus may range from social and economic to physicochemical and ecological. In this volume are many contributions emphasising spatial or visualisation approaches to modelling land use change with a view to decision support. These have tremendous potential as humans are visual creatures and landscapes are inherently spatial.

Envisioning the outcomes of land use decisions could have massive potential for supporting better decisions. Our point of departure in this chapter is to ask what revegetated landscapes might look like *for biodiversity* 20, 50, 100 or 200 years in the future. We don't know how other organisms see the world (Smith Chapter 7), but through habitat modelling we can try to project the suitability of a landscape for supporting persistence of one or more species.

Most of the contributions to this volume that are concerned with visualisation are necessarily simplistic in their treatment of the ecological aspects; areas are selected for revegetation, trees grow, animals return and thrive. In this contribution we emphasise the ecological detail and necessarily ignore spatial and visualisation aspects. Importantly, we point out how thinking about what species need in landscapes and the time course of change helps to understand the outcomes of land use decisions and to point to making effective decisions for improving the status of biodiversity in agricultural landscapes. In time, the ecological and visualisation approaches will inevitably become better integrated, for the benefit of understanding landscape change.

The limited extent of native vegetation in the 'wheat-sheep' regions of south-eastern and south-western Australia is insufficient to sustain native biodiversity (Barrett et al. 2003; Mac Nally 2007; Recher 1999; Robinson and Traill 1996), and extensive revegetation is needed (Mac Nally 2008; Saunders and Hobbs 1995; Vesk and Mac Nally 2006). While some progress toward broad-scale revegetation has been made, areas revegetated are orders of magnitude too small (Freudenberger et al. 2004; Huggett 2007). If we assume that responsible agencies must act to counter the potential meltdown of terrestrial biodiversity in these regions of southern Australia, we now must consider how to plan for 'landscape-scale' restoration, which we call *landscape reconstruction* (Bennett and Mac Nally 2004; Huxel and Hastings 1999; Lambeck 1997; Westphal and Possingham 2003; Westphal et al. 2007).

In most existing approaches to landscape reconstruction, vegetation is represented as a binary variable (present/absent), revegetation is assumed to be feasible, and new vegetation appears 'immediately' in a mature state (Mac Nally 2008). Timelags in reconstruction can translate to substantial declines of biota in fragments (Martínez-Garza and Howe 2003; Vesk and Mac Nally 2006). Explicitly factoring maturation time into restoration planning becomes imperative when we contemplate areas where regrowth of native vegetation is likely to be very slow (due to poor nutrients or little water) and where existing vegetation is senescing. These conditions are widespread in southern Australia, as well as many parts of Africa, Eurasia and North and South America. Empirical models of habitat resource provision demonstrate century-scale timelags in development of some key resources in Victorian woodlands, such as large boughs for nesting, tree hollows and fallen timber (Vesk et al. 2008).

Here, we recognise explicitly that timelags and species-specific habitat requirements strongly influence the success of landscape reconstruction for biodiversity conservation. Therefore, we examine the temporal dimension of habitat restoration through revegetation on potentially retirable agricultural land. We use a simple model of habitat suitability for woodland bird species defined by breeding and foraging resources provided by *Eucalyptus* woodland habitats during revegetation. We believe birds are a useful taxon for this approach because:

- birds respond to the amount and arrangement of vegetation at the scales of human activity, for example, clearing, retention and revegetation (10–10,000 ha) (Radford et al. 2005; Trzcinski et al. 1999)
- birds are the most species-rich vertebrate taxon in these regions
- knowledge of ecological requirements of bird species is rich
- data on species' distributions is extensive (Barrett et al. 2003; Radford and Bennett 2007; Thomson et al. 2007).

We define 'habitat' as a set of resources provided by vegetation (Dennis et al. 2003), focussing on vegetation structure through time rather than floristic succession. We construct a generalised 'Habitat Suitability Index' (O'Connell et al. 2000), which we parameterise for different species individually based on literature accounts of species ecology. We then use this model to identify temporal dynamics that are crucial for planning revegetation actions in box-ironbark (*Eucalyptus* spp.) woodlands of northern Victoria, Australia.

10.2 Methodology

Model details are extensive so we have gathered these together in the Appendices. This means that the minutiae will not detract from the general flow. We are generally concerned with the agricultural landscapes of southern Australia which are primarily woodlands and open forests dominated by eucalypt trees. Specifically, here we model the responses of grazed or cropped grassy box-ironbark forests and woodlands of northern Victoria (Lunt and Bennett 2000). These were extensively cleared for grazing and cropping over the past 200 years of European agriculture. More information about the study system can be found in Vesk and Mac Nally (2006), and about historical modification of vegetation structure in Lunt et al. (2006).

10.2.1 Model Description

As vegetation grows it provides required habitat resources for birds (Fig. 10.1). The better these requirements are satisfied by the resources produced by vegetation, the higher the suitability for the bird species. The model is focused on local resource provision through time. Spatial configuration was not considered, although clearly it is important for some species (Radford et al. 2005). The 'unit' modelled was a site of approximately 1 ha, which frequently is the area of revegetation sites (Freudenberger et al. 2004).

We modelled two cohorts of vegetation, an extant mature cohort and a cohort recruiting from revegetation actions on land formerly cleared for agriculture. Site conditions reflect management legacies through effects on soil structure (including compaction) and fertility (including sodicity and acidity), seed banks and weed loads. Recruitment of new vegetation results from two possibilities: active revegetation (direct seeding, planting tubestock) and from natural recruitment. These were combined into a recruiting cohort. Ongoing management ('business as usual', exclusion of stock) also modifies recruitment rate (Vesk and Dorrough 2006). The detailed model scenarios for sites and management are outlined in Appendix 10.1.

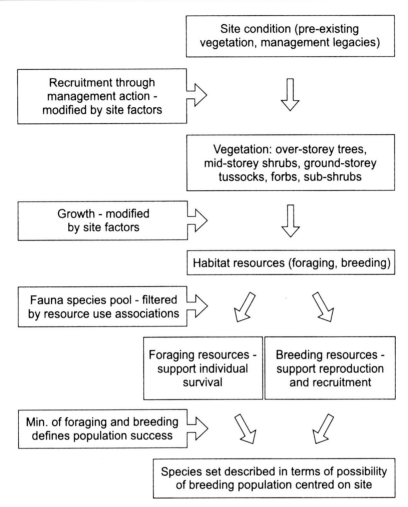

Fig. 10.1. Flow chart of model of structural dynamics of revegetation and suitability for bird species use

Vegetation was modelled as three layers: ground-storey, shrubs and trees. Each layer grows and provides consumable resources for birds, such as nectar and seeds, and structural resources, such as boughs and tree hollows for nesting (Fig. 10.1). The same resources could act as both foraging and breeding resources (e.g. fallen timber). Prey resources emerging from vegetation growth — invertebrates, small mammals and birds — also were modelled. Resources were depicted on a proportional scale and accumulated as simple ramp functions of time (delay, linear increase and maximum, Fig. 10.2). As we accumulate further empirical data on habitat resources, these functions can be replaced with ones resulting from statistical modelling (Vesk et al. 2008).

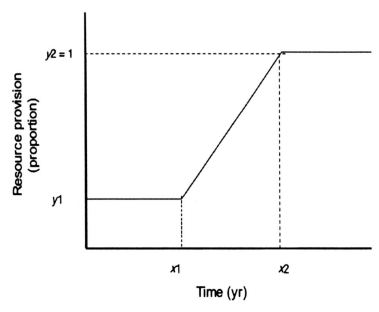

Fig. 10.2. Diagram of habitat resources through time provided by revegetation. Resources are provided at some initial level, $y1$ (usually 0), until time $x1$, when they increase linearly to a maximum, $y2$ (usually 1), at time $x2$, thereafter remaining at the maximum

Table 10.1 shows the modelled relationships for habitat resources supplied by new vegetation in the default case, equivalent to seeding of bare ground. The delay is equivalent to $x1$ in Fig. 10.1, time to maximum is time from $x1$ to $x2$ in Fig. 10.1. Maximum provision of resources is 1 for all resources; open ground decreases with vegetation growth. Initial provision of all resources is 0; open ground is initially 1.

Ground dwelling and aerial invertebrate abundances do not change with different vegetation and land use, despite compositional changes, and are maintained at 1.

Revegetation and Bird Resources   189

Table 10.1. Modelled relationships for habitat resources supplied by new vegetation in default case, equivalent to seeding of bare ground

Resource	Delay [yr]	Time to max [yr]	Slope of increase	Basis for relationship[a]
Ground-storey foraging				
Leaves, flowers, seeds	0	4	2.00	
Fallen timber	114	305	0.043	= small hollows
Soil and litter	0	10	0.797	2°, Σ(gl, ml, ol)
Open ground	1	24	-0.357	
Mid-storey foraging				
leaves, flowers, seeds	0	29	0.286	
Invertebrates	0	29	0.286	2°, (ml)
Over-storey foraging				
Leaves	0	77	0.106	
Flowers/seeds	29	257	0.036	
Canopy invertebrates	0	115	0.071	2°, Σ(ol, ofs)
Stem invertebrates				2°, Σ(ol, ob)
Higher prey				
Small ground vertebrates				2°, Σ(gf, gi, gsl, gl)
Medium arboreal verte-brates				2°, Σ(ol, ofs, ml, mf, oi)
Small insectivorous birds				2°, Σ(mi, ml, ol, of, oi)
Ground-storey breeding				
Tussocks	0	4	2.0	= gl
Fallen timber	114	305	0.043	= small hollows
Mid-storey breeding				
Shrub branches	0	29	0.286	= ml
Over-storey breeding				
Leaves and twigs	0	77	0.106	= ol
Tree boughs	43	233	0.043	
Stem bark and fissures	29	138	0.074	
Small hollows <10 cm	114	305	0.043	
Large hollows >10 cm	171	362	0.043	

[a] Relationship codes: g=ground-storey, m=mid-storey, o=over-storey, l=leaves, f=flowers/seeds, i=invertebrates, 2°=secondarily supplied from, Σ=sum of.

There are few data on vegetation growth rates and fewer for resource provision over time in box-ironbark forests and woodlands. We used diameter at breast height (*DBH*) data from empirical surveys of revegetation plantings (Vesk et al. 2008) to derive power relationships between age (*AGE*) and size (*DBH*):

$$\log(DBH) = 0.61[0.58, 0.67] \times \log(AGE) + 0.56[0.33, 0.72] \qquad (10.1)$$

where closed brackets define parameters for slow and fast growth sites respectively. We then used relationships between tree diameter and resource provision to predict resource provision over time. For example, an abundance of tree hollows has been related to tree diameter in several studies (Gibbons and Lindenmayer 2002). Floral production in several species of *Eucalyptus* is strongly, positively related to tree diameter (Wilson and Bennett 1999). Resources and their inferred dynamics are shown in Table 10.1. We made no distinction between resources allocated between few or many individuals. For example, few scattered large old trees may provide a moderate total number of hollows, while many younger trees may provide the same total number. We modelled two cohorts of vegetation: extant and revegetated (or recruiting). Available resources of the two cohorts were summed. Senescence dynamics in the existing vegetation is described in Appendix 10.2.

The frequency with which bird species use the resources was expressed proportionally. We coded resource use, u (0: never, 0.15: rarely or occasionally, 0.65: regularly, 1: most frequently or obligate) from autecological accounts in Marchant and Higgins (1990–2002) (see Appendix 10.3). These values for resource use frequency were derived using a set of simple rules (see Appendix 10.4).

For a given point in time (t) for species j and a set of resources r_i, we derived a 'sufficiency' (S_{jt}) as the summed product of the resources r_{it} provided by the vegetation and the frequency of use u_{ij} over the set of resources. This was constrained to a maximum of unity. For tractability, we assume that foraging resources are substitutable. Thus, sufficiency is defined by:

$$S_{jt} = \sum_i \left(u_{ij} \times r_{it} \right) \leq 1 \qquad (10.2)$$

Sufficiencies for foraging (S_{fjt}) and for breeding (S_{bjt}) were estimated separately. Foraging sufficiency was modified by known preferences of bird species for foraging in open or closed vegetation (Appendix 10.5). Nest site availability is described in Appendix 10.6. Summed foraging sufficiency S_{fjt} was multiplied by the foraging cover modifier, m_j.

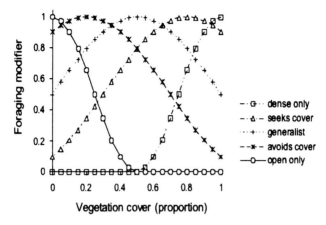

Fig. 10.3. Modification of foraging suitability by vegetation cover

A breeding population can occupy the site at a given time only if its minimum requirements for foraging and breeding resources are both provided. That is, we expect foraging and breeding resources to be non-substitutable. Hence, we used the minimum of the values for sufficiency of foraging and breeding resources to estimate the habitat suitability (*HS*) for a species *j* as:

$$HS_{jt} = \min(S_{fjt}, S_{bjt}) \quad (10.3)$$

The model is implemented as a series of tables in a *Microsoft Excel®* spreadsheet.

Three forms of pre-existing vegetation were used: over-grazed, transformed pastures; croplands; and lightly grazed, grassy woodlands. Cropping is thought to increase soil erosion and deplete soil seed banks, but increase soil nitrogen from fertilization (Prober et al. 2002). Over-grazing is thought to reduce soil fertility because of biomass off-take and to degrade soil structure via compaction by livestock (Prober et al. 2002).

10.3 Case Study

We consider the case of intensive revegetation on grazed grassy box-ironbark (*Eucalyptus* spp.) woodland of northern Victoria, Australia (Lunt and Bennett 2000). The landscape between 150 m and 350 m above sea level, mean annual rainfall ranges from 400 mm to 750 mm, and mean daily maximum temperatures vary from approximately 30°C in summer to 12–13°C in winter. We assumed the site initially had remnant, scattered, mature trees following clearing (approximately 1 to 2 trees/ha), but no juvenile trees or shrubs, as typical of the region. Ground-storey was assumed to be initially a mixture of native perennial and annual grasses and forbs and annual exotic grasses. We assumed intense restoration effort (Schirmer and Field 2000), that is, sites are ripped to improve soil structural characteristics, herbicides are applied, dominant native species in ground-storey, shrubs and trees are direct seeded or planted as tubestock, and management is ongoing (weed control, replacement plantings, thinning, maintenance of shrub and grass populations). Existing mature trees were assumed to die by 100 years from now (Appendix 10.2; Gibbons and Boak 2002). We modelled 133 bird species recorded during surveys of the box-ironbark woodlands (Mac Nally et al. 2000; Mac Nally and Horrocks 2002). Species nomenclature follows Marchant and Higgins (1990–2002).

10.3.1 Results

Bird Species Differ in Their Ability to Use Revegetation

We identified several broad patterns of bird species responses to vegetation maturation. Some species benefited from revegetation because both foraging and breeding resources were provided quickly by the maturation of shrubs and trees. These included the yellow thornbill and rufous whistler (Fig. 10.4a). For other species (e.g. striated pardalote and southern boobook Fig. 10.4b), foraging requirements were met soon after revegetation but breeding resources, such as tree hollows, developed only after long delays (approximately 150 years). Foraging resources for birds tolerant of open country, such as the Australian magpie and wedge-tailed eagle, were already satisfied, but the suitability of those resources declined as vegetation matured and cover increased (Fig. 10.4c). Availability of large boughs for nesting decreased as mature trees senesced, but gradually increased with growth of the new cohort of trees.

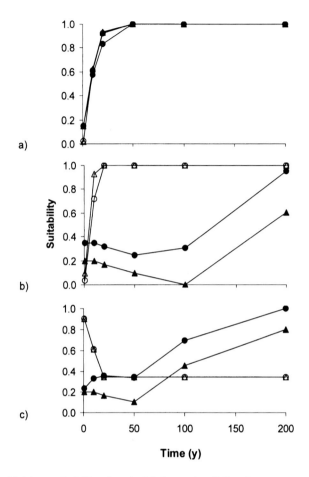

Fig. 10.4. Habitat suitability for six bird species following revegetation. Suitability of resources for foraging (open symbols) and for breeding (filled symbols). Examples are for a lightly grazed site with 1–2 trees/ha planted with tubestock. (**a**) Species for which suitability for both foraging and breeding initially is low and increases rapidly; yellow thornbill (△,▲), rufous whistler (○,●). (**b**) Species for which foraging resources initially are poor but increase rapidly, whereas breeding resources initially decline before a slow increase; southern boobook (△,▲), striated pardalote (○,●). (**c**) Species characteristic of open country for which foraging suitability initially is high but declines as vegetation matures and cover increases, whereas breeding resources gradually increase after an initial delay; wedge-tailed eagle (△,▲), Australian magpie (○,●)

Summaries over the whole avifauna show that foraging sufficiency increased rapidly for most species, but suitability increased more slowly for a small number of species (Fig. 10.5a). Breeding suitability was much slower to improve over the entire avifauna (Fig. 10.5b). The minimum of foraging and breeding broadly reflected the change in suitability for breeding (Fig. 10.5c).

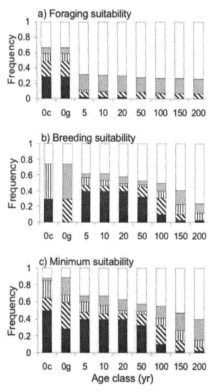

Fig. 10.5. Habitat suitability for 134 bird species of box-ironbark forests and woodlands over time after intensive revegetation on an overgrazed site, lacking shrubs and trees, summarised in classes. Suitability values classed as follows. Black: very low, S<0.2; diagonal hatching: low, $0.2 \leq S<0.4$; vertical hatching: medium, $0.4 \leq S<0.6$; grey: high, $0.6 \leq S<0.8$; white: very high, $0.8 \leq S<1.0$. Age classes 0c and 0g are cropped site and grazed grassy woodland, respectively

Representations of habitat suitability for selected functional groups demonstrated that expected suitability is more variable for breeding than for foraging, and the minimum suitability reflects suitability for breeding (Fig. 10.6a, 10.6c, 10.6e). This echoes the pattern in overall suitability and highlights the importance of breeding resources. Breeding suitability for hollow-nesting birds and frugivores is slow to increase, while suitability for canopy foragers and ground foragers improves more rapidly. The variation of breeding suitability within functional groups tends to be greater for groups defined by foraging than by groups defined by breeding (Fig.

10.6b), whereas foraging suitability is more variable within a group defined by breeding than those groups defined by foraging (Fig. 10.6d).

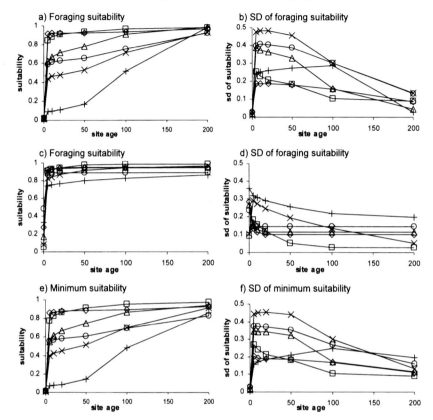

Fig. 10.6. Average suitability (**a, c, e**) and standard deviation of suitability (**b, d, f**) for breeding, foraging and their minimum for selected functional groups of 96 species of box-ironbark woodland birds. The modelled scenario was for intensive revegetation on an overgrazed site, lacking shrubs and trees. Symbols as follows: 41 species of ground foragers ◇; 26 species of small birds (<30 g) foraging amongst canopy □; 18 species of small birds that forage on bark △; 18 species of mid-storey foragers ○; 18 species of frugivores ×; 13 species of hollow-nesters +

Comparing Strategies over Time: Species that Require Old Trees

Here we consider the decision whether to actively revegetate a cropland site or to destock a lightly-grazed site (see also Dorrough et al. 2008). In cleared lands being grazed, there are few trees and therefore few large boughs and tree hollows; large boughs and tree hollows are absent from cropland. Many bird species nest in large boughs or tree hollows that develop after 50–150 years (see Fig. 10.6; Gibbons and Lindenmayer 2002). These birds are of conservation concern because boughs and hollows upon which they depend are few and declining, and quick replacement is unlikely. Availability of these resources declines as mature trees senesce, even were grazing livestock removed ('destocking', Fig. 10.7a), because of low recruitment and long timelags in resource development. The aim of planting trees is to replace senescing trees and supplement low rates of natural recruitment. However, breeding resources from revegetated cropland sites will not be available for a long time. A comparison of the provision of breeding resources for large-bough- and hollow-nesters by destocking a grazed site (cheap) against intensively planting a cropland site (expensive) is shown in Fig. 10.7a. Destocking a grazed site is superior over the first century before old trees senesce, but intensively planting a cropland site is superior after 100 years when the planted cohort matures (Fig. 10.7a).

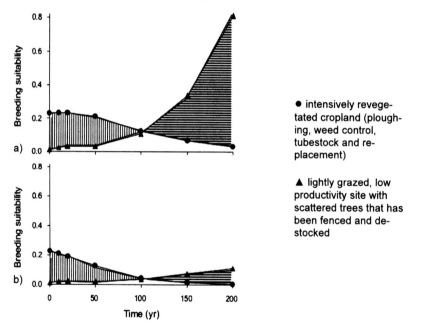

Fig. 10.7. Average breeding suitability for 26 bird species that nest in either large boughs or in tree hollows. (**a**) Integrating over time (the areas under the curves) demonstrates that revegetation on the broad-scale crop site has greater value (horizontal hatching) than destocking the grazed site (vertical hatching). (**b**) Discounting future value at 1% p.a. reduces the value of the planting on broad-scale cropland relative to vegetation on destocked grazing land

We believe that for assessing the value of vegetation, or for comparing revegetation alternatives, integration of resource suitability over time (e.g.

200 years) is better than calculating suitability at any single time (Gibbons and Lindenmayer 2007; Mac Nally 2008). A trajectory of habitat quality over time provides a curve, the area beneath which can be used to compare strategies (Fig. 10.7a). In this example, intensive replanting provides a greater time-integrated value than destocking alone (Fig. 10.7a). However, future value should be discounted because populations cannot take advantage of the resources unless they persist until the resources become available. Bottlenecks in the medium term may lead to local extinction irrespective of the quality of habitat in the more distant future. Using a future discount rate of 1% per annum, the distant future value of intensive planting is substantially decreased whereas the near future value of destocking is little changed (Fig. 10.7b) (Drechsler and Burgman 2004). Under this logic, destocking is preferred.

10.3.2 Discussion

When envisioning landscapes for natural resource management, vegetation is a surrogate for native biodiversity. In essence this requires that future landscapes support viable populations of species. For the general public, biodiversity concepts often translate most strongly to mammals and birds, rather than insects, for example. Species' responses to changing landscapes can be modelled through a rules-based approach as used here, with population demographic models, or statistical models of spatial distribution. Our work highlights the opportunity to use extensive existing knowledge on bird species' ecology together with a simple model of habitat suitability, that is, animals need to feed and to breed to maintain viable populations. This is appropriate when there is no sound reason to select one focal species over others.

Explicit population demographic modelling, which explicitly addresses population viability, requires intensive study to estimate demographic parameters, and thus is likely to be limited to one or few species. Statistical modelling of species distribution requires substantial surveys of species occurrence across the landscape and is limited in extrapolation beyond the sample population, that is, we can't project to 100-year-old revegetation if we have none in the landscape currently. Also, occurrence does not necessarily imply population viability. All these approaches are useful for projecting biodiversity benefits of revegetation and constraints thereupon, and all move us away from a simplistic conceptualisation where biodiversity restoration is assumed possible, instant and perfect. The challenge then is to integrate this into envisioning of spatial land use scenarios.

In this work, we have demonstrated how important ecological processes and principles can be when analysing land use change scenarios. Our model is heuristic so we would not have great confidence in the precision or accuracy of predictions about any particular species and scenario. However, by making reasonable assumptions about how vegetation develops, and how bird species use habitat resources based on textbook ecology, we have drawn out some implications of temporal change in vegetation resulting from revegetation activities that we believe are quite general. We see this as contributing to a future integration of spatial models of vegetation change and ecological responses into spatial scenario analyses.

We emphasise five main ecological points based on these analyses:

- First, as revegetation matures, availability of habitat resources increases and the proportion of open ground decreases, so suitability of revegetation for different native species changes through time.
- Second, species differ in responses to habitat change, and so, particular revegetation actions will favour species differentially.
- Third, at any given time, and at any given site, vegetation may not provide both foraging and breeding resources for a single species. If foraging resources are limited at some times, and breeding resources at others in the maturation sequence, the population may remain continuously limited (Fig. 10.4c).
- Fourth, time lags in development of some resources, such as peak floral production by eucalypt trees and tree-hollow formation, are considerable, in the order of decades to centuries (Gibbons and Lindenmayer 2002; Wilson and Bennett 1999; Vesk et al. 2008).
- Fifth, assessing suitability of revegetation for target species requires integration over time-to-maturation and perhaps should include consideration of future discounting, or some form of complex future benefit function, to avoid solutions that may be good in the distant future but that pass through a prior (deleterious) bottleneck.

As restored vegetation matures, its suitability shifts for different species of birds. This leads to management compromises between species. Summing of suitability values for many species can be unweighted, or weighted by species' conservation concern. Models based on species occurrences may lead to planning of revegetation that satisfies foraging needs but not breeding requirements, and so will allow survival of an individual but not population persistence (Thomson et al. 2007). Long time-lags and bottlenecks for crucial habitat resources, coupled with future-discounting, requires acceleration of the development of certain resources, such as hollows, or artificial supplements. The latter option may be very expensive or, in some cases, infeasible (Spring et al. 2001).

10.4 Caveats and Extensions

We have assumed successful revegetation, yet this is not certain. Many efforts are only partially successful and the ensuing suitability for birds would likely reflect that. Anecdotally, the least successful restoration is of the ground-storey. We have also assumed replacement of ground- and mid-storey species after senescence, which will be over relatively short terms (approximately 30 years). Replacement may require planned disturbance treatments to encourage recruitment. We also have not accounted for recruitment of a second generation of trees after the initial revegetation action. However, in practice it always is desirable to establish self-recruiting populations because subsequent management may not occur and a key aim should be sustainable systems. Observations suggest limited regeneration occurs in extant plantings (Vesk et al. 2008). We have not modelled disturbances that reset succession, such as wildfires. Although these disturbances would complicate the model, their omission is unlikely to invalidate our main conclusions.

The chosen future discount rates will greatly affect the preferred options (Drechsler and Burgman 2004). It seems necessary to avoid allocation strategies whereby revegetation actions that provide high value habitat in the long-term are proposed to offset loss of current habitat and which lead to short-term population bottlenecks from a paucity of key resources such as tree hollows. Moilanen et al. (in press) highlight the importance of future discounting in calculating habitat offsets.

There are several ways to extend this work. Our spreadsheet model needs to be migrated to computational GIS layers that can be combined with knowledge about individual species' spatial requirements (minimum areas, connections, spatial configurations of breeding and non-breeding habitats) to make better assessments of probable population viabilities through time. We currently are working towards this goal. Some compromises between foraging and breeding resources can be addressed in a spatially explicit model. For example, some birds may use a dense patch of vegetation for breeding and adjacent more open areas for foraging (Law and Dickman 1998). However, other birds and less mobile organisms, such as small mammals and reptiles, may have to use the same patch for foraging and breeding. The use of different patches of habitat implies that scheduling of revegetation activities on sites might be required to maintain a matrix of successional stages (Richards et al. 1999). Existing approaches to scheduling of silvicultural operations and fire management may be helpful (Bevers and Hof 1999; Hansen et al. 1993; Richards et al. 1999). Size and isolation of revegetated patches will clearly influence colonisation, population size and time to local extinction (Huxel and Hastings 1999; Lambeck 1997).

Another way forward would be to introduce a vegetation-dynamic element to the spatially efficient landscape reconstruction approach developed by Westphal and Possingham (2003). That work used statistical species-occurrence models based on landscape metrics coupled with estimates of bird densities to model population sizes (and hence persistence) in a reconstructed landscape (Westphal and Possingham 2003; Westphal et al. 2007). The species-occurrence probabilities for each patch could be modified by values for vegetation resources, which in turn, would be dependent on time. Desirable landscapes would have high values for time integrated (future discounted), landscape-level bird populations aggregated across species.

Further extension would be to account for interspecific interactions such as predation and competition, which are known to have strong effects on bird community composition. Given that habitat restoration involves decade- and century-scale ecological processes, we must also consider climate change (Hobbs and Hopkins 1990), which will shift climatic envelopes of plants and animals across the landscape (Root et al. 2003). In 50 years, the climatic envelopes of many species of *Eucalyptus* — the dominant trees in south-eastern Australian woodlands — may not overlap with their current distributions (Hughes 2003; Hughes et al. 1996). Future climates should be considered when selecting long-lived species for revegetation.

In conclusion, we see much scope for the development of spatial analysis and visualisation tools that incorporate ecological processes. This will enable planners and stakeholders to make better informed decisions about land use to accommodate social, economic and ecological concerns. We hope that this chapter contributes to this goal.

Acknowledgements

We appreciate the contributions of Josh Dorrough and Erica Fleishman to the development of ideas presented here and for comments on drafts, for which we also thank David Duncan and an anonymous reviewer. We acknowledge funding from the Monash University Research Fund: New Research Areas and from the Australian Research Council LP 0560518 with co-funding from the Victorian Department of Sustainability and Environment, Victorian Department of Primary Industries, North Central CMA, Goulburn-Broken CMA, Mallee CMA and North East CMA, and contributes to the Key Project 1.1 under the Our Rural Landscapes Initiative of the Victorian Department of Primary Industries. This is contribution number 57 from the Australian Centre for Biodiversity: Analysis, Policy and Management.

References

Barrett G, Silcocks A, Barry S, Cunningham R, Poulter R (2003) The new atlas of Australian birds. Birds Australia (Royal Australian Ornithologists Union), Melbourne

Bennett AF, R Mac Nally (2004) Identifying priority areas for conservation action in agricultural landscapes. Pacific Conservation Biology 10:106–123

Bevers M, Hof J (1999) Spatially optimising wildlife habitat edge effects in forest management linear and mixed integer programs. Forest Science 45:249–258

Blakers M, Davies SJJF, Reilly PN (1984) Atlas of Australian birds. Melbourne University Press, Melbourne

Boland DJ, Brooker MIH, Chippendale GM, Hall N, Hyland BPM, Johnston RD, Kleinig DA, Turner JD (1984) Forest trees of Australia, 4th edn. Nelson/CSIRO, Melbourne

Dennis RLH, Shreeve TG, Van Dyck H (2003) Towards a functional resource-based concept for habitat: a butterfly biology viewpoint. Oikos 102:417–426

Dorrough J, Vesk PA, Moll J (2008) Integrating ecological uncertainty and farm-scale economics when planning restoration. Journal of Applied Ecology 45:288–295

Drechsler M, Burgman MA (2004) Combining population viability analysis with decision analysis. Biodiversity and Conservation 13:115–139

Emison W, Bennett S, Beardsell C, Norman F, Loyn R (1987) Atlas of Victorian birds. Birds Australia, Melbourne

Freudenberger D, Harvey J, Drew A (2004) Predicting the biodiversity benefits of the Saltshaker Project, Boorowa, NSW. Ecological Management and Restoration 5:5–14

Gibbons P, Boak M (2002) The value of paddock trees for regional conservation in an agricultural landscape. Ecological Management and Restoration 3:205–210

Gibbons P, Lindenmayer DB (2002) Tree hollows and wildlife conservation in Australia. CSIRO Publishing, Melbourne

Gibbons P, Lindenmayer DB (2007) Offsets for land clearing: No net loss or the tail wagging the dog? Ecological Management and Restoration 8:26–31

Hansen AJ, Garman SL, Marks B, Urban DL (1993) An approach for managing vertebrate diversity across multiple use landscapes. Ecological Applications 3:481–496

Hobbs RJ, Hopkins AJM (1990) From frontier to fragmentation: European impact on Australia's vegetation. In: Saunders D, Hobbs RJ, Hopkins AJM (eds) Australian ecosystems: 200 years of utilisation, degradation and reconstruction. Ecological Society of Australia, Chipping Norton, NSW, pp 93–114

Huggett A (2007) A review of the focal species approach in Australia. Land and Water Australia, Canberra

Hughes L (2003) Climate change and Australia: trends, projections and impacts. Austral Ecology 28:423–443

Hughes L, Cawsey EM, Westoby M (1996) Climatic range sizes of Eucalyptus species in relation to future climate change. Global Ecology and Biogeography Letters 5:23–29

Huxel GR, Hastings A (1999) Habitat loss, fragmentation, and restoration. Restoration Ecology 7:309–315

Lambeck RJ (1997) Focal species: a multi-species umbrella for nature conservation. Conservation Biology 11:849–856

Law BS, Dickman CR (1998) The use of habitat mosaics by terrestrial vertebrate fauna: implications for conservation and management. Biodiversity and Conservation 7:323–333

Lunt ID, Bennett AF (2000) Temperate woodlands in Victoria: distribution, composition and conservation. In: Hobbs RJ, Yates CY (eds) Temperate Eucalypt woodlands in Australia: biology, conservation, management and restoration. Surrey Beatty and Sons, Chipping Norton, NSW, pp 17–31

Lunt ID, Jones N, Spooner PG, Petrow M (2006) Effects of European colonisation on indigenous ecosystems: post-settlement changes in tree stand structures in Eucalyptus-Callitris woodlands in central New South Wales, Australia. Journal of Biogeography 33:1102–1115

Mac Nally R (2007) Use of the abundance spectrum and relative-abundance distributions to analyze assemblage change in massively altered landscapes. American Naturalist 170:319–330

Mac Nally R (2008) The lag dæmon: hysteresis in rebuilding landscapes and implications for biodiversity futures. Journal of Environmental Management doi:10.1016/j.jenvman.2007.1006.1004

Mac Nally R, Horrocks G (2002) Relative influences of patch, landscape and historical factors on birds in an Australian fragmented landscape. Journal of Biogeography 29:395–410

Mac Nally RC, Bennett AF, Horrocks G (2000) Forecasting the impacts of habitat fragmentation. Evaluation of species-specific predictions of the impact of habitat fragmentation on birds in the box-ironbark forests of central Victoria. Australia. Biological Conservation 95:7–29

Marchant S, Higgins PJ (1990–2002) Handbook of Australian, New Zealand and Antarctic birds. Oxford University Press, Melbourne

Martínez-Garza C, Howe HF (2003) Restoring tropical diversity: beating the time tax on species loss. Journal of Applied Ecology 40:423–429

Moilanen A, van Teeffelen A, Ben-Haim Y, Ferrier S (in press) How much compensation is enough? Explicit incorporation of uncertainty and time discounting when calculating offset ratios for impacted habitat. Restoration Ecology

O'Connell TJ, Jackson LE, Brooks RP (2000) Bird guilds as indicators of ecological condition in the central Appalachians. Ecological Applications 10:1706–21

Pizzey G, Doyle R (1980) A field guide to the birds of Australia. Collins, Sydney

Prober SM, Thiele KR, Lunt ID (2002) Identifying ecological barriers to restoration in temperate grassy woodlands: soil changes associated with different degradation states. Australian Journal of Botany 50:699–712

Radford JQ, Bennett AF (2007) The relative importance of landscape properties for woodland birds in agricultural environments. Journal of Applied Ecology 44:737–47

Radford JQ, Bennett AF, Cheers GJ (2005) Landscape-level thresholds of habitat cover for woodland-dependent birds. Biological Conservation 124:317–37

Recher HF (1999) The state of Australia's avifauna: a personal opinion and prediction for the new millennium. Australian Zoologist 31:11–27

Richards SA, Possingham HP, Tizard J (1999) Optimal fire management for maintaining community diversity. Ecological Applications 9:880–892

Robinson D, Traill BJ (1996) Conserving woodland birds in the wheat and sheep belts of southern Australia. Royal Australasian Ornitholoists Union Conservation Statement 10, Supplement to Wingspan 6(2), Royal Australian Ornithological Union, Melbourne, p 16

Root TL, Price JT, Hall KR, Schneider SH, Rosenzweig C, Pounds JA (2003) Fingerprints of global warming on wild animals and plants. Nature 421:57–60

Spring DA, Bevers M, Kennedy JOS, Harley D (2001) Economics of a nest-box program for the conservation of an endangered species: A reappraisal. Canadian Journal of Forest Research 31:1992–2003

Schirmer J, Field J (2000). The cost of revegetation, final report to Environment Australia. Australian National University Forestry, and Greening Australia, Canberra

Saunders DA, Hobbs RJ (1995) Habitat reconstruction: the revegetation imperative. In: Bradstock RA, Auld TD, Keith DA, Kingsford RT, Lunney D, Siversten DP (eds), Conserving biodiversity; threats and solutions. Surrey Beatty and Sons, Chipping Norton, NSW, pp 104–112

Thomson JR, Mac Nally R, Fleishman E, Horrocks G (2007) Predicting bird species distributions in reconstructed landscapes. Conservation Biology 21:752–66

Trzcinski MK, Fahrig L, Merriam G (1999) Independent effects of forest cover and fragmentation on the distribution of forest breeding birds. Ecological Applications 9:586–593

Vesk PA, Dorrough JW (2006) Getting trees on farms the easy way? Lessons from a model of eucalypt regeneration on pastures. Australian Journal of Botany 54:509–19

Vesk PA, Mac Nally R (2006) Changes in vegetation structure and distribution in rural landscapes: implications for biodiversity and ecosystem processes. Agriculture, Ecosystems and Environment 112:356–366

Vesk PA, Nolan R, Thomson JR, Dorrough JW, Mac Nally R (2008) Time lags in provision of habitat resources through revegetation. Biological Conservation 141:174–186

Watson J, Freudenberger D, Paull D (2001) An assessment of the focal-species approach for conserving birds in variegated landscapes in south-eastern Australia. Conservation Biology 15:364–1373

Westphal MI, Possingham HP (2003) Applying a decision-theory framework to landscape planning for biodiversity: follow-up to Watson et al. Conservation Biology 17:327–329

Westphal MI, Field SA, Possingham HP (2007) Optimising landscape configuration: a case study of woodland birds in the Mount Lofty Ranges, South Australia. Landscape and Urban Planning 81:6–66

Wilson J, Bennett AF (1999) Patchiness of a floral resource: Flowering of red ironbark *E. tricarpa* in a box and ironbark forest. Victorian Naturalist 116:48–53

Woodgate P, Black P (1988) Forest cover changes in Victoria 1869–1987. Victorian Department of Conservation, Forests and Lands, Melbourne

Appendices

Appendix 10.1: Scenarios for Sites and Management Activities

Business-as-usual Scenario

A good condition site assumes 'moderate' grazing history. Initial vegetation: 50% of native ground-storey vegetation, 20% of scattered mature shrubs and trees. Propagule supply is proportional to density, but recruitment, and thus rate of increase, is halved by ongoing grazing. Potential maximum native vegetation components are limited by ongoing grazing and weed competition: 50% in ground-storey and 20% in each of shrub- and tree-storeys.

Cropped sites have no native vegetation. Propagule supply from distant vegetation is 5%, as is recruitment. Maximum native vegetation for each stratum is 0%. Land is subjected to ongoing cropping, removing all native vegetation periodically.

Grazed sites have no native shrubs or trees, but have 20% native ground-storey. Propagule supply is proportional to vegetative density, with 10% of possible tree seeds (assuming well dispersed). Rate of increase (recruitment) is very low, 0.1%, due to continuing intense grazing. Potential maximum native vegetation components are limited by ongoing grazing and weed competition: 20% in ground-storey and 5% in each of shrub- and tree-storeys.

Fencing and Destocking

Fencing and destocking on a good site result in checks of recruitment being removed, so rate of increase is 50% in ground-storey and 20% in each of shrub- and tree-storeys. Fencing and destocking a cropped site results in the cap on native vegetation being relaxed, allowing up to 50% native ground-storey and 80% of shrub- and tree-storeys. Weeds comprise the remainder. Propagule supply still is minimal, as is resultant rate of increase, 5% each stratum. Destocking benefits grazed sites and the rate of increase climbs to 20% in ground-storey, shrubs are unchanged as few propagules arrive and none are present in the seedbank. Trees increase at 10%, matching propagule supply of small seeds.

Direct Seeding

Maximum propagule supply is achieved, as is rate of increase on a good site. On a cropped site, although maximum propagules are supplied, recruitment is reduced by competition with weeds. Rate of increase is 50% in native ground-storey and 80% in shrub and tree strata. Maximum native vegetation cover is capped at 50% ground-storey and 80% shrub and tree strata by weeds abetted by residual fertilizers. On grazed sites, direct seeding saturates propagule supply, but rate of increase is reduced by poor soil structure and fertility due to years of grazer hooves and production off-take (Prober et al. 2002).

Planting Tubestock after Site Preparation and Follow-up Weed Control

Good sites have saturated propagule supply, achieve maximum rate of increase and have enhanced initial vegetation due to planting advanced seedlings: 60% ground-storey, 30% shrub- and 25% tree-storeys. Weed control removes cap on native vegetation in cropped sites, maximum growth rate is achieved, and planting advanced seedling enhances the initial vegetation: 20% ground-storey, 10% shrubs and 5% trees. Grazed sites have soil structure improved by ripping and spot fertilization under planting. So maximum growth rate is achieved and planting tubestock enhances initial vegetation: 25% ground-storey, 10% shrubs and 5% trees.

Appendix 10.2: Tree Senescence

We follow the reasoning of Gibbons and Boak (2002). Extensive clearing (tree and shrub removal) of woodlands and forests of south-eastern Australia occurred between approximately 1850 and 1900 (Woodgate and Black 1988). We assume that scattered trees remaining in paddocks today were retained because of their value as shade for livestock (Gibbons and Boak 2002). This would mean that the trees were large and mature ($\geq$ 100 years old) when clearing was occurring (1850–1900, at latest). This would result in trees of 200 to 400 years of age in 2000. If we also assume that the lifespan of *Eucalyptus* spp. trees is approximately 400 to 500 years (Boland et al. 1984), then all trees would be expected to die between 2000 and 2300. Most trees would be dead by 2150. Gibbons and Boak (2002) estimated annual mortality rate of paddock trees from a number of studies was 0.54–2.5%. We assume a mid-range value loss of 1.5% of trees per year. For scattered trees at 20% of capacity, this means that decline to 5% of original density will occur by 2090. Even where scattered trees at 50% of tree capacity, a fall to 5% occurs by 2150. Taking these two approaches, we assume that trees will die at 2150.

For shrub- and ground-storeys, we assume that management has established a new equilibrium, that is, population and structure are maintained at current levels by ongoing recruitment.

Appendix 10.3: Bird Species' Resource Use

A list of bird species and their resource use compiled from autecological accounts in the Handbook of Australian, New Zealand and Antarctic Birds (Marchant and Higgins 1990–2002) is available from the senior author upon request. Volume 7 (Yellow-breasted Boatbill to Starlings) was unavailable, so for some *Passerinidae* (fam. *Artarmidae, Pardalotidae*), accounts in field guides and atlases were used (Blakers et al. 1984; Emison et al. 1987; Pizzey and Doyle 1980).

Appendix 10.4: Weights

Resource use sufficiency, S_j, for species j was estimated using the formula:

$$S_{jt} = \sum_i \left(u_{ij} \times r_{it} \right) \tag{10.4}$$

for resources $r_{i=1,...,n}$ with resource use frequency, u_i, at time t. Values for the u_i were derived by first classifying resource use frequency from bird species' autecological accounts, such that a = obligate or most frequently used, b = used regularly, c = used occasionally or rarely. Then, values of a = 1, b = 0.65, c = 0.15 were derived such that the following conditions were met (Table 10.2).

Table 10.2. Rules for determining weightings for resources

Verbal rule (to be read as 'when the resource/s is/are maximally abundant ...')	Equation
one most frequently used or obligate resource should suffice	$a_i \times r_i = 1$, $r_1 = 1$
two regular resources should suffice	$(b_1 \times r_1) + (b_2 \times r_2) \geq 1$, $r_1 = r_2 = 1$
one regular resource and one-half abundance of a second regular resource should suffice	$(b_1 \times r_1) + (b_2 \times r_2) \geq 1$, $r_1 = 1, r_2 = 0.5$
less than four occasional resources should not suffice	$\sum_{i=1}^{3} c_i \times r_i < 1$, $r_1 = r_2 = r_3 = 1$
ten occasional resources should suffice	$\sum_{i=1}^{10} c_i \times r_i \geq 1$, $r_1 = ... = r_{10} = 1$
One regular and two occasional resources should almost suffice	$0.7 \leq (b_1 \times r_1) + (c_2 \times r_2) + (c_3 \times r_3) \leq 0.95$, $r_1 = r_2 = r_3 = 1$
One regular and one occasional resource should not suffice	$(b_1 \times r_1) + (c_2 \times r_2) < 1$, $r_1 = r_2 = 1$
One regular should be better than three occasional resources	$(b_1 \times r_1) > \sum_{i=2}^{4} (c_i \times r_i)$, $r_1 = r_2 = r_3 = r_4 = 1$
Five occasional resources should be at least as good as one regular resource	$(b_1 \times r_1) \leq \sum_{i=2}^{6} (c_i \times r_i)$, $r_1 = ... = r_6 = 1$

Appendix 10.5: Effects of Cover on Foraging Resource Use

Because birds display varied preferences for foraging in different cover environments, the availability of a resources to a bird should be modified accordingly. Bird species were categorised according to foraging responses to shrub and tree cover. We considered whether they require vertical structure, like perches, from which to pounce. Effect of cover on foraging was assigned by Peter Vesk using data from abovementioned literature sources, and Gregory Horrocks using expert opinion. Conflicts were resolved by re-examining bases for the categorisation. If conflict remained and assignments differed by two categories, the middle category was used. If assignments differed by one category, the category closer to the middle (occasional use of open areas) was used. Sine curves were used to characterise foraging responses to vegetation cover. Open areas refer to open grassy areas or bare ground with patchy ground cover (> 100 m x 100 m). Dense cover ($C \sim 1.0$) refers to areas of dense shrub and or tree cover (e.g. closed heath and closed forest, >70% foliage projective cover). Summed foraging sufficiency was multiplied by the foraging cover modifier, m. These functions appear in Fig 10.3.

Table 10.3. Formulae used to modify foraging suitability due to vegetation cover

Foraging response to mid- and over-storey vegetation cover	Formula for foraging modifier m to account for cover C
Forages exclusively in open areas or bare ground with patchy grass.	If $C > 0.5$ then $m = 0$, otherwise $$m = \frac{1}{2}\left[\sin \pi(2C + \frac{1}{2}) + 1\right]$$
Regularly forages in open areas but requires some over-storey structure.	$$m = \frac{1}{2}\left[\sin \pi C + 1\right]$$
Occasionally forages in open areas and in dense cover but neither actively avoids nor actively seeks those areas.	$$m = \frac{1}{2}\left[\sin \pi(C - 0.3) + 1\right]$$
Actively avoids open areas.	$$m = \frac{1}{2}\left[\sin \pi(C + 0.3) + 1\right]$$
Forages exclusively in dense cover. If sighted in open areas, only crossing to suitable habitat.	If $C < 0.5$ then $m = 0$, otherwise $$m = \frac{1}{2}\left[\sin \pi(2C + \frac{1}{2}) + 1\right]$$

Appendix 10.6: Nesting Requirements

Ground Nesting

For species that nest on the ground in open areas, increases in vegetation cover reduce nest-site availability. Otherwise, ground nesting sites are assumed to be unlimited. For species that nest in earthen banks, the model assumes there is a bank nearby.

For species that nest on the ground but often under/next to/among tussocks/logs/trees, then nest site availability is not to be determined solely by ground (i.e. unlimited) but by the supplementary resource too. We use 0.65 contribution from ground, which means 0.5 of the supporting resource will suffice, but that 0.25 of the supporting resource will not suffice.

Some species that nest on or near the ground respond either positively or negatively to vegetation cover. Therefore, we used a modifier for the summed nesting sufficiency from ground, tussocks and fallen timber. Species were categorised for their use of open ground nest sites:

- restricted to very open sites (<10% mid- and tree-storey), $g = 1$.
- nesting is favoured by some mid- and tree-storeys but is less likely in <10%, less likely at >70% cover, $g = 0.5$.
- nesting is reduced at <70% cover and is best at high cover (> 70%), $g = 0$
- unaffected, not ground nester.

The following formula was used to modify nesting sufficiency for ground nesting birds by the amount of vegetation cover:

$$O = 1 - 1.8 \times |1 - C - g| \qquad (10.5)$$

where O is the open ground modifier for nesting, C is cover, g is the value appropriate for the class of use of open ground nest sites.

Cuckoos

Cuckoos should be not be limited by nest sites providing hosts are present. Hosts are diverse, but particular cuckoos have specific suites of host species:

- pallid and fan-tailed cuckoos parasitise honeyeaters
- Horsefield's bronze, shining bronze and black-eared cuckoos parasitise thornbills, fairy wrens and speckled warblers.

11 The Application of Genetic Markers to Landscape Management

Paul Sunnucks and Andrea C Taylor

School of Biological Sciences and Australian Centre for Biodiversity, Monash University, Victoria, Australia

Abstract: There is great concern about how landscape change will affect the persistence of native biota, and the services they provide to human wellbeing. Of fundamental concern is the effect on population processes — birth and death rates, migration and genetic exchange. Molecular ecology and landscape genetics make contributions to the analysis and monitoring of landscape change that are otherwise inaccessible. They can yield powerful and unambiguous information about where individuals and species move, with whom they reproduce, and what landscape features and arrangements impact their movement. These data can feed into modelling of population persistence. It is a common misconception that genetic signatures of landscape change take many generations to be detectable. Recent methods detect changes in population processes over very short timeframes, such as years to decades. It is surprising to many non-geneticists that these methods are: (a) available now and wholeheartedly applied by management agencies elsewhere in the world, and (b) highly cost-effective. We summarise the application of selectively neutral genetic markers in landscape management.

11.1 Introduction

11.1.1 The Need for Information on How Biota Occupies and Moves through Landscapes

Following massive habitat loss and degradation, conservation and restoration of native biodiversity in natural and production landscapes is a priority in Australia, as in much of the world (Young and Clarke 2000). Retention or re-establishment of natural levels of connectivity among groups of organisms is likely to improve survival probability, leading to retention of genetic and species diversity, and thus ecosystem function, resilience and persistence (Frankham 2005; Hansson 1991). However, for most organisms there is little information on dispersal abilities, how these relate to features and structure of the landscape, and how these are likely to change with ongoing environmental and climate change.

Most studies of the pattern of occupation of habitat are conducted at relatively small scale; however, increasingly strong arguments are being made that we need to characterise biotic responses at the level of landscapes (Bennett et al. 2006; Radford et al. 2005). Vital information on biotic mobility is labour-intensive and expensive to collect by traditional field approaches, and thus its inclusion in landscape restoration and management has been problematic. This challenge is intensified by the realisation of the critical importance of rare migration events in modelling persistence (Davies and Melbourne 2007; Hoehn et al. 2007). If we add to this the fact that most biotic surveys are 'snap-shots' that do not capture the change of patterns of mobility and demographic trajectories, and that these may be changed under future environments, understanding the relationship between structural (landscape features) and functional (demographically relevant movements of individuals) connectivity seems an even greater challenge.

To improve our knowledge of how species respond to landscape change, we need to augment currently available descriptions of *patterns* with new understanding of *processes* — movement of individuals, gene flow and population trends — in landscapes with different spatial characteristics. For example, a particularly important aspect of modelling population persistence and planning landscapes is an understanding of dispersal and colonisation abilities of organisms. Such information is extremely difficult to obtain by classical methods, yet is critical input into models (Debain et al. 2007; Epps et al. 2007; McIntire et al. 2007). As we outline, molecular population biology can supply such information. Nonetheless, to date rather few studies have incorporated genetic estimates of dispersal into models of population persistence, despite the potentially extremely

The Application of Genetic Markers to Landscape Management   213

positive outcomes of doing so in terms of evaluating functional connectivity of landscapes and enabling future planning (Epps et al. 2007).

11.1.2 A Spectrum of 'Genetics' in Landscape Management and Planning

Functional genetics/evolution and molecular ecology are at either end of the spectrum of genetic approaches which can contribute to landscape planning and management. Understanding the genetics of key functions can be critical to understanding natural distributions and abilities of organisms for inclusion in models of demographic trajectories and projected persistence of populations.

Such an approach has been applied to management of the grassland forb *Rutidosis leptorrhynchoides* (Young et al. 2000). Characterisation of the genetic self-incompatibility system and genetic dissection of the mating system was essential for understanding the relationship between habitat fragmentation, effective population size and declining demographic performance. Functional and evolutionary genetics can also provide the most meaningful insights into predicting how organisms will respond to changing environments. For example, there have been rapid advances in appreciation of the speed of adaptation of organisms to warming climates and the limits of those changes (Kellerman et al. 2006; Reusch and Wood 2007; van Heerwaarden and Hoffmann 2007), and invasiveness may be strongly linked to simple genetic traits (e.g. in *Solenopsis* fire ants, Krieger 2005). More broadly, negative impacts on genetic integrity are strongly implicated in extinction risk and ecosystem resilience after environmental perturbation (Frankham 2005; Reusch et al. 2005; Spielman et al. 2004).

At the other end of the spectrum is molecular population biology ('molecular ecology' and 'landscape genetics', see below), wherein DNA that is predominantly not subject to natural selection is used to estimate individual and population attributes and phenomena. This latter end of the spectrum is the main focus of this article.

11.1.3 Molecular Population Biology Supplies Information Essential for Landscape Planning and Management

The advent of powerful and innovative genetic analyses based on highly-resolving neutral genetic markers means that functional connectivity through landscapes and population processes (e.g. mating systems, dispersal, kin associations) of organisms at various temporal and spatial scales can now be assessed routinely and cost-effectively. This field, molecular population biology (often referred to as 'molecular ecology' or 'landscape genetics'), has matured in the last decade or two, owing to key discoveries and innovations that have allowed synthesis of theory, molecular population genetics, numerical analyses and increasingly accessible software. A full review is beyond the scope of this article, but readable accounts can be found in the ecological and general biological literature (Banks and Taylor 2004; DeSalle and Amato 2004; DeYoung and Honeycutt 2005; Manel et al. 2003, 2005; Pearse and Crandall 2004; Selkoe and Toonen 2006; Sunnucks 2000), and several recent books on molecular ecology and conservation genetics, such as Allendorf and Luikart (2006). A recent special issue of *Landscape Ecology* (see Holderegger and Wagner 2006), focussed on landscape genetics which potentially informed a different audience of the kind of research that is routinely published in more genetics-orientated biology journals such *Molecular Ecology, Animal Conservation, Proceedings of the Royal Society Series B, Conservation Genetics,* and *Evolution.*

The effectiveness of molecular population biology in supplying information that is otherwise prohibitively expensive, too difficult or impossible to obtain, is reflected in its strong uptake by North American and European land management and wildlife agencies. Its use in wildlife and landscape management is mainstream — many dozens of significant private and public sector molecular ecology/conservation genetics laboratories in the USA (including the US Fish and Wildlife Service, and the US Geological Survey) are involved in relevant work. Applications span an enormous range including:

- making a census of thinly-distributed animal species without the need for captures
- understanding population organisation and dispersal
- identifying management units
- planning landscape development and reconstruction
- enhancing knowledge of species diversity and relationships.

Although 'non-invasive' sampling (e.g. collection of faeces, hair or feathers) may be the most feasible method for collection of sufficient molecular data for some rare and/or cryptic species (Piggott and Taylor 2003), where possible, sampling should be via capture of individuals. The individual and population life history data (e.g. age and condition) routinely obtained during capture increases the inferential power of incorporating genetic approaches. By the same token, molecular population biology analyses add substantial value to information gained during biotic surveys, but the full potential of this may not be realised unless a culture of routine genetic sampling is developed.

Our objective here is to summarise molecular ecology with application in landscape planning and management, with particular emphasis on population structure, dispersal and gene flow. Some tasks and questions directly associated with this are outlined in the Appendix, with indications of the sorts of analyses that might be applied. We also provide some illustrative case studies from our own recent research, and conclude with some suggestions for increasing the uptake of molecular population biology analysis in Australian landscape management.

While we focus on population structure, dispersal and gene flow in this chapter, that is not to underestimate the importance of the many other applications of molecular ecology (see Tables 11.1–11.4 in Appendix), including delineation of taxonomic units for management consideration, monitoring of individual species and estimation of population size. For example, recent genetic estimates of historic population sizes of several whale species and under-reporting of minke whale by-catch demand a complete rethinking of management of cetacean populations (Alter et al. 2007; Baker et al. 2007).

11.2 Background

Molecular population biology takes advantage of information encoded in the DNA of individual organisms reflecting their identity and that of their parents, peers, or more distantly related individuals including ancestral species (Avise 2004). As DNA passes through this hierarchy, from parent to offspring to grand-offspring and so on, it accumulates differences that demonstrate links between individuals in the chain of ancestry. These differences in genetic 'markers' are detected by molecular genetic assays, and can be analysed numerically in three major ways, each reflecting different spatial and temporal scales (Sunnucks 2000). Fuller details of the approaches highlighted in this section are given in the references above.

11.2.1 Three Levels of Analysis Assess Three Levels in Time and Space

There are three levels of analysis, providing information at different levels in time and space:

- genotypic analyses provides information at the finest spatial and shortest temporal (contemporary) scales

- genic analyses provides information at medium spatial and recent past temporal scales
- genealogical analyses provides information at large spatial and distant past temporal scales.

Genotypic Analysis

Genotypic analyses (sensu Sunnucks 2000) provide information at the finest spatial and shortest temporal scales. They are based on DNA 'signatures' of individual organisms that are reshuffled in every generation of a sexual species. For most animals and many plants, these signatures are typically investigated via a suite of one or two dozen different highly-resolving genetic markers of a type called microsatellites (reviewed by Selkoe and Toonen 2006). Taken together, the information from a sufficient array of microsatellites is such that all individuals effectively are born with a unique tag that can be attributed to them, to their parents (via parentage estimation), relatives, and population of birth (by analyses loosely termed 'assignment tests') (Manel et al. 2005).

Thus these approaches can be used to identify individuals, and detect if and how far they have dispersed since birth. A major advantage over traditional capture-mark-recapture or radio-tracking is that all individuals are born with their 'tag' and thus at most a single capture is necessary for many applications, so relieving a major bottleneck in data collection. As Waser and Strobeck (1998) pithily note, '…few birds have bands but all have genotypes.'

These approaches are typically extremely powerful because sample sets can be very large, all individuals in a sample set can have important inferences made about them (say postnatal dispersal distance), and analyses can distinguish individual movements from gene flow — the latter is demographically and evolutionarily far more important. The power to track large numbers of individuals means that, rather than average values, we obtain the full distribution of dispersal distances — extremely useful in demographic modelling — and have high individual replication to investigate the factors contributing to differences in dispersal tendency.

Since they can be applied to large numbers of individuals, approaches are able to detect rare dispersal events that might be critical in predicting population persistence in highly altered environments (Davies and Melbourne 2007; Hoehn et al. 2007) that can be infeasible to detect by traditional methods (e.g. long distance dispersal in rock-wallabies: Eldridge et al. 2001). These sources of investigative power, coupled with the ability to detect inhibited dispersal before serious demographic effects result (Schmuki et al. 2006), mean that molecular population biology has the potential to inform effective intervention while situations are relatively less critical.

Genic Analyses

The very same datasets analysed from a genotypic perspective (see above) can also be analysed in a 'genic' framework, where the focus is on frequencies of genetic variants, rather than on individuals. Gene frequencies change more slowly than the generation-to-generation shuffling of genotypes, so genic approaches register biological events that occur over a few, or tens or hundreds of generations. Thus they tend also to reflect processes at larger spatial scales. In landscape biology, genic frequency-based analyses are often used to investigate medium-term structuring of populations associated with pervasive, ongoing restriction of mobility and gene flow.

Genealogical Analyses

DNA sequences change by spontaneous mutation. A new variant arises at a point in time and space, and will have a trajectory that will dictate how it comes to be distributed over landscapes. This is a relatively slow process, therefore genealogical analyses, based on DNA sequences, register the deepest view back in time, and usually reflect patterns and processes at the largest scales of the three approaches outlined here. As DNA evolution progresses with an order and often at an approximately consistent rate, it contains information about relative and even absolute timelines. In landscape biology these approaches are perhaps most useful for understanding the natural patterning of organisms, as a pre-requisite for understanding human impacts, say since the industrial revolution. They are also essential for assessments of species level (and above) biodiversity and evolutionary relationships among biota.

11.2.2 Main Molecular Tools in Landscape Molecular Population Biology

The mainstay of molecular population biology as applied to landscape management is the genotyping of a suite of microsatellites, often coupled with sequencing and haplotype analysis of mitochondrial DNA (DeYoung and Honeycutt 2005). Between them, these tools offer resolution at the broad range of temporal and spatial scales required for most landscape management applications, and are suitable for use on a range of field-collected samples. The review articles mentioned in the previous section should be consulted for detailed appraisals of the attributes of other marker systems. Microsatellites are a type of highly-resolving molecular marker located in nuclear DNA and are thus inherited from both parents in sexual species. These markers are becoming increasingly readily available for much of the biota, although applying them to more obscure or understudied taxonomic groups can require some development effort. A *Current Contents* search up to May 2007 located 991 peer-reviewed publications with 'microsatellite' and 'management' in the text, indicating that this area is a major area of activity, albeit an area not often encountered by landscape managers.

Mitochondrial DNA complements microsatellites for landscape-level molecular population biology because it reflects longer term, larger scale biological processes, and typically becomes diagnostic of populations and species more readily (Avise 2004). Mitochondrial DNA markers have been used in many thousands of studies worldwide, and are extremely readily applied to almost all animals for a range of purposes including species identification, delineation of geographic ranges of management units, and estimating history of populations. They are not so useful in plants, where chloroplast genes are often used in their place.

Main analyses in Landscape Molecular Population Biology

Many diverse analyses and computer software products used to carry out molecular population analyses have matured over the last two decades (DeSalle and Amato 2004; DeYoung and Honeycutt 2005; Manel et al. 2005; Pearse and Crandall 2004; Selkoe and Toonen 2006). The previous section and Tables 11.1–11.4 (Appendix) present overviews of the general approaches to choosing amongst the differing analyses. As with any research program, landscape management agencies need carefully to frame research questions, since this typically will guide sampling strategies (nonetheless, in the absence of convincing ethical or logistical contraindications, we would advocate opportunistic sampling even in the absence of solid plans or framed research). As with any other research approaches and management applications, there are many complexities and caveats associated with selecting among alternatives to gain the most benefit per unit investment. Rather than using the many published analyses of the efficiencies and power of the key systems (e.g. Bernatchez and Duchesne 2000; Lancaster et al. 2006) prescriptively, these can be used to assist in planning a sensible approach that can be fine-tuned adaptively as a project progresses.

The main concepts are that microsatellite and mitochondrial DNA from individuals can be used to characterise those individuals, identify their parents (parentage analysis), allocate them to a prior group or cluster of like-individuals (assignment tests and 'STRUCTURE' analysis), analyse the spatial patterns in their relationships (often using spatial autocorrelation and spatially-explicit pattern analysis), and estimate effective population sizes. These forms of information can be used to estimate if and how far individuals have dispersed, what life stages disperse (e.g. estimation of sex-biased dispersal is very important in many population modelling situations), whether they have bred at their destination, and what terrain and conditions they must have traversed to get from location of birth to point of capture.

Individual data of these kinds can be assembled into estimates of key demographic and ecological parameters or data arrays (e.g. matrix of dispersal distances for all individuals in a sample) for populations or other groupings — perhaps those based on different landscape contexts (e.g. intact versus modified habitat). As the three levels of analysis (outlined above) give views into different temporal windows (contemporary, recent past, distant past), many of these operations can be conducted for different time scales, perhaps most usefully in the current contexts for comparing current processes to those arising from human impacts over 200 years to those that existed from the last glaciations. Critically for increased understanding of human impacts and the role of landscape features in realised ecological connectivity, these analyses can yield estimates of how rates of dispersal and gene flow have changed over time and what the impacts have been on local effective population sizes.

A strong example involving land management in southern Australia is a program investigating human impacts on yellow-footed antechinus (*Antechinus flavipes*) (Lada et al. 2007a, 2007b). This work applied sophisticated co-analysis and modelling of demographic, environmental and genetic data to estimate the time scales of changes in migration patterns in response to diverse human impacts such as forest clearance and utilisation, establishment of towns and alteration of river flows. On small and large scales, the genetic data indicate landscape features that facilitate or inhibit functional connectivity of habitat *from the perspective of the species.* These approaches are becoming increasingly widespread — the assessment of landscapes from the perspective of bighorn sheep (*Ovis canadensis)* is a recent example in which genetic data significantly improved the ability to plan landscapes (Epps et al. 2007).

At a higher level of synergy, parameter estimates and conclusions from single species can be analysed alongside those from many other species, to facilitate ever broader inferences about the landscape structures that support ecological processes. As part of this process, suites of analyses are available that explicitly analyse population genetic data in terms of landscape scale information such as vegetation, climate, soil and hydrology (e.g. ecological distance approaches exemplified by Geffen et al. (2004) directed investigation of potential dispersal barriers in Schmuki et al. (2006)), and test for statistical congruence among patterns in different biota. It is well recognised that species have variable responses to habitat alteration (Radford et al. 2005). Comparative analysis of the responses of species can be a powerful tool in understanding differences and similarities, for example, via exploration of ecological or physiological differences among species (Schmuki et al. 2007) and reasons underpinning different persistence in altered landscapes (Hoehn et al. 2007).

Under certain conditions of preservation and storage, DNA is quite robust. Consequently it is often possible to obtain estimates of past population processes directly from museum collections and other stored biological material. Obvious applications include 'before and after' assessment of habitat fragmentation.

11.3 Case Studies

Our laboratory has applied molecular population biology to dozens of species that are of direct conservation concern, or that are models for other species. We present three brief case studies from published work, which illustrate some of the ideas outlined above.

11.3.1 Impacts of Habitat Fragmentation on Cunningham's Skinks

At two sites, rock-outcrop-dwelling Cunningham's skinks (*Egernia cunninghami*) were sampled in natural habitat and adjacent habitat with vegetation removed. Assignment tests showed that in unmodified habitat, males and females usually disperse similar distances, while habitat fragmentation (de-vegetation) reduces dispersal in both sexes, but more so in females, creating a sex-bias in dispersal that was not present previously (Stow et al. 2001; Stow and Sunnucks 2004a, 2004b). The brake on dispersal is so great that habitat patches only 100 m apart have started to reach significantly different allele frequencies in less than 100 years — hence they are now on different demographic trajectories, and potentially different evolutionary trajectories.

Populations in de-vegetated habitat experience extremely elevated relatedness — which could change kin interactions and increase potential for breeding between close relatives — with expected strongly harmful demographic effects through inbreeding depression. Parentage analysis revealed that Cunningham's skinks at the two sites have not yet experienced elevated inbreeding because they are extremely effective in avoiding mating with close relatives — relatedness coefficients of pairs that did mate and produce offspring were 10 times lower than for pairs in the same rock outcrops that could have mated but did not. Such inbreeding avoidance is not without its costs, which could be demographically important. In the fragmented sites, a significantly higher proportion of males were unmated than in the natural sites (about one skink per fragmented group, as opposed to one skink per 10 natural groups).

11.3.2 Dispersal and Gene Flow of Greater Gliders through Forest Fragmented by Pine Plantation

It was predicted in the 1960s, as native forest was cleared to make way for pine plantation near Tumut in southern New South Wales, that greater gliders (*Petauroides volans*) would not persist in the plantation, based on aspects of their ecology and observations of their response to the forest clearance. However, over eight glider generations later, the species still occurred in many remnant native vegetation patches retained within the plantation boundary, albeit at a lower occupancy rate than at matched continuous forest control sites. To determine the role of patch connectivity in persistence of the greater glider in remnants, we gathered microsatellite genotypes from individuals from 11 remnant patches and nearby continuous native eucalypt forest sites.

Patch samples retained substantially more genetic diversity than expected under an isolation model, suggesting that patches have experienced some immigration despite the expected hostility of the pine matrix. Genetic parentage and assignment analyses identified five putative patch immigrants (among 81 sampled patch animals): two that immigrated from sampled patches 1 km and 7 km distant, and three from unresolved or unsampled localities. Samples from the oldest and most geographically-isolated patches showed lower admixture than those from more recent patches closer to continuous forest, suggesting the former have experienced relatively little immigration. Nonetheless, evidence of at least some immigration into most patches may explain why the greater glider has persisted contrary to expectation in heavily fragmented habitat (Taylor et al. 2007). This study highlights that existing species information may be inadequate for predicting or explaining processes and patterns in biodiversity distributions, and that rapid approaches to filling these important data gaps are vital.

Greater gliders are just one of a suite of species residing in the Tumut plantation for which population genetic data have been collected, revealing comparative dispersal patterns through an exotic pine matrix. Unlike the greater gliders, indirect and direct genetic estimates suggest dispersal by agile antechinus (*Antechinus agilis*) is only minimally affected (Banks et al. 2005a, 2005b). Impacts of fragmentation on bushrats (*Rattus fuscipes*) are unclear because genetic connectivity is more a function of the presence or absence of gullies, which may provide corridors for dispersal even in continuous forest (Lindenmayer and Peakall 2000). Differential species responses have been revealed at an even finer scale within the plantation, in a study showing that dispersal was more impacted for habitat specialist beetles than for generalist ones (Schmuki et al. 2006). Such multi-species data have the potential to promote planning and management that will maximise retention of biodiversity in production landscapes.

11.3.3 Catchments Catch All: Congruent Patterns in Diverse Invertebrate Fauna in Decaying Wood at a Landscape Scale

Tallaganda Forest in montane New South Wales comprises five hydrological catchments lying along 100 km of the Gourock Range, an off-shoot of the Great Dividing Range. The area is a complex of state forests and national parks, and has been subject to forestry activities that have intensified over the last four decades. We have been investigating its patterns in biodiversity and population processes in representatives of the ecologically-critical diverse log-dependent animals: 'giant' springtails (Collembola), velvet worms (Onychophora), funnel web spiders (Mygalomorpha) and terrestrial flatworms (Terricola) (Garrick and Sunnucks 2006; Garrick et al. 2004, 2006, 2007; Sunnucks et al. 2006). Before collecting biological data, we created a series of prior expectations based on expert opinion regarding physical data including topography, climate and history of vegetation change, against which we would test the biological inferences from molecular population biology (Garrick et al. 2004).

A combination of nuclear genetic markers and mitochondrial DNA has revealed extremely high levels of unrecognised species-level endemism in most of the fauna analysed. A powerful combination of analyses at all three levels (genotypic, genic and genealogical) was applied where possible (Garrick and Sunnucks 2006).

For most species, some or all of the five catchments hold distinct forms of each 'species' that are probably themselves species — genotypic and genic analyses show very little dispersal or admixture among catchments. Coalescent-based estimators of past population growth and spatial change in geographic ranges were used in combination with phylogeny-based approaches to investigate the changes in population size and locations over time. Impacts of Pleistocene or earlier climatic cycles are detected on multiple time scales, and multiple putative moist forest refuges were identified. Water catchment divisions predict population patterning and present day population structure with high precision, and may serve as an excellent surrogate for biodiversity indication in sedentary arthropods from topographically heterogeneous montane temperate forests.

The analytical comparison between the two Collembola provides an interesting example of how such a historical approach can be useful in understanding landscape-scale patterns relevant to management and planning. While these two species have quite different historical demographies and somewhat different biology, landscape processes have acted on them sufficiently strongly that they show concordant spatiotemporal patterning, which should simplify landscape-scale solutions for their management (Garrick et al. 2008).

Comparative studies of animals with low mobility and/or high habitat specificity remain rare, yet such organisms may hold fine-grained palaeoecological signal. This research program has driven a novel appreciation of the fine scale at which invertebrate organisms 'view' the world. Specifically, planning and management at an individual catchment scale should retain much more biodiversity (including species level biodiversity) than management that does not cater for different catchments. Log-dependent creatures in Tallaganda Forest nearly all show evidence of extremely long (millions of years) localised (catchment scale) residency, highlighting the critical role of areas within each catchment that acted as refuges from vegetation loss during Quaternary vegetation reorganisation. The data have been used in the management plans for the new Tallaganda National Park.

11.4 Future Trends

Although they are mature, effective and available, approaches of molecular population biology are yet to be routinely incorporated into the landscape manager's toolbox (the exceptions being in North America, Europe and South Africa) even though they represent the best and most appropriate science for many purposes (Forest et al. 2007; Scribner et al. 2005). Properly applied and conducted, the approaches are extremely cost-effective — operating costs are typically around 20% of payroll costs. These costs are comparable with other disciplines associated with landscape management, and for some applications they are the only feasible approach to providing critical information.

An example of successful uptake of the approaches in Australia is the case of the northern hairy-nosed wombat, a sensitive endangered species in Queensland. Managers have elected to employ non-invasive genetic techniques based on collecting hairs at burrow entrances to conduct multiple censuses along with demographic/biological monitoring. The genetic 'hair census' produces more precise population estimates, is less disruptive, costs less than the traditional techniques they replaced, and yields considerable extra information (Banks et al. 2003). Each genetic census costs approximately $15,000 to $20,000. In 2000, a genetic census produced 95% confidence intervals of 96 to 150 individuals, compared to 42 to 186 for the 1993 trapping census. The hair census revealed twice as many individuals as trapping immediate before, thus doubling the minimum population estimate.

There is a critical need for continuing to build global capacity and investment in the field of landscape genetics/molecular ecology. In Australia where we, the authors, have most direct experience, genetic research in landscape and species management (at least of animals) is done mainly by university researchers typically with support from the Australian Research Council or philanthropic trusts, rather than management-oriented funding sources. Australian management agencies, with rare exceptions, lack the necessary infrastructure and expertise to directly carry out such research, and they thus have little direct control over the research and its reporting. Management agencies currently have two main options:

- engagement of commercial genetic labs (relatively expensive and typically without broad and deep expertise to interpret data, which is also largely lacking in agency staff themselves)
- collaboration with university research laboratories.

University research laboratories: have expertise in data interpretation; are able flexibly to design species- and application-specific assays rather than being restricted to off-the-shelf assays; and produce research findings that may feed into new projects and typically involve dedicated and cost-efficient research students. However in order for university laboratories to increase their involvement in management oriented research there needs to be a culture of collaboration at all stages, including the funding application, project design, data interpretation and management recommendation phase. Fruitful interactions between management agencies and good quality laboratories with proven, peer-reviewed research publication records will be greatly facilitated by a culture of understanding and engagement with the potential offered by population genetics (Scribner et al. 2005).

The Application of Genetic Markers to Landscape Management   225

Uptake may be further facilitated by molecular population biology training of management staff, which is offered at a number of Australian Universities (e.g. the Applied Conservation Genetics Workshop, at the University of Western Australia: http://www.invasiveanimals.com/images/pdfs/ACG_Flyer_UWA_December_3-5.pdf).

11.5 Conclusion

Where appropriately applied by proficient practitioners, molecular population biology makes irreplaceable contributions to landscape management. Molecular population biology detects and estimates key population processes either more efficiently than competing approaches or by enabling estimations that are otherwise impossible. The field is fully mature, and is being applied routinely in many jurisdictions. Its greater uptake in landscape management would be highly beneficial in many circumstances. The main impediments to this uptake appear to be inertia and low prioritisation of resource availability, which is limiting the ability to build capacity. Thus there is considerable potential for landscape managers to gain the substantial benefits of molecular population biology.

11.6 Future Research Directions

Molecular population biology has the potential to interface extremely effectively with other fields and create important synergies for landscape and species management. While progress has been made in several key areas, there are still plenty of opportunities for improving efficiency of workflows and accuracy of estimates. Integration of habitat and environmental information with genetic estimates of population processes (notably dispersal) has begun to reveal the nature of capabilities and limits of organisms in modified landscapes (Hoehn et al. 2007; Scribner et al. 2005), but software and analyses are still emerging, and datasets are rare. 'Landscape friction analyses' would benefit from more sophisticated search routines for identifying optimal least-cost paths through landscapes (i.e. those that best explain measured genetic structures). One of the most appealing outcomes of genotypic estimates of dispersal is that large numbers of individuals can have attributes such as dispersal ability attributed to them. This makes them capable of feeding directly into spatially explicit, agent-based models of population function and persistence, which increasingly are chosen for modelling of population persistence and landscape design (e.g. McIntire et al. 2007).

Due to an increased focus on the organismal capabilities that interact with landscape structures, there have been incredible advances in the sophistication and precision of ecophysiological trait-based research — in technology, analysis and modelling (Kearney and Porter 2006). So far, studies are relatively few, and this research field offers many great opportunities for important science with strong management applications. In the arena of preparedness for climate change, these integrated approaches will provide radical improvements in forecasting how the abilities of organisms will interact with human-induced changes to environments under different scenarios of industrial activities and habitat alteration. This information will be critical for planning how we want our future landscapes to look.

Finally, the genomics revolution has provided the technology, skills and knowledge to efficiently find large numbers of genes that are directly responsible for the dispersal, thermal and other ecological properties of organisms (Feder and Mitchell-Olds 2003; Joost et al. 2007; Lee and Mitchell-Olds 2006). These approaches have moved forward from being applicable only to model organisms, and their costs are falling extremely rapidly. We are progressing towards being able to integrate biochemical function, phenotypic properties, genetic estimates of dispersal and landscape features, with environmental overlays and demographic modelling. These will provide excellent predictions of what organisms will be capable of under different conditions and which environments are the highest priority for maintenance and restoration to maximise population persistence.

Acknowledgements

We thank our numerous collaborators on the research programs outlined here, and in our other molecular population biology programs. Funding has come from diverse sources, including the Australian Research Council, the Holsworth Wildlife Trust, and Monash University. Thanks to David Duncan and the Landscape Futures Alliance for the invitation and support to attend the conference on which this volume draws — 'Place and Purpose: Spatial Models for Natural Resource Management and Planning' (May 2007, Bendigo, Victoria). This is a publication of the Australian Centre for Biodiversity. Two anonymous reviewers made valuable suggestions that improved the manuscript.

References

Allendorf FW, Luikart G (2006) Conservation and the genetics of populations. Blackwell, Oxford

Alter SE, Rynes E, Palumbi SR (2007) DNA evidence for historic population size and past ecosystem impacts of gray whales. In: Proceedings of the National Academy of Sciences of the United States of America 104:15162–15167

Avise JC (2004) Molecular markers, natural history, and evolution, 2nd edn. Sinaeur, Sunderland, Massachusetts

Baker CS, Cooke JG, Lavery S, Dalebout ML, Ma YU, Funahashi N, Carraher C, Brownell RL Jr (2007) Estimating the number of whales entering trade using DNA profiling and capture-recapture analysis of market products. Molecular Ecology 16:2617–2626

Banks SC, Taylor AC (2004) The utility of genetic analysis in forest fauna conservation. In: Lunney D (ed) Conservation of Australia's forest fauna, 2nd edn. Royal Zoological Society of NSW, Mosman, NSW, pp 576–590

Banks SC, Hoyle SD, Horsup A, Sunnucks P, Wilton A, Taylor AC (2003) Demographic monitoring of an entire species by genetic analysis of noninvasively collected material. Animal Conservation 6:101–108

Banks SC, Finlayson GR, Lawson SJ, Lindenmayer DB, Paetkau D, Ward SJ, Taylor AC (2005a) The effects of habitat fragmentation due to forestry plantation establishment on the demography and genetic variation of a marsupial carnivore, *Antechinus agilis*. Biological Conservation 122:581–597

Banks SC, Lindenmayer DB, Ward SJ, Taylor AC (2005b) The effects of habitat fragmentation via forestry plantation establishment on spatial genotypic structure in the small marsupial carnivore, *Antechinus agilis*. Molecular Ecology, 14:1667–1680

Bennett AF, Radford JQ, Haslem A (2006) Properties of land mosaics: implications for nature conservation in agricultural environments. Biological Conservation 133:250–264

Bernatchez L, Duchesne P (2000) Individual-based genotype analysis in studies of parentage and population assignment: how many loci, how many alleles? Canadian Journal of Fisheries and Aquatic Sciences 57:1–12

Davies KF, Melbourne BA (2007) The tails of two geckos tell the story of dispersal in a fragmented landscape. Molecular Ecology 16:3289–3291

Debain S, Chadaeuf J, Curt T, Kunstler G, Lepart J (2007) Comparing effective dispersal in expanding population of *Pinus sylvestris* and *Pinus nigra* in calcareous grassland. Canadian Journal of Forest Research-Revue Canadienne De Recherche Forestiere 37:705–718

DeSalle R, Amato G (2004) The expansion of conservation genetics. Nature Reviews Genetics 5:702–712

DeYoung RW, Honeycutt RL (2005) The molecular toolbox: genetic techniques in wildlife ecology and management. Journal of Wildlife Management 69:1362–1384

Eldridge MDB, Kinnear JE, Onus ML (2001) Source population of dispersing rock-wallabies (*Petrogale lateralis*) identified by assignment tests on multilocus genotypic data. Molecular Ecology 10:2867–2876

Epps CW, Wehausen JD, Bleich VC, Torres SG, Brashares JS (2007) Optimising dispersal and corridor models using landscape genetics. Journal of Applied Ecology 44:714–724

Excoffier L, Heckel G (2006) Computer programs for population genetics data analysis: a survival guide. Nature Reviews Genetics 7:745–758

Feder ME, Mitchell-Olds T (2003) Evolutionary and ecological functional genomics. Nature Reviews Genetics 4:651–657

Forest F, Grenyer R, Rouget M, Davies TJ, Cowling RM, Faith DP, Balmford A, Manning JC, Proches S, van der Bank M, Reeves G, Hedderson TAJ, Savolainen V (2007) Preserving the evolutionary potential of floras in biodiversity hotspots. Nature 445:757–760

Frankham R (2005) Genetics and extinction. Biological Conservation 126:131–140

Garrick RC, Sunnucks P (2006) Development and application of three-tiered nuclear DNA genetic markers for basal hexapods using single-stranded conformation polymorphism coupled with targeted DNA sequencing. BMC Genetics 7:11. doi:10.1186/1471-2156-7-11

Garrick RC, Sands CJ, Rowell DM, Tait NN, Greenslade P, Sunnucks P (2004) Phylogeography recapitulates topography: very fine-scale local endemism of a saproxylic 'giant' springtail at Tallaganda in the Great Dividing Range of south-east Australia. Molecular Ecology 13:3329–3344

Garrick RC, Sands CJ, Sunnucks P (2006) The use and application of phylogeography for invertebrate conservation research and planning. In: Grove, SJ, Hanula JL (eds.) Insect biodiversity and dead wood: Proceedings of a Symposium for the 22nd International Congress of Entomology. General Technical Report, US Department of Agriculture Forest Service, Southern Research Station, Asheville, pp 15–22, available at http://www.srs.fs.usda.gov/pubs/

Garrick RC, Sands CJ, Rowell DM, Hillis DM, Sunnucks P (2007) Catchments catch all: long-term population history of a giant springtail from the southeast Australian highlands — a multigene approach. Molecular Ecology 16:1865–1882

Garrick RC, Rowell DM, Simmons CS, Hillis DM, Sunnucks P (2008) Fine-scale phylogeographic congruence despite demographic incongruence in two low-mobility saproxylic springtails. Evolution (online acceptance) doi:10.1111/j.1558-5646.2008.00349.x

Geffen E, Anderson MJ, Wayne RK (2004) Climate and habitat barriers to dispersal in the highly mobile grey wolf. Molecular Ecology 13:2481–2490

Hansson L (1991) Dispersal and connectivity in metapopulations. Biological Journal of the Linnean Society of London 42:89–103

Hoehn M, Sarre SD, Henle K (2007) The tales of two geckos: does dispersal prevent extinction in recently fragmented habitat? Molecular Ecology 16:3299–3312

Holderegger R, Wagner HH (2006) A brief guide to landscape genetics. Landscape Ecology 21:793–796.

Joost S, Bonin A, Bruford MW, Despres L, Conord C, Erhardt G, Taberlet P (2007) A spatial analysis method (SAM) to detect candidate loci for selection: towards a landscape genomics approach to adaptation. Molecular Ecology 16:3955–3969

Kearney M, Porter WP (2006) Ecologists have already started rebuilding community ecology from functional traits. Trends in Ecology and Evolution 21:481–482

Kellermann VM, van Heerwaarden B, Hoffmann AA, Sgro CM (2006) Very low additive genetic variance and evolutionary potential in multiple populations of two rainforest *Drosophila* species. Evolution 60:1104–1108

Krieger MJB (2005) To b or not to b: a pheromone-binding protein regulates colony social organisation in fire ants. Bioessays 27:91–99

Lada H, Mac Nally R, Taylor AC (2007a) Distinguishing past from present gene flow along and across a river: the case of the carnivorous marsupial (*Antechinus flavipes*) on southern Australian floodplains. Conservation Genetics 10.1007/s10592-007-9372-5

Lada H, Thomson JR, Mac Nally R, Horrocks G, Taylor AC (2007b) Evaluating simultaneous impacts of three anthropogenic effects on a floodplain-dwelling marsupial *Antechinus flavipes*. Biological Conservation 134:527–536

Lancaster ML, Gemmell NJ, Negro S, Goldsworthy S, Sunnucks P (2006) Ménage à trois on Macquarie Island: hybridisation among three species of fur seal (*Arctocephalus* spp.) following historical population extinction. Molecular Ecology 15:3681–3692

Lee CE, Mitchell-Olds T (2006) Preface to the special issue: ecological and evolutionary genomics of populations in nature. Molecular Ecology 15:1193–1196

Lindenmayer DB, Peakall R (2000) The Tumut experiment-integrated demographic and genetic studies to unravel fragmentation effects: a case study of the native bush rats. In: Young A, Clark G (eds.) Genetics, demography, and the viability of fragmented populations. Cambridge University Press, Cambridge, pp 173–201

McIntire EJB, Schultz CB, Crone EE (2007) Designing a network for butterfly habitat restoration: where individuals, populations and landscapes interact. Journal of Applied Ecology 44:725–736

Manel S, Gaggiotti OE, Waples RS (2005) Assignment methods: matching biological questions with appropriate techniques. Trends in Ecology and Evolution 20:136–142

Manel S, Schwartz MK, Luikart G, Taberlet P (2003) Landscape genetics: combining landscape ecology and population genetics. Trends in Ecology and Evolution 18:189–197

Otis DL, Burnham KP, White GC, Anderson DR (1978) Statistical inference from capture data on closed animal populations. Wildlife Monographs 62:1–135

Peakall R, Smouse PE (2006) GENALEX 6: genetic analysis in Excel. Population genetic software for teaching and research. Molecular Ecology Notes 6:288–295

Pearse DE, Crandall KA (2004) Beyond FST. Analysis of population genetic data for conservation. Conservation Genetics 5:585–602

Piggott MP, Taylor AC (2003) Remote collection of animal DNA and its applications in conservation, management and understanding the population biology of rare and cryptic species. Wildlife Research 30:1–13

Radford JQ, Bennett AF, Cheers GJ (2005) Landscape-level thresholds of habitat cover for woodland-dependent birds. Biological Conservation 124: 317–337

Reusch TBH, Wood TE (2007) Molecular ecology of global change. Molecular Ecology 16:3973–3992

Reusch TBH, Ehlers A, Hammerli A, Worm B (2005) Ecosystem recovery after climatic extremes enhanced by genotypic diversity. In: Proceedings of the National Academy of Sciences of the United States of America 102:2826–2831

Schmuki C, MacEachern S, Runciman D, Vorburger C, Sunnucks P (2006) Beetles on islands of bush in a sea of pine: impacts of habitat fragmentation on two species of *Adeliini* at Tumut, SE Australia. Molecular Ecology 15:1481–1492

Schmuki C, Woodman J, Sunnucks P (2007) Physiology complements population structure of two endemic log-dwelling beetles. Environmental Entomology 36:524–530

Scribner KT, Blanchong JA, Bruggeman DJ, Epperson BK, Lee CY, Pan YW, Shorey RI, Prince HH, Winterstein SR, Luukkonen DR (2005) Geographical genetics: Conceptual foundations and empirical applications of spatial genetic data in wildlife management. Journal of Wildlife Management 69:1434–1453

Selkoe KA, Toonen RJ (2006) Microsatellites for ecologists: a practical guide to using and evaluating microsatellite markers. Ecology Letters 9:615–629

Stow AJ, Sunnucks P (2004a) High mate and site fidelity in Cunningham's skinks (*Egernia cunninghami*) in natural and fragmented habitat. Molecular Ecology 13:419–430

Stow AJ, Sunnucks P (2004b) Inbreeding avoidance in Cunningham's skinks (*Egernia cunninghami*) in natural and fragmented habitat. Molecular Ecology 13:443–447

Stow AJ, Sunnucks P, Briscoe DA, Gardener MG (2001) The impact of habitat fragmentation on dispersal in Cunningham's skink (*Egernia cunninghami*): evidence from allelic and genotypic analyses of microsatellites. Molecular Ecology 10:867–878

Spielman D, Brook BW, Frankham R (2004) Most species are not driven to extinction before genetic factors impact them. Proceedings of the National Academy of Sciences of the United States of America 101:15261–15264

Sunnucks P (2000) Efficient genetic markers for population biology. Trends in Ecology and Evolution 15:199–203

Sunnucks P, Blacket MJ, Taylor JM, Sands CJ, Ciavaglia SA, Garrick RC, Tait NN, Rowell DM, Pavlova A (2006) A tale of two flatties: different responses of two terrestrial flatworms to past environmental climatic fluctuations at Tallaganda in montane south-eastern Australia. Molecular Ecology 15:5413–4531

Tallmon DA, Koyuk A, Luikart G, Beaumont MA (2008) ONESAMP: a program to estimate effective population size using approximate Bayesian computation. Molecular Ecology Resources 8:299–301

Taylor AC, Tyndale-Biscoe H, Lindenmayer DB (2007) Unexpected persistence on habitat islands: genetic signatures reveal dispersal of a eucalypt-dependent marsupial through a hostile pine matrix. Molecular Ecology 16:2655–2666

van Heerwaarden B, Hoffmann AA (2007) Global warming: fly populations are responding rapidly to climate change. Current Biology 17:R16–R18

Waser PM, Strobeck C (1998) Genetic signatures of interpopulation dispersal. Trends in Ecology and Evolution 13:43–44

Young AG, Clarke GM (eds) (2000) Genetics, demography and viability of fragmented populations. Cambridge University Press, Cambridge

Young AG, Brown AHD, Murray BG, Thrall PH, Miller CH (2000) Genetic erosion, restricted mating and reduced viability in fragmented populations of the endangered grassland herb *Rutidosis leptorrhynchoides*. In: Young AG, Clarke GM (eds) Genetics, demography and viability of fragmented populations. Cambridge University Press, Cambridge, pp 335–359

Appendix

Some Key Tasks, Questions and Analyses in Molecular Population Genetics

Table 11.1. Questions and analyses employed in the definition and mapping of populations, management units and taxa

Question	Analyses	Example analytical programs[a]
What is taxonomic status?	Molecular systematics/ population membership	MODELTEST PAUP* PHYLIP MEGA
How many populations are there?	Clustering approaches	STRUCTURE BAPS
Where are they?	Spatially explicit landscape genetics	GENELAND
How long have they been there?	Comparisons of outcomes of genotypic, frequency-based and DNA sequence-based analyses	(relevant combinations)

[a]Complete list and citations contained in Excoffier and Heckel (2006)

Table 11.2. Questions and analyses applicable to between-population processes

Question	Analyses	Example analytical programs[a]
What model of population structure applies?	Spatial autocorrelation, linear Mantel tests	GenAlEx (Peakall and Smouse 2006) SPAGeDi
How different are populations?	Frequency differentiation tests of genotypes (contemporary differentiation), gene frequencies and DNA sequences (longer term differences)	GENEPOP GenAlEx ARLEQUIN
What is the rate and pattern of dispersal and gene flow? What kinds of individuals disperse, with what probability?	Assignment/ parentage tests (contemporary effects), medium-term estimates from gene frequencies, long-term estimates from DNA sequences and coalescence analyses	BAYESASS GENECLASS CERVUS GENEPOP FSTAT MIGRATE IM
What proportion of individuals is from different sources?	Assignment and mixed stock analyses	STRUCTURE GENECLASS BAYESASS
How admixed are individuals?	Assignment tests	(See above)

[a]Complete list and citations contained in Excoffier and Heckel (2006)

Table 11.3. Questions and analyses applicable to within-population processes

Question	Analyses	Example analytical programs[a]
What is the effective population size?	See Pearse and Crandall (2004)	Ne ESTIMATOR, ONESAMP (Tallmon et al. 2008)
Has demography changed recently and/or in the past?	Tests for recent or past changes in effective population size, loss of genetic variation, coalescence tests	BOTTLENECK FLUCTUATE DNASP (See above)
Have basic population processes changed?	Assignment, parentage and kinship tests→ local dispersal, social/ mating systems and kin structure	KINSHIP

[a]Complete list and citations contained in Excoffier and Heckel (2006)

The Application of Genetic Markers to Landscape Management   233

Table 11.4. Questions and analyses applicable to specific intensive management tasks

Question	Analyses	Example analytical programs[a]
Is a species of conservation concern being used inappropriately?	Molecular systematics Assignment of individuals to stocks	See Table 11.1 See Table 11.2
How many individuals in population? Births, deaths, reproductive success, migration, sex ratio, space use etc.	Non-invasive sample collection, genotype-matching, capture-mark-recapture analysis	GenAlEx and Tables 11.2 and 11.3 CAPTURE (Otis et al. 1978)

[a]Complete list and citations contained in Excoffier and Heckel (2006)

12 Scenario Analysis with Performance Indicators: a Case Study for Forest Linkage Restoration

David V Pullar and David Lamb

Geography, Planning and Architecture, The University of Queensland, Australia

Abstract: Regional conservation programs have as a goal the integration of protected areas into broader ecological networks. Creating landscape corridors and buffers within a land mosaic is one way to support the functional viability of landscapes. Implementing these programmes on the ground is problematic, however, especially in areas with a mix of land uses and fragmented forests. In such cases it is necessary to prioritise restoration decisions to balance the benefits against the cost of restoration efforts. We approach this as a planning problem to work with stakeholders to explore options that find a balance between the environmental benefits and costs. The options are mapped onto the landscape as scenarios and then evaluated in a GIS planning support system. This chapter describes a methodology for prioritising restoration efforts and applies this to a hypothetical study for northern Australia. Spatial indicators are used to assess alternate options for vegetation corridors linking forest remnants. The indicators mainly measure spatial structure, but also aim to incorporate knowledge on provision of landscape ecological functions. The indicators assess both site level and landscape benefits. Diagnostic spatial indicators and scenarios are developed generically within GIS so the approach may be tuned to different applications and planning problems.

12.1 Introduction

In north-eastern Australia large tracts of tropical rainforests on good soils and gentle topography in the wet tropics have been cleared for agriculture. This has resulted in high levels of ecosystem fragmentation which causes detrimental changes to the physical environment from local to bio-geographical scales (Fahrig 2003; Saunders et al. 1991). These landscapes now consist of isolated forest remnants located within predominately agricultural land uses. The long-term survival of species that occupy these remnants is threatened if they are unable to access or utilise resources in the landscape (e.g. Vesk et al. Chapter 10), for instance if lowland areas providing access to water are isolated from highland areas utilised for foraging. Bioregional plans stipulate that a percentage of the landscape should be set aside for conservation and protection of native ecosystems — however these plans do not account for finer scale spatial configuration and interactions within the landscape. More detailed landscape planning is needed to adequately protect areas that support wildlife movement and dispersal and other landscape functions.

Most the remaining large tracts of forests in north-eastern Australia are found within state forests and inaccessible mountainous areas (Erskine 2002). There are some smaller areas of forest located on agricultural land, but they are highly fragmented and, on their own, provide minimal ecological benefits. Where cleared land and agricultural land uses dominate, the best opportunity to improve landscape functions is to create corridors between residual patches on private lands and the larger protected state forests. The task of prioritising areas for reforestation is daunting. What areas are best suited for forest re-establishment? How do you quantify the benefits of restoring areas at the site and landscape levels? What are the trade-offs between biodiversity gains and lost agricultural production?

Forman (1995) gives simple guidelines for understanding ecology organised around four geographical scales: landscapes and regions, patches and corridors, mosaics, and efficient landscape patterns. The objective of this paper is to translate these general guidelines into a simple quantitative analysis to prioritise the best areas for restoration. We are focused on a modified landscape where natural forest patches exist in a background matrix of agricultural land. We assume that key landscape functions can be improved by reforestation that buffers existing remnants or connects smaller remnants with wide vegetated corridors, and that there are important trade-offs between the quality, costs and spatial configuration of restoration options. Our aim is to see if quantitative measures from landscape ecology are able to discriminate between different restoration plans and to assess the best plan. We explore measures that quantify the:

- capacity and potential for restoring sites
- potential to improve overall landscape connectivity
- cost of site in terms of foregone agricultural production.

We use a scenario planning approach (Ahern 1999; Botequilha Leitão et al. 2006) where different restoration options are proposed, and the task is to analyse the options based upon comparative scores. This is a normative approach to landscape planning (Nassauer and Corry 2004) where the focus is on the trade-offs between factors and the relationships between landscape ecology and decision options. We are particularly interested in exploring the trade-offs for benefits attributable to a site (or restored patch) compared to its landscape context as measured by landscape connectivity.

The outline of this chapter is as follows. The next section expands on general landscape ecology guidelines that we use to explore reforestation options. It puts these guidelines into a planning framework to evaluate alternative restoration options as scenarios. Scenarios provide a useful way to explore alternatives and to assess how well they meet objectives and other requirements. We next describe specific rules and indicators used to evaluate scenarios to quantify their performance to achieve benefits. The subsequent section presents a hypothetical case study for north-eastern Australia. Four scenarios are described for reforesting land to improve forest connectivity. Scenario analysis is implemented by a custom tool in GIS. The conclusion discusses results from this implementation, and the use of scenario analysis using indicator rules for assessing landscape functions.

12.2 Background

Ecological networks are seen as an important pillar for conservation and sustainable use of biological diversity (UN 2002). An ecological network provides a link to connect and buffer an assemblage of core areas. They are particularly relevant for species that require very large habitat areas. For instance in the wet tropics of Australia, some marsupials utilise multiple parts of the landscape for foraging and breeding. In the short-term, species need to move to avoid periodic natural disturbances, such as droughts, and longer term survival may require moving large distances away from a threat and recolonising more favourable habitat areas (Jongman et al. 2004). Ecological networks are planned in different countries as biosphere reserves, reserve networks, bioregional planning schemes, biological or conservation corridors and ecoregion-based conservation. Despite the differences in terminology, these approaches share a core vision on how best to integrate conservation of biodiversity with sustainable development (Bennett 2004).

Lamb (2005) expands on the types of landscape patterns and planning interventions desirable for restoring forests. The context for these interventions is assumed to be a rural landscape where land clearing has resulted in highly fragmented native vegetation, which in turn threatens the long-term biodiversity of species and soil degradation in the region. The imperative is to propose interventions to protect existing forests and restore landscape functions. Table 12.1 lists different types of interventions along with the dominant landscape threats they address. The next section discusses how these are assessed with metrics from landscape ecology.

Table 12.1. Alternative points of intervention or priorities for restoration in 'degraded' landscapes

Types of restoration interventions	Reason
Protect areas of degraded or secondary forest and facilitate natural regrowth (perhaps by enrichment planting)	Many degraded forests are capable of recovering much of their original biodiversity if protected and if residual natural forest is nearby. The cost of doing this is likely to be less than replanting
Enlarge small residual forest fragments by planting species-rich forest around their margins (i.e. restoration plantings)	To diminish the role of the 'edge effect' and increase the effective inner core habitat area
Create buffer zones around small residual natural forest fragments using commercial timber plantations (i.e. monocultures or rehabilitation plantings)	To provide buffers against fires, weeds etc. and to reduce 'edge effects'
Create habitat corridors between residual forest areas (restoration or rehabilitation plantings)	To facilitate the movement and genetic interchange of poorly dispersed species
Establish new forest patches in the agricultural matrix between residual forest areas	To act as 'stepping stones' that facilitate the movement and genetic interchange of more easily dispersed species
Create protective zones along large riverine buffer areas	To act as filters limiting the movement of soils into streams
Reforest erosion areas on hill slopes	To stabilised hill slopes and limit erosion
Reforest recharge areas in landscapes prone to salinity	To restore key hydrological processes

12.3 Linkage restoration

Landscape ecology, with its focus on spatial structure and landscape function (Turner 2005), is suited to analysis of landscapes at multiple scales. Structure describes the composition and configuration of natural elements to support landscape functions, and function encompasses broad ecological services provided by the landscape, such as, regulation, protection and production (de Groot 2006). Landscape planning utilises structure–function relationships at the design stage to explore environmental plans and monitoring to validate the assumptions (Botequilha Leitão et al. 2006). While approaches do vary in different landscape regions (Hawkins and Selman 2002), ideally some generic principles emerge that are transferable between similar landscapes. Forman (1995) defines generic landscape elements in terms of the spatial structure for patches and landscape mosaics. Patches support habitat and local resources for biodiversity. They may be assessed on their quality and shape. A landscape mosaic consists of an arrangement of patches and corridors to support movement and dispersal of species. In our case study the patches are closed forest systems against a background matrix of agricultural land. There are many ways to restore habitat patches and introduce corridors in the matrix. Figure 12.1 shows the landscape elements of interest. A patch has local habitat benefits. A corridor links remnant patches and forest appendages to create a larger and more influential element connecting the landscape. This also contributes to a larger landscape network.

Here we have created a number of generic rules for structure–function relationships to assess restoration objectives for alternate plans in terms of the:

- capacity of degraded sites to recover without the need for external intervention
- value of the restored area for in situ biodiversity conservation
- biodiversity value of a linkage corridor
- context of the corridor within the landscape.

While a number of studies have identified the ecological importance of shape–configuration metrics in landscape planning conceptually (Bennet et al. 2006; Hersperger 2006; Turner et al. 2001), the challenge remains to translate metrics into a form that will permit an analysis of trade-offs in achieving planning objectives. The performance indicators we propose do this. We develop indicative rules for each of the landscape restoration objectives identified above. While a more comprehensive study would be required to understand the specific impacts of scenarios upon the biodiversity and ecological values, our rules demonstrate an approach that may help to ensure that restoration plans conform to basic rules of thumb from landscape ecology. Our four indicator rules are elaborated on below.

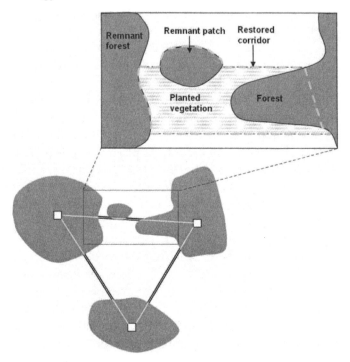

Fig. 12.1. Spatial elements forming forest configuration and connectivity through the establishment of restored vegetation, restored corridors and landscape networks

12.3.1 Indicator Rule 1: Site Recovery Capacity

Gibbons and Freudenberger (2006) review a number of approaches for rapid assessment of vegetation condition for restored areas. The methods are based upon concepts of landscape functions and they combine indicators for vegetation structure and composition. They acknowledge that not all aspects of landscape function are represented by these indicators, but they do capture enough information to inform decision making. One indicator described by Gibbons and Freudenberger (2006), which we will use in our study, is a measure the capacity of a vegetation community to recover after a disturbance is removed. This indicates the likelihood that a replanted area recovers to a natural state. They describe decision criteria based upon past disturbance (clearing and soils modified by cultivation) and the composition of existing vegetation (exotic or native). This is shown in Table 12.2 as a decision classification table to illustrate the combination of possible attributes.

Table 12.2. Decision classification table for the recovery capacity of degraded site (adopted from Gibbons and Freudenberger 2006)

Disturbance		Recently cleared (<2 years)		Historically cleared (>2 years)	
Soils		Unmodified	Cultivated	Unmodified	Cultivated
Vegetation	Native	High	Moderate	Moderate	Low
	Exotic dominated	Moderate	Low	Low	Very low

12.3.2 Indicator Rule 2: Site Biodiversity Value

Lamb (2005) discusses some general principles for biodiversity used to prioritise restoration activities. These principles assume familiarity with the ecology for an area, including knowledge of the spatial patterns of threatening land uses and forest types. The biodiversity for sites are scored based upon landscape characterisation and simple indicative rules. In our case study we interpret site biodiversity as having a direct relationship with the spatial structure of forested patches and proximity to neighbouring forests. Table 12.3 illustrates a decision classification using a number of spatial criteria to construct a biodiversity index; the values in the table may be changed to suit local circumstances.

Table 12.3. Rating values 0–1 (low to high) for biodiversity based on the type of forest, its size, and landscape context. The numbers give a subjective value for biodiversity based upon interpreted benefits for forest spatial structure and composition

Context		Neighbouring forest			Isolated		
Type		Mono	Mixtures	Regrowth	Mono	Mixtures	Regrowth
Size	Small (<10 ha)	0.1	0.2	0.5	0.1	0.1	0.3
	Large (>10 ha)	0.4	0.7	1.0	0.2	0.5	0.8

12.3.3 Indicator Rule 3: Landscape Linkage Qualities

Tucker (2000) examines issues in a linkage restoration project in northeastern Australia. The issues relate to threats from disturbances and viable linkage configurations. Many of the restoration issues are related to forest shape and configuration. A high edge to area ratio of a corridor increases the susceptibility of the area to weed establishment. Thin corridors are likely to leave wildlife prone to predation compared to broad corridors (Tucker 2000). However, it has been suggested that the disadvantages of narrow corridors with large edge to area ratios can sometimes be reduced if species with dense canopies that persist to ground level are used to 'seal' the boundary (Tucker and Murphy 1997). In the case of the Lake Barrine corridor this was done using rows of *Araucaria cunninghamii* which has a deep crown with branches to the bottom of the tree. A decision table of the shape and configuration rules are shown in Table 12.4.

Table 12.4. Ranked values for linkage quality based on the forest width, shape and core area

Shape		Elongated		Regular	
Core area		Small	Large	Small	Large
Width	Narrow (<200 m)	Very low	Low	Very low	Low
	Wide (>200 m)	Low	Moderate	Moderate	High

12.3.4 Indicator Rule 4: Landscape Connectivity

The most significant anticipated benefit for a corridor is that it connects patches in a landscape and provides benefits beyond its patch size. Connectivity is critical for animals to forage for food and to utilise the forest for protection, but longer term population survival may require larger ecological neighbourhoods than the immediate home range of an animal (Ricketts 2001; Turner et al. 2001). Two methods used to measure connectivity are graph structure and pattern indices.

Graph structures (Urban and Keitt 2001; Vos et al. 2001) represent a landscape as a set of nodes that are assumed to have a habitable extent, and are connected by edges that are assumed to allow transmission (Moilanen and Nieminen 2002). The graph structure is constructed as a minimal spanning tree or triangulation to reflect importance of nearby patches with cut-off values for distances above the species' home range. The significance of edges is that they provide a dispersal mechanism — also called a dispersal flux — which is measured as a distance between the nodes. The distance may be weighted based on an impedance for species movement. This is expressed as a distance–decay function (Urban and Keitt 2001) or as a least-cost pathway (Marulli and Mallarach 2005). Pascual-Hortal and Saura (2006) review a number of graph-based indices and found ones that combine patch size, habitat quality and dispersal flux are more effective. Pattern indices also combine habitat quality as a carrying capacity and connectivity. Cohesion of patches is expressed as a ratio between the patch area–edge areas ratio and total area (Opdam et al. 2003). Connectivity is then indirectly represented as an aggregate of connectivity linkages between distinct patches.

A graph-based index is preferred as it provides a holistic representation of the habitat and connectivity relationships between landscape elements (Chetkiewicz et al. 2006). The connectivity index C is computed for patch i to patches $j=1,n$ using:

$$C_i = \sum_{j=1,n} A_j \exp(-\alpha \cdot d_{ij}) \qquad i \neq j \qquad (12.1)$$

where j is the index to adjoining patches with area weight A_j and distance d_{ij}. The influence of distance is controlled by the parameter α, if $\alpha=0$ then distance has no effect. This parameter is normally specific to the degree of dispersal of a species so a higher value, that is, a value of 5 means a species is sensitive to fragmentation (Vos et al. 2001). Another way of looking at this is that larger C indices (larger A_j and short distance d_{ij}) are less sensitive to animal dispersal. The index in Eq. 12.1 may be made sensitive to forested corridors by adding a linkage width. This is shown in Eq. 12.2 by dividing the distance by width (note that a nominal minimum width is needed to avoid division by zero).

$$C_i^* = \sum_{j=1,n} A_j \exp(-\alpha \cdot d_{ij} / w_{ij}) \qquad i \neq j \qquad (12.2)$$

where w_{ij} is the width of the corridor. A larger C^* index (larger A_j, short distance d_{ij} and large w_{ij}) is less sensitive to animal dispersal. This modification does change the experimental interpretation of α (Vos et al. 2001) but our intention is to use C^* in a relative sense.

12.4 Atherton Tablelands Case Study

Our study area is based in the Atherton Tablelands in the wet tropics region of north-eastern Australia. Our case study involves two forest elements, namely the parks for Lake Barrine and Lake Eacham, which are separated from the main forest, namely Gadgarra State Forest (Fig. 12.2). The remnants contain crater lakes and are surrounded by dense rainforest, but are isolated within the landscape due to agriculture. Amongst the rich biodiversity of the area, we chose to use the southern cassowary[1] as a focal species to assess likely biodiversity benefits of the restoration options. The iconic cassowary is a large flightless bird that lives in dense tropical rainforest. While it can utilise open eucalypt woodlands to move between rainforest patches, it prefers the protection afforded by dense forest. Hence land clearing is recognised as a major threat to the species survival. Biological records from the WildNet database maintained by the Queensland EPA show that cassowaries have been sighted in each of the crater lake parks and within the state forests.

Fig. 12.2. Map showing restoration scenarios for crater lakes in the Atherton Tablelands

In 1995 a restoration project was undertaken to link the northern lake, Lake Barrine, to Gadgarra State Forest in the east. The objective was to allow the movement of species such as the southern cassowary. A 70–120 m wide and 1 km long corridor was restored along a creek to link the forests. Details of the project, issues and restoration concepts are discussed in Tucker (2000). There is another crater lake with a fragmented rainforest to

---

[1] A description for the biology of the southern cassowary may be obtained from http://www.epa.qld.gov.au/nature_conservation/wildlife/threatened_plants_and_animals/endangered/cassowary/

the south, namely Lake Eacham. As an exercise we investigated four options for linking the Lake Eacham fragment to the main state forest. These are explained in the next section.

12.4.1 Restoration scenarios

As a hypothetical situation we suggest four scenarios for forest restoration in the study area (Fig. 12.2). These scenarios would normally be explored as part of a community engagement process where local stakeholders want to see demonstrated benefits for conservation and sustainable development actions (Opdam et al. 2006; UN 2002). The scenarios and the evaluation with indicator rules provide a simple scientific analysis, but the indicators are easy to communicate to stakeholders so they can see where, how and what ecological benefits are possible. We quantify ecological benefits at a site scale as this has greatest impact on land owners, and also at a landscape scale as this has greatest impact for ecological benefits. Another important consideration for stakeholders is land costs. The costs for land restoration on private land can be high, but it may be warranted for critical areas that are strategically located within the landscape. The scenarios allow us to evaluate the opportunity costs for removing land from agricultural use. We include a very simple indicator for opportunity cost based upon the product of the estimated yield per hectare for a particular land use and the area for restoration.

Scenario 1: Establish a Forest Linkage between Lake Eacham and the Gadgarra State Forest

The area to be planted is not large as it mainly fills in gaps between existing forests. However these areas are cultivated so the capacity for native vegetation recovery is low and they have a high opportunity cost. The proposed corridor area is sufficiently large and near other forest so the biodiversity is rated as good. The landscape connectivity index is also high (i.e. supports dispersal for many species) because it links the park to the larger state forest.

Scenario 2: Establish a Forest Linkage between Lake Barrine and the Gadgarra State Forest

The area to be planted is not large and mainly covers pastoral land with unmodified soils that were cleared a number of years ago. Therefore it would have a moderate propensity for recovery of native vegetation. The area is sufficiently large and near other forest so the biodiversity indicator is rated as good. Again the landscape connectivity index is high because it links the park to the larger state forest. The restored area also covers streams and would enhance riparian functions.

Scenario 3: Establish a Forest Linkage between Lake Eacham and Lake Barrine

The reforested area is relatively large and mainly covers pastoral land with unmodified soils that were cleared a number of years ago. The area is sufficiently large and near other forest so the biodiversity is rated as good. The landscape connectivity index is moderate as the restored area connects two smaller parks, each of which has low values for open space connectivity to the larger state forest.

Scenario 4: Establish a Forest Linkage between Lake Barrine and the Remaining Forests Next to the Reservoir

The reforested area is relatively large and covers a mix of pastoral land that was recently cleared and a number of smaller remnant patches. The area has a good propensity for recovery of native vegetation. The landscape connectivity index is not high as it only links two smaller forest patches, and the state forest is far enough away that it does not contribute to connectivity in a large way.

12.4.2 Scenario Evaluation

Figure 12.3 shows the results of the evaluation. Each of the scenarios are evaluated with indicator rules and scored to a uniform scale of 1 to 100 for comparison purposes. In general we use a linear scaling of raw indicator values to scores. A more comprehensive approach would be applied based upon stakeholder consultation to specify the utility function for scoring. The scores we present in Fig. 12.3 are mainly intended to show that trade-offs do occur and distinctions can be quantified for the scenarios. At a local scale the capacity for a site to be restored favours scenario 4 where there is less historical clearing and cultivation. The area encompassed in scenario 4 is considered more suited to restoration treatments and monitoring (Gibbons and Freudenberger 2006). The areas for all scenarios had the same biodiversity benefit as they are similar in size and are all replanted with native forest. This would change if we considered other planting options such as monocultures and mixed plantation. In terms of linkage values, scenarios 3 and 4 are more favoured by our rules. This is because they are the widest and encompass the largest areas to reduce edge effects

(Tucker 2000). At a landscape scale, network connectivity is improved with scenarios 1 and 2, as both these scenarios link to the larger state forest over short distance corridors.

There are trade-offs to be decided between these scenarios — to undertake restoration in an area (scenario 4) that could be rapidly restored, or to improve the overall landscape context with a corridor connecting the state forest (scenarios 1 and 2). We do not resolve these trade-offs, except to say that the indicator rules do provide a way to measure and quantify the benefits for each of the scenarios to inform decision making.

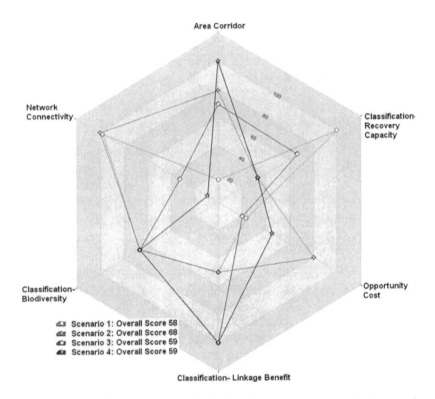

Fig. 12.3. Result of scenario analysis displayed as radar chart for the four options using six indicator rules

12.5 Conclusion

The chapter has explored the use of performance indicators to evaluate landscape forest restoration scenarios. A review of the literature on landscape forest restoration shows that we need to look at the problem at different geographical scales. We identified indicator rules for the local site area, its relationship as a corridor and nearby forest, and the larger context within the landscape. Applying this to a case study on the Atherton Tablelands in the wet tropics region of north-eastern Australia shows that there are potential trade-offs for each of the indicators. A more comprehensive analysis might include other indicators. For instance, how recovery capacity is improved with seed dispersal around existing rainforest (White et al. 2004), reforesting riparian areas (Lawson et al. 2007) and forestation utilising plantation species (Lamb et al. 2005) would add further scenarios for consideration. The decision of what scenario to test and with what indicators should be decided as part of a consultative process with stakeholders. Given that this process does take place and plausible scenarios for landscape forest restoration are proposed, then we have shown that these scenarios can be quantified and compared to assess their benefits and to help prioritise decisions on restoration.

All the analysis described in this chapter was performed in a GIS with a custom extension[2] to automate the analysis and management of scenario evaluation. Scenarios are created in a GIS using conventional editing to create input spatial layers. Indicators are implemented as spatial analysis procedures or classifications within the custom extension. The indicators compute values for the scenarios which are then converted to scores using monotonic utility functions set by the user. Our study applied linear transformations to map indicator values to scores, but these utility functions should be understood and agreed upon by stakeholders (Malczewski 2004). The classification rules for site recovery capacity, biodiversity value, and landscape linkage qualities are not overly sensitive to scoring as these are already defined via an expert classification. However, the indicator for landscape connectivity is sensitive to scoring and should be explored and understood by stakeholders so they can interpret the relationship between connectivity values and their benefit as an ecological network.

Despite the shortcomings of scoring procedures we see the importance of teasing out likely trade-offs between indicators. We have demonstrated that landscape structure and function should influence the location and configuration of reforestation decisions. Being able to understand trade-offs on local and landscape qualities for site restoration planning is important for optimising outcomes.

---

[2] The extension runs in the *ArcGIS* software from Environmental Systems Research Institute. The extension is called *Scenario Analysis* and may be obtained by contacting the author.

References

Ahern J (1999) Spatial concepts, planning strategies and future scenarios: a framework method for integrating landscape ecology and landscape planning. In: Klopatek J, Gardner R (eds) Landscape ecological analysis: issues and applications. Springer, New York, pp 175–201

Bennett G (2004) Integrating biodiversity conservation and sustainable use: lessons learned from ecological networks. The World Conservation Union (IUCN), Cambridge

Bennett AF, Radford JQ, Haslem A (2006) Properties of land mosaics: implications for nature conservation in agricultural environments. Biological Conservation 133(2):250–264

Botequilha Leitão A, Miller J, Ahern J, McGarigal K (2006) measuring landscapes: a planner's handbook. Island Press, Washington

Chetkiewicz C-L, St Clair C, Boyce M (2006) Corridors for conservation: integrating pattern and process. The Annual Review of Ecology, Evolution and Systematics 37:317–342

de Groot R (2006) Function-analysis and valuation as a tool to assess land use conflicts in planning for sustainable, multi-functional landscapes. Landscape and Urban Planning 75(3–4):175–186

Erskine PD (2002) Land clearing and forest rehabilitation in the Wet Tropics of north Queensland, Australia. Ecological Management and Restoration 3:135–137

Fahrig L (2003) Effects of habitat fragmentation on biodiversity. Annual Review of Ecology and Systematics 34:487–515

Forman RT (1995) Some general principles of landscape and regional ecology. Landscape Ecology 10(3):133–142

Gibbons P, Freudenberger, D (2006) An overview of methods used to assess vegetation condition at the scale of the site. Ecological Management and Restoration 7(1):10–17

Hawkins V, Selman P (2002) Landscape scale planning: exploring alternative land use scenarios. Landscape and Urban Planning 60(4):211–224

Hersperger AM (2006) Spatial adjacencies and interactions – neighborhood mosaics for landscape ecological planning. Landscape and Urban Planning 77:227–239

Jongman R, Külvik M, Kristiansen I (2004) European ecological networks and greenways. Landscape and Urban Planning 68:305–319

Lamb D (2005) Scenario modelling to optimise outcomes. International Tropical Timber Organisation Technical Series 23:117–124

Lamb D, Erskine P, Parrotta J (2005) Restoration of degraded tropical forest landscapes. Science 310:1628–1632

Lawson T, Gillieson D, Goosem M (2007) Assessment of riparian rainforest vegetation change in tropical north Queensland for management and restoration purposes. Geographical Research 45(4):387–397

Malczewski J (2004) GIS-based land-use suitability analysis: a critical overview. Progress in Planning 62(1):3–65

Marulli J, Mallarach JM (2005) A GIS methodology for assessing ecological connectivity: application to the Barcelona metropolitan area. Landscape and Urban Planning 71:243–262

Moilanen A, Nieminen M (2002) Simple connectivity measures in spatial ecology. Ecology 83(4):1131–1145

Nassauer JI, Corry RC (2004) Using normative scenarios in landscape ecology. Landscape Ecology 19(4):343–356

Opdam P, Verboom J, Pouwels R (2003) Landscape cohesion: an index for the conservation potential of landscapes for biodiversity. Landscape Ecology 18(2):113–126

Opdam P, Steingröver E, van Rooij S (2006) Ecological networks: a spatial concept for multi-actor planning of sustainable landscapes. Landscape and Urban Planning 75:322–332

Pascual-Hortal L, Saura S (2006) Comparison and development of new graph-based landscape connectivity indices: towards the priorisation of habitat patches and corridors for conservation. Landscape Ecology 21:959–967

Ricketts TH (2001) The matrix matters: effective isolation in fragmented landscapes. The American Naturalist 158:87–99

Saunders DA, Hobbs RJ, Margules CR (1991) Biological consequences of ecosystem fragmentation: a review. Conservation Biology 5(1):18–32

Schumaker N (1996) Using landscape indices to predict habitat connectivity. Ecology 77:1210–1225

Tucker N (2000) Linkage restoration: interpreting fragmentation theory for the design of a rainforest linkage in the humid wet tropics of north-eastern Queensland. Ecological Management and Restoration 1(1):35–41

Tucker NIJ, Murphy TM (1997) The effect of ecological rehabilitation on vegetation recruitment: some observations from the wet tropics of north Queensland. Journal of Forest Ecology and Management 99:133–152

Turner MG (2005) Landscape ecology: What is the state of the science? Annual Review of Ecology, Evolution, and Systematics 36(1):319–344

Turner M, Gardner R, O'Neill R (2001) landscape ecology in theory and practice: pattern and process. Springer, New York

UN (2002) Report of the World Summit on Sustainable Development, Johannesburg, South Africa, August 2002. United Nations, New York

Urban D, Keitt T (2001) Landscape connectivity: a graph-theoretic perspective. Ecology 2(5):1205–1218

Vos C, Verboom J, Opdam P, Ter Braak C (2001) Toward ecologically scaled landscape indices. The American Naturalist 157(1):24–41

White E, Tucker N, Meyers N, Wilson J (2004) Seed dispersal to revegetated isolated rainforest patches in North Queensland. Forest Ecology and Management 192:409–426

PART 3
SOCIOECONOMIC DIMENSIONS TO LANDSCAPES

13 Strategic Spatial Governance: Deriving Social–Ecological Frameworks for Managing Landscapes and Regions

David J Brunckhorst, Ian Reeve, Phil Morley and Karl Bock

Institute for Rural Futures and UNESCO Centre for Bioregional Resource Management, University of New England, NSW, Australia

Abstract: Since the 1980s, watersheds or catchments have been the primary regions used for natural resource management in many countries. Catchments however, often do not represent the range of biophysical and social characteristics of importance for effective resource governance. The requirements for spatial definition of resource governance regions have received little analysis in science or policy. Research on 'place' attachment and community participation in resource management provides a grounding to re-examine such regional arrangements. This chapter describes three characteristics considered to be of priority importance in identifying regional boundaries for resource governance. Firstly, the boundaries of resource governance regions should enclose areas of high interest and importance to local residents. Secondly, the biophysical characteristics of a resource governance region should be as homogenous as possible. Thirdly, the nature and reach of environmental externalities of resource use should determine the size and nesting of resource management regions. An example of the application of these concepts to derive a hierarchy of nested regions for the State of New South Wales, Australia is provided. The results have been used by the New South Wales Government and the Electoral Commission to reconsider local government, natural resource management regions, and representative democracy within the State. Wider applications might include review of municipal boundaries and regional governance frameworks in other nations and across the European Union.

13.1 Introduction

Around the world, an increasing number of governments face rising social and environmental costs of resource use. There is increasing understanding amongst both scientists and policy makers that many resource governance issues relate to the complex interdependencies of social and ecological systems operating at various scales. The emergent patterns and properties from social–ecological interactions across landscapes provide further evidence supporting the growing emphasis on efficient and effective community engagement and civic action at multiple scales (Beckley 1995; Berkes and Folke 1998; Brown and MacLeod 1996; Brunckhorst 2000; Slocombe 1993). Planning for resource management at multiple scales of biophysically similar landscapes or ecoregions is considered important because they reflect characteristics influencing land and other resource use (Bailey 1996; Omernik 1987, 1995). Federated or nested administrative arrangements and spatial planning units have been employed by various governments in natural resource management (Frey and Eichenberger 1999; McGinnis, 1999; Waldo 1984). However, these arrangements have not always been effective in ensuring natural resources are used sustainably (Barham 2001; Blomquist and Schlager 2005; Carpenter and Gunderson 2001; Johnson et al. 1999).

The placement of boundaries to define regions for integrated resource governance warrants more careful analysis than it has been accorded in the past. What actors with an interest exist in an area, what spatial civic representation and networks, and what landscapes of ecological patterns, function and ecosystem services are included in a resource governance region? These considerations are vitally important to the success or failure of strategies, plans and actions towards more resilient and sustainable social–ecological systems. We put forward three basic principles that need to be considered in defining resource governance regions and, using the State of New South Wales in Australia, demonstrate an empirical method of deriving a nested hierarchy of such regions that are consistent with these principles.

13.2 A Potted History of Catchments for Resource Governance

Watersheds and catchments have become the dominant form of regionalisation for natural resource governance in many countries. Modern integrated catchment management has its roots in early 20th century progressivism in the United States (Margerum 1995; Muskingum Water

Conservancy District 2002; Waldo 1984). In the 1960s, new social movements concerned with environmental and civil rights issues led to increased demands for direct citizen participation in public policy making. Together with other areas of public policy, integrated catchment management responded with a shift from technocratic planning to various forms of participative planning. This shift took place in the late 1980s and early 1990s, with little consideration either of the implications for the definition of resource governance regions, or of the considerable body of theory in the social sciences that is relevant to regionalisation — such as Central Place Theory (Christaller 1933), gravity modelling (Carrothers 1956), theories of place attachment (Altman and Low 1992; Cuba and Hummon 1993; Kemmis 1990) and Hierarchy Theory (McGinnis 1999; Pattee 1973).

Research on public participation in natural resource governance grew through the 1990s (Buchy et al. 2000), as has the development, empirical testing and refinement of theories of place attachment (Feld and Basso 1996; Stedman 2003; Wilkinson 2000). There has also been increasing understanding of the role that place and community play in influencing natural resource politics and management (Beckley 1995; Carr 2004; Cheng et al. 2003; Field et al. 2003; Parisi et al. 2004; Shannon 1998). While this conceptual and theoretical development was being assembled over the last decade, catchments have nevertheless remained the dominant administrative unit for regional natural resource governance in Australia and elsewhere (Phelps 2003; Reeve et al. 2002). Within the integrated catchment management literature, most authors accept unquestioningly that catchments should form the areal units within which natural resource governance takes place. Others make a case that river catchments can also form a natural unit encompassing cultural and social commonalities (McGinnis et al. 1999; Webler and Tuler 1999).

There is however, a growing weight of argument against the assumption that catchment-based regions automatically incorporate all resource governance issues and their communities of interest (Blomquist and Schlager 2005; Getches 1998; Omernik and Bailey 1997; O'Neill 2005). Brunckhorst (2000, 2002), Parisi et al. (2003, 2004) and Johnson et al. (1999) pointed out that regions of similar biophysical attributes and climate have little correlation to either watershed topography or areas of interest to land use communities. Barham (2001) argued that processes of democratic deliberation that have evolved over long periods of time prior to the emergence of modern environmentalism do not often fit with catchment boundaries. Other authors have argued that physical catchment boundaries rarely coincide with the boundaries of communities that usually form natural units within which resource governance issues are negotiated and resolved (Ewing 2003; Lane et al. 2004; O'Neill 2005). Syme et al. (1994) went so far as to suggest that organisation of community involvement on catchment boundaries would act against the achievement of the stated goals and purposes of integrated catchment management.

13.3 Defining Regions for Resource Governance

Although there have been mounting criticisms of catchments as natural resource governance regions, and the growing conceptual and theoretical development in socio–spatial aspects of natural resource governance, there has been surprisingly few attempts to propose and apply empirical techniques of regionalisation that might address some of these criticisms and build on this growing body of theory (see Cheng et al. 2003; González and Healey 2005; Omernik and Bailey 1997). One attempt described as an example here, was a major study by the Institute for Rural Futures to derive a nested hierarchy of resource governance regions for the non-metropolitan part of New South Wales in Australia. To underpin the spatial analysis however, it was necessary to distil from the growing literature on socio–spatial aspects of natural resource governance some principle characteristics that could inform methodological and analytical development. The three key principles chosen are described below.

13.3.1 Principle 1

> Principle 1: Resource management regions should reflect the area of most interest to local resident communities.

People are quite capable of identifying the locality of their 'place attachment' or the area they regard as their community (Altman and Low 1992; Cheng et al. 2003; Cuba and Hummon 1993; Hillery 1955; Hobbs et al. 2002; Kemmis 1990; Stedman 2003). Place-based territorial development for local to regional governance is considered important for a variety of purposes and processes (Albrechts et al. 2003; Brandenburg and Carroll 1995; Wilkinson 2000). Parisi et al. (2004) and Brunckhorst et al. (2006) have found that the 'place geography' of residents of communities corresponds with their area of local civic interest, their social networks and the area for which they want representation. A spatially representative social survey and initial methodological trials demonstrated that there is a high degree of spatial conformity between the areas regarded as the location of one's community, the areas regarded as acceptable for the residential location of one's elected representative in local government, the area of one's local social networks and interactions, and the areas within which one would wish to be consulted about resource governance decisions affecting those areas (Brunckhorst et al. 2006). Such an area is referred to here as a 'community area'. While people will have interests in distant places too, their local community area is the locus of substantial social and economic interaction with other residents, and of interaction with the natural resource base.

A position in the landscape will lie within one or more community areas belonging to the people living in the vicinity of that point. A point in the landscape that lies within a large number of overlapping community areas is a point in which a correspondingly large number of people have an interest. Resource governance decisions affecting this point in the landscape will have to consider the interests of this large number of people. If the boundaries of natural resource governance regions cut through such an area, local community participation and engagement will be greatly compromised (Fig. 13.1). Indeed, it is likely that many residents will feel dissatisfied with consultative processes and the representation of their interests (Knight and Landres 1998; Parisi et al. 2002, 2003; Reeve et al. 2002; Shannon 1998). Other points in the landscape will lie within relatively few community areas. It is preferable therefore, if the boundaries of natural resource governance regions pass through areas of minimum collective interest to local people. If the boundaries of natural resource governance regions pass through these parts of the landscape, then a minimum of people will be in a situation in which their community area is divided between one or more resource governance regions (Fig. 13.1). For this reason, this first principle proposes that resource governance boundaries should pass through points that lie within relatively few areas of shared interest to local communities.

13.3.2 Principle 2

Principle 2: The administrative region within which natural resource management occurs should contain a relatively homogeneous set of landscapes with similar climate, ecological and geophysical characteristics.

The biosphere can be divided into continents and oceans, and the former further subdivided into broad continental regions.

Fig. 13.1. The 'place geography' of community residents corresponds with their area of local civic interest, their social networks and the area for which they want representation or to be involved in decision making. It is more desirable for residents to have their community of interest wholly within local and regional governance and administrative boundaries (A) than the boundary intersecting and dividing their community of interest (B)

These can be subdivided into ecoregions and landscapes, and landscapes into ecosystem components, and further subdivided into patches or structural units and so on (Bailey 1996; Brunckhorst 2000; Omernik 1987, 1995; Wiken 1986). Across broad continental regions, patterns are generally observable at various spatial scales where similar organisms or biophysical attributes occur together. These mosaics are composed of units within which internal homogeneity is relatively high. When similar recurring ecological communities are replaced by a different set of recurring natural units, landscape boundaries can be observed and their underlying causes inferred fairly accurately (Forman 1995; Forman and Godron 1981, 1986; Hansen and di Castri 1992). Efficiency and effectiveness in resource governance is likely to be considerably enhanced if planning, priority setting, and management actions and monitoring take account of these boundaries (Field et al. 2003; Johnson et al. 1999; McGinnis et al. 1999; Omernik and Bailey 1997; Reid and Murphy 1995). Resource management planning and actions are improved when dealing with similar contexts of soils, local climate, elevation and topography. Infrastructure capital expenditure and maintenance is also more efficient when understood in terms of similar local to regional biophysical conditions (Brunckhorst 2002; Slocombe 1993).

13.3.3 Principle 3

> Principle 3: The nature and reach of the environmental externalities of resource use determine the size and nesting of resource governance regions.

Collective decision making and collaboration in natural resource management is necessary because one person's use of natural resources impacts upon other parts of the landscape and people. The spatial extent of these environmental externalities can range from the local (e.g. noise pollution), to the regional (e.g. groundwater extraction from regional aquifers), to the national or global (e.g. carbon dioxide emissions). If those who create, and those who are affected by these externalities, are to be represented in shared decision making, then the resource governance region within which this takes place has to be of a similar scale as the reach of the externalities (Cole 2002; Reeve 2003).

Many environmental externalities operate simultaneously across a range of scales. For example, vegetation clearing for agriculture on a farm might result in outbreaks of salinised land on adjacent farms, and an increase in salinity of surface waters which has impacts on urban water users some distance away. For this reason, it is likely that in most areas resource governance regions will need to be nested, with smaller regions (dealing with local problems) nested within larger regions (dealing with environmental externalities with a longer reach). The institutional design principles by which nested resource governance regions might operate are beyond our focus herein, but should be carefully considered in application and development of governance arrangements (see Marshall 2005; McGinnis 1999).

13.4 Application of Principles to Spatial Analysis

The regionalisation approach consisted of three major components, and required the formulation of the concept of a 'social surface' — a topography of the areas of highest collective community identity and interest (described further below). The three steps were:

- Step 1: Derivation of a social surface and a hierarchy of 'civic' regions defined by the 'valleys' in social surface (to satisfy Principles 1 and 3).
- Step 2: Derivation of a hierarchy of biophysical regions (to satisfy Principles 2 and 3).
- Step 3: Optimisation of the boundaries of the two hierarchical regionalisations so that all three principles are satisfied to the maximum degree possible.

The following sections describe the methods followed and results for each of these three components, with an emphasis on the social surface and civic regions.

13.4.1 Delineating Civic Regions from a Social Surface

The methods currently being used in Australia for the derivation of social catchments are highly dependent on the acquisition of primary data from surveys of residents and, for economic reasons, are infeasible to apply on a large scale (see Hugo et al. 2001).

A modelling approach was developed that could use mostly secondary data, and which utilised insights from theories of place and cognitive mapping (Altman and Low 1992; Austin 1994; Cheng et al. 2003; Cuba and Hummon 1993; Hillery 1955; Hobbs et al. 2002; Kearney and Kaplan 1997; Tuan 1974). This modelling approach was founded on the observation from the primary data from a study of northern New South Wales (see Brunckhorst et al. 2006) that the community areas that people drew on a map of their region approximated ellipses in outline, with sizes ranging from a few kilometres across the shortest dimension to over a hundred kilometres. For the majority of rural residents, the ellipse was defined by their place of residence (home point) at one end of the ellipse and a town at the other end. For residents in smaller towns or villages, the elliptical community area generally included the nearest larger town, while for residents in larger towns, the community area included one or more smaller towns in the region, usually along major highways. Community areas tended to be larger in the more sparsely settled regions of northern New South Wales, and smaller in the more densely settled coastal regions. This suggested that it would be possible to model community areas by populating the State of New South Wales (NSW) with simulated home points, and attaching an elliptical simulated community area to each home point, appropriately sized and orientated according to the location of towns of various sizes in the vicinity.

Simulating Home Points

A spatial resolution of 1 km had been set for the study which led to a spacing of simulated home points at intervals of 500 m or less. The Census Collection Districts (CCDs) for NSW were ranked by population density and the population fraction for simulation for the least dense CCD set to a value that would provide for distances of 500 m between simulated home points when that fraction of the population of the CCD was uniformly distributed across the geographical extent of the CCD. The required population fraction for the least dense CCD was found to be 0.66. However, if this value were to be used in densely settled areas, this would result in far more simulated home points than needed to generate the social surface described below. Accordingly, a continuously variable population fraction was used, where the fraction was an inverse function of population density. This resulted in one simulated home point per CCD in population dense metropolitan areas and large cities. The procedure described above resulted in 14,339 simulated home points spread across NSW.

Simulating Community Areas

Simulated elliptical community areas, sized and oriented according to the factors described above, were placed on each of these home points. NSW was divided into five regions, each region having a different mean community area size. These mean sizes were chosen to reflect the variation in community area size known from our previous study. As community areas were generated by the model in each region, they were randomly varied in size to give a size distribution similar in shape to that found in the earlier social survey of residents of north-east NSW, with a mean community area size equal to that set for the region (Brunckhorst et al. 2006). The next transformation of the simulated community areas was to orientate them such that they included one or more towns in the vicinity of the home point. To avoid boundary effects in regions close to the NSW border, towns in Queensland, South Australia and Victoria were included among the towns influencing the orientation of generated community areas.

The final step in the modelling procedure was to assign each simulated community area a height of one unit in a third dimension at right angles to the north-south and east-west dimensions of the map of NSW. Working in this three-dimensional space, the simulated community areas were summed to produce a 'social surface'. The social surface obtained by summing the elliptical community areas on each of the 14,339 simulated home points is shown in oblique view in Fig. 13.2. High points on this surface corresponded to points that lay within the community areas of relatively large numbers of people (strictly, large numbers of simulated home points). Low points on the surface corresponded to points that lay within the community areas of relatively few people. As proposed in Principle 1, above, it is these low points in the social surface that are suitable areas through which resource governance region boundaries might pass.

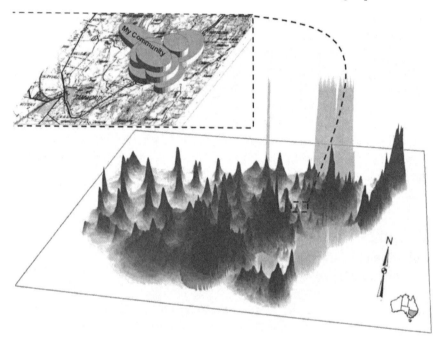

Fig. 13.2. Areas of shared interest to local communities overlap and stack on each other creating a social–civic topography. This figure shows the simulated social surface for non-metropolitan New South Wales. The peaks representing Sydney and Canberra have been truncated and rendered semi-transparent to avoid obscuring the parts of the surface behind. With the exception of the Sydney and Canberra regions, darker areas indicate higher elevations of the surface

13.4.2 Deriving a Hierarchy of Civic Regions

A hierarchy of regions based on the simulated social surface can be produced by locating major and minor 'valleys' in the social topography of the surface. Boundaries based on the major 'valleys' will define larger level 1 regions, and boundaries following the 'valleys' within these regions will define the smaller level 2 sub-regions. At the next level, boundaries on minor 'valleys' within the level 2 sub-regions will define the yet smaller level 3 sub-regions.

Hierarchies and tributary levels of river watersheds and catchments are derived by the height and position of valleys in the landscapes topography. Likewise, a social surface can be treated as a topography where the hills and peaks represent spatially defined areas of high shared community interest, and valleys at various lower levels indicate areas of less and lesser collective community of interest. The Hydrological Modelling Tool in ESRI, *ArcView* 3.2, was used to produce such a 'drainage network' on the topography of the modelled social surface. 'Valleys' at the lower 'altitudes' of the modelled social surface, indicate possible locations for level 1 boundaries, those in the middle 'altitudes' — level 2 boundaries and those at the upper 'altitudes' — level 3 boundaries.

In some areas, the 'topography' of the social surface did not necessarily give a strong indication as to the placement of boundaries. This was a consequence of broad shallow 'valleys' in the surface, or the presence of several 'valleys' in close proximity that were equally good candidates for the location of a boundary. For this reason, a telephone survey of a number of community organisations with hierarchical structures of local, regional and state branches was undertaken. Use of 'key informants' is an efficient way of gathering surrogate data or for 'ground truthing' as used here (Cheng et al. 2003; Parisi et al. 2002). A total of 403 interviews with office bearers in the Country Women's Association, the Hockey Association, the Soccer Association and the Netball Association were completed. Interviewees were asked about the localities in their region where their organisation interacted with similar organisations as part of social activities and/or sporting competitions. Information from the telephone survey of community organisations and the spatial arrangement of 'valleys' in the social topography was combined to produce a three-level hierarchical regionalisation of the modelled social–civic surface, shown in Fig. 13.3.

Validation of the Hierarchy of Civic Regions

Our earlier study provided an empirically measured social surface and associated set of civic regions for north-eastern NSW, against which the modelled civic regions could be compared. In that study, a classification matrix was used to record, for each civic region, the proportion of home points that were assigned to the same civic region when the modelled surface is used to derive the boundaries between the regions. The accuracy of placement or agreement between the modelled boundaries and the measured boundaries in north-eastern NSW was extremely good (Carletta 1996), with correct classifications of more than 98.6% of the 1,973 home points in the region for which measured data was available (Kappa=0.982, p<0.0005).

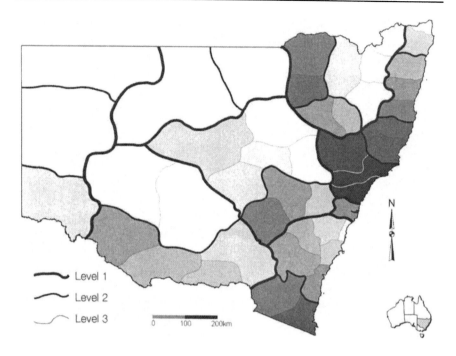

Fig. 13.3. Hierarchy of civic regions derived from the simulated social surface

13.4.3 Deriving Ecoregions

The biophysical regionalisation was based on elevation, soil moisture, soils, and climate data at scales of 1 km or finer, using the ERDAS Imagine 8.5 classification routine (for details, see Brunckhorst et al. 2006). Vegetation data, as a surrogate for environmental attributes, were also classified separately for comparative purposes and to confirm nesting of ecoregions. The result was a hierarchical biophysical regionalisation comprising eight major regions (level 1), each of which was divided into sub-regions (level 2). The level 2 sub-regions were further subdivided into level 3 sub-regions. While a fourth level might have been derived for some areas, the scale of mapping and spatial accuracy might be compromised without finer scale social survey data. The hierarchical ecoregional boundaries of the biophysical regionalisation are shown in Fig. 13.4.

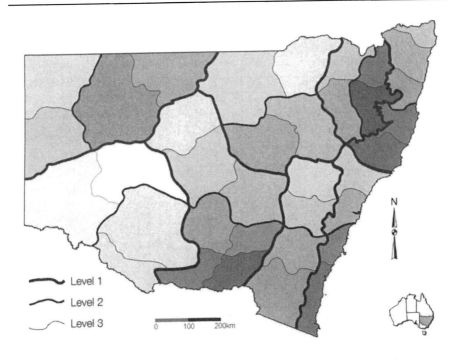

Fig. 13.4. Nested ecoregions — the hierarchical biophysical regionalisation of the State of New South Wales

13.4.4 Integrating Ecoregions and Civic Regions through Boundary Optimisation

The boundaries that define the civic regions (Fig. 13.3) do not necessarily coincide with the boundaries of the ecoregions derived from the multi-attribute biophysical regionalisation (Fig. 13.4). There is coincidence along the eastern escarpment of the northern and southern tablelands, because a sparsely settled area coincides with a major climatic, floral and faunal discontinuity in the landscape.

At the three different scales of the hierarchy across the regions, it is necessary to adjust the boundaries of the civic regions to bring them into closer coincidence with the boundaries of the biophysical regionalisation. This is possible by the fact that the 'valleys' in the social surface can be quite broad. This is particularly so for the 'valleys' at lower 'altitudes' (level 1) in the social surface. This means that the boundary can be moved reasonable distances within the valley, without causing a significant increase in the number of community areas that are intersected by the boundary. At broader scales (levels 1 and 2) therefore, the optimisation routine can give more weight to the biophysical boundaries. However, at finer scales (level 3) it is necessary to ensure that the optimisation routine does not shift boundaries into relatively high areas on the social surface. Following this routine for boundary placement boundaries when integrating biophysical and civic regions is termed 'eco–civic optimisation'. The resulting set of regions is termed an 'eco–civic regionalisation'. The eco–civic regionalisation for NSW is shown in Fig. 13.5.

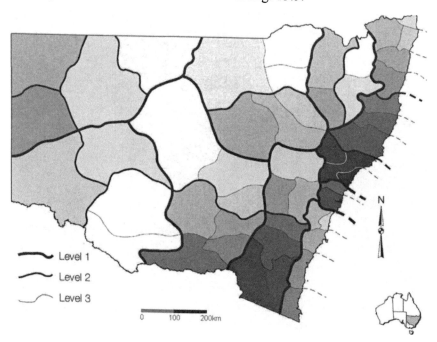

Fig. 13.5. Eco–civic regionalisation of New South Wales, following optimisation of boundaries through integration of biophysical and civic regionalisations

13.4.5 Comparing the Performance of Regions

If the catchment boundaries of existing Catchment Management Authorities (CMAs) are overlaid on the boundaries of the level 2 eco–civic regions (Fig. 13.6), it can be seen that these catchment-based boundaries are a poor fit with both the areas of community interest and with ecoregions. The eco–civic regions appear to be more representative of homogeneous social and ecological characteristics.

Before eco–civic regions are implemented as planning and administrative areas for resource governance, policy makers may wish to know how well existing administrative regions perform in comparison. For any given administrative region, some community areas will be wholly within the region boundary, while others will be intersected by the region boundary. The proportion of people's community areas that are wholly within a region boundary, compared to the total number of people living within that boundary, provides an index of the performance of the particular resource governance region in terms of its ability to include the areas that are of most civic interest to residents. This index is termed the 'Community Capture Index' (CCI). The CCI provides a means of comparing the performance of different regions in terms of the extent to which people's community areas are intersected by region boundaries. In conformity with Principle 1, above, a regionalisation with boundaries that intersect fewer community areas (higher value of the CCI), is preferable to a regionalisation that intersects a greater number of community areas (lower value of CCI, see Fig. 13.1).

Figure 13.7 shows the plot of CCIs for the three levels of the eco–civic regionalisation, and for a range of current administrative regions in NSW, including Local Government Areas (LGAs), and Catchment Management Authority (CMA) regions (Fig. 13.6) which are based on catchment boundaries. The figure demonstrates that the current administrative boundaries including CMAs are in sub-optimal locations because they intersect or divide up areas of shared collective concern and interest to local residents and communities. The eco–civic regions minimise the number of people for which a regional governance boundary might intersect an area of interest to them. Of note is the very poor performance of existing LGAs in their representation of communities of shared interest and identity (Fig. 13.7). Indeed previous work has shown that the most populous LGAs represent only around 10% of the area of social and civic interest to resident communities, and perform worse in their representation of communities of interest than would a random allocation of areas (Brunckhorst et al. 2006).

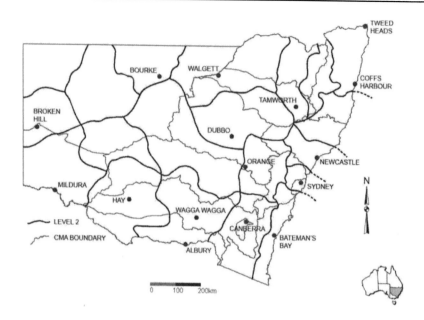

Fig. 13.6. Boundaries of Catchment Management Authorities (CMAs) and the level 2 eco–civic regionalisation

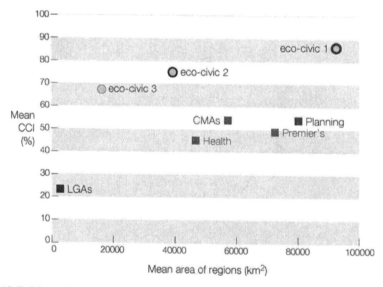

Fig. 13.7. Mean Community Capture Index (CCI) plotted against mean area of regions for a range of administrative regions. LGA: Local Government Areas; CMAs: Catchment Management Authority regions; Health, Planning and Premier's refer to government departments for administration and service delivery

13.5 Conclusion: Past, Present and Future Resource Governance

The past three decades have seen the emergence of catchments and watersheds as the dominant resource governance for planning and management regions. The assumption has been that soils, vegetation, other biodiversity, land use, and groundwater, along with community engagement and collective action occur within catchment boundaries. In practice, catchment management has a history of inefficiency, inappropriate monitoring and high transaction costs (Barham 2001; Brunckhorst 2000; Getches 1998; Lane et al. 2004; Minami and Tanaka 1995; Omernik and Bailey 1997; O'Neill 2005; Stokols and Shumaker 1981; Syme et al. 1994). Catchments however generally do not represent either the 'place attachment', the communities' area of interest or the ecological resource base very well.

This chapter has described three principles which can underpin the development of regionalisations for government administration of, and community participation in, natural resource governance. The principles relate to the spatio–social context representing communities of interest, optimised for homogeneity of the ecological landscape, and spatially bounded in a nested hierarchy to facilitate scaling of institutional arrangements for management of externalities. While some small catchments and watersheds might reflect these characteristics, most do not. The approach illustrated involves identifying where boundaries between resource governance regions should pass so as to minimise the fragmentation of the areas of the landscape with which local people identify, and in which they have an interest. Boundary placement is further optimised to ensure that natural resource issues and ecosystem functions are as homogenous as possible within the regions defined by the boundaries. The Community Capture Index provides a numerical measure of the extent to which boundaries cut through areas of the landscape with which resident communities identify and have an interest.

Utilisation of eco–civic regions is more likely to increase and maintain civic interest and engagement in local governance issues, including the planning and monitoring of natural resource management, while reducing transaction costs and externality effects. Application of the eco–civic methodology to the design of local to regional institutions for resource governance would be valuable in Australia and other nations, for reassessing federalism and regionalism governance issues — including the restructuring of local government areas to regional government — while integrating appropriate scales of regional environmental and development planning, and other government surface delivery. Nesting at broader scales would also enhance collaboration and cross-jurisdictional management as dictated by externalities, and efficiently integrate sustainability policy and actions across multiple scales of social–ecological systems interactions.

13.6 Future Directions

The science–policy dialogue must become increasingly responsive, each to the other. Understanding and identifying windows of opportunity to change direction in policy, planning and governance need to be strategic priorities of science–policy and community interactions (Brunckhorst 2005). The concepts and applications of the advances described in this chapter suggest not only a practical and sensible breakthrough in resource governance — they provide a radical new direction in strategic spatial governance. There are challenges in the practice of forging such systems-based integrative solutions. Changing trajectories in policy and planning for integrated resource management is not likely to be easy, as entrenched practices have been institutionalised beyond their capacity to deliver sustainable resource management.

Future applications of the eco–civic approach to establishing nested regional governance at international levels include contributing solutions and options to address the challenges faced by the European Union with regionalism issues (see for example, Albrechts et al. 2003) and social–ecological systems contexts, which sometimes transgress nation state boundaries. Plans for use of the eco–civic methodology are being developed to better understand scales of social–ecological interactions across Idaho and Montana for conservation and sustainable land use cross-boundary programs. At other scales of application in strategic spatial governance of resources, future applications include finer scales of local policy communities, and nested institutions for water sharing, trade and management.

References

Albrechts L, Healey P, Kunzmann KR (2003) Strategic spatial planning and regional governance in Europe. Journal of the American Planning Association 69(2):113–129

Altman I, Low SM (eds) (1992) Place attachment. Plenum Press, New York

Austin DE (1994) Incorporating cognitive theory into environmental policy making. Environmental Professional 16:262–274

Bailey RG (1996) Multi-scale ecosystem analysis. Environmental Monitoring and Assessment 39:21–24

Barham E (2001) Ecological boundaries as community boundaries: the politics of watersheds. Society and Natural Resources 14:181–191

Beckley TM (1995) Community stability and the relationship between economic and social well-being in forest-dependent communities. Society and Natural Resources 8(3):261–266

Berkes F, Folke C (eds) (1998) Linking social and ecological systems: management practices and social mechanisms for building resilience. Cambridge University Press, Cambridge

Blomquist W, Schlager E (2005) Political pitfalls of integrated watershed management. Society and Natural Resources 18:101–117

Brandenburg AM, Carroll MS (1995) Your place, or mine: the effect of place creation on environmental values and landscape meanings. Society and Natural Resources 8:381–398

Brown J, MacLeod N (1996) integrating ecology into natural resource management policy. Environmental Management 20(3):289–296

Brunckhorst DJ (2000) Bioregional planning: resource management beyond the new millennium. Taylor and Francis, Amsterdam

Brunckhorst DJ (2002) Institutions to sustain ecological and social systems. Ecological Management and Restoration 3(2):109–117

Brunckhorst DJ (2005) Integration research for shaping sustainable regional landscapes. Journal of Research Practice 1(2), article M7. Retrieved January 2008, http://jrp.icaap.org/content/v1.2/brunckhorst.html

Brunckhorst D, Reeve I, Coop P (2006) Eco-civic optimisation: a nested framework for planning and managing landscapes. Landscape and Urban Planning 75(3–4):265–281

Buchy M, Ross H, Proctor W (2000) Enhancing the information base on participatory approaches in Australian natural resource management. In: Land and Water Australia. Natural resources management: people and policy. Land and Water Australia, Canberra

Carletta J (1996) Assessing agreement on classification tasks: the kappa statistic. Computational Linguistics 122:249–254

Carpenter SR, Gunderson LH (2001) Coping with collapse: ecological and social dynamics in ecosystem management. Bioscience 51(6):451–457

Carr AJ (2004) Why do we all need community science? Society and Natural Resources 17(9):841–849

Carrothers G (1956) A historical review of the gravity and potential concepts of human interaction. Journal of American Institute of Planners 22:94–102

Cheng AS, Kruger LE, Daniels SE (2003) 'Place' as an integrating concept in natural resource politics: Propositions for a social science research agenda. Society and Natural Resources 16(2):87–104

Christaller W (1933) Central places in southern Germany (translated by CW Baskin 1966). Prentice-Hall, Englewood Cliffs, NJ

Cole DH (2002) Pollution and property. Comparing ownership institutions for environmental protection. Cambridge University Press, Cambridge

Congalton R (1991) A review of assessing the accuracy of classifications of remotely sensed data. Remote Sensing of the Environment 37:35–46

Cuba L, Hummon DM (1993) A place to call home: identification with dwelling, community and region. Sociology Quarterly 34:111–131

Ewing S (2003) Catchment management arrangements. In: Dovers S, Wildriver S (eds) Managing Australia's environment. Federation Press, Sydney, pp 393–412

Feld S, Basso KH (1996) Senses of place. School of American Research Press, Santa Fe

Field DR, Voss PR, Kuczenski TK, Hammer RB, Radeloff VC (2003) Reaffirming social landscape analysis in landscape ecology: A conceptual framework. Society and Natural Resources 16(4):349–362

Forman RT (1995) Land mosaics: the ecology of landscapes and regions. Cambridge University Press, Cambridge

Forman RT, Godron M (1981) Patches and structural components for landscape ecology. BioScience 31:733–740

Forman RT, Godron M (1986) Landscape ecology. J. Wiley and Sons, New York

Frey BS, Eichenberger R (1999) The new democratic federalism for Europe. Edward Elgar, Cheltenham

Getches D (1998) Some irreverent questions about watershed-based efforts. Chronicles of Community 2:28–34

González S, Healey P (2005) A sociological institutionalist approach to the study of innovation in governance capacity. Urban Studies 42(11):2055–2069

Hansen AJ, di Castri F (eds) (1992) Landscape Boundaries: consequences for biotic diversity and ecological flows, Springer-Verlag, New York

Hillery GA (1955) Definitions of community: areas of agreement. Rural Sociology 20:111–124

Hobbs BF, Ludsin SA, Knight RL, Ryan PA, Biberhofer J, Ciborowski JJ (2002) Fuzzy cognitive mapping as a tool to define management objectives for complex ecosystems. Ecological Applications 12:1548–1565

Hugo GP, Smailes P, Macgregor C, Fenton M, Brunckhorst D (2001) Defining social catchments in non-metropolitan Australia. Bureau of Rural Sciences, Canberra

Johnson K, Swanson F, Herring M, Greene S (1999) Bioregional assessments: science at the crossroads of management and policy. Island Press, Washington DC

Kearney AR, Kaplan S (1997) Toward a methodology for the measurement of knowledge structures of ordinary people: the conceptual content cognitive map (3CM). Environmental Behavior 29:579–617

Kemmis D (1990) Community and the politics of place. University of Oklahoma Press, NormanKnight RL, Landres PB (eds) (1998) Stewardship across boundaries. Island Press, Washington DC

Lane MB, McDonald GT, Morrison TH (2004) Decentralisation and environmental management in Australia: a comment on the prescriptions of the Wentworth Group. Australian Geographical Studies 42(1):103–115

Margerum RD (1995) Examining the practice of integrated environmental management: towards a conceptual model. Unpublished PhD dissertation. University of Wisconsin-Madison

Marshall GR (2005) Economics for collaborative environmental management: renegotiating the commons. Earthscan, London

McGinnis MD (ed) (1999) Polycentric governance and development. University of Michigan Press, Ann Arbor

McGinnis VM, Wooley W, Gamman J (1999) Bioregional conflict resolution: rebuilding community in watershed planning and organising. Environmental Management 24(1):1–12

Minami H, Tanaka K (1995) Social and environmental psychology: transaction between physical space and group-dynamic processes. Environmental Behavior 27:43–55

Muskingum Water Conservancy District (2002) Frequently asked questions about the Muskingum Water Conservancy District. Retrieved 16 April 2007, http://www.mwcdlakes.com/about.htm

Omernik JM (1987) Ecoregions of the conterminous United States (Level II). Annals of the Association of American Geographers 77(1):118–125

Omernik JM (1995) Ecoregions: a spatial framework for environmental management. In: Davis W, Simon T (eds) Biological assessment and criteria: tools for water resource planning and decision making. Lewis Publishing, Boca Raton, Florida, pp 49–62

Omernik JM, Bailey RG (1997) Distinguishing between watersheds and ecoregions. Journal of the American Water Resources Association 33(5):1–15

O'Neill KM (2005) Can watershed management unite town and country? Society and Natural Resources 18:241–253

Pattee HH (1973) Hierarchy theory: the challenge of complex systems. Braziller, New York

Parisi D, Grice SM, Taquino M, Gill DA (2002) Building capacity for community efficacy for economic development in Mississippi. Journal of Community Development and Society 33(2):19–38

Parisi D, Taquino M, Grice SM, Gill DA (2003) Promoting environmental democracy using GIS as a means to integrate community into the EPA-BASINS approach. Society and Natural Resources 16(3):205–220

Parisi D, Taquino M, Grice SM, Gill DA (2004) civic responsibility and the environment: linking local conditions to community environmental activeness. Society and Natural Resources 17(2):97–112

Phelps A (2003) Total catchment management in New South Wales: governance rhetoric and reality. Unpublished thesis, University of New South Wales, Sydney

Reeve I (2003) Principles for the nested governance of water resources (Occasional Paper 2003/1), Institute for Rural Futures, University of New England, Armidale. Retrieved 15 August 2007, http://www.ruralfutures.une.edu.au/publications/occpapers/occpapers.htm

Reeve I, Marshall G, Musgrave W (2002) Resource governance and ICM. Published report to the Murray-Darling Basin Commission. Institute for Rural Futures, University of New England, Armidale, and Murray-Darling Basin Commission, Canberra

Reid TS, Murphy DD (1995) Providing a regional context for local conservation action: a natural community conservation plan for the southern California coastal sage scrub. BioScience Supplement 1995:84–90

Shannon MA (1998) Understanding social organisations and institutions. In: Naiman RJ, Bilby RE (eds) River ecology and management: lessons from the Pacific Coastal ecoregion. Springer Verlag, New York, pp 529–551

Slocombe DS (1993) Implementing ecosystem-based management: development of theory, practice and research for planning and managing a region. BioScience 43(9):612–622

Stedman RC (2003) Is it really just a social construction? The contribution of the physical environment to sense of place. Society and Natural Resources 16(8):671–686

Stokols D, Shumaker SA (1981) People in places: a transactional view of settings. In: Harvey JH (ed) Cognition, social behaviour, and the environment. Lawrence Erlbaum Associates, Hillsdale, NJ, pp 441–488

Syme GJ, Butterworth JE, Nancarrow BE (1994) National whole catchment management: a review and analysis of processes (Occasional Paper 01/94), Land and Water Resources Research and Development Corporation, Land and Water Australia, Canberra

Tuan YF (1974) Topophilia: a study of environmental perception, attitudes and values. Prentice Hall, New Jersey

Waldo D (1984) The administrative state: a study of the political theory of American public administration. Holmes and Meier, New York

Webler T, Tuler S (1999) Integrating technical analysis with deliberation in regional watershed management planning: applying the National Research Council Approach. Policy Studies 27(3):530–543

Wiken E (1986) Terrestrial ecozones of Canada (Ecological land classification series No. 19). Environment Canada, Ottawa, Ontario

Wilkinson KP (2000) The rural community in America. Social Ecology Press, Middleton, WI

Additional Reading

Berkes F, Folke C (eds) (1998) Linking social and ecological systems: management practices and social mechanisms for building resilience. Cambridge University Press, Cambridge

Brunckhorst DJ (2000) Bioregional planning: resource management beyond the new millennium. Taylor and Francis, Amsterdam, Netherlands

Forman RT (1995) Land mosaics: the ecology of landscapes and regions. Cambridge University Press, Cambridge

Langmore J (2007) To firmer ground: restoring hope in Australia. University of New South Wales Press, Sydney

Lyle JT (1999) Design for human ecosystems: landscape, land use, and natural resources. Island Press, Washington, DC

McHarg IL (1992) Design with nature. John Wiley & Sons Inc., New York

McGinnis MD (ed) (1999) Polycentric governance and development. University of Michigan Press, Ann Arbor

Steinitz C, Arias H, Basset S, Flaxman M, Goode T, Maddoch III T, Mouat D, Peiser R, Shearer A (2003) Alternative futures for changing landscapes: The Upper San Pedro River Basin in Arizona and Sonora. Island Press, Washington, DC

14 Placing People at the Centre of Landscape Assessment

Patricia J Fitzsimons[1] and Donald Cherry[2]

[1] Department of Primary Industries, Parkville Centre, Victoria, Australia
[2] Department of Primary Industries, Epsom Centre, Victoria, Australia

Abstract: Environmental indicators assist governments to fulfil legislative requirements to report on environmental condition and trends, but they also play a broader role in the management of natural and human resources. To ensure an assessment of environmental condition and trends digestible to policy makers, an integrated approach is advocated that gives recognition to conflict and uncertainties surrounding environmental impacts whilst developing the capacity to consider cultural values and economic variables.

This chapter provides a review of three conceptual frameworks used to identify indicators and to guide an integrated assessment of socioecological processes. A case study approach is used to apply the Millennium Ecosystem Assessment (MA) framework to two Victorian landscapes.

14.1 Introduction

Governments and other agencies use environmental indicators to assist in fulfilling legislative requirements for reporting on the state and condition of the environment along with its natural resources. Much of the information derived from indicators is not digestible by policy makers (who are at arm's length from the science). Indicator-based research often has an emphasis on scientifically rigorous processes involving the collection of individual datasets, rather than on ensuring that the data and information is not only robust in its analysis of the condition of the environment, but is able to tell an integrated story of changing landscapes.

Assessing progress towards sustainable development is a challenge for governments at all levels. Asking the right questions is an essential component to being able to measure progress against policy documents or organisational goals. Even if there is adequate data and information available, an integrated assessment of performance cannot be achieved without the guidance of a framework that integrates sociocultural, economic, environmental and governance information. A framework ensures information and indicators used to assess performance function within an analytical problem solving logic (Niemeijer and de Groot 2008). In addition, a triple bottom line approach would give consideration to an assessment that integrates social and economic dimensions to gauge progress towards sustainable development (Henriques and Richardson 2004).

14.2 Background

The Brundtland Commission report (WCED 1987) placed environmental issues firmly on the political agenda and sought to clarify the meaning of sustainable development by linking the environment to economic and social development. The report highlighted the importance of engagement at all levels of society to achieve the aims of a sustainable future, explained through the much quoted definition:

'Sustainable development is development that meets the needs of the present without compromising the ability of future generations to meet their own needs' (WCED 1987).

Increased public awareness of environmental issues and a greater understanding of their linkages to economic and social issues continue to rise throughout the world. In Australia the debate on climate change is enunciated daily in newspapers, on television and the Internet and through public forums. Ensuring information produced is responsive to policy needs as well as public information requirements can be challenging. The collection of data through a range of monitoring programs is often conditional on domestic (political) requirements (i.e. State of the Environment Reporting) as well as fulfilling international commitments (i.e. The Convention on Biological Diversity or the Ramsar Convention on Wetlands). Data collection practices can be fragmented and may not be supported during changes in political leadership. However, the development of indicators provide valuable tools of assessment and aid strategic planning (OECD 2003). The focus of this chapter is the development of indicators applied to the Victorian context.

Since the inception of indicator-based environmental reporting many of the drivers for reporting remain the same:

- to respond to international obligations
- to assist in determining policy and investment decisions
- to identify the major causal factors influencing environmental degradation
- information provision to a public increasingly concerned about environmental issues.

However, much environmental information is not able to be fully utilised by policy makers or the general public due to the technical nature of the work, the overwhelming amount of information produced, and because it provides information that is disconnected from people's holistic experience of the environment. The research on which this chapter is based seeks to provide a more integrated approach to environmental reporting and, importantly, places people at the centre of assessment.

14.3 Methodology

The first phase of the research involved the selection of an indicator framework. A review was undertaken that focused on the Pressure–State–Response (PSR) model (OECD 1991, 1994, 2003), its ongoing development through the Driving Forces–Pressure–State–Impact–Response (DPSIR) model (Bosch 1999) and the Millennium Ecosystem Assessment (MA) framework (Capistrano et al. 2005; Carpenter et al. 2005; Chopra et al. 2005; Hassan et al. 2005). The assessment involved a review of each framework's capacity to guide the selection of indicators that could initiate an integrated story of land and biodiversity, agriculture and sociodemographic change across Victorian landscapes. The research also sought to address current gaps in the ability to assess climate change and greenhouse impacts on regional Victoria.

14.3.1 Pressure–State–Response Model

The PSR model was initially proposed by Friend and Rapport in the 1970s and subsequently adapted by the Organisation for Economic Co-operation and Development (OECD) for environmental reporting. Key environmental issues were identified for each of the OECD countries and a series of indicators defined for each issue. The indicators focused on environmental pressures both direct and indirect, environmental conditions and societal responses (OECD 2003). The logic put forth in the PSR asserts that humans cause damage to the environment, which in turn affects the quality and quantity of natural resources (Hughey et al. 2004; OECD 2003). The PSR developed the causal chain concept to identify environmental indicators around cause and effect. For example, a direct pressure to the environment is defined as the use of resources or consumption patterns. This establishes an opposing relationship between the environment and the economic benefit derived from the use of natural resources and the consequences for the wellbeing of its people.

Developments in modern technology have provided scientists with opportunities to evaluate the way in which science is conducted. A more recent approach sought a more holistic or systems approach to transcend the compartmentalised perspective proposed by Descartes in the 1600s (Descartes 1993). Descartes developed the analytic framework known as 'reductionism' which sought to breakdown a problem into its component parts and therefore to see each component of knowledge as being isolated into causal chains (Sposito et al. 2007a). General System Theory responded by placing emphasis on the importance of understanding the logic and organisation between different elements of a problem through the use of 'hardware' — computers — and 'software' — systems science (von Bertalanffy 1968). Recent literature on 'systems thinking' (Checkland 1981; Checkland and Poulter 2006; Midgley 2000) considers both hard and soft systems. For example, Checkland (1981) challenges our thinking by proposing a more abstract approach to analysis through a focus on human activity. This offers a number of challenges to scientists whilst assisting our understanding of the role played by people within an integrative system.

Whilst it is not the purpose of this chapter to provide a detailed overview of the history and philosophy of systems thinking, we argue that systems thinking provides an important theoretical basis to understanding the inter-relationships of a system, as opposed to reductionism which seeks to study its component parts. As the MA framework focuses on the linkages between ecosystem services and human wellbeing it is therefore informed by a systems thinking approach to the analysis of ecosystem changes and policy responses (Alcamo et al. 2003)

By way of contrast, the PSR applies a reductionist approach to establish 'cause and effect' relationships — this is one of its greatest weaknesses. Whilst the PSR framework has been successful in highlighting causal chains it oversimplifies the linkages between cause and effect. The model assumes linearity which has reduced applicability for decision making and its ability to 'see the big picture'. The addition of scenarios, a more recent tool to guide planning for the future, is a valuable addition to strategic planning whilst providing a systematic way to analyse policy options. The lack of capacity to weigh-up policy options limits the usefulness of the PSR framework (Lempert et al. 2003; Wolfslehner and Vacik in press).

14.3.2 Driving Forces–Pressure–State–Impact–Response Model

An enhanced PSR framework, the Driving Forces–Pressure–State–Impact–Response (DPSIR) model, was developed by the European Environment Agency to report on its activities (Bosch et al. 1999). DPSIR sought to describe the relationship between the origins and consequences of environmental problems whilst identifying causal networks. Whilst the PSR is focused on identifying environmental indicators (physical, biological and chemical), the DPSIR can be used as a starting point to answer a variety of research questions, as well as to identify a set of indicators (USGAO 2004). However, criticism of the DPSIR has focused on its inability to support the development of practical long-term policy making due to its focus on analysing past performance and predetermined environmental stresses rather than on its capacity to highlight future environmental impacts (Spangenberg et al. 1998).

The DPSIR, like its precursor the PSR, remains concerned with pressure indicators, in particular the role of humans in placing pressure on the environment. An alternative perspective is to place the socioeconomic dimension at the centre of ecological research. This is essential when addressing questions such as:

- What environmental improvements or deterioration occur in relation to ongoing economic growth?
- What will be the impact on human wellbeing?

Placing people at the centre of assessment through a consideration of human development is put forward in the Millennium Ecosystem Assessment framework (Ohl et al. 2007).

14.3.3 Millennium Ecosystem Assessment Framework

The Millennium Ecosystem Assessment (MA) framework was developed between 2001 and 2005 under the auspices of the United Nations to provide an integrated assessment of the consequences of ecosystem change for human wellbeing (Capistrano et al. 2005; Carpenter et al. 2005; Chopra et al. 2005; Hassan et al. 2005). The MA framework highlights the richness of multiple causality and interactions across landscapes as it provides a move away from principally reporting on the biophysical environment through a focus on human wellbeing. The MA framework provides the capacity to undertake multi-scale assessments — one that incorporates global, national, regional and local contexts. Its focus on human wellbeing gives recognition to the idea that degraded ecosystems are placing a burden on human wellbeing and economic development, and that better managed ecosystems can meet the goals of sustainable development.

Gaining an understanding of the way in which people function in the landscape and their relationship with ecosystems can assist in identifying improved outcomes for environmental managers. This focus on ecosystems delivers an investigation of the dynamic interactions between people, plants, animals, micro-organisms and the built environment. As a consequence of its capacity to guide an integrated assessment of ecosystem services and human wellbeing, the MA framework was selected to identify a series of social, environmental and economic indicators and to provide an integrated assessment of the condition and trends across a range of Victorian landscapes.

14.3.4 Indicator Selection

A set of indicators were identified and developed to measure human wellbeing. These include basic considerations for a good life such as employment and income. Good social relations were measured by the quality of social networks and security was measured by land tenure arrangements. Once a measurement for human wellbeing was established, ecosystem services of significance to Victoria were identified. These included supporting services, such as diversity of land uses, soil characteristics (structure and decline, salinisation, acidification) and biology. As water plays an important role in life cycle processes and is a critical resource to both urban and rural activities, indicators for groundwater trends and water use efficiency were identified. Provisioning services included the quantity and value of food produced, forest extent and condition, as well as the carbon stocks and value of production from forests, wood and wool. Measuring the quality and quantity of fresh water and consumption patterns were also important indicators. Regulating services included changes in temperature and rainfall over time, the impact of weeds and pest plants and animals, along with annual greenhouse emissions. Cultural services included heritage sites, recreational use of parks and reserves and the social value of rivers.

Significant direct and indirect drivers of change were identified and included changes in demographics, land use and land cover changes, technological adaptation and use, as well as biophysical drivers such as soil erosion by water and wind. Whilst there were initially 100 indicators identified, a process of consultation with key stakeholders, including a number of state government agencies and Catchment Management Authorities, the list was refined to some 40 indicators. The focus was on the capacity of the indicators to tell an integrated story of landscape condition and change.

14.4 A Landscape Approach for Victoria

Victoria lies in the south-eastern corner of Australia and includes a section of the Great Dividing Range which extends some 3,000 km along the eastern coast. Victoria's other main geographic feature is its 1,800 km of coastline, which ranges from sandy beaches and rugged cliffs to mangrove-fringed mudflats.

Victoria is the most densely populated state in Australia with a current population of 5 million, of which 3.4 million live in Victoria's capital city of Melbourne. Its increasing density is a consequence of its history, geography and temperate climate. Land and water resources in Victoria are variable in condition and subject to a range of ongoing and new pressures. Long-standing pressures to catchment health include soil erosion, salinity, and pest plants and animals. Two more recent pressures on land and water resources include socioeconomic and climate change (VCMC 2002, 2007).

Demographic and economic changes are accelerating, particularly in regional Victoria, along with land ownership and land use changes presenting both risks and opportunities. Changes in the rate of natural increase, the ageing of the population and the movement of people between regions presents significant challenges for regional Victoria in the coming years (DSE 2005b; Barr et al. 2005).

A wide range of approaches have been used to describe and map Victorian landscapes. This research takes the approach of understanding landscapes as an association and interaction of various factors that influence and shape the land. These factors include physical, social, human, economic, political and cultural aspects, all of which interact with each other and any representations or analysis are subjective (Johnston et al. 2000). Forman (1995) recognised that boundaries are often determined by natural processes such as bioregions or drainage basins, but suggests that:

'to accelerate the use of ecology in design, planning, conservation, management and policy, we must use regions and landscapes that balance and integrate natural processes and human activities' (Forman 1995, p.14).

This places emphasis on landscapes as social constructs that result from the interactions between people, their communities and their environment (Brunckhorst 2005).

The use of landscapes provides a flexible method for an analysis of condition and trends. Rural landscapes are a combination of:

- people and communities
- resource production and related industries
- economies and political institutions
- biodiversity and ecological systems.

Therefore it is critical to understand the forces of economic, social and environmental change to ensure the sustainable management of natural resources as well as the social sustainability of its people (Brunckhorst 2005).

The MA framework, for pragmatic reasons, defines a series of 'systems' for reporting on the condition of ecosystems across the globe. Whilst not being ecosystems themselves, each system represents easily recognisable landscape categories which include both the human and natural environments. Ten distinct systems were characterised representing themes of management or areas of interest and include: marine, coastal, inland water, forest, dryland, island, mountain, polar, cultivated and urban (Hassan et al. 2005). A subset of five landscapes was selected for the Victorian study and these are described below. The definitions do not always conform to those developed in the MA framework as they were adapted and modified to ensure understanding within the Victorian context.

14.4.1 Definitions of Five Victoria Landscapes

Landscapes chosen for the Victorian study include:

- Semi-arid: Lands where plant production is limited by water availability. The dominant uses are livestock grazing, and cultivation.
- Forest: Lands dominated by trees. Often used for timber, fuelwood, and non-timber forest products.
- Coastal: Interface between ocean and land, extending seawards to about the middle of the continental shelf and inland to include all areas strongly influenced by their proximity to the ocean.
- Mountain: Steep and high lands.

- Cultivated: Lands dominated by domesticated plant species. Cultivated land is used for, and substantially changed by, cropping, agroforestry, or aquaculture production.

14.4.2 The Role of Indicators

The series of indicators, developed as part of the research, provided data and a statewide analysis of the condition of Victoria's ecosystem services, human wellbeing and drivers of change. Australia is generally a dry continent with a harsh climate and the coastal region plays a significant role in enabling Australians to deal with these extremes. Case studies of two Victorian landscapes (Semi-arid and Coastal) follow. The aim is to show how the MA framework can be employed to integrate the selected indicators through a focus on the condition of ecosystems services, the drivers of change and their impact on human wellbeing. Two landscapes were used to enable a comparison of the methodology whilst highlighting causes and effects that occur across two very different Victorian landscapes. The statewide analysis of condition and trends was undertaken by scientific experts and the integrated analysis was undertaken by the authors, with emphasis placed on policy relevance.

14.5 Case Study 1: Semi-arid Landscape

Semi-arid is defined by the United Nations Convention to Combat Desertification (UN 1996), as lands where annual precipitation is less than two-thirds of potential evaporation. Semi-arid systems include cultivated lands, scrublands, grasslands, semi-deserts, and true deserts.

Scarcity of water is internationally defined with an Index of Aridity (IA) expressed as the ratio of annual rainfall to annual potential evapotranspiration. The ratio was developed by Thornthwaite (1948) for climate classification. Based on both the UN Convention (above) and Thornthwaites Index of Aridity (IA of less than 0.66 indicates a dry to arid landscape), Fig. 14.1 shows an adaptation of this ratio spatially mapped across Victoria. Much of northern Victoria (approximately 41%) falls into the Semi-arid (IA=0.2–0.5) and Dry subhumid (IA=0.5–0.65) classes with some smaller areas to the south. The data for Fig. 14.1 is based on a long-term average for rainfall divided by potential evapotranspiration. If the IA were calculated for the last 20 to 30 years only, the Semi-arid Landscape would be significantly larger. This is due to a prolonged period of low rainfall recorded across southern Australia.

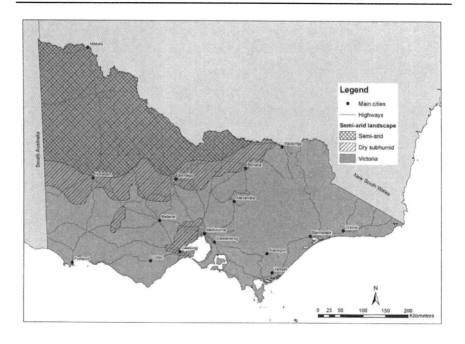

Fig. 14.1. Semi-arid Landscape as defined by the Index of Aridity

14.5.1 Overview

Semi-arid Landscapes are the most vulnerable to loss of ecosystem services. Internationally, much of this loss is due to high human populations in relation to the productive capacity of the system. However, in Victoria the Semi-arid Landscape has a relatively low population that continues to decline in comparison to coastal areas which are experiencing significant growth (DSE 2005b). The main threats to catchment health are the consequences of land clearing and cultivation practices leading to erosion and soil salinity, in addition to the introduction of pest plants and animals. Demographic change and economic adjustments will lead to significant changes in land ownership and use over the next decade and will provide both risks and opportunities (VCMC 2007). In addition, the capacity of agriculture to adapt to climate change will vary according to location and social and environmental vulnerability. Agricultural industries in the Semi-arid Landscape will rely upon the capacity of communities to generate new economic activities, with a greater emphasis on science and technology (Sposito et al. 2007b).

A declining trend in rainfall over much of eastern Australia has been observed since the wet years of the mid 1950s, and dry conditions have persisted since October 1996. It is noteworthy that for Victoria's cropping zone (93% of Victoria's wheat and 55% of Victoria's barley is grown in the Semi-arid Landscape: Sposito et al. 2007b), the six-year period from 2001 to 2007 was more than 80 mm drier than the record 1939–45 drought, and clearly the driest on record (BOM 2007). Drought and loss of land productivity are dominant factors that cause people to migrate from the Semi-arid Landscape to areas that are more productive and better serviced.

Significant pressures and key indicators for the Semi-arid Landscape were identified from a range of studies (Barr et al. 2005; Biggs et al. 2004; Chopra et al. 2005; CSIRO 2003; DNM 2002) and key policy documents for the State of Victoria, and are outlined in Table 14.1. These seek to identify the most dominant pressures on ecosystem services and the types of indicators that could be used to measure their impact. Whilst there are obvious overlaps in ecosystem services and drivers of change across each of the five landscapes, some are more important for each system. For the Semi-arid Landscape, water availability and its efficient use, soil degradation, and farm production are important considerations, with primary production a significant service provided. However, of the forty or so indicators studied, three key areas stood out as fundamentally important in this landscape — water availability and the condition of the water, land use and land use change, and changes in demographics highlighted through changing patterns of employment. Land use and employment are identified as key driving forces whilst the condition of streams is an important indication of the health of this resource.

Table 14.1. Ecosystem services, pressures and indicators for the Victorian Semi-arid Landscape

Ecosystem services	Important pressures	Key indicators
Food production from cropping and grazing	Soil degradation Rainfall variability Water quality	Agriculture tourism Air quality Drought (Aridity Index)
Water resources	Water use	Employment
Production of fibre	Urbanisation Nutrient cycling	Fresh water (access to clean drinking water)
Recreation	Pest infestation	Genetic resources
Cultural services	Carbon emissions	Groundwater trends
Biodiversity	Farm production Rural restructuring	Index of stream condition Land use change Native vegetation cover and condition Nutrient dispersal (fertiliser use) Population growth/decline Production and consumption of wool Rainfall distribution Real income Social capital/social cohesion Soils: salinisation, structure decline, erosion by wind and water, biology Tree plantings Water use efficiency Impact of weeds and pest plants

14.5.2 Employment Indicator

An indicator of employment was derived from monthly estimates of the Labour Force Survey conducted by the Australian Bureau of Statistics (ABS) and ABS Census of Population and Housing spatially mapped (ABS 2007). Figure 14.2 highlights a number of industry sectors with significant changes in employment over the last 20 years. Most significant is the reduction in employment within the manufacturing sector, the rise in property and business services, and a gradual reduction in employment in agriculture, fishing and forestry. This trend reflects a shift in Victoria away from manufacturing and agriculture, forestry and fisheries. The sharp drop during 2002 could be related to an exodus of farmers from the land, as Victoria was in the grip of a severe drought (80% of the state was declared drought affected in July 2002). Although most of Victoria's land is in agricultural production and the state exports much of what it produces, Victorian agriculture is moving from being labour intensive to capital intensive (Barr et al. 2005).

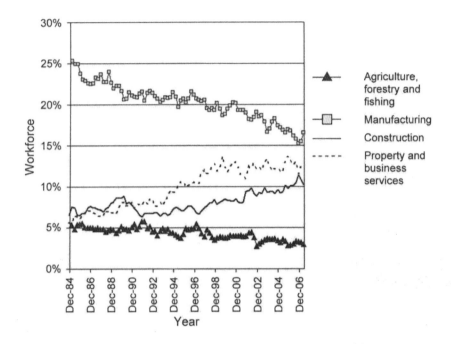

Fig. 14.2. Trends in full-time employment by industry (ABS 2007)

The spatial maps presented in Fig. 14.3 along with the graph presented Fig. 14.2 identify a significant reduction in employment in the agriculture, fishing and forestry sector. The spatial map in Fig. 14.3 highlights the important role of this sector in the Semi-arid Landscape as agriculture is consistently the highest employer and any reduction in this sector will have a significant impact on the wellbeing of the community.

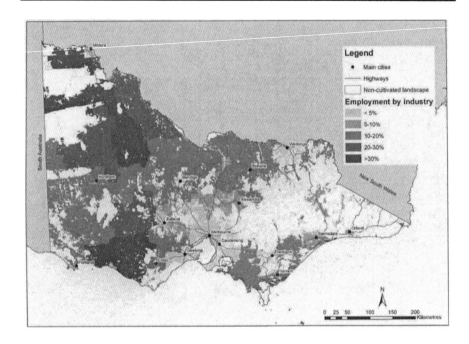

Fig. 14.3. Employment by industry (agriculture, forestry and fisheries) in Victoria

14.5.3 Index of Stream Condition Indicator

The Index of Stream Condition (ISC) is an integrated measure of the overall state of a stream reach, based on the assessment of five component subindices: hydrology, water quality, physical form, riparian zone and aquatic life. It is a measure of changes to the stream from a natural or ideal condition. The ISC has been applied to all major rivers in Victoria between 1999 and 2004. Whilst there was no general improvement detected, importantly, the overall deterioration in stream condition appears to have been controlled. The water quality reflects the degree of modification of the surrounding environment and as a consequence of extensive land clearing in the Semi-arid Landscape; ratings in this region reflect consistently low quality water which are highlighted in Fig. 14.4. The area along the northern boundary reflects a major river system (VCMC 2007).

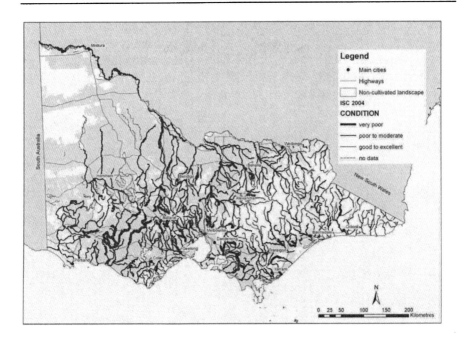

Fig. 14.4. Index of Stream Condition in Victoria

Results from the ISC confirm international trends that inland water ecosystems are in a worse condition than any other broad ecosystem type. The reasons are diverse: land conversion within catchments, over-harvesting of water, the introduction of exotic species, climate change, infrastructure development (dams, diversions, etc.), to name a few. Demand for water from aquatic ecosystems is strong. Changes to flow regimes as a consequence of modifications to habitat or land use (as outlined above) as well as the addition of chemical pollutants reduces the condition of inland water. The declining condition of inland waters is putting the services derived from these ecosystems at risk and this is most severe in the Semi-arid Landscape (Hassan et al. 2006).

14.5.4 Land Use Diversity Indicator

Land use is one of the most crucial drivers of biodiversity change and lies at the interface between social and ecological processes. Changes in land use reflect changes in economic and demographic structure and the interplay between economic diversity, demography and policy goals. The resilience of a community may sometimes be determined by the diversity of industries that supports it. In an agricultural sense, the mapping in Fig.

14.5 identifies the statewide diversity of land uses and highlights a broadacre non-irrigated cropping and grazing mosaic across the Semi-arid Landscape. Major irrigation regions in the Semi-arid Landscape are located on the Murray River which can be identified along the northern border of Victoria and highlighted by the dark shading.

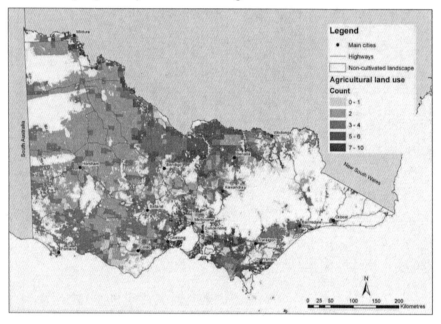

Fig. 14.5. Agricultural land use diversity in Victoria

It is generally regarded that a community with greater diversity is more robust than one dependant on one or two industries. Three main forces determine this diversity: the availability of irrigated water, increased cost of land pressuring land owners into intensification of agriculture, and the advent of hobby/tourist farms. Only one of these determinants is evident in the Semi-arid Landscape — the availability of irrigated water. The prolonged period of drought and a policy of increasing environmental flows to river systems limit the capacity to increase production from irrigation. The pressure then to draw water and use it from existing sources impacts on stream condition, opening debate between stakeholders over the best policy direction to follow on the status of a limited resource. Increasing agricultural production has an impact on ecological sustainability and this will need to be measured and assessed at a range of scales (Dale and Polasky 2007).

14.5.5 Management Response

The Semi-arid Landscape is reliant upon broadacre cropping and crop pasture for its livelihood. With increasing demands for food and potential new markets in ethanol, it plays an important role in the Victorian economy. Employment in agricultural industries continues to decline and the capacity to diversify land uses is restricted due to the availability of water infrastructure. The challenge is in developing an adaptive management strategy that is economically sound and specific to the uniqueness of the Semi-arid Landscape and particularly, to the culture of its people (Hassan et al. 2005).

14.6 Case Study 2: Coastal Landscape

The Coastal Landscape is generally defined as the area between 50 m below mean sea level and 50 m above the high tide level, or extending landward to a distance of 100 km from shore, whichever is nearer the shore (Hassan et al, 2005). It typically includes coral reefs, intertidal zones, estuaries, coastal aquaculture and sea grass communities.

In Victoria however, the 100 km buffer does not apply since in some areas 100 km inland from the coast falls beyond the coastal river catchments. Instead a 50 m boundary works well as it encompasses the majority of ecological systems along the coast as highlighted in Fig. 14.6. It is also more meaningful than trying to use local government areas, catchment boundaries or a set distance, as along the coast many estuaries, mudflats and sea grass beds are included as well as coastal lakes and offshore reefs. Much of the offshore areas are included within the state's marine parks and reserves. While most of the coastline faces the open ocean to the south, Port Phillip Bay (lying roughly in the centre) provides a protected harbour, which acts as a focus for industry, urban development and relatively safe recreational waters.

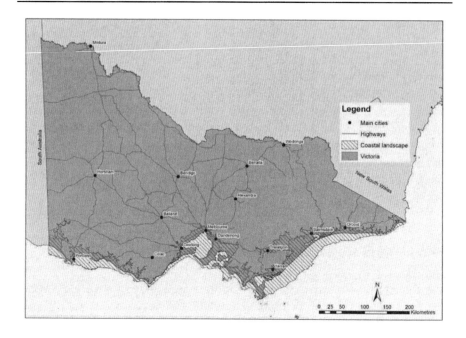

Fig. 14.6. The Coastal Landscape in Victoria, as defined by the +50 m to −50 m rule

14.6.1 Overview

Australians share a love affair with their coast. Rich in natural and cultural resources, the coastal zone is the focus of Australia's economic, social, tourism and recreational activity and supports about 86% of Australia's population (DEWHA 1995). Victoria's coastline is some 1,800 km in length and supports a diverse range of ecosystems. As it is the only major south-facing coastline in the world (it has been isolated for 65 million years), many distinct species have evolved. Whilst these ecosystems are dynamic, due to the buffering from currents and winds, human activity continues to cause serious decline in marine and terrestrial fauna, and the impacts of climate change will increase the effects of human pressures (VCC 2007).

Table 14.2 highlights important ecosystem services, significant pressures and key indicators of the condition of the Coastal Landscape. Significant values are placed on the Victorian coast from the perspective of lifestyle, recreation and tourism, all of which rely heavily on its biodiversity, climate and natural beauty. The main threats to the Coastal Landscape

Placing People at the Centre of Landscape Assessment   295

are therefore related to pressures placed by population growth, and infrastructure development, which may ultimately cause degradation of the natural beauty. Key indicators must be related to the services provided and associated pressures. Significant pressures are derived from the attractiveness of being near the coast and the resulting changes in land use impacts on land value. The indicators chosen: visitors to parks and reserves, land value, and land use diversity, highlight these related factors and assist us in telling an integrated story of the coast.

Table 14.2. Ecosystem services, pressures and indicators in the Coastal Landscape of Victoria

Ecosystem services	Important pressures	Key indicators
Biodiversity Tourism Recreation	Population growth Development Loss of natural environment Lack of infrastructure	Employment Extent and conservation status of native vegetation Index of Stream Condition Acid sulfate soils Land use diversity Land value Population Real income levels Total food production Visitors to parks and reserves Welfare dependency

14.6.2 Visitors to Parks and Reserves Indicator

As an indicator of cultural ecosystems, 'visits to parks in Victoria' highlights the linkages between people and the environment. Increasing visitor numbers to the state's parks shows the high value that the community places on these protected places and an increasing awareness in the sanctuary they provide. However, the increasing demand from recreational activities, whilst developing social capital and enhancing wellbeing through increased employment and business opportunities, has an impact on the provision of other ecosystem services, such as quality of water or competition for food and fibre production. This is particularly apparent along much of Victoria's coastline and highlighted in Fig. 14.7 by the consistent visitor nodes.

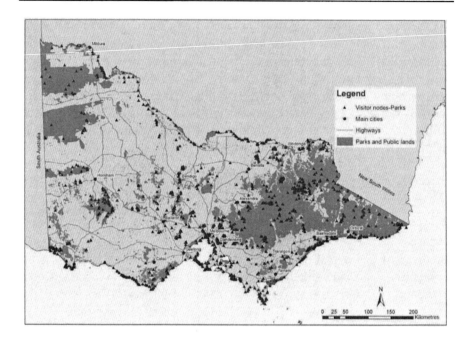

Fig. 14.7. Visitor nodes indicating recreation in Victoria's parks

14.6.3 Ratio of Land Value to Production Value Indicator

The growth of urban populations and increasing affluence has stimulated a growing interest from urban Australians in buying rural land. Land that may have once been valued as an agricultural resource is now valued as a housing site. The result is an escalation in land values that bears no relationship to the agricultural capacity of the land or the state of the agricultural economy.

Data presented in the 'ratio of land values to production values' indicator in Fig. 14.8 is for the period 1996 to 2001 (Barr et al. 2005). This shows high property values close to Melbourne, provincial cities, the coast and along major transport corridors. In these areas farm businesses are under greater pressure to intensify operations if they wish to remain viable.

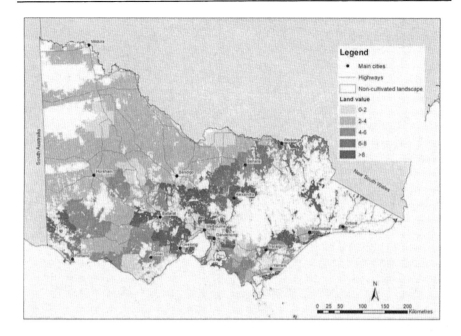

Fig. 14.8. Ratio of land value to production value in Victoria

14.6.4 Land Use Diversity Indicator

In comparison to the Semi-arid Landscape, the Coastal Landscape has all three significant pressures: increased cost of land, the advent of hobby or tourist farms, and water infrastructure enabling a diversity of land uses. Rather than present a statewide map of land use diversity which is unable to highlight more localised land uses, the map in Fig. 14.9 highlights the diversity of land use along one section of the coast, with the addition of a close up of the Geelong portion of Port Phillip Bay.

The diversity of land uses in the Coastal Landscape includes: urban development, industrial sites, parks and recreation spaces, and agriculture (particularly forestry and dairying). Eucalypt plantations have increased in the coastal hinterland and offer both environmental and production values. Their development impacts employment and offers reduced nutrient loss into the watertable, whilst also affecting water flows and groundwater recharge rates (DSE 2005c).

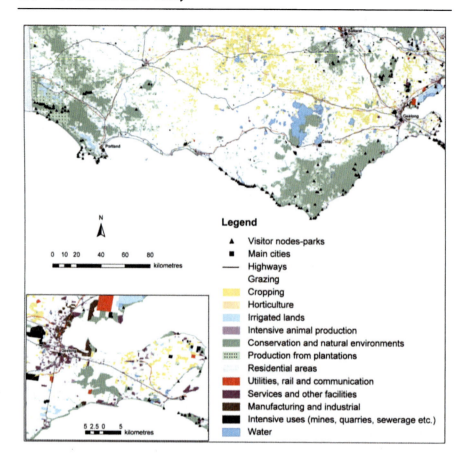

Fig. 14.9. Land use diversity in the Coastal Landscape

14.6.5 Policy Response

The enlarged portion of the Port Phillip coastal area highlights a wide range of land uses: hobby farms, intensive agriculture, industrial sites, public land used for military infrastructure, and tourism associated with vineyards. The city of Geelong is a major shipping port and therefore requires land for its associated infrastructure. Whilst placing enormous pressures on the natural environment, there are a range of impacts on human wellbeing that require consideration, such as increasing land prices and cultural amenity due to overuse. This range of activities provides increased employment and business opportunities, but if this area reflects the potential pressures faced by many other parts of the coastal landscape, then an integrated policy response is required that represents a diverse set of interventions and approaches to ecosystem services and human wellbeing.

14.7 Overview of Results

A comparison of two Victorian landscapes has assisted in identifying the key issues for each of the landscapes and the consequences for sustainable development. Dramatic changes in population growth along the coast places pressure on a number of Coastal Landscape ecosystem services, while land traditionally used for agricultural industry becomes increasingly diversified in order to survive.

In the Semi-arid Landscape, where the quality and quantity of water is critical to survival for both the natural environment and the humans living in it, there is every reason to assume that the situation will not improve in the short-term, and may well get worse before it stabilises. Whilst the drivers of change have been identified by policy makers in a series of policy documents, what is not so well understood is the consequences of long-term impacts of these drivers on the condition of ecosystem services and an analysis of the consequences for the wellbeing of the community. Indicators (as outlined in Tables 14.1 and 14.2) that are consistently developed over a period of time can evaluate the condition and trends of ecosystem services on human wellbeing.

14.8 Conclusion

This research evaluated the Pressure–State–Response (PSR) model, its further development through the Driving Forces–Pressure–State–Response (DPSIR) model and the Millennium Ecosystem Assessment (MA) framework through its capacity to report on the condition and trends of Victorian landscapes. The MA framework was selected to guide the development of a series of indicators, classify a series of Victorian landscapes and provide a framework to conduct an integrated assessment of Victorian landscapes.

A case study of two Victorian landscapes — Semi-arid and Coastal — was undertaken to highlight the methodology developed to assess the consequences of ecosystem change for human wellbeing. This study highlighted inequities in the distribution of losses and gains for ecosystem services and the consequences for human wellbeing through a comparison of two distinct Victorian landscapes. Some of the changes in ecosystems that can be measured have contributed to substantial gains in human wellbeing and economic development, but these gains were achieved at a cost to the sustainable health of ecosystems through exploitation, degradation or loss.

The differences highlighted between the two Victorian landscapes through an examination of some indicators, emphasises this loss/gain paradigm. Significantly, these changes have occurred over a short time-frame. Rapid growth in demands for more resources have drained other ecosystem services and run the risk of causing irreversible damage. An integrated assessment of the consequences of the impact of changes in the condition of ecosystems services and their consequences for human well-being is fundamentally important to ensure policy makers are provided with information based on scientific evidence presented in an integrated format to highlight the impacts for the sustainable development of Victoria and, importantly, the wellbeing of its people.

The MA framework provides a paradigm shift away from an emphasis on the pressures humans exert on the environment to one that provides an integrated assessment of the dynamic interactions between people, living organisms and the built environment. A landscape analysis highlights the inter-dependencies and interactions between ecosystem services and human wellbeing at a scale that seeks to balance and integrate natural processes and human activities, whilst enhancing strategic planning.

Information products, which include maps, graphs and an analysis of condition and trends, were developed to provide policy makers with an evidence-based approach to evaluate progress against policy objectives, particularly progress towards environmentally sustainable development and strategic investment decisions. Application of the MA framework guides assessments at different scales and assists in problem definition through the linking of condition and trends of ecosystem services to human wellbeing.

14.9 Future Research Directions

Policy makers and agencies with a responsibility to report on the state of land and water resources in Victoria are currently sourcing the suite of indicators to tell their own story of landscape condition and change. They are being used:

- as inputs into State of the Environment Reporting
- to undertake an assessment of the condition of land and water resources
- as a guide to policy development for sustainable agriculture
- to review progress against key policy documents such as 'Our Environment Our Future' (DSE 2005a).

It is intended to make the products available to the broader research community and the general public through an online portal, Victoria Resources Online (VRO) <http://www.dpi.vic.gov.au/dpi/vro/vrosite.nsf/pages/vro home>, thus providing a publicly accessible knowledge base which highlights the interactions and interdependencies between people and their landscapes. Whilst the indicator products on VRO will be in a PDF format, several of the indicators have been further developed using visualisation software to highlight changes over time. These temporally animated visualisations of landscape change provide decision makers with an enhanced understanding of changes occurring to land and water resources and socio-ecological processes.

Monitoring biophysical and socioeconomic phenomena through the collection of data with broad spatial coverage and time series attributes requires tremendous resources. Poor data monitoring, collection, and curation of fundamental datasets which underpin indicator products remains an ongoing challenge. Further research is required in developing data collection standards which have sufficient historical persistence and comparability to identify location and temporal changes. Further research should consider the role of spatial modelling to assist in managing gaps in data and information, in addition to the role of scenario development to assess a range of response options. The challenge remains to develop a suite of indicator products that will enable trend analysis between collection years and allow comparison across the state. In this way knowledge can be scaled up and down to satisfy local to global reporting requirements.

Acknowledgements

The research for this paper was carried out as part of the project 'Triple Bottom Line Indicators for Victorian Landscapes' which was funded by the Victorian Government's Department of Primary Industries and Department of Environment and Sustainability. The authors would like to thank the project's Steering Group for their continued review and support — particularly its former chairperson Kimberley Dripps, the current chairperson Angela Avery, and the project team. Thanks are due to the support of Dr Christopher Pettit in encouraging the further development of the methodology presented in this chapter, and the two anonymous referees for valuable suggestions on improving the manuscript.

References

Alcamo J, Bennett EM (2003) Ecosystems and human wellbeing: a framework for assessment, Millennium Ecosystem Assessment. Island Press, Washington

ABS (2007) Labourforce, Australia. Detailed Quarterly (cat.no. 6291.0.55.003), Australian Bureau of Statistics, Canberra

Barr N, Wilkinson R, Karunaratne K (2005) The changing social landscape of rural Victoria. Department of Primary Industries, Melbourne

Biggs R, Bohensky E, Desanker PV, Fabricius C, Lynam T, Misselhorn AA, Musvoto C, Mutale M, Reyers B, Scholes RJ, Shikongo S, van Jaarsveld AS (2004) Nature supporting people – the South African Millennium Ecosystem Assessment integrated report. Council of Scientific and Industrial Research, South Africa

Bosch P, Buchele M, Gee D (1999) Environmental indicators: typology and overview, Technical Report No. 25. European Environment Agency, Copenhagen

Brunckhorst D (2005) Integration research for shaping sustainable regional landscapes. Journal of Research Practice 1(2) Article M7, available at http://www.icaap.org

BOM (2007) Six years of widespread drought in southern and eastern Australia November 2001 – October 2007 (Special Climate Statement 14) Commonwealth of Australia, Bureau of Meteorology, Docklands, available at http://www.bom.gov.au/climate/current/special-statements.shtml

Capistrano D, Samper CK, Lee MJ, Raudsepp-Hearne C (2005) Ecosystems and human wellbeing: multiscale assessments. Volume 4: Findings of the Subglobal Assessments Working Group, Millennium Ecosystem Assessment. Island Press, Washington

Carpenter SR, Pingali PL, Bennett EM, Zurek MB (2005) Ecosystems and human wellbeing: scenarios. Volume 2: Findings of the Scenarios Working Group, Millennium Ecosystem Assessment. Island Press, Washington

Checkland P (1981) Systems thinking, systems practice. John Wiley & Sons, New York

Checkland P, Poulter J (2006) Learning for action, John Wiley & Sons, Chichester

Chopra K, Leemans R, Kumar P, Simons H (2005) Ecosystems and human wellbeing: policy responses. Volume 3: Findings of the Responses Working Group, Millennium Ecosystem Assessment. Island Press, Washington

CSIRO (2003) Natural values: exploring options for enhancing ecosystem services in the Goulburn Broken catchment. CSIRO, Sustainable Ecosystems, Canberra

Dale VH, Polasky S (2007) Measures of the effects of agricultural practices on ecosystem services. Ecological Economics 64(2):286–296

DEWHA (1995) The Commonwealth Coastal Policy, Commonwealth of Australia. Department of the Environment, Water, Heritage and the Arts, Canberra

DSE (2005a) Our environment, our future: Victoria's environmental sustainability framework. Department of Sustainability and Environment, East Melbourne

DSE (2005b) Regional matters: an atlas of regional Victoria. Department of Sustainability and Environment, East Melbourne

DSE (2005c) Victoria's state of the forests report. Department of Sustainability and Environment, East Melbourne

Descartes R (1993) Meditations on first philosophy: in which the existence of God and the distinction of the soul from the body are demonstrated. Hackett Publishing, Indiana

DNM (2002) Norwegian Millennium Ecosystem Assessment – Pilot Study 2002. Directorate for Nature Management, Trondheim

Forman RTT (1995), Land mosaics: the ecology of landscapes and regions. Cambridge University Press, Cambridge

Hassan R, Scholes R, Ash N (2005) Ecosystems and human wellbeing: current state and trends. Volume 1: Findings of the Condition and Trends Working Group, Millennium Ecosystem Assessment. Island Press, Washington

Henriques A, Richardson J (Eds) (2004) The triple bottom line: does it all add up? Earthscan, London

Hughey KFD, Cullen R, Kerr GN, Cook AJ (2004) Application of the Pressure–State–Response framework to perceptions reporting of the state of the New Zealand environment. Journal of Environmental Management 70(1):85–93

Johnston RJ, Gregory D, Pratt G, Watts M (2000) Dictionary of human geography, 4th edn. Blackwell, USA

Lempert RJ, Popper SW, Bankes SC (2003) Shaping the next one hundred years: new methods for quantitative, long-term policy analysis. RAND, Santa Monica

Midgley G (2000) Systemic intervention – philosophy, methodology and practice. Kluwer Academic/Plenum Publishers, New York

Niemeijer D, de Groot RS (2008) A conceptual framework for selecting environmental indicator sets. Environmental Indicators 8(1):14–25

OECD (1991) Environmental indicators. A preliminary set. Organisation for Economic Co-operation and Development, Paris

OECD (1994) Using the Pressure–State–Response model to develop indicators of sustainability: OECD Framework for Environmental Indicators. Organisation for Economic Co-operation and Development, Environment Directorate-State of the Environment Division, Paris

OECD (2003) OECD environmental indicators: development, measurement and use reference paper. Organisation for Economic Co-operation and Development, Environment Directorate, Paris. Available at www.oecd.org/env/.

Ohl C, Drauze K, Clemens G (2007) Towards an understanding of long-term ecosystem dynamics by merging socioeconomic and environmental research criteria for long-term socioecological research sites selection. Ecological Economics 63:383–391

Spangenberg JH, Femia A, Hinterberger F, Schutz H (1998) Material flow-based indicators in environmental reporting. European Environment Agency, Luxenberg

Sposito VA, Faggian R, Aurambout JP (2007a) Systems thinking: foundations and perspectives, technical report 1 on systems thinking. Department of Primary Industries, Melbourne

Sposito VA, Hossain H, Anwar M, Sietchiping R, Fitzsimons P, Rees D, Wu Y, Evans C (2007b) Facing the challenge of climate change impacts and adaptation in regional and agricultural systems in Victoria. Victorian Government, Melbourne

Thornthwaite CW (1948) An approach toward a rational classification of climate. In: Glossary of meteorology, Geographic Reviews 38:55–94, available at http://amsglossary.allenpress.com/glossary

UN (1996) United Nations Convention to Combat Desertification in countries experiencing serious drought and/or desertification, particularly in Africa. United Nations, Bonn

USGAO (2004) Environmental indicators: better co-ordination is needed to develop environmental indicator sets that inform decisions: report to Congressional Requesters. United States Government Accountability Office, Washinton, DC

VCMC (2002) The health of our catchments: a Victorian report card. Victorian Catchment Management Council, East Melbourne

VCMC (2007) Catchment condition report. Victorian Catchment Management Council, East Melbourne

VCC (2007) Draft Victorian coastal strategy. Victorian Coastal Council, East Melbourne

von Bertalanffy L (1968) General system theory: foundations, development, applications. George Braziller, New York

Wolfslehner B, Vacik H (in press) Evaluating sustainable forest management strategies with the analytic network process in a Pressure–State–Response framework. Journal of Environmental Management

WCED (1987) Our common future. World Commission on Environment and Development, Oxford University Press, Oxford and New York

Additional Reading

DPI (2006) Landscape analysis models and frameworks – a review, (Our Rural Landscape 1.1– New Dimensions for Agricultural Landscapes). Department of Primary Industries, Victoria

Moran M, Rein M, Goodin RE (2006) The Oxford handbook of public policy. Oxford University Press, New York

OECD (2007) OECD environmental indicators. Organisation for Economic Co-operation and Development, Environment Directorate, Paris. Retrieved January 2008, http://www.oecd.org/findDocument/0,3354,en_2649_34283_1_119 699_1_1_37465,00.html

Victoria Resources Online (2007) Victoria's geomorphological framework. State of Victoria, East Melbourne, available at http://www.dpi.vic.gov.au/dpi/vro/vrosite.nsf/pages/vrohome

15 The Social Landscapes of Rural Victoria

Neil Barr

Department of Primary Industries, Epsom Centre, Victoria, Australia

Abstract: The restructuring of Victoria's farm industries is conceptualised as an outcome of the competition for land between the farm sector and migrants seeking to purchase land for amenity purposes. Nine statistical indicators are used to build a spatial representation of the restructuring of the rural sector based upon two orthogonal factors. The first factor represents a continuum between localities where rural land is purchased for farm aggregation and localities where rural land is purchased for rural amenity. The second factor distinguishes between localities where agricultural enterprises are intensifying land use and those where agriculture is on a trajectory towards less intense land use. The two factors are used to create five 'social landscapes' each with a divergent trajectory of rural restructuring.

15.1 Introduction

The latter half of the twentieth century saw a transformation of the rural economies and communities of Europe, North America, Australia and New Zealand. The economic explanation is that this transformation was driven by the increasing efficiencies of agricultural production and the diminishing position of agriculture within national economies. Increasing efficiencies lower the price of agricultural commodities, putting price pressure on the farm sector. A diminishing position within the national economy ensures the agricultural sector experiences greater competition for resources from other sectors of the economy. An alternative conceptualisation developed within rural sociology is that this is a 'post-productivist transition' driven by changing social values that are only partially expressed through farm commodity markets (Wilson 2001).

This chapter sets out a conceptual framework of the rural transformation as experienced in Australia. The chapter proceeds in five steps:

- A narrative of Australian restructuring is provided, based upon the interaction between the forces of productivism and migration driven by a search for non-production values from farm land (henceforth referred to as amenity migration).
- This narrative is used to guide the selection of nine statistical indicators of the steps in the narrative. These indicators are the basis of a data matrix of nine indicators for 125 Statistical Local Areas of rural Victoria.
- The data matrix is simplified using Principal Components Analysis. This reduces the nine variables into two significant factors.
- The 125 Statistical Local Areas are segmented into five 'social landscapes' on the basis of the derived factor scores.
- The five 'social landscapes' are examined to reveal unique characteristics and divergent trajectories of structural change reflecting different balances in production and amenity-based demand for land.

15.2 A Narrative of Rural Transformation in Australia

15.2.1 International Agricultural Competition

Agricultural industries have a tendency to create chronic oversupply in agricultural markets. This oversupply is an outcome of the farm sector capturing productivity increases greater than any matching increased demand from the market. The result is a long-term decline in the prices of farm products in relation to the prices of inputs required to produce those products. This is referred to as a decline in the farm sector's terms of trade (Productivity Commission 2005).

Australia has a relatively small domestic market. The farm sector exports 70% of its produce. Farm industries have received only limited protection from trade competition. Government has been gradually withdrawing this protection over the past two decades. Government policy has increasingly emphasised primary producer self-reliance and risk management (Botterill 2003a, 2003b; Lockie 1997). This emphasis has in part been driven by the need to appear consistent in World Trade negotiations (Argent 2002; Botterill 2003a). Most Australian agricultural industries now operate relatively unsheltered from international agricultural competition (Stoeckel and Reeves 2005).

Those that supply domestic markets have not been shielded from the economic pressures experienced in the export sector. A combination of import competition and supply chain restructuring has in some industries intensified the terms of trade squeeze. For example, the processing tomato sector has experienced a doubling of production and a 90% decline in farm numbers in the past 20 years (Pritchard et al. 2007). Food processors and supermarket purchasers use their contractual negotiations to facilitate the expansion of the largest and most efficient farms to increase production and shave margins. This strategy excludes increasing numbers of smaller farms from the dominant outlets in the domestic food supply chain, leaving only residual market structures or new niche opportunities (Griffith 2004).

15.2.2 Agricultural Restructuring

In response to continuing price pressure, Australian farmers seeking to remain competitive have little option but to continuously strive for farm productivity improvements. If they achieve improvements greater than the long-term average 2% decline in their terms of trade, they will improve their business position. If they achieve less than the 2% benchmark, they will fall behind. In striving to keep ahead of the productivity benchmark, they are doing their own part in stoking agricultural over-capacity (Owen 1966).

In many agricultural industries a long-term increase in farm productivity can only be achieved by increasing the scale of farm operations. A larger farm allows the business to take advantage of larger machinery, and to spread labour costs and fixed farm costs over a greater volume of production. Over time it is inevitable that farms will increase in area and decrease in number. In the Australian grain-belt, farms tend to double in size and halve in number every 30 to 40 years. The number of farmers decreases by 1% to 2% per year (Barr 2004, 2005). In a regional rural economy where farming is the dominant industry, increasing farm scale is matched by declining farm numbers and a decline in the population beyond the farm.

15.2.3 Amenity Values in the Rural Land Market

The growth of urban populations and increasing affluence has stimulated a growing interest in buying rural land by urban Australians (Burnley and Murphy 2003). Some purchasers are seeking a quiet weekend retreat. Some seek a lifestyle change. Some aspire to become farmers, though perhaps with little understanding of the economic realities faced by the modern farmer. Land that may have once been sought as an agricultural resource is now purchased as a housing site. Characteristics once little valued by the farm sector become important in setting land value. Hills provide opportunities for an elevated house site. Native bushland provides an attractive landscape. The result is an escalation in land values that bears no relationship to the agricultural capacity of the land or the state of the agricultural economy (Barr 2005).

The competition for land from amenity purchasers makes it more difficult for farmers to capture productivity increases by increasing farm scale. Those in high amenity locations are no longer able to purchase farm land and expand their farm holdings. To remain competitive they must intensify their farm operations within the existing land area. Not all farm managers can, or wish to, implement this option. This means they must choose between selling and moving where land is cheaper, taking off-farm work, or living within the constraints of declining farm income. The outcome is a gradual long-term decline in the competitiveness and financial scale of many farms and a gradual increase in the prevalence of off-farm work. The trade-off for decreasing farm competitiveness is increasing community sustainability. The population living in the rural domain increases. The local economy becomes more diversified as the new migrants bring new businesses and income streams to the area (Barr 2005).

15.2.4 Indicators Derived from the Narrative

The narrative recounted above is a representation of the interaction between the pressure for farm adjustment and the demand for land based on its amenity characteristics. Nine statistical indicators were chosen to represent this narrative as the first step in exploring the geographic variation in agricultural restructuring in the State of Victoria. Each indicator represents a step in the logic of the restructuring narrative. Table 15.1 describes the construction of each indicator, the transformation used to normalise the data, the source of data and the link between the indicator and the narrative. The indicators are measured for Statistical Local Areas (SLA) within the State of Victoria. The SLA is a standard geographical subdivision used by the Australian Bureau of Statistics for providing regional census data (ABS 2001). Statistical land areas were excluded from the analysis if they contained fewer than 10 farmers and fewer than 10 agricultural establishments as defined by the Australian Bureau of Statistics. This reduced the influence on the analysis of extreme ratio scores associated with urban and close peri-urban SLAs.

Table 15.1. Indicators representing geographic variation in Victorian agricultural landscapes

Indicator	Indicator construction	Normalisation transformation	Data source	Place within the narrative
INCREASE_SCALE	Farm scale measured using the ABS EVAO. This is a proxy for the value of farm production. The indicator is the proportional change in median farm EVAO between 1986 and 2001	Ln(x)	ABS Australian Agricultural Census	In response to declining farm terms of trade, farmers increase the scale of farm output
POPULATION	Proportional change in total population between 1996 and 2001	Ln(x)	ABS Census of Population and Housing	In a rural economy where farming is the main industry, population declines as farms increase in size and decrease in number
LAND_MARKET	Purchasers of rural land who have addresses inside the region as a percent of all purchasers (over the period 1996–2001)	Ln(x)	Land Victoria PRISM database of land transactions	In some locations farmers seeking to expand their farm holdings face competition from purchasers from outside their locality seeking property for amenity purposes
LAND_VALUE	Median value of land per hectare as a proportion of the median value of agricultural production per hectare. The data is averaged over the period 1996–2001.	√x	Land Victoria PRISM database of land transactions, ABS Australian Agricultural Census. Land value derived from records of all land transactions for rural blocks larger than 10 hectares. Value of production per hectare created using the ABS EVAO indicator and farm area reported in the Agricultural Census	Competition for land from amenity purchasers raises land value beyond that which would be expected to be paid for agricultural purposes

Table 15.1. (cont.)

FARM_SIZE	Median farm size in 2001 measured using EVAO	$1/x^2$	ABS Australian Agricultural Census	Higher rural land prices constrain the capacity of farms to increase scale by purchasing land. Farms in areas where land values are inflated by amenity purchasers will become progressively smaller
AREA_LARGE	Proportion of farm area managed as part of farms falling within the highest quartile of national farm EVAO in 2001	$Arcsin(\sqrt{x})$	ABS Australian Agricultural Census	Where there is little competition for land, large farms will be the norm. Where there is competition from amenity purchasers, small farms will become the norm
OFF_FARM	Proportion of farm managers who nominate farming as a minor occupation in 2001	$\sqrt{x}$	ABS Australian Agricultural Census	Where small farms become the norm, farmers will often supplement income by seeking off-farm work
ECONOMY	Agricultural employment as a proportion of total employment	$Ln\,(x/(1-x))$	ABS Census of Population and Housing	A more diversified local economy is more attractive to migrants. Migration helps further diversify the economy
IN-MIGRATION	Proportion of population that migrated into the region in the previous 5 years	$Ln\,(x/(1-x))$	ABS Census of Population and Housing	New residents move into rural areas can mitigate the impacts of agricultural consolidation on population

ABS: Australian Bureau of Statistics *EVAO*: Estimated Value of Agricultural Operations

15.3 From Indicators to Social Landscapes

15.3.1 Factor Analysis Using the Principal Components Method

The first step in a geographic exploration of agricultural restructuring was to simplifying the nine indicator variables to a smaller number of underlying factors. 'Principal Components Analysis' is the most commonly used multi-variate technique of exploratory statistical analysis (Everitt and

Dunn 1983). The technique describes a dataset of n cases in a p-dimension environment as a series of uncorrelated variables that are linear combinations of the original variable set. Each new variable has a mean of zero and a standard deviation of one. The new variables are derived in decreasing order of importance. The objective of the Principal Components Analysis is to discover whether a small number of underlying factors can summarise most of the variation in the original data. The 'scree plot' (Fig. 15.1) shows the relative importance of the nine orthogonal factors derived from the data set as measured by the factor's Eigenvalue. Eigenvalues can be taken as a measure of the relative importance of a factor. The first two factors account for 65% of the variance within the data. The remaining factors accounted for only small amounts of variance. The use of the first two factors alone is an efficient strategy to summarise the dataset and is consistent with the recommendation of Catell (1965) who advised selecting those factors that fall before the break of slope in the scree plot.

Principal Components Analysis is often followed by rotation of the factor solution to improve the interpretability. This step was not necessary in this analysis. The un-rotated solution provided clearly interpretable factors. Various rotations of these orthogonal factors were explored, but none yielded any improvement in the interpretability of the first two factors. The factor component matrix (Table 15.2) can be considered as the correlation between the two factors and each of original variables. This matrix underlies the following interpretation of the factors.

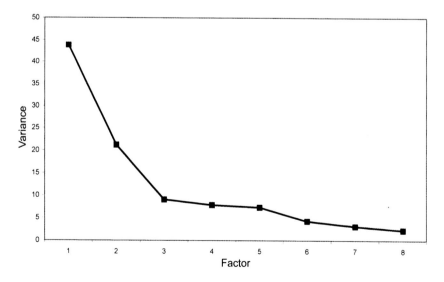

Fig. 15.1. Scree plot of extracted factor Eigenvalues

Table 15.2. Factor component matrix

Indicator	Factor 1	Factor 2
ECONOMY	0.773	-0.421
LAND_VALUE	-0.564	-0.458
LAND_MARKET	0.553	-0.019
FARM_SIZE	0.796	0.526
AREA_LARGE	0.726	0.074
OFF_FARM	-0.797	-0.020
POPULATION	-0.566	0.652
IN-MIGRATION	-0.590	0.630
INCREASE_SCALE	0.511	0.647

Factor 1: Amenity–Production

This factor accounted for 44% of the variance in the dataset. All nine of the variables loaded strongly on this factor. This factor portrays the relative position of amenity and agricultural purchasers in the land market. Those SLAs scoring low on this factor have a more diversified employment base, significant in-migration leading to population increase, smaller farms and higher rural land prices. The land market is dominated by non-agricultural land purchasers. Those SLAs scoring high on the factor generally have high levels of employment in agriculture, declining populations, larger farms and lower land prices. Most rural land transactions involve sales of land between farmers.

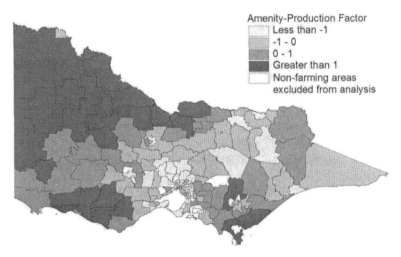

Fig. 15.2. Spatial distribution of Factor 1: Amenity–Production

The city of Melbourne lies in the centre of the main cluster of SLAs scoring low on this factor (Fig. 15.2). Other low scoring SLAs lie along the Hume Freeway (the main arterial link between Melbourne and Sydney), around major regional towns, or in popular coastal or mountain holiday destinations. High scoring SLAs are generally found distant from Melbourne, particularly in the flatter cropping country to the north-west.

Factor 2: *Extensification–Intensification*

The second factor accounted for 21% of the variance in the dataset. Six of the nine variables contributed to this factor. The SLAs that scored high on this factor had larger farms (as measured by value of farm output) and had increased the scale of production over the previous 15 years. However, there was no relationship between this factor and farm size as measured by area. One way of interpreting this factor is that it represents intensification of production on a fixed area as a response to high land values.

The highest scoring SLAs included a cluster to the east of Melbourne that host some intensive peri-urban agricultural industries including vineyards, vegetable production, chicken meat production and mushroom production (Fig. 15.3). These industries have in common a capacity to produce a high value of output on a relatively small area of land. Moderately high scores are associated with districts in southern Victoria with reliable rainfall, and areas to the north with access to irrigation water.

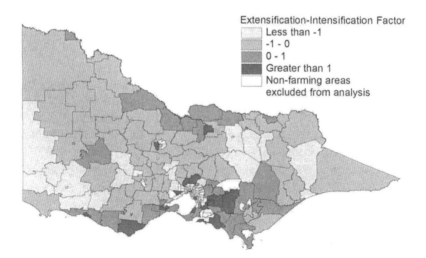

Fig. 15.3. Spatial distribution of Factor 2: Extensification–Intensification

15.3.2 Creating a Geography of Amenity and Intensification

The next task was to integrate the scores from two orthogonal factors to create a spatial geography of rural restructuring. The obvious approach was to use hierarchical cluster analysis to seek further patterns within the two factor scores. Cluster analysis is recommended as an exploratory analytic tool when there is evidence of clustering within the data (Everitt and Dunn 1983). With only two factors to consider, any underlying clustering should be evident in a scattergram of these two factors (Fig. 15.4). There is no evidence of gross clustering in this Cartesian representation. In this situation, the use of clustering algorithms is likely to produce unstable solutions, with the choice of clustering method greatly influencing the solution. Attempts to use various clustering methodologies produced a wide variety of solutions and often results that were not easily interpreted.

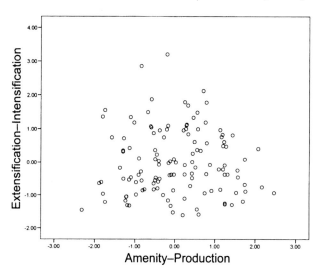

Fig. 15.4. Scattergram of factor scores

With no clear clustering apparent in the data, the decision was taken to divide SLAs into groups according to objective but arbitrary thresholds. This strategy makes it simpler to monitor future changes in group membership.

The first factor should provide the basis for the mapping of rural restructuring processes. It accounts for more than twice the data variance summarised in the second factor. The first factor was divided into three ranges with the break points set at -1, 0 and 1. The choice of these standard deviation ranges was not entirely arbitrary, as these also coincided with some minor natural breaks within the score distribution. This subdivision created four groups of SLAs that correspond with high amenity influence, low amenity influence and a transitional zone between the two.

Though the second factor may be less important, it provides significant information about those districts where intensification has replaced land purchase as the main strategy to ensure farm business viability. It is also evident from the scattergram of factor scores that high scores on the second factor are associated with less variation in first factor scores. Based on these two arguments, it was decided to allocate SLAs with a factor score greater than 1 into a separate grouping. These arbitrary thresholds are represented in Table 15.3.

Table 15.3. Thresholds used for allocating SLAs to social landscapes

		Factor 2 score			
		< -1	-1–0	0–1	> 1
Factor 1 score	< -1		Production		
	-1–0		Transitional		Intensive Agriculture
	0–1		Amenity Farming		
	> 1		High Amenity		

There is a degree of arbitrariness in allocating SLAs into these five groups on the basis of a continuous score distribution. Despite this simplification, the five groupings of SLAs presented have quite distinct characteristics, and together they are instructive in explaining the geographic variation in rural restructuring within Victoria.

15.4 Five Social Landscapes

In this section the five groupings of SLAs, which are referred to as 'social landscapes' are presented. This phrase is chosen to convey the pervasive influence of the transformation under study on many aspects of land occupancy (Holmes 2006), demographic structure and agricultural production. These wider influences are discussed in other work (Barr 2005; Barr and Karunaratne 2002; Barr et al. 2003). The landscapes are portrayed spatially in Fig. 15.5. The indicator characteristics of each landscape are shown in Table 15.4.

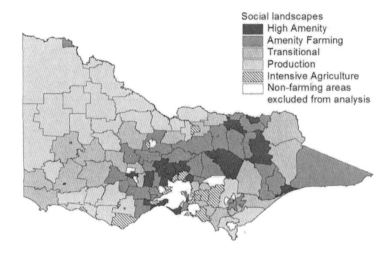

Fig. 15.5. Five social landscapes of rural Victoria

15.4.1 The Production Landscape

Of the 26 SLAs classified into the Production Landscape, 17 are in northwest Victoria (Fig. 15.6). This is a landscape that is generally flat and lacking in scenic amenity and has few major towns. These characteristics dampen the interest of amenity-driven land purchasers. This has provided an economic advantage to successful farmers by providing greater opportunity to purchase farm land. In much of this landscape the predominant broadacre agricultural industries (mainly cropping and grazing) have achieved increases in productivity that match or exceed the long-term decline in the farm terms of trade (Barr 2005).

Today larger farms cover much of the landscape and few small farms exist. Part-time farming is relatively uncommon. The obverse of a history of farm aggregation is a matching history of farm business extinction and population decline. If current trends continue, half of all farm businesses will disappear within a 30 to 40 year span (Barr 2004). Small towns are caught between the continuing decline in the number of farm businesses and the growth of a small number of major regional centres. Population has been gradually falling, and through much of the 1990s, services were gradually closing. Population projections suggest a future of both continued population decline and increasing aged dependency ratios because of a falling natural birth rate and out-migration of young adults (DSE 2004).

15.4.2 The Transitional Landscape

The Transition Landscape is comprised of 25 SLAs that fall along the boundary between areas of high and low amenity influence. Half of these SLAs comprise that part of Victoria that was once regarded as 'prime wool country'. In the past two decades the wool industry has experienced difficult times. Until recently wool prices have remained flat. Production costs have risen and the industry has found it difficult to produce productivity increases to match its terms of trade decline (Ha and Chapman 2000). The industry is constrained by the labour intensive nature of wool production.

The response of the wool industry has been a gradual and ongoing decline. The volume of wool production has halved, and the sheep flock has shrunk since a peak in the late 1980s. Few young farmers are being recruited to the industry and an increasingly aged population of wool producers gradually retiring from farming (Barr et al. 2003; Wilkinson 2003; Wilkinson et al. 2002).

A transitional landscape is being created by this decline. The obvious question is what is the landscape transiting towards? There is no single answer to this question. In some parts of the transitional landscape more profitable forms of broadacre agriculture are supplanting the traditional wool industry. Farmers with the capacity to change the balance of production on their farms have diversified from wool production into other agricultural outputs that give a better chance of keeping pace with the declining terms of trade. The easiest shift for a wool grower is from wool to prime lamb production. In the longer term, those who are successful in increasing the intensity of cropping on their farms will out-compete specialist wool producers in the market for arable land. To the south of the state, the dairy industry has been expanding, fuelled in part by water shortage in the northern irrigation areas and by an influx of migrating dairy farmers from New Zealand seeking cheaper land. They have been displaced by the escalation of land values in rural New Zealand.

More novel land uses are also supplanting traditional agricultural pursuits. In the south-west of the state many sheep properties have been purchased for blue gum plantations. These purchasers have raised the price of land well above local sheep farmers' expectations of its worth. This has allowed many older wool producers to leave the industry with a lump sum that exceeded their expectations. The replacement of wool farms and traditional patterns of land ownership and management with industrial forests is seen as a threat by those remaining as residents in afforesting communities (Petheram et al. 2000).

Not all of the region is arable or has reliable rainfall or wind. A new breed of settler is purchasing land in these less favoured parts of this landscape. Many of these new buyers seek escape from the city, sharing the same dreams of an idyllic rural lifestyle. But these new arrivals are not always as wealthy as the new migrants to the amenity landscapes. The attraction of the transitional landscape is the greater affordability of the land as compensation for its less dramatic amenity. The new settlers are bringing untraditional and diverse aspirations for their land. Amongst these migrants are small lot farmers who might aspire to grow boutique crops or animals or to re-create the original native vegetation and landscape.

The future for many small towns in the transitional landscape is uncertain. Residents of these towns generally have low incomes (Barr 2002; DOI 2003). Less than a decade ago there was debate over residents 'trapped' in these towns. There were few job opportunities and no opportunity to move because no one was interested in purchasing houses in those towns (Forth 2000). In the past five years there has been a new migration to many of these towns, driven by limited access to affordable housing in the state capital, Melbourne. The price of metropolitan housing rose rapidly over a five-year period. This precipitated a housing affordability crisis. During this same period the median value of rural housing fell from 65% to 53% of the median metropolitan value. This price difference was most pronounced in the Transitional and Production landscapes. For low income-earners this was the only opportunity to purchase affordable housing within the state.

15.4.3 The Amenity Farming Landscape

The Amenity Farming Landscape is comprised of 35 SLAs in locations with high scenic qualities but beyond the Melbourne commuting zone. The trajectory of agriculture in the Amenity Farming Landscape is dominated by the land purchases of migrants for whom the agricultural potential of land is a minor consideration. Sixty percent of all rural land purchases are made by persons who do not, at the time of purchase, live within the local region. Most land is purchased for its value as a residence rather than its agricultural value. Many of these districts lie too far from Melbourne to allow a commuting lifestyle. As a consequence, many land purchasers use their newly acquired blocks for weekend living. Many local governments in this landscape report between 30% and 50% of properties having non-resident ratepayers (McKenzie 2007).

Rural land prices are high in terms of their capacity for agricultural production. For many years this has acted as a constraint upon the capacity of farm businesses to expand by purchasing land. Farm businesses have become economically smaller as the shifting benchmark of a viable farm size has gradually increased (Barr and Karunaratne 2002). In this social landscape, farming is being slowly gentrified as an occupation to be enjoyed by those who can afford the entry price and who are not then driven to increase productivity to pay their mortgage. In 2001 the average farmer age was 54 years, five years older than the median age in the Production Landscape (Barr and Karunaratne 2003). The small number of agricultural businesses that have prospered and grown within this landscape have often done so by following the intensification path. The ossified state of rural restructuring within this landscape is balanced by the positive outlook for regional population. Farm businesses are being extinguished at only one-third of the rate of the Production Landscape. Many farms that cease operation are not absorbed by other farms, but are fragmented into a number of sub-economic units. This fragmentation provides the basis for population increase.

15.4.4 The High Amenity Landscape

The 19 SLAs that comprised the High Amenity Landscape are mostly on the outskirts of Melbourne or major provincial centres. A small number are centred on major mountain tourism destinations. In this landscape the influence of amenity purchasers in the land market is even greater than in the Amenity Farming Landscape. Two-thirds of rural property purchases are made by people living outside the district. Land is extremely expensive in comparison with its agricultural potential. Land price is determined by location. Scenic qualities and proximity to urban locations are the main factors determining land price. Farms are smaller than in the other landscapes and have been shrinking in size over the past two decades. Agriculture plays a minor role in the local economy, accounting for less than 4% of the local workforce. Commuting to work in major centres is common (Buxton et al. 2007).

15.4.5 The Intensive Agriculture Landscape

The 20 SLAs that comprise the Intensive Agriculture Landscape are mostly located on the outskirts of Melbourne or the major regional cities of Bendigo and Ballarat. This landscape mirrors some aspects of the High Amenity Landscape and some aspects of the Production Landscape. It has high population growth, high land values and a significant part-time farming segment in common with the High Amenity Landscape. Despite these constraints, the statistics of this social landscape portray a farm sector that is flourishing. Farm financial-scale increased by 47% between 1986 and

2001. The median farm Estimated Value of Agricultural Operations (EVAO) is similar to that of the Production Landscape. These apparently contradictory characteristics can be explained by the highly bifurcated nature of agriculture within this landscape.

Much private land is occupied by a highly gentrified grazing sector. Most managers of the smaller properties are either heavily engaged in off-farm employment or are retired. There is little pressure or capacity to adjust to a declining farm terms of trade. Over 15 years there has been little change in the number or scale of these properties. Much farm activity is little more than grazing to keep the grass down. Long ago these businesses stepped off the terms of trade race. Despite the relatively minor economic value of these businesses, this industry is an important component of the landscape that is valued by the local community and that the planning authorities seek to protect through controls on subdivision.

At the other end of the farm spectrum is the intensive sector that produces chicken meat, vegetables, mushrooms, pork and eggs. These businesses are tightly integrated into the supermarket food supply chain. The median scale of operations of the intensive agriculture sector increased 400% in a 15-year period while the number of operations declined by 15% (Parbery 2007). The peri-urban location provides some advantages. One attraction is the proximity of a large labour pool. The other is proximity to supply chain partners. This agricultural sector has been able to prosper and expand at a rapid rate despite high land values and not insignificant opposition from sections of the community opposed to potential loss of amenity from intensive animal industries. This is ironic — the successful agricultural industries are unwelcome, while those that are welcome are unable to remain viable.

The other very successful agricultural business in this landscape is the vineyard and winery combination. This industry can take advantage of its proximity to the metropolis through a combination of intensification (winery investment) and through cellar door sales, restaurant catering and accommodation services. In part the business strategy is to transform farming into a commoditised experience. Some smaller farm businesses are taking advantage of the opportunities of the peri-urban location with similar strategies. They are developing new marketing methods or re-opening older marketing forms that are based upon a personal relationship between the consumer and producer. These include direct marketing by mail, regular food basket deliveries, pick your own, roadside stalls, and farmers markets. The major advantage of these forms of business to the farmer is the higher share of the return that is captured by the farmer. These strategies are a logical response to the gradual exclusion of small farms from the major food supply chains.

The Social Landscapes of Rural Victoria   321

Table 15.4. Median SLA characteristics of the five social landscapes

Indicator	High Amenity	Amenity Farming	Transition	Production	Intensive Agriculture	Graphical representation
Farm scale measured using the ABS EVAO. The indicator is the proportional change in regional median EVAO 1986–2001	-17.9	-4.7	21.0	34.7	47.3	
Percent change in total population 1996–2001	10.8	2.5	.365	-3.8	10.4	
Percent of population that migrated into the region in the previous 5 years	42.6	37.0	34.6	30.9	44.8	
Agricultural employment as a proportion of total employment.	3.2	10.4	22.1	33.0	3.3	
Purchasers of rural land who have addresses inside the region as a percentage of all purchasers (over the period 1996–2001)	33.6	40.9	47.1	61	39.4	
Value of land per hectare as a percentage of value of agricultural production per hectare (averaged over the period 1996–2001)	25.9	10.5	6.5	3.8	5.4	
Median farm size in 2001 measured using EVAO	77.0	96.0	138.0	184.0	158.0	
Percentage of farm area managed as part of farms falling within the highest quartile of national farm EVAO in 2001	14.1	19.7	36.6	50.0	33.9	
Percentage of farm managers who nominate farming as a minor occupation in 2001	37.8	26.1	18.0	13.5	25.5	

15.5 Conclusion

Rural Australia is highly vulnerable to the economic and social forces driving the transformation of agriculture across the world. This makes the Australian context an interesting choice for a spatial exploration of the rural transformation. A narrative of the transformation was used as the basis for the selection of a suite of indicators. These indicators were simplified into two orthogonal factors. The first factor portrays the landscape transformation Potter (2006) described as the 'bifurcated rural space' that European policy makers appear to be increasingly willing to accept as the price of progress in World Trade Organisation negotiations.

However, this single factor was not sufficient to account for the geographic heterogeneity apparent in the data. A second factor emerged from the analysis, consistent with the tentative observation of Smailes et al. (2005) that it was not possible to rank their Australian locality clusters according to a 'single polarisation between two dichotomous groups'.

These two factors were used to create a five landscape classification. The five social landscapes in Victoria have quite different characteristics and appear to be following divergent trajectories of transformation. This heterogeneity makes clear the dangers of drawing conclusions about the nature of the rural transformation on the basis of geographically limited case studies.

The phenomenon that this analysis has explored is not unique to Victoria or even Australia. United States researchers have identified the influence of amenity upon the population growth of rural United States, though the interaction with agricultural restructuring is less explicitly explored (McGranahan 1999). The amenity influence has been represented in the development of typology of US counties (Parker 2005). European scholars have created an extensive literature debating the nature of Europe's 'postproductivist transition'. Scholars argued over the existence of this transformation and the factors driving the transformation. Underlying the debate has been a gradual emergence of evidence that transformation has been spatially heterogeneous, rendering as questionable many of the generalisations based upon small-scale geographic case studies (Wilson 2001). However, there has been little research into the spatial aspects of rural restructuring within Europe (Holmes 2006).

15.6 Future Research Directions

There are two clear directions for future research. The first is to extend the social landscapes analysis to other regions. The task of extending to other geographic domains will be challenging due to problems of incomparable data. For Australia, Australian Bureau of Statistics datasets are nationally available. However land transaction data is not easily obtainable for the rest of Australia. Factor scores would need to be calculated using a subset of the indicator variables regressed against the full Victorian dataset. This is not an insurmountable challenge. Initial exploration has demonstrated the feasibility of this approach, and also suggested that the Australian rangelands may require an extension of the indicator suite to encompass the influences of indigenous land tenure and growth in the mining industry.

The second challenge is to transform this static analysis into a methodology for monitoring the progress of the rural transition (Mather et al. 2006). The current analysis has been undertaken with a suite of indicators derived from data for the period for 1986 to 2001. At the time of writing a comparable suite of data for the period 2001 to 2006 was being released by the Australian Bureau of Statistics. Once again data comparability will create difficulties. There have been significant changes in the 2006 Agricultural Census procedures that make comparability between this and previous collections challenging.

References

ABS (2001) Australian standard geographic classification 2001. Australian Bureau of Statistics, Canberra

Argent N (2002) From pillar to post? In search of the post-productivist countryside in Australia. Australian Geographer 33(1):97–114

Barr N (2002) Going on the land and getting off it: farm income and farm adjustment. Retrieved July 2002, from http://www.agrifood.info/Connections/Winter2002/barr.htm

Barr N (2004) The micro-dynamics of occupational and demographic change in Australian agriculture: 1976–2001 (No. 2055.0). Australian Bureau of Statistics, Canberra

Barr N (2005) The changing social landscape of rural Victoria. Department of Primary Industries, Tatura, Victoria

Barr N, Karunaratne K (2002) Victoria's small farms. Natural Resources and Environment, Bendigo, Victoria

Barr N, Karunaratne K (2003) Victorian small farms: an update. Department of Primary Industries, Bendigo, Victoria

Barr N, Wilkinson R, Karunaratne K (2003) The changing social landscape of the Victorian wool industry: 1976–2001. Department of Primary Industries, Bendigo, Victoria

Botterill L (2003a) From Black Jack McEwan to the Cairns Group (No. 86). Australian National University, Canberra

Botterill L (2003b) Uncertain climate: The recent history of drought policy in Australia. Australian Journal of Politics and History 49(1):61–74

Burnley I, Murphy P (2003). Sea change: movement from metropolitan to arcadian Australia. University of NSW Press, Sydney

Buxton M, Tieman G, Bekessy S, Budge T, Butt A, Coote M (2007) Peri-urban case study: Bendigo Corridor. RMIT University, Melbourne

Catell R (1965) Factor analysis: an introduction to the essentials. Biometrics 21:190–225

DOI (2003) Regional matters. Department of Infrastructure, Melbourne

DSE (2004) Victoria in future. Department of Sustainability and Environment, Melbourne

Everitt B, Dunn G (1983) Advanced methods of data exploration and modelling. Heinemann, London

Forth G (2000) Following the Yellow Brick Road and the future of Australia's declining country towns. Paper presented at the First National Conference on the Future of Australia's Country Towns, 28–39 June 2000, Bendigo, Victoria

Griffith GR (2004) The impact of supermarkets on farm suppliers. The Australian Economic Review 37(3):329–336

Ha A, Chapman L (2000) Productivity growth trends across Australian broadacre industries. Australian Commodities 7(2):334–340

Holmes J (2006) Impulses towards a multifunctional transition in rural Australia: Gaps in the research agenda. Journal of Rural Studies 22(2):142–160

Lockie S (1997) Beyond a 'good thing': political interests in the meaning of Landcare. In: Lockie S, Vanclay F (eds) Critical Landcare. Charles Sturt University, Wagga Wagga, pp 1–8

Mather AS, Hill G, Nijnik M (2006) Post-productivisim and rural land use: cul de sac or challenge for theorization. Journal of Rural Studies 22(4):441–455

McGranahan DA (1999) Natural amenities drive rural population change (No. 781). Economic Research Service, US Department of Agriculture, Washington

McKenzie F (2007) Non-resident ratepayers by Local Government Area 2007. Unpublished map

Owen WF (1966) The double development squeeze on agriculture. American Economic Review 56(1):43–70

Parbery P (2007) Rural land use and sustainable green wedges: Interim report. Department of Primary Industries, Melbourne, Victoria

Parker T (2005) Measuring rurality: 2004 county typology codes: methods, data sources, and documentation. Economic Research Service, US Department of Agriculture, Washington

Petheram RJ, Patterson A, Williams K, Jenkin B, Nettle R (2000) Socioeconomic impact of changing land use in south west Victoria. Institute of Land and Food Resources, University of Melbourne, Parkville

Potter C (2006) Competing narratives for the future of European agriculture: the agri-environmental consequences of neoliberalization in the context of the Doha Round. The Geographical Journal 172(3):190–196

Pritchard B, Burch D, Lawrence G (2007) Neither 'family' nor 'corporate' farming: the restructuring of Australian tomato growing under conditions of neo-liberal agriculture. Journal of Rural Studies 23:75–87

Productivity Commission (2005) Trends in Australian Agriculture. Productivity Commission, Canberra

Smailes P, Griffin T, Argent N (2005) The changing social framework. In: Cocklin C, Dibden J (eds) Sustainability and change in rural Australia. University of New South Wales Press, Sydney, pp 80–102

Stoeckel A, Reeves G (2005) Agricultural trade policy made easy. Rural Industries Research and Development Corporation, Canberra

Wilkinson R (2003) Entry, retirement and succession strategies of Victorian woolgrowers: a qualitative study. Department of Primary Industries, Bendigo, Victoria

Wilkinson RL, Barr N, Karunaratne K (2002) 'The kids don't want to take over the farm': what's happening to the demographics of Victoria's wool industry? Wool Technology and Sheep Breeding 50(3):295–301

Wilson GA (2001) From productivism to post-productivism ... and back again? Exploring the (un)changed natural and mental landscape of European agriculture. Transactions of the Institute of British Geographers 26(1):77–102

Additional Reading

Barr N, Karunaratne K, Wilkinson R (2005) Australia's farmers: past, present and future. Land and Water Australia, Canberra

Holmes J (2002) Diversity and change in Australia's rangelands: a post-productivist transition with a difference? Transactions of the Institute of British Geographers 27(3):362–384

Marsden T (1998) New rural territories: Regulating the differentiated rural spaces. Journal of Rural Studies 14(1):107–117

Marsden T (1999) Rural futures: the consumption countryside and its regulation. Sociologia Ruralis 39(4):501–521

McGranahan DA, Beale C (2002) Understanding rural population loss. Rural America 17(42):2–11

16 A Decision Aiding System for Predicting People's Scenario Preferences

Ray Wyatt

School of Resource Management and Geography, University of Melbourne, Victoria, Australia

Abstract: This chapter introduces a potentially profitable addition to the methods used by present-day Spatial Decision Support Systems (SDSS). It is best described as a Decision Aiding System (DAS). It predicts how different sorts of people will score different scenarios being evaluated for any problem. The first section speculates why most current SDSS researchers have, so far, failed to address this vital preference prediction part of the decision-support process. Subsequent sections then clarify the DAS' mechanisms using a real-world spatial planning case study. The conclusion is reached that the DAS has exciting potential for increasing SDSS' level of community consciousness, especially in the future when it morphs into an Internet-based application, thereby enabling it to 'learn' decision-making priorities from a broader cross-section of users.

16.1 Introduction

The system described in this chapter predicts how various groups within the community will rate the alternative spatial scenarios being considered within any decision-support exercise. It is true that human evaluation behaviour cannot be predicted completely, but if a system provides just an approximate guide, then decision makers will be grateful. This is because decision makers fear opting for scenarios which few people within the community will want, and they will always welcome guidance as to which scenarios are likely to be more highly thought of.

16.2 Background

The current impact of SDSS, upon real-world policy-making practice, is not as great as some might expect it to be. Uran and Janssen (2003) explored why this is so, and they grouped all of the possible reasons into four categories:

- SDSSs sometimes inflict severe information overload upon hapless users who, therefore, give up on interpreting SDSSs' messages.
- Information overload is often inevitable anyway because of the sheer complexity of the phenomena being modelled.
- The paucity of strong theory within many SDSSs can sometimes cause users to doubt the logical correctness of SDSS' conclusions.
- Some users have doubts about SDSS' intrinsic suitability for providing guidance within real-world problem settings anyway.

The first three categories — information overloads, intractable complexity and paucity of background theory — are not addressed here, but the fourth, real-world irrelevance, is. The software described has been labelled a Decision Aiding System (DAS) even though this tag is frequently misused to describe any system which has potential to support real-world practitioners. For example, Holloway and White (2003) describe their algorithm as a DAS simply because it generates suitable questions for trimming scenarios into a manageable number. Moreover, Kuchar et al. (2002) refer to their system as a DAS just because it handles risk assessment and management. Also, Spada et al. (2005) supposed 'decision-aiding methodology' is actually just a program for optimising school bus routes.

In actual fact, 'decision aiding' is a broad process that thoroughly and comprehensively reviews the alternatives on offer. Gregory et al. (2001) claim it is a five-step process, as follows:

- Step 1: Find stakeholders' values
- Step 2: Develop alternatives
- Step 3: Simulate alternatives' effects
- Step 4: Identify trade-offs between alternatives
- Step 5: Summarise and explain disagreements about alternatives' levels of merit.

Since current SDSS practice already covers the first four of these steps, this chapter contributes towards the fifth step — summarise and explain disagreements ... As such, it advances SDSS towards becoming comprehensive 'decision aiding', as defined by Gregory et al. (2001), rather than just 'decision support'.

16.3 An Extra Step for the SDSS Discipline

It needs to be remembered that current efforts by SDSS researchers to boost the field's technical excellence are nothing short of spectacular (Mennecke et al. 2000). Detailing them is outside the scope of this chapter, but examples include the system by Dymond et al. (2004) which exploits extensive cross-disciplinary modelling and web-based interaction, the SDSS by Makropoulos et al. (2003) which incorporates the extra flexibility afforded by fuzzy logic, and the program contributed by Sengupta et al. (2005) which uses intelligent agents as part of its simulation.

Less emphasis has been placed upon increasing SDSSs' political awareness. The reasons for this are probably complicated and historical, and although they are tangential to the thrust of this chapter it is useful to speculate about some of them briefly. For instance, most developers of SDSS are employed either in academic or government institutions where they are far removed from both the diversions of having to make money for shareholders and enduring the 'cut and thrust' of real-world, politically sensitive policy making. Hence most SDSS researchers take a theoretical stance towards policy making rather than a practical one. They tend to focus on what sorts of information might be useful for decision takers, and how to best present such information. They usually have little interest in how their system is actually used in the real-world.

Put differently, concentration on technique tends to make the SDSS discipline unaware of the plight of real-world decision takers. The latter are almost always under many pressures, many of them political in nature and all of them time critical. The result is that SDSS outputs are sometimes going to have little impact on the observed behaviour of policy makers — unless an SDSS also incorporates a quick mechanism for better seeing through the smokescreen of complications that always swirls around politically delicate situations. The contribution made by the DAS presented in this chapter is one such mechanism.

Although it is offered squarely within the SDSS management tradition, the DAS has taken some guidance from theoreticians like Maslow (1971), who pondered people's basic motivations, and Glasser (1988), who used Maslow's work to develop 'Choice Theory'. This states that humans' choices are driven by five genetically defined needs — survival (e.g. food, clothing, breathing, safety), belonging (connection/love), power (significance), freedom (responsibility) and fun (learning) — and so people will rank highest those scenarios which best satisfy such basic needs.

In economics, the equivalent to Choice Theory is known as Rational Choice Theory (Swedborg 1990), and it says that humans will usually try to maximise utility which, presumably, is related to how well basic drives are satisfied. Moreover, it assumes that all actions (alternative scenarios) can be ranked, and it also assumes transitivity between rankings. Nevertheless, the Rational Choice Theory has been criticised by authors like Green and Shapiro (1994), for actually being a weak predictor of human (political) behaviour. Also, from a management perspective these theories are too unspecific for precisely predicting people's ratings of concrete scenarios. This is because it is often difficult, if not impossible to measure exactly how well or otherwise each concrete scenario satisfies people's drives for survival, belonging, power, freedom and fun.

Turning, therefore, to psychology for help, it seems that although this discipline has been quite ambitious in trying to explain and predict human behaviour exactly, it has tended to become suffocated by the sheer complexity of the phenomena that it is still trying to unravel. Even so, some useful pointers towards practical prediction of people's preferences have emanated from psychologists' laboratories, such as Fishbein and Ajzen's (1975) 'Theory of Reasoned Action', which says that human motivation increases whenever a behaviour is positive (attitude) and whenever significant others want it performed (subjective norm). Hence the higher that a scenario scores for attitudes and subjective norms, the more likely it is to be ranked highly overall.

Ajzen (2005) later added a third factor — 'perceived behavioural control' (self efficacy) to derive his 'Theory of Planned Behaviour' which maintains that scenario rankings are based on:

- behavioural beliefs (consequences)
- normative beliefs (expectations of others)
- control beliefs (facilitating and impeding factors within the environment).

Despite this approach having predicted accurately within many problem domains, many researchers have claimed that it is severely hampered by its non consideration of factors like fear, threat, mood, and negative and positive feelings. Hence others have tried to concretise some of the latter, for example Nuttin (1984), who maintained that the valence (importance) of any desire is a function of things like temporal distance, perceived instrumentality, reality character and difficulty.

16.4 Description of the *Preference Prediction* Software

The *Preference Prediction* software extends the insights of researchers like Nuttin (1984) into the realm of management so as to make them more operational. For instance, Nuttin's temporal distance, perceived instrumentality, reality character and difficulty respectively are loosely translated into the positive criteria of speed, effectiveness, acceptability and ease. This enables any scenario's merit to be predicted once its scores on such criteria have been specified.

16.4.1 Finding a Larger Set of Criteria

The author has used his long experience within urban planning and management (Wyatt 1989), to identify nine key plan-evaluation criteria (Table 16.1). There is some overlap between such criteria, but each has a distinct meaning of its own. Collectively, they are assumed to cover all of the attributes that evaluators ever look for when comparing alternative scenarios.

Indeed, one can relate these nine criteria to Ajzen's Theory of Planned Behaviour described above, by noting that:

- acceptable, permissive and autonomous approximate normative beliefs
- timely, easy and responsive approximate behavioural beliefs
- safe, fast and effective approximate control beliefs.

However, since the nine criteria in Table 16.1 are more measurable than Ajzen's or Nuttin's concepts they are more usable.

16.4.2 Finding Relationships between Criterion Scores and Overall Scenario Merit

At the many community workshops at which the *Preference Prediction* software has been used, nobody has been allowed to use it until they have worked through at least three past problems that are part of the software package. When users address such problems the software asks them to score the alternative scenarios on the nine evaluation criteria listed in Table 16.1. It then asks the user to score each scenario for overall merit, and all users' responses are stored on the computer's hard drive.

Table 16.1. The nine evaluation criteria used by the *Preference Prediction* software

Criterion	Definition	Synonyms	Antonyms
Acceptable	capable, worthy, likely to be accepted or gladly received	tolerable, admissible, welcome, agreeable, fitting, apposite, agreeable, gratifying	objectionable, intolerable
Permissive	non preventative	indulgent, lenient	blockading, hindering, impeding, obstructive, deterring, thwarting, preclusive, obviating, pre-emptive, prohibitive, prohibitory, non indulgent
Autonomous	free from external control and constraint	independent, self-governing, self-sufficient, self-directed, self-reliant	dependent, governed, directed
Timely	suitable given the current conditions or circumstances	opportune, due, expedient, well timed, convenient, judicious, adapted	untimely, inexpedient, inopportune, injudicious
Easy	posing no difficulty	effortless, unproblematic, user-friendly	challenging, arduous, awkward, knotty, problematic, problematical, tricky
Responsive	responding readily to effort	cost-effective, productive, rewarding, amenable, tractable, susceptible	unrewarding, unproductive
Safe	free from danger or the risk of harm	harmless, innocuous, risk-free, riskless, unhazardous, sound, secure	chancy, chanceful, dicey, dodgy, harmful, hazardous, risky, insidious, mordacious, parlous, perilous, treacherous
Fast	capable of acting or moving quickly	quick, swift, hurrying	dilatory, laggard, drawn-out, sluggish, slow-moving
Effective	produces its intended effect	effectual, successful, capable, functional, helpful, deft, adept, adroit, competent, efficacious'	ineffectual, unsuccessful, incompetent

In this way the system 'learns' what overall merit ratings the user tends to associate, on average, with each score (between -5 and +5) for each criterion. Eventually, therefore, if there are a large number of problems/users, the software is able to average out what scenario merit rating is associated with each possible score on each of the nine criteria. For example, a score of say, -3 for 'responsiveness' might be associated, on average, with an overall scenario merit of say, 1.2, whereas a score of say, +4 for 'effectiveness' might be associated, on average, with a scenario merit of say, 2.7.

A Decision Aiding System for Predicting People's Scenario Preferences   333

It then becomes easy to translate any person's or group's set of criterion scores directly into scenario merit ratings.

Note that the software also asks users to tick boxes indicating their personal characteristics such as age, sex and occupation. This enables the software to find not only relationships between criterion scores and scenario merit for that user, and for the total user group as a whole, but also for sub-groups such as males, females, younger people, older people, managers and administrators. Hence the software is able to predict scenario preferences on behalf of either an individual, or the total group of users or each of the sub-groups.

When last tested it was found that the *Preference Prediction* software's success rate at predicting a person's top-scoring option was around 90%, with a p-value of ~0.16. Moreover, its performance in predicting that the actual top scenario is in the predicted top two scenarios was higher, with a p-value of ~0.11. This is impressive, especially given that the software is not really designed to make predictions of individuals' scores; it is designed to make (averaged) predictions for aggregated groups of individuals, which is almost certainly a lot easier.

16.4.3 The Underlying Assumption

Many will argue, however, that this seems all too glib. They will assert that the relationships between scenario merit and criterion scores vary with circumstance, which makes it impossible to predict people's scenario merit ratings at any one time. Yet the DAS presented in this chapter makes the opposite assumption — people's behaviour is sufficiently consistent to allow prediction of how they will score scenarios in every circumstance.

It is true that sometimes the emphases that people place on some criteria will change, for example the emphasis that they put on scenarios' (financial) safety will be different when they stand to lose their own money rather than somebody else's. Yet across many instances within many problem domains, it is expected that any person's basic values will begin to reveal themselves. For example, some people will show themselves to be risk adverse whereas other people will reveal that they are less so.

Such a belief in the intrinsically consistent 'bounded rationality' of human behaviour has considerable currency in the literature, especially within applied disciplines like economics. As McFadden (1998) so thoroughly points out, the jury is still out on whether people act more like what he calls 'Chicago man' — consistently because of innate rationality — or whether they act more like what he designates as 'K-T man' — inconsistently because of illusions and context. Nevertheless, the first assumption has been adopted here in the interests of uncovering some consistent principles of evaluation behaviour, and it seems to be valid. That is, the software works; it predicts people's preferences fairly precisely, which could only happen if people exhibit relationships between criterion scores and scenario merit that remain reasonably consistent across all sorts of problem domains.

16.5 An Urban Planning Case Study Application of the *Preference Prediction* Software

The case study problem addressed here has been outlined by the Santa Barbara Region Economic Community Project (2003). This is a non-profit organisation of business, government and community activists who wish to preserve the quality of life for residents within the Santa Barbara region, which is on the southern coast of California about 300 km north-west of Los Angeles (Fig. 16.1).

Fig. 16.1. The location of Santa Barbara in southern California

This area is plagued by many of the problems encountered by the natural resource management (NRM) discipline within the peri-urban regions. It has attractive beaches, tourists, retirees, immigrants, a university, a fertile hinterland and high mountains constraining spatial expansion, as well as all of the resulting conflicts. Accordingly, the Economic Community Project (ECP) has designated an urban growth boundary (UGB) and generated some standard planning scenarios, as shown in Table 16.2.

The *Widespread Development* scenario is not analysed below because its plausibility is highly questionable and it is undesirable. The *All Affordable* scenario may be just as unrealistic, but it at least has merit through its

A Decision Aiding System for Predicting People's Scenario Preferences   335

social equity ideals. This leaves five scenarios — *Existing Policies, No Growth, New Neighbourhoods, Infill* and *All Affordable.*

Table 16.2. Spatial scenarios considered in the Santa Barbara case study

Scenario	Description
Existing Policies	Market forces determine the locations of new development, almost all of which is kept to within the existing built up area
No Growth	Only one new building on each vacant residential and small commercial lot; a UGB; maximum preservation of open space; down zoning of commercial and selected residential areas
New Neighbourhoods	Twice the present growth rate but located only in new, peripheral neighbourhoods, with no major changes to existing urbanised areas; reduced commercial development; UGB expanded to include Naples and the Carpinteria Valley; some government involvement in development
Infill	Rural open space and agriculture preserved; a UGB; denser development near employment centres and along built up transportation corridors; 50% more housing; a little commercial development; some government involvement in development
All Affordable	The same as Infill but residential development restricted to affordable housing
Widespread Development	Twice the present growth rate; no UGB; a weakening of most agricultural, coastal and open space protection policies

16.5.1 Assigning Criteria Scores to the Scenarios

In previous research, a SDSS known as the South Coast Outlook and Participation Experience (SCOPE) simulated the five scenarios' impacts upon the following measurable parameters (Onsted 2002):

- housing affordability
- the jobs–housing balance
- traffic congestion and commuting
- preservation of open space and agricultural land
- other quality of life factors.

Such impacts enabled the author to sensibly score each scenario in terms of the nine criteria in Table 16.1 above, and to then enter such scores into the *Preference Prediction* software, as shown (partially) in Fig. 16.2. The scenarios are listed down the left side and each column represents a criterion, with the score for each scenario on each criterion shown by a number in a box and by a dot on the x-axis of a graph. Such scores are allowed to range from -5 to +5.

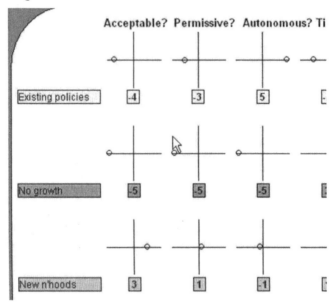

Fig. 16.2. Screen shot of some criterion scores for alternative spatial scenarios in the Santa Barbara case study

16.5.2 Predicting Scenario Ratings for Overall Merit

When a user clicks on the 'ratings' button the program takes the scores, as partially shown in Fig. 16.2, plus the relationships between criterion scores and scenario merit that it has learned on behalf of different sorts of people amongst its 288 past users, and makes predictions of each scenario's merit.

For example, Fig. 16.3 shows predictions of scenario merit levels according to people who are less than 31 years old, and for people who are older than 30. The *Preference Prediction* software has calculated that the younger people will probably rate the *Infill* scenario the highest (1.6), followed by the *New Neighbourhoods* (1.4), *All Affordable* (0.9), *Existing Policies* (0.8) and *No Growth* (0.3) scenarios.

A Decision Aiding System for Predicting People's Scenario Preferences   337

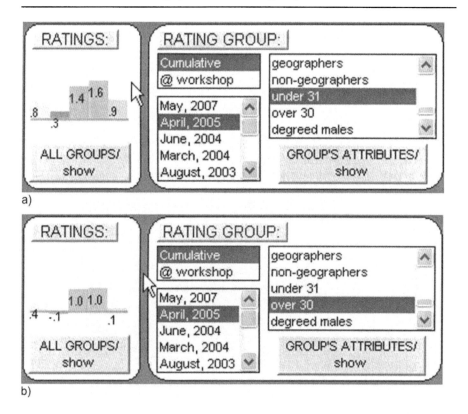

Fig. 16.3. Screen shot of predicted merit ratings for the five scenarios according to younger people (**a**) and older people (**b**)

By contrast, the older people are likely to rate the *New Neighbourhoods* scenario just as highly as the *Infill* scenario (1.0), they will rate the *Existing Policies* scenario at a low 0.4, and the *All Affordable* scenario at a barely tolerable 0.1. Moreover, the older people will actually assign a negative rating of -0.1 to the *No Growth* scenario.

This contrast between younger and older people's preferences seems plausible. It possibly reflects the young's fewer responsibilities, as well as their idealistic favouring of the *Infill* scenario as a stance against suburban sprawl. By contrast, older people have a more pressing need for at least some, and preferably optimal living conditions for child-rearing, and so they are extremely likely to actually oppose the *No Growth* strategy and to prefer the *New Neighbourhoods* scenario just as much as *Infill* scenario.

16.5.3 Checking the Personal Characteristics of the Advisors

Clicking on the 'group's attributes/ show' button will generate an output like that shown (partially) in Fig. 16.4. Here the attributes of some of the older people — those who will probably favour the *New Neighbourhoods* scenario — are shown. It can be seen that several of them have children, and they tend to be well educated; some are geographers whereas others are interested in sailing, sociology or English, and most either have professional or managerial jobs, or are tertiary level students.

GROUP'S ATTRIBUTES				
PERSON -	2	28	30	72
AGE:	31-40	41-50	51-60	31-40
SEX:	Male	Male	Male	Male
CHILDREN:	none	2	4	none
FIELD:	geography	marketing	sailing	geography
EDUCATION:	Tertiary	Tertiary	Secondary	Tertiary
OCCUPATION:	Other	Sales	Professional	Student
PROBLEM:	reduction	better	better	eating pizza
CLIENT:	soil erosion	yacht club	yacht club	friends
ROLE:				
PERSON -	113	114	116	123
AGE:	31-40	41-50	41-50	31-40
SEX:	Female	Female	Female	Male
CHILDREN:	none	2	2	none
FIELD:	empowermei	sociology	english	development
EDUCATION:	Tertiary	Tertiary	Tertiary	Tertiary
OCCUPATION:	Managerial	Professional	Managerial	Student
PROBLEM:	productivity	less drugs	precinct	less drugs
CLIENT:	company	Australia	Carnegie	Australia
ROLE:	Bystander	Bystander	Solver	Legitimate

Fig. 16.4. Screen shot of the attributes of some of the older people who have used the *Preference Prediction* software in the past

16.5.4 Predicting Scenario Merit Ratings on Behalf of Past Workshops

The *Preference Prediction* software will also predict the scenario merit ratings of past collections of users, such as those attending a workshop held at a university in Brazil, even though the participants have, presumably, never heard of the Santa Barbara problem. Their probable ratings of the Santa Barbara scenarios' merit levels are shown in Fig. 16.5, and their most favoured one will probably be the *Infill* strategy, which is the scenario that will be the most favoured by the totality of young people (less than 31 years old) who have used the software so far.

This seems reasonable, since most of the Brazilian students were also young. However, the Brazilians will probably be more in favour of the *No Growth* and *Existing Policies* scenarios, as well as slightly more against the *All Affordable* scenario than the totality of young people will be. The possible cultural reasons behind such differences are, of course, a fascinating topic, and if they are discussed at a stakeholder workshop running the *Preference Prediction* software, perhaps new insights into the merits and demerits of the respective scenarios will emerge.

Fig. 16.5. Screen shot of predicted merit ratings for the five scenarios according to a group of Brazilian students

16.5.5 Exploring How Scenario Ratings Were Derived

Clicking on the 'derivation' button makes the *Preference Prediction* software explain how it calculated advisors' scenario merit ratings, normally using histograms like those shown in Figs. 16.6 and 16.7. The x-axes represent criterion score and the heights of the columns up the y-axes are proportional to the average scenario merit that has been found to be associated with that score for that criterion. Accordingly, any scenario's merit is the average height of the selected columns from each of the nine criteria, selection of column, of course, being dictated by what that scenario scores for that criterion, as shown by the dots on the x-axes. Columns so selected are shown on the left, above each scenario, together with their average height/scenario rating.

340   R Wyatt

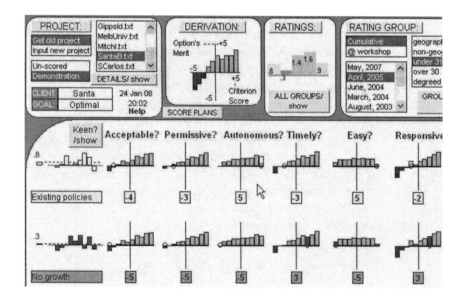

Fig. 16.6. Screen shot of how the software derived scenarios' merit ratings according to younger people

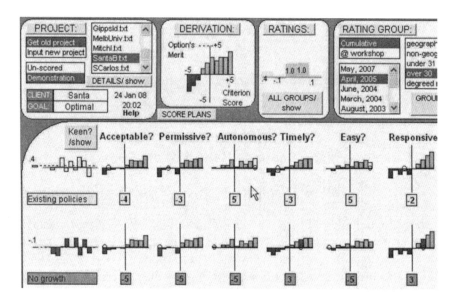

Fig. 16.7. Screen shot of how the software derived scenarios' merit ratings according to older people

Yet this is just one method that users can employ to predict scenario merit ratings. There are others, such as the regression line method, which is more suitable whenever advisory group numbers are small. This is because, for such groups, the software has not gathered enough instances of certain scores for certain criteria, so it has been unable to derive a stable average scenario merit rating that is associated with such scores. Consequently, if the manual method, as illustrated in Figs. 16.6 and 16.7 is used, predictions of scenario merit ratings become unreliable.

To illustrate this point, compare Fig. 16.6 with Fig. 16.7. The histograms in Fig. 16.6 are coherent, for example, as the scores for 'Timely?' increase along the x-axis so too does scenario merit (y-axis). By contrast, the older people's equivalent histogram in Fig. 16.7 is much more unstable. This is because there were far fewer older people within the total of 288 past users compared to the number of younger people. Hence whenever there is a small sample size, like there is for older people, it is probably more accurate to plot regression lines through their histograms and use such lines to predict their scenario merit ratings. This is shown in Fig. 16.8. Ratings are no longer at the mercy of idiosyncratic data that can plague small samples.

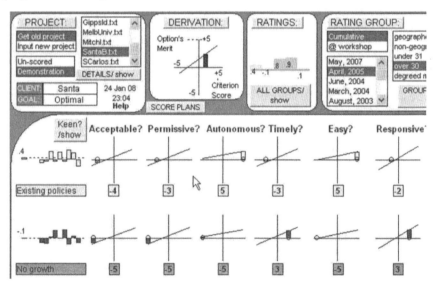

Fig. 16.8. Screen shot of how the software derived scenarios' merit ratings according to older people using the regression line method

16.5.6 Searching for Reasons behind Each Scenario Merit Rating

Users of the software can get some clues about possible, more subtle reasons for scenario ratings by clicking on the 'reasons' button. This generates a number of face charts, which holistically summarise both the nine-dimensional way that advisory groups differentially emphasise the nine criteria, and the nine-dimensional way that scenarios achieve high and low scores on these same criteria. Consequently, if the advisory group's face chart is overlaid onto the scenarios' face charts, this gives clues as to which emphasised criteria scored well, or otherwise, in each scenario. Since such holistic comparisons are difficult to make using a series of conventional charts like histograms or star plots, face charts are used by the software in an attempt to bring greater clarity.

Nevertheless, it is important to realise that comparing face charts is an inexact and subjective method rather than a detailed and objective analytical technique. Indeed, comparisons between face diagrams take the form of 'feeling' which faces seem to bear a 'family resemblance' to which other faces. This is hardly a rigorous process, but by taking advantage of humans' spectacular ability to recognise and compare faces quickly, face charts are probably the best way to proceed within any complicated, nine-dimensional search space (Wyatt 2003). In any case, this part of the *Preference Prediction* software is meant to stimulate workshop discussion about the possible, deeper reasons behind why scenarios rate the way that they do, not to fully document such reasons, which this is probably impossible.

Figure 16.9 shows how to interpret the face charts. The left side of the figure refers to those past users of the software who indicated their occupation as 'managerial', and the right hand side refers to those who designated their occupation as 'administrative'. Moreover, the top diagrams show that, given the five Santa Barbara scenarios' scores for the nine criteria as partially shown above in Fig. 16.2, managers will probably prefer *Infill* (rated at 1.4) whereas administrators will probably prefer *New Neighbourhoods* (rated at 1.9). Possible clues as to why this is so are implicit within the face charts.

A Decision Aiding System for Predicting People's Scenario Preferences   343

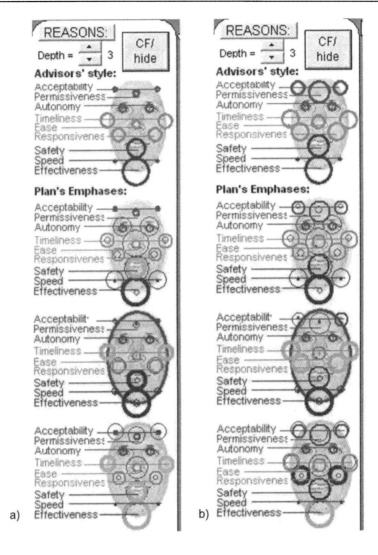

Fig. 16.9. Screen shot of face charts showing the reasons behind managers' (a) and administrators' (b) scenario merit ratings

The top two face charts contain circles that are proportional to the slopes of each criterion's regression lines, as shown in Fig. 16.8. This is because steep slopes indicate that scenario merit rating grows strongly as the criterion's score increases, and so this criterion must be important to those advisors. Hence the top left face chart shows that the three criteria of 'acceptability', 'permissiveness' and 'autonomy', as represented by the hair, forehead and eyes respectively and collectively corresponding to Ajzen's *normative beliefs*, are not very important to managers. However, taking the top three criteria, it seems that one of the middle three, 'timeliness' (the ears), which is part of Ajzen's *behavioural beliefs*, is important to managers, as are two of Ajzen's *control beliefs*, 'safety' and 'effectiveness' represented by the mouth and neck respectively.

Contrast this with the top right face chart. Taking the three most important criteria again, the bottom circles show that administrators, like the managers, emphasise 'effectiveness' but they put more importance than the managers do on two of the *behavioural beliefs* — 'timeliness' and 'responsiveness'. Note also that they assign relatively more importance than the managers do to two of the *normative beliefs* at the top — 'acceptability' and 'permissiveness'. That is, administrators seem to be more concerned with community acceptance than managers are and, therefore, they will tend to prefer different scenarios.

Users can quickly see this by, for example, comparing the managers' face chart with face chart of each scenario and noting that the face charts of the *Infill* and *New Neighbourhoods* scenario look the most like the managers' face chart, which partly explains why these two scenarios were rated the highest by the managers. Moreover, the situation can be further clarified if the user clicks on the 'CF/ show' button, causing the advisors' face chart to be overlaid onto each of the scenario's face chart and the highlighting of each important criterion for which that scenario scored a high or low merit rating.

For example, moving down the two columns at once, the second pair of face charts show that the *Existing Policies* scenario scores poorly for 'effectiveness', which is important to both managers and administrators, and so its merit rating is low. However, this scenario also scores highly for 'safety' which is important to managers, and so the managers do not rate this option so low relatively low as the administrators do.

Also, the third pair of face charts show that the *No Growth* scenario, is being rated lowest of all by the managers because, even though it is 'timely', it scores poorly for both 'effectiveness' and 'safety'. By contrast, the administrators do not rate this option so poorly, both because 'safety' is not as important to them and because this scenario scores strongly for two of what they see as important criteria — 'timeliness' and 'responsiveness'.

Moreover, the fourth pair of face charts refers to the *New Neighbourhoods* scenario. This scenario is the managers' second favourite and the administrators' favourite. Yet it scores strongly on all three of the managers' top three criteria, but strongly on only one of the administrators' top three, and it even scores poorly for 'responsiveness', which is also very important to administrators. This shows that using face charts is not infallible and it only serves as a rough guide as to why certain scenarios are favoured and disfavoured by some groups of advisors.

A Decision Aiding System for Predicting People's Scenario Preferences   345

Note that it is from such findings that various prescriptions can be formulated. For example, one way to boost the merit of say, the *Existing Policies* scenario in the eyes of the managers would be to alter it so that it becomes more 'effective'. Also, if the 'responsiveness' of say, the *New Neighbourhoods* strategy could be improved just slightly, then this scenario would become even more superior in the eyes of the administrators.

It is such speculation, about possible improvements to the scenarios in order to make them more desirable in the eyes of different groups within the community, which demonstrates the consciousness-expanding potential of the *Preference Prediction* software. Users are unlikely to think of such possibilities, let alone discuss them, without the software first nudging them in this direction.

16.5.7 Predicting All Groups' Preferences Simultaneously

Finally, users can obtain the scenario merit ratings on behalf of all groups simultaneously, simply by clicking on the 'for all groups/show' button. The information that this generates is shown in Fig. 16.10, but note that the numbers are sometimes slightly different to predictions shown above because the regression line method, rather than the manual graphical method, is used throughout. The rating for the scenario with the highest merit, according to each advisory sub-group, is shown in bold font.

It can be seen that according to the total group of 288 past users the *Infill* scenario rates the most highly at 1.53, followed by the *New Neighbourhoods* scenario on 1.52, and most sub-groups will also favour the *Infill* scenario.

But there are exceptions. For instance, the *New Neighbourhoods* scenario and not the *Infill* strategy is the best according to those aged between 11 and 20, females, parents with one child, administrators, powerful stakeholders, geographers, females with a degree and people with a primary school level of education. Although there is likely to be considerable overlap between such groups, listing them in this way gives a reasonable idea of where within the community the strongest support for the second-ranking *New Neighbourhoods* scenario is likely to come from. Moreover, the *Existing Policies* scenario is rated as top by tradespersons and by local councillors, and there are no groups who rate as highest either the *No Growth* scenario or the *All Affordable* scenario.

RATINGS ACCORDING TO EACH GROUP					
ACCORDING TO:	Existing	No	New	Infill	All
Everyone	.80	.59	1.52	**1.53**	.83
11-20	1.61	.93	**2.18**	2.11	1.30
31-40	.45	-.27	.80	**.90**	-.05
51-60	.22	-.86	.71	**.71**	-.40
Females	.84	.80	**1.71**	1.69	1.05
1 child	.61	.04	**1.23**	1.16	.28
3 children	1.22	.89	1.44	**1.54**	.92
5 children	.01	-.10	.17	**.22**	-.02
Secondary ed.	1.18	.59	1.77	**1.80**	.92
Professionals	.48	.16	.83	**.89**	.25
Administrators	.88	.94	**1.87**	1.83	1.19
Tradespersons	**.12**	-.51	-.04	-.05	-.25
Others	-.02	-.56	.62	**.65**	-.23
Cluster 2	.73	.69	**1.51**	1.46	.99
Cluster 4	.54	1.01	1.41	**1.46**	.97
Powerful stake.	1.26	.47	**1.85**	1.75	.94
Urgent stake.	1.03	.38	1.39	**1.48**	.64
Councillors	**.49**	.10	.17	.28	.12
Geographers	.60	.66	**1.46**	1.43	.92
< 31	.85	.67	1.60	**1.60**	.92
Degree. males	.49	.25	1.09	**1.13**	.45
Degree. females	.77	.78	**1.67**	1.63	1.02
21-30	.69	.60	1.48	**1.49**	.84
41-50	.54	.34	.93	**1.05**	.48
Males	.76	.33	1.31	**1.34**	.57
Non-parents	.82	.62	1.56	**1.56**	.85
2 children	.52	.28	.94	**1.08**	.38
4 children	.24	-.03	.88	**1.14**	.04
Primary ed.	1.52	.13	**1.58**	1.38	.65
Tertiary ed.	.65	.56	1.41	**1.41**	.76
Managers	.97	.44	1.32	**1.40**	.67

Fig. 16.10. Screen shot showing predicted scenario merit ratings according to all advisory sub-groups

16.6 Future Trends

There are many research efforts and software packages which seek to anticipate people's preferences, but invariably these approaches ask the user (before they start) to actually specify the criteria on which they would like their scenarios to be scored. This even applies to the well known *Expert Choice* software package based on Saaty's (1996) Analytic Hierarchy Process (AHP) and Friend and Hickling's (1997) *Strategic Decision Making (STRAD)* software. However, asking the user to specify the criteria always ruins any chance that the system has of 'learning' different types of past users' relationships between criterion scores and scenario merit. If the criteria that are used alter from problem to problem, then no relationships between criterion scores and scenario desirability can ever emerge.

Put differently, the *Preference Prediction* software is the only known package that 'learns' to make more and more precise predictions of a scenario's merit as the number of users grows. Moreover, the software is sufficiently generic to facilitate predictions of scenario ratings from groups of people who have no experience of the particular problem being addressed. With higher numbers of users, further research will investigate how the software performs in terms of its capability for making consistent and coherent predictions.

16.7 Conclusion

This chapter has argued that the SDSS discipline should try to improve its political awareness by adding a preference-prediction component to its armoury, thereby making the discipline more community-sensitive and closer to genuine 'decision aiding' rather than simple 'decision support'. The DAS presented here raises many questions about decision Choice Theory because it takes a macro and approximate managerial approach rather than one that deeply explores human behaviour (although it does make some attempts at the latter). However, its crucial contribution is its ability to make accurate preference predictions, which is attractive to real-world decision makers because it appends political intelligence to the SDSS field for the first time.

The suggestions made about probable sub-community responses to the alternative scenarios in the Santa Barbara case study are certainly valid, as least in so far as they apply to the past users of the *Preference Prediction* software. This means that such insights can serve as spurs for further and deeper thinking, which constitutes one way of taking the SDSS discipline towards a new level of (political) savvy.

16.8 Future Research Directions

The *Preference Prediction* software requires substantial data input in order to increase the reliability of its estimates. Moreover, collecting such data requires users to attend a facilitated workshop session for the mechanical harvesting of the relationships data which each of them leaves on their computer's hard drive. To make this process easier, future research will build a web-based version of the software, complete with server-side automatic processing and amalgamation of the data supplied by each user. This will enable the software to run interactively with, and learn from its on-line users, who should be present in greater numbers since the software will become more convenient to access. The *Preference Prediction* software can then base its learning upon a wider range of problem types and across a wider range of people.

There also needs to be some investigation into how to incorporate into the software neural and Bayesian Belief Network-based methods for predicting scenario merit. Past versions of the software did include a neural network-based prediction option, but because it could not clearly show how it arrived at its ratings it tended to be mistrusted by users at community workshops — so it was removed. Yet neural network simulation is likely to be far more accurate than either the mechanical or the straight-line regression approach given that scenario rating is almost certainly a non-linear process. Therefore in order to counter such suspicion in the future, research into why it occurs would seem worthwhile.

References

Ajzen I (2005) Attitudes, personality and behavior, 2nd edn. Open University Press/McGraw-Hill, Milton-Keynes

Dymond RL, Regmi B, Lohani VK, Dietz R (2004) Interdisciplinary web-enabled spatial decision support system for watershed management. Journal of Water Resources Planning and Management 130(4):290–300

Fishbein M, Ajzen I (1975) Belief, attitude, intention and behavior: an introduction to theory and research. Addison-Wesley, Reading, Massachusetts

Friend JK, Hickling A (1997) Planning under pressure: the strategic choice approach. Elsevier, Oxford

Glasser W (1998) Choice theory: a new psychology of personal freedom. Harper Collins, New York

Green DP, Shapiro I (1994) Pathologies of rational choice theory: a critique of applications in political science. Yale University, New Haven, Connecticut

Gregory R, McDaniels T, Fields D (2001) Decision aiding, not dispute resolution: creating insights through structured environmental decisions. Journal of Policy Analysis and Management 20(3):415–432

Holloway HA, White CC (2003) Question selection for multi-attribute decision-aiding. European Journal of Operational Research 148(3):525–533

Kuchar JK, Walton DS, Matsumoto DM (2002) Integrating objective and subjective hazard risk in decision-aiding system design. Reliability Engineering and System Safety 75(2):207–214

Makropoulos CK, Butler D, Maksimovic C (2003) Fuzzy logic spatial decision support system for urban water management. Journal of Water Resources Planning and Management 129(1):69–77

Maslow AH (1971) The farther reaches of human nature. Viking, New York

McFadden D (1998) Rationality for economists? Department of Economics, University of California, Berkeley. Retrieved 2 December 2007, http://emlab.berkeley.edu/eml/nsf97/mcfadden.pdf

Mennecke BE, Crossland MD, Killingsworth BL (2000) Is a map more than a picture? The role of SDSS technology, subject characteristics, and problem complexity on map reading and problem solving. Management Information Systems Quarterly 24 (4):601–629

Nuttin J (1984) Motivation, planning and action. Lawrence Erlbaum/Leuven University Press, Hillside, New Jersey

Onstead JA (2002) SCOPE: a modification and application of the Forrester Model to the south coast of Santa Barbara County. M.A. thesis, Department of Geography, University of California, Santa Barbara, available at http://zenith.geog.ucsb.edu/title.html

Saaty R (1996) The analytic hierarchy process and utility theory: ratio scales and interval scales. In: Proceedings of the Fourth International Symposium on the Analytic Hierarchy Process, 12–15 July 1996, Simon Frasier University, British Columbia, Canada, pp 22–27

Santa Barbara Region Economic Community Project (2003) South Coast Regional Impacts of Growth Study, Santa Barbara, California

Sengupta R, Lant C, Kraft S, Beaulieu J, Peterson W, Loftus T (2005) Modelling enrolment in the Conservation Reserve Program by using agents within spatial decision support systems: an example from southern Illinois. Environment and Planning B 32(6):821–834

Spada M, Bierlaire M, Liebling TM (2005) Decision-aiding methodology for the school bus routing and scheduling problem. Transportation Science 39(4):477–491

Swedborg R (1990) Economics and sociology. Princeton University Press, New Jersey

Uran O, Janssen R (2003) Why are spatial decision support systems not used? Some experiences from the Netherlands. Computers, Environment and Urban Systems 27(5):511–526

Wyatt R (1989) Intelligent planning. Unwin Hyman, London

Wyatt R (2003) Face diagrams. Journal of Machine Graphics and Vision 12(3):335–352

Additional Reading

Ferranti M (1993) Decision makers turn to Expert Choice for help. PC Week 10(22):37–39

Friend J, Hickling A (2005) Planning under pressure – the strategic choice approach (3rd ed). Elsevier, Amsterdam

Maxwell DT (2004) Decision analysis. Institute for Operations Research and the Management Sciences 31(5):44–56

Nagel SS (ed) (1992) Applications of decision-aiding software. McMillan, London

Salo AA (1995) Interactive decision aiding for group decision support. European Journal of Operational Research 84(1):134–149

PART 4
LAND USE CHANGE AND
SCENARIO MODELLING

17 Mapping and Modelling Land Use Change: an Application of the *SLEUTH* Model

Keith C Clarke

Department of Geography, University of California, Santa Barbara, USA

Abstract: SLEUTH *is a cellular automaton model of urban growth and land use change. The model uses two tightly coupled cellular automata models, one for urban growth and the second for land use change. It includes self-modification of control parameters, and has a self-calibrating capacity built into the computer code for the model. Over a decade of model development, refinement, sensitivity testing and experiment has now gone into the model, and* SLEUTH *has accumulated over 100 applications in the USA and worldwide. In this chapter, the theme of why the land use component of the model has been less applied and tested than the urban growth part is examined. The impact of the inclusion of land use in the model on the processes involved is discussed. Two revisions of the model are thought desirable beyond additional testing: one allowing land transition probabilities to change as a function of time, and one allowing change within urban areas.*

17.1 Introduction

SLEUTH is a cellular automaton model of urban growth and land use change that has been applied extensively in the geographic simulation of future planning scenarios (Clarke and Gaydos 1998; Clarke et al. 1997, 2007). *SLEUTH* is an acronym for the spatially explicit input data layers required by the model, that is Slope, Land use, Exclusion, Urban extent over time, Transportation, and the Hill-shaded backdrop used for visualisation. The model is actually a tightly coupled blend of two cellular automata models: the original Urban Growth model and the Deltatron Land

Use Change model. The structure of the coupled model and its many applications, its sensitivity testing, and methods for improving its performance have been documented extensively elsewhere (Candau et al. 2000; Clarke et al. 2007; Silva and Clarke 2002).

SLEUTH started as the Urban Growth model, a cellular automaton model that uses weighted maps as inputs, divides the study area into square cells, and applies from year to year in sequence a set of cellular automaton rules that determine whether or not cells will change from non-urban to urban. Behaviour of the cellular automaton is regulated by five parameters that control diffusive growth, outward spread, the degree of new centre creation, and the influence of roads on the growth pattern. The Deltatron model is a second cellular automaton that takes as its input the number of new urban cells in each time period, and creates and perpetuates change among land uses other than the urban class, for example from wetlands to agriculture. *SLEUTH* has been comparatively examined among the many alternative cellular automata and other models of land use change (e.g. Agarwal et al. 2002; Gaunt and Jackson, 2003). The model has been shown to produce both convincing and statistically valid results, and has been integrated into methods for scenario planning (e.g. Jantz et al. 2003). Recent work has experimented with *SLEUTH* to test models and theories of complex future urban forms (Gazulis and Clarke 2006; Goldstein 2007; Silva 2004).

Many studies have been devoted to testing *SLEUTH* to ensure the calibration is most effective, and therefore the model's results are optimal, or at least, the best under the circumstances (Dietzel and Clarke 2007). Candau (2000) tested *SLEUTH*'s temporal sensitivity, and proved that the model gives better results using recent data and short-term forecasts than either long-term historical data or long-term forecasts. Goldstein et al. (2005) tested assumptions about sensitivity to the number of Monte Carlo simulations used in the model, and experimented with genetic algorithms as alternatives to brute force methods (Goldstein 2005).

While attention to testing the model and its assumptions is welcome as a means to assure rigor and 'honesty' in modelling, far less attention has been paid to the impact of including land use in the model, curious given the attention devoted to the Urban Growth model part of *SLEUTH*. The definitive descriptive paper by Candau et al. (2000) diagrams how the deltatron cellular automata works and lists the parameters that control the behaviour. Dietzel and Clarke (2004a, 2004b) examined the effects of aggregation of land uses in hierarchical classification systems, and found that the model does somewhat better with more aggregated land use classifications than with more detailed subclasses. Yet, no paper to date explicitly examines the land use component, nor discusses the impact of using the model to simulate land use change versus urban growth.

In this chapter, *SLEUTH*'s land use behaviour is examined, the assumptions used are critiqued, and the consequences of modelling both land use change and urban growth are discussed. In general, more applications that employ the Deltatron model are necessary, and refinements to the model are desirable that (a) allow a more dynamic process for land use change forecasting, and (b) allow change within urban areas.

17.2 Methodology

In this section, *SLEUTH*'s Deltatron model is more closely examined. *SLEUTH* assumes a cellular or gridded world, and model layers equivalent to the inputs. Two land use layers are necessary, and from these a transition matrix is created that controls how class-to-class transitions take place. An important first step is to realise that the two cellular automata are run in sequence, and the output of newly urbanised cells determines how many times the deltatron code executes. Thus when urban growth is stagnant, land change pressure is reduced. Alternately, when large areas are being consumed by rapid urban growth, more inter-class transitions are also created.

The land use change dynamics are created through a three-phase process:

- *Initiate and cluster* change, in which pixels are selected for potential change, two possible new classes are drawn, and the transition probabilities used to draw a random number that decides whether change takes place based on which new class has the closest average slope to the pixel in question (Fig. 17.1a). Clustering is done by applying the same potential class change to the changed cell's neighbouring pixels.
- *Age the deltatrons* by one time period each cycle, adding any new changes to the grid, and killing any that have aged too far (Fig. 17.1b).
- *Propagate change*, by spreading the deltatrons and imposing them back onto the land use map grid (Fig. 17.1c).

These stages will be examined in sequence, but of course in the model, they are applied at every time step.

The land cover change dynamics are assumed to have been active for at least three update cycles so that besides the land cover grid, a corresponding 'aged Deltatron space' also exists.

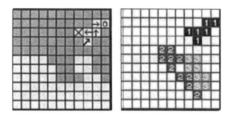

Fig. 17.1a. Deltatron model phase 1: initiate and cluster land use change. Colour images at: http://www.colorado.edu/research/cires/banff/pubpapers/94/

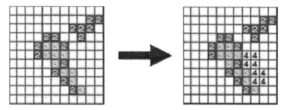

Fig. 17.1b. Deltatron model phase 2: age deltatrons. Colour images at: http://www.colorado.edu/research/cires/banff/pubpapers/94/

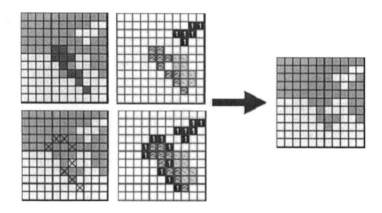

Fig. 17.1c. Deltatron model phase 3: propagate change. Colour images at: http://www.colorado.edu/research/cires/banff/pubpapers/94/

Deltatrons are 'bringers of change', they track the spatial and temporal effects of land transitions. Depending upon its age, a deltatron's locally associated land class may be available for propagating change, or for holding it in its current state. This means that change can be persistent, something impossible in classical cellular automata. To *initiate change*, each newly urbanised cell induces a potential change in land cover and, as a result, can produce a deltatron in a randomly selected non-urbanised cell. In the code, pixels are selected at random, and the selected cell either will stay the same or transition to another land use type. Change is determined by a probability weighted by: the average topographic slopes, values pre-computed for each land cover class, the historical land cover changes, and the slope of the current cell. These probabilities are defined such that more frequent land use changes, as well as historically occurring correlations between slopes and land uses, are weighted appropriately. However, these values are not updated, and so simulated transitions do not influence future change.

Random growth and change would result if there was no aggregation effect. *Change clustering* is simulated using the deltatrons, which store the location and class history of cells. A parameter — 'cluster size' — controls how large each new deltatron cluster can grow. Each new cell that is picked within the cluster started by a random change is changed to the same land cover class as the new deltatron, or it remains unchanged, taking into account the weighted slope and the precomputed class transition probabilities. Unlike in the initial condition phase, in this phase the cells can only change to the land use class that the associated deltatron has, or remain unchanged.

Each newly transitioned cell now acts as the land cover change aggregation centre. Again a random cell from its neighbourhood is tested for land cover change with the same probabilities (to remain unchanged or change to the same new land cover class). To encourage clustering, a probability exists that the aggregation centre will be moved back to the original (first) deltatron location. The effect is to allow change to drift both outwards and directionally, should conditions be suitable.

The dynamics of the *propagate change* phase are very similar to the organic growth step in the Urban Growth model dynamics. All non-deltatron cells which are neighbours to at least two deltatron cells with an age of two or more (i.e. they were created at or prior to time *t-1*) are tested against the cell's weighted probability. Cells can either remain unchanged or change to the same land cover type as a neighbouring deltatron's land cover type. Finally, in the *age deltatrons* phase, all deltatrons are aged into the next time step. The number of cycles a deltatron persists is defined by a constant. If they become 'older' than this maximum deltatron age, they 'die' and can, in principle, be recruited as a new potential deltatron in the next growth cycle.

Fundamental to the deltatron is the transition or Markov matrix. In the model, at least two land use layers are necessary so as to be able to compute this matrix. Classes must be in common, and are signified by a set of indexes to which names and colours are assigned in the scenario file that controls each model execution. Users can assign invariant classes that by definition cannot change, such as roads and water. Users can also exclude pixels and classes from change, so that for example, an area can be masked and change can take place only within the study area, a county with irregular boundaries for example. The program then does a complete pixel count of transitions by class. This raw pixel count is divided by the number of time steps between the two land use layers. Of course the assumption that 100 pixels changing from agriculture to forest (for example) happen at five pixels a year if the two layers are 21 years apart may not be valid, but was made because a more complex change model was not thought necessary. A superior model would allow a sub-model of probability change over time, perhaps calibrated using a large number of land use layers. Thus non-linear change could be accommodated.

To investigate the consequences of these modelling assumptions over time, two experiments were undertaken. These did not involve model runs, but they did examine the latter assumption in detail — that is, does the probability of land use transition change as a linear or a non-linear function of time? If the former is true, then *SLEUTH*'s assumptions about land use change modelling are reasonable. If not, then how can modifications be made that allow *SLEUTH* to model such non-linear changes among land use classes?

17.3 Results and Discussion

Land use, unlike land cover which can be assessed directly by measurement, can be defined as a human function applied to, or dominant over, a defined zone of the earth's surface. Choice of a land use classification scheme implies defining and naming a set of classes that have domain-specific meaning and that is exhaustive (covers all of the extent), definitive (unambiguous) and temporally invariant (does not change over time). Unfortunately, all of these factors are frequently violated in land use mapping, and are often violated in the case of land cover mapping. Major problems are that the classification scheme is approximate, such as in remote sensing and image classification, and the error becomes type one or type two errors artificially over time, even when there are no changes.

To empirically examine how *SLEUTH* deals with land use change, two datasets were examined. First, the *SLEUTH* website contains a test calibration dataset for a hypothetical city called Demo City. This set contains land use defined exhaustively in seven classes, loosely approximating the so-called Anderson Level I classes: Urban (class 1, red), Agriculture (class 2, yellow), Rangeland (class 3, grey), Forest (class 4, green), Water (class 5, cyan), Wetlands (class 6, blue) and Barren (class 7, black). Colour im-

Mapping and Modelling Land Use Change: an Application of the *SLEUTH* Model   359

ages are available on the project website <http://www.ncgia.ucsb.edu/projects/gig>.

Computer code was written that, unlike *SLEUTH*, allowed an arbitrary number of land use images. The code computed the raw and annually standardised class transition probabilities, and plotted them by sequences of transitions. In Fig. 17.2, for example, the five land use images are shown at the top of the figure, and below are maps showing the 'from' and the 'to' class for changed pixels only.

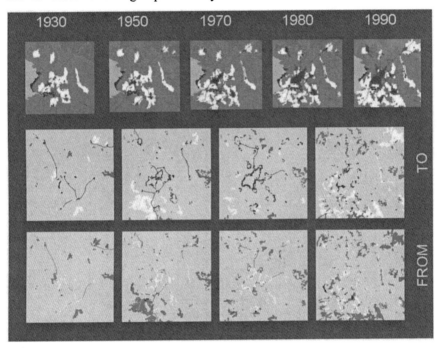

Fig. 17.2. Demo City land use change: five datasets and their 'from' and 'to' transitions

Figure 17.3 illustrates the four transition time periods and the changes in each probability trajectory over time. Note that different length time transitions are normalised by using the *SLEUTH* method of distributing pixel transitions evenly over the time step. Only the water-to-water transition is invariant, all others show variations, some very significant, over time. Furthermore, they tend to form clusters of change types, with some changing little at all (and so being suited to *SLEUTH*), and others changing broadly.

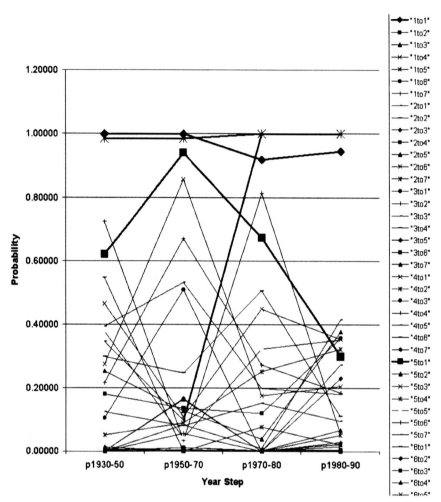

Fig. 17.3. Demo City data, all land use changes, normalised transition probabilities over time. Transitions are land use codes, based on the Anderson Level I schema (Anderson 1976)

Yet this is only a simulated dataset. Do these sorts of transitions also occur in the real world? To test this assertion, the equivalent dataset was compiled for a portion of the Santa Barbara, California region. The compilation process was particularly demanding, and was conducted as follows. First, a highly detailed, six inch resolution colour infra-red aerial photo of the area was registered accurately using the USGS Digital Orthophoto Quadrangle base at 1:12,000 equivalent. Using local knowledge, field visits and photo interpretation expertise gained over many years, the 1998 image was classified into the Anderson Level II classes, with the only addition being an extra class for riparian forest, which is important in Santa Barbara's Mediterranean climate for wildlife habitat and urban setback protections. Then the vectorised hand-digitised land use polygons were kept, the previous equivalent image imported and registered, and the vector coverage edited in place to remove changes that had not yet taken place. This level of matching was essential to avoid problems of image mis-registration and ambiguous or statistically biased classification. When complete, the vector layers were rasterised at the same resolution and extent, and brought into the same software used for Demo City. The final ground resolution was approximately one metre. One of the final land use maps is shown as Fig. 17.4.

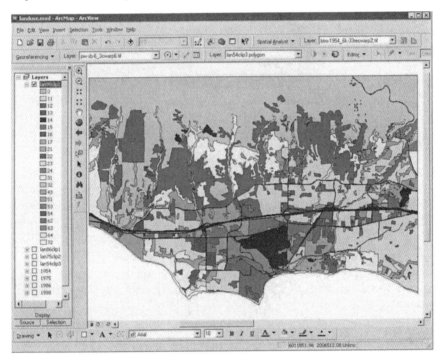

Fig. 17.4. Screen shot of Santa Barbara: classified Anderson Level II land use 1998 (classes are numbered as in Fig. 17.6) (map by Jeff Hemphill, UCSB)

Data were compiled from imagery acquired in 1954, 1975, 1986 and 1998. Since this allowed only one intervening change record per class change, deviation from a straight line in the resulting plots of the transitions (Fig. 17.5) represents changes unlikely to be well modelled in *SLEUTH*. Figure 17.5 is rather congested as it shows all of the classes (including many invariant classes); however, some class transitions do seem to follow a straight line.

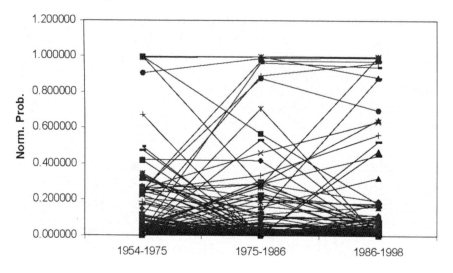

Fig. 17.5. Santa Barbara data: all land use changes over three time periods, 1954–98

In Fig. 17.6, the top 15 change classes that are least linear were extracted and shown separately. This figure shows that many classes do indeed violate the Markov assumption of linear transition. In fact, class changes seemed to fall into groups with similar patterns — for example, rise slowly then rapidly, or fall rapidly then rise slowly. Some of these change sets may be equivalent, that is, they are subsets of classes that exchange space not with the land use pattern as a whole but only with specific other classes. This implies that specific trajectories (for example, forest to agriculture to urban) may lead or persist in the land use change, both qualitatively and spatially. A specific example is class 23 'Combined feed operations' to class 15 'Industrial and commercial', which changes from high to low probability and class 23 'Combined feed operations' to class 11 'Urban and residential' (low to high), with the break coming between 1975 and 1986. In Santa Barbara almost all listed class 23 land use is horse farms. So a pattern exists in that once horse farms started converting to business properties, the next step was for them to be converted into residential developments. The change in the period 1954–1975 seems to have triggered the following change in 1986–1998. It is possible that there are many similar highly complex class dependencies in the transitions.

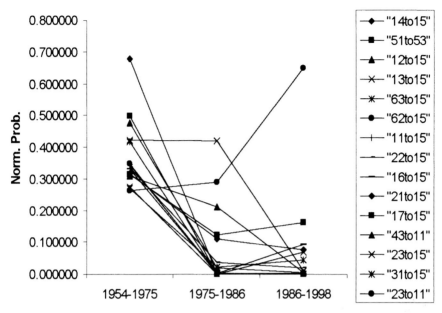

Fig. 17.6. Santa Barbara land use change: top 15 change classes ('from-to'). Land use codes are: 11=urban–residential; 12=urban–commercial; 13=urban–industrial; 14=urban–transportation; 15=urban–industrial/commercial; 16=mixed urban; 17=urban–other; 22=agriculture–orchards; 23=agriculture–ornamental horticultural; 31=herbaceous rangeland; 43=forest–mixed; 51=water–streams; 53=water–reservoirs; 62=non-forested wetlands; 63=wetlands–riparian

While the class transition matrix — when computed from only two datasets and using Markov assumptions — is an approximation of reality and of use in *SLEUTH* modelling, it remains only part of the picture. The matrix can be computed from raw pixel counts, and the class transition probabilities must be normalised over time for modelling, since they are measured over uneven time extents. These matrices are expected to be symmetrical about their leading diagonals, at least over short periods. However, as we have seen from the examples above, if the matrices are raised to a power, we would expect subsets of absorbing class transitions that simply exchange space over time, and whole sets of zones that undergo an identical set of multiple-step transitions, or at least converge on a single final class. These likely reflect very long-term land use trends. In Santa Barbara, for example, rangeland was used for beef cattle ranching until its conversion to citrus orchards starting in about 1890. After World War II, by far the dominant trend has been urban conversion.

An additional assumption of *SLEUTH* is that once a cell becomes urban — and it can do so only within the Urban Growth model code, since urban is defined as an invariant class during the Deltatron's phases — it remains so. Yet in the real world, urban land moves among urban land use classes (residential to commercial, for example) as a normal part of urban internal dynamics. A major improvement to *SLEUTH* would be to incorporate a third model of interior land change. A first step would be to 'age' urban pixels the same way that the Deltatron model ages pixels. Different classes could have expected lifetimes (say 100–200 years for residential), after which the land becomes available for change, or even de-urbanisation. Should de-urbanisation be conducive to modelling, much could be added to *SLEUTH* modelling, including the changes expected with population decline, now already common in Europe, and long-term decay of civilisations. In the long-term, all cities are either completely rebuilt on their foundations (e.g. modern Istanbul), or abandoned (e.g. Tikal). A versatile urban model should be able to handle each of these phases, not simply urban growth.

Quite clearly, the more land use change data that are available, the more complex a possible model of changes can be. With only two periods, the *SLEUTH* linear model is all that is available. With three periods, diversions from linearity can be measured. With four or more periods, quite sophisticated models, perhaps incorporating cross dependencies and non-linear time models could be possible. Unfortunately, uncertainty, class consistency, and mis-registration issues become more of a problem as more time periods are used. Just as uncertainty increases with forecasts into the future, they also increase in uncertainty going backwards in time, when data on land use were sparse and maps inaccurate. Similarly, there is scope to measure the spatial autocorrelation of changes in land use. Counting polygon contact faces is one method, but there are many others that would allow an empirical model of land use dependencies in space. *SLEUTH*'s assumption of spatial autocorrelation, and association with slope, remains untested and a possibility for future research once suitable datasets become available.

17.4 Conclusion

In conclusion, we find the assumption of linear change in land use class transition probabilities, the Markov assumption, to be only partially valid for land use transitions. We tested this assertion using both hypothetical and real data for California. Good test data to further examine the complexity of land use transitions is difficult and expensive to create, and suffers from the inconsistencies of class definition, inaccurate maps, and misclassification and registration of multi-temporal land use maps. A more complex model should also examine non-linear changes, change sequences, and path dependencies in land use transitions. After further testing of these findings, it is hoped to incorporate them into a new and more accurate cellular automaton model of land use change.

Acknowledgements

SLEUTH was developed with funding from the United States Geological Survey, the Environmental Protection Agency, and the National Science Foundation in the USA. This chapter was written while the author was a visiting Distinguished Professor at the University of Trieste, Italy, supported by the Fulbright Foundation.

References

Agarwal C, Green GM, Grove JM, Evans TP, Schweik CM (2002) A review and assessment of land use change models: dynamics of space, time, and human choice. General Technical Report NE-297. US Department of Agriculture, Forest Service, Northeastern Research Station, Newton Square, Pennsylvania

Anderson JR, Hardy EE, Roach JT, Witmer, RE (1976) A land use classification system for use with remote-sensor data: U.S. Geological Survey Professional Paper 964. US Government Printing Office: Washington, DC

Candau J (2000) Calibrating a cellular automaton model of urban growth in a timely manner. In: Parks BO, Clarke KC, Crane MP (eds.) Proceedings of the Fourth International Conference on Integrating Geographic Information Systems and Environmental Modelling: problems, prospects and research needs. 2–8 September 2000, Banff, Alberta, Canada. Online at http://www.geog.ucsb.edu/~kclarke/ucime/banff2000/529-jc-paper.htm

Candau J, Rasmussen S, Clarke KC (2000) A coupled cellular automaton model for land use/land cover dynamics. Paper 94, 4th International Conference on Integrating GIS and Environmental Modelling (GIS/EM4): problems, prospects and research needs. 2–8 September 2000, Banff, Alberta, Canada. Online at http://www.geog.ucsb.edu/~kclarke/ucime/banff2000/533-jc-paper.htm

Clarke KC, Gaydos L (1998) Loose-coupling a cellular automaton model and GIS: long-term urban growth prediction for San Francisco and Washington/Baltimore. International Journal of Geographical Information Science 12:699–714

Clarke KC, Gazulis N, Dietzel C, Goldstein NC (2007) A decade of *SLEUTH*ing: lessons learned from applications of a cellular automaton land use change model. In: Fisher P (ed) Classics from IJGIS. Twenty years of the International Journal of Geographical Information Systems and Science. Taylor and Francis, CRC Boca Raton, Florida, pp 413–425

Clarke KC, Hoppen S, Gaydos L (1997) A self-modifying cellular automata model of historical urbanisation in the San Francisco Bay area. Environment and Planning B 24:247–261

Dietzel C, Clarke KC (2004a) Replication of spatio-temporal land use patterns at three levels of aggregation by an urban cellular automata. Lecture Notes in Computer Science 3304:523–532

Dietzel C, Clarke KC (2004b) Spatial differences in multi-resolution urban automata modelling. Transactions in GIS 8:479–92

Dietzel C, Clarke KC (2007) Toward optimal calibration of the *SLEUTH* land use change model. Transactions in GIS 11(1):29–45

Gaunt C, Jackson L (2003) Models for assessing the effects of community change on land use patterns. In: Geertman S, Stillwell J (eds) Planning support systems in practice. Springer, Berlin

Gazulis N, Clarke KC (2006) Exploring the DNA of our regions: classification of outputs from the *SLEUTH* model. In: El Yacoubi S, Chapard B, Bandini S (eds) Cellular Automata. Proceedings from the 7th International Conference on Cellular Automata for Research and Industry (ACRI 2006). 20–23 September 2006, Perpignan, France Lecture Notes in Computer Science. No. 4173. Springer, New York

Goldstein N (2005) Brains versus brawn – comparative strategies for the calibration of a cellular automata-based Urban Growth model. In: Atkinson PM (ed) Geodynamics. CRC Press, Boca Raton, Florida, pp 249–272

Goldstein N (2007) Coupled spatiotemporal simulation modelling explorations of co-evolving systems International Journal of Sustainable Development and World Ecology 14:37–51

Goldstein N, Dietzel C, Clarke K (2005) Don't stop 'til you get enough – sensitivity testing of Monte Carlo iterations for model calibration. In: Proceedings of the 8th International Conference on GeoComputation, 1–3 August 2005, University of Michigan, Ann Arbor. Available from http://igre.emich.edu/geocomputation2005/abstract%5Flist/

Jantz CA, Goetz SJ, Shelley MK (2003) Using the SLEUTH urban growth model to simulate the impacts of future policy scenarios on urban land use in the Baltimore–Washington metropolitan area. Environment and Planning B 30:251–271

Silva EA (2004) The DNA of our regions: artificial intelligence in regional planning. Futures 36:1077–1094

Silva E, Clarke KC (2002) Calibration of the *SLEUTH* urban growth model for Lisbon and Porto, Portugal. Computers, Environment and Urban Systems 26:525–552

18 Uncertainty in Landscape Models: Sources, Impacts and Decision Making

Kim E Lowell

Primary Industries Research Victoria, Parkville Centre, Carlton, Victoria, Australia
Cooperative Research Centre for Spatial Information, Carlton, Victoria, Australia

Abstract: This chapter focuses on impacts on decision making of uncertainty associated with outputs of complex systems-based models. Fundamental underlying sources of uncertainty that can impact quantitative model outputs are discussed — these sources include: model structure; natural variability of phenomena; spatial, temporal, and taxonomic scales of data; and the fundamental nature of data used for modelling. These are discussed relative to map complexity, data characteristics, how they impact model calibration and validation, and how model structure and usage is impacted by them.

The potential impact of uncertainty information on decision making is discussed. It is suggested that the importance of uncertainty information in the decision-making process depends on how 'obvious' a decision is and the amount of influence that model outputs have on the decision being made. Knowing the uncertainty of model outputs is most useful in decision making when several different options appear to have comparable acceptability, and model outputs have a high influence on the decision being made. Conversely, uncertainty is much less important in decision making if one option is clearly superior to all others, or if factors such as policy considerations are given more weight than model outputs.

In situations where uncertainty information would be useful, the high level of complexity of the uncertainty in outputs of spatial systems-based models makes it difficult for such information to be used efficiently by decision makers. An example is presented to demonstrate that having uncertainty information will not necessarily change a decision made, nor will it provide more confidence that a correct decision has been made.

18.1 Introduction

Prior to the late 1960s, information used to manage landscapes was available primarily in static forms such as tables, graphs and alignment charts. Subsequent widespread availability of high speed computing led to the creation of more sophisticated tools. Coupled with this demand was the maturation and consolidation of knowledge about various landscape processes. The result was that in the early 1970s, quantitative modelling of landscape processes became an area of study in its own right.

With the emergence of specialists trained in landscape process modelling, models have increased in complexity and sophistication, and the diversity of processes modelled has increased. As confidence in such models has increased, models have been adopted for use by landscape managers. This has changed the landscape management decision-making process. If a choice is required among competing landscape management options, in addition to using accumulated scientific knowledge that is specific to the problem/region of interest, models now provide decision makers an opportunity to more thoroughly explore different options.

Users of model outputs have also become more sophisticated. An increasing reliance on models to inform decisions has resulted in increased demand from decision makers to be able to assess the value of the information provided by a given model. In short, model users want to know how good the information produced by a model is and its associated limits, that is, they want to know the quality of the model and its outputs.

While this awareness of model limitations is a healthy trend in modelling use and development, it presents difficulties. In increasingly sophisticated models, an increasing number of components that describe different processes are coupled in order to describe entire systems. Unfortunately, individual components of such models may not be completely validated, the interactions among system components may be poorly understood, and the increased model complexity may cause excessive data demands for model validation and calibration. Moreover, modellers of landscape processes generally focus on the uncertainty related to numerical calibration of their models rather than on uncertainties associated with the fundamental nature of the data used to create and apply models, or the basic structure of the model itself.

The result is that it is extremely difficult to provide a comprehensive statement of the quality of a complex, systems-based model and its outputs. This in turn makes it difficult to provide an answer to a model user's relatively simple question, 'How good are the model and its outputs?' While a satisfactory answer to this question can be an important input into some de-

Uncertainty in Landscape Models: Sources, Impacts and Decision Making   369

cision-making processes, there are situations in which there is actually little need for model developers to provide an answer to this question.

This chapter has two objectives. The first is to provide model developers with a better understanding of underlying, fundamental sources of uncertainty in their models so that they will be better able to explain to users the difficulty of answering the question, 'How good is the model and its outputs?' The second objective is to provide guidelines about situations in which uncertainties associated with models are of high, medium or low importance. It is ultimately demonstrated in this chapter that knowledge of model uncertainty will not necessarily provide a decision maker with elevated confidence that a correct decision was made.

18.2 Models, Variability and Sources of Uncertainty

For the purposes of this chapter, models are defined as mathematical constructs used to describe some physical process. Hydrological models will be used as the primary example, but the concepts presented extend into many other domains. For illustrative purposes, a simplistic model is formulated whereby the amount of water that flows from a certain fully stocked (i.e. high canopy closure) forested area is described as:

$$outflow = rainfall^a \div (stems\_per\_ha^b \times soilperm)^c \qquad (18.1)$$

where a, b, and c are model coefficients.

Though fictitious, this model does make a certain amount of sense:

- As *rainfall* increases, *outflow* increases at a rate controlled by a.
- As the number of trees (*stems_per_ha*) in the fully stocked forest increases, *outflow* decreases at a rate controlled by b. Since this model only applies to fully stocked stands, *stems_per_ha* is in reality a surrogate for tree size — more stems per hectare in a fully stocked stand means smaller trees.
- As soil permeability (*soilperm*) increases, *outflow* decreases at a rate controlled by c.

In formulating this model, the relationships of rainfall and soil permeability to outflow have been scientifically proven. However, suppose that though it has not been proved conclusively that more trees in fully stocked forests impacts outflow in the way described for all species and geographical regions, what limited scientific literature exists on the subject suggests that this occurs.

To produce such a model, the coefficients *a*, *b*, and *c* can be derived empirically or can be constants known to describe some process; the speed at which an object falls when dropped, or the temperature at which water freezes are examples of known constants.

The accuracy and precision of outflow estimates from this model will be affected by numerous sources of uncertainty — some that are fairly obvious and others that are more subtle — whose impacts can be trivial or extreme. The impact of each source of uncertainty on model outputs depends on many factors. Moreover, it is not possible to say that one or another will always, or never, have an extreme/trivial impact.

18.2.1 Model Structure

Though erroneous model structure will generally not cause imprecision, it can cause either extreme inaccuracy (i.e. bias), or no inaccuracy at all. If an increase in *stems_per_ha* in a fully stocked forest does decrease *outflow* for the tree species in question, and coefficient *b* (Eq. 18.1) is formulated as a 'known constant' that is positive, then the model is correctly specified and *outflow* is correctly modelled as the bottom-most line in Fig. 18.1. However, suppose that for the species/area in question, an increase in *stems_per_ha* actually increases *outflow* as described by the 'truth' line in Fig. 18.1. If so, Eq. 18.1 is improperly structured and the inaccuracy resulting from modelling it using the positive known constant is relatively large. However, if coefficient *b* is empirically determined to be negative (fitted line in Fig. 18.1), then the flaw in model structure will not be apparent. Hence, the mathematical formulation of the model and the way that coefficient values are determined can cause highly inaccurate model outputs, or perversely may have no effect on model outputs.

Another way that model structure can impact uncertainty is the form of the relationship. In Eq. 18.1, the relationship between *outflow* and *stems_per_ha* is modelled as being curvilinear — a relationship that the fictitious 'truth' line in Fig. 18.1 indicates is correct. If this relationship is modelled as being linear (Fig. 18.1), however, inaccuracy results with its magnitude being dependent on the *stems_per_ha* — that is, greatest error occurs with *stems_per_ha*=100 to 150 and the least with *stems_per_ha* =475.

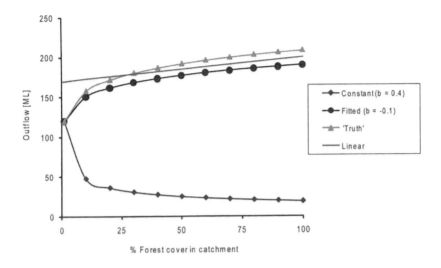

Fig. 18.1. Estimates of catchment outflow from a fictitious hydrological model (Eq. 18.1 in text) with different coefficient values associated with rainfall

An additional element of model structure can affect the precision of outputs rather than their accuracy. If a model is missing a critical element, then model estimates will be less precise because an important source of model variability has not been captured. For example, it may have been appropriate to include in Eq. 1 some term that describes the topography of the area of interest, or the sub-surface geology. Not including such terms might mean that outputs from Eq. 1 have a precision of $\pm 20\%$ whereas the inclusion of topography/sub-surface geology would have increased the precision to $\pm 10\%$. Note that if an important component has not been included in a model, the accuracy of outputs is not likely to be impacted — that is, outputs will be unbiased provided the 'incomplete' model is correctly structured.

18.2.2 Natural Variability, Temporal Resolution and Spatial Resolution

Natural variability is a relatively obvious source of model uncertainty that has the effect of decreasing the precision of model outputs. *Rainfall* clearly is a variable in our model that has a large amount of natural variability. Of greater importance, however, is that all rainfall events of a given magnitude, say 10 mm, will not produce the same outflow — even if the *stems_per_ha* and *soilperm* are identical for those rainfall events.

There are two conflicting conceptual solutions for decreasing the impacts of natural variability on model output uncertainty; increase model complexity, or decrease model complexity.

If one believes that a model is missing an important variable, then the inclusion of that variable will improve model precision and decrease uncertainty — thus increasing model complexity. Though this may help resolve the problems associated with natural variability, it might do so at a cost that would make the use of the model impractical. Suppose that our model (Eq. 1) is known to be imprecise because the range of soil moisture at a depth of 1 m across an area is not known. Adding soil moisture to the model in some manner would therefore improve the model, but would require that a model user exhaustively sample soil moisture at a depth of 1 m across the entire area. This would improve model estimates, but is not likely to be economically feasible.

The alternative solution — decreasing model complexity — requires an examination of spatial and temporal scale. In our simplistic model, no time period has been specified for rainfall. Suppose that someone wanted to use the model to know how much outflow is likely to occur on a specific future date — the day the fishing season starts, for example. To do this, they may use the long-term average rainfall measured for the date of interest. However, due to the higher variability of daily rainfall (Fig. 18.2), model estimates for daily outflow will be much worse than those for monthly outflow. This real world variability interacts with the fact that the model is not 'perfect' which means that higher variability in rainfall translates into less precision in model outputs. The inclusion of a variable — *rainfall* — that has a high amount of natural variability will harm model performance. Removing it and instead incorporating the 100-year average rainfall would considerably decrease the impact of rainfall as a source of variability. It would do so, however, at the potentially high cost of having a model that could not be used reliably for any time period less than 100 years.

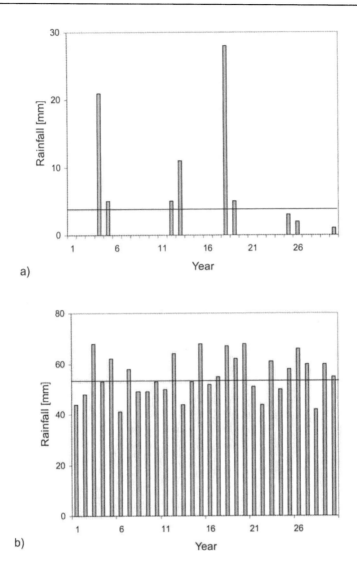

Fig. 18.2. Illustration of (**a**) high variability in rainfall for a single calendar day over 30 years (coefficient of variation = 240), and (**b**) lower variability for a single month over 30 years (coefficient of variation = 15). The solid line indicates the average daily or monthly rainfall

As for spatial scale, a similar dynamic operates, but there is an added complication related to the way in which spatial data are produced. Our model in Eq. 1 requires a statement of the *stems_per_ha* in the forested area of interest. Such information is usually collected via small plots located on the ground that are considered representative of the forest type of interest as identified on existing forest maps. Such maps are generally produced by human interpretation of aerial imagery. Therefore, they are subject to constraints of human cognition, image resolution, and a minimum mapping unit. Simply put, less detail is visible on 1:80,000 aerial photographs than on 1:15,000 aerial photographs, and on the former, one might only map forests that are at least 10 ha in size whereas on the latter one might map forests as small as 4 ha.

In the example presented in Fig. 18.3, different estimates of the amount of forest result from different photographic scales and minimum mapping units. Notably, the change in estimates is dependent on the level of fragmentation of the forest type. A similar effect occurs if 'soil permeability', as used to calibrate the model, was based on ground-based point estimates — but the model is applied areally using soil permeability as represented on a soils map. The previous suggestion to add a topographic factor such as slope could also cause a similar problem for similar reasons — in hilly terrain, 'slope' measured in the field is not the same as 'slope' produced using algorithmic treatment of a digital terrain model.

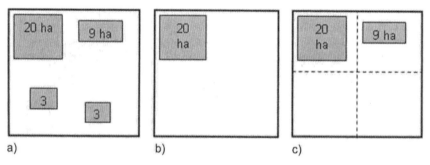

Fig. 18.3. Illustration of how spatial scale affects the quality of information. (**a**) Truth (35 ha of forest or 35% of the catchment). (**b**) Interpretation based on one single small scale photograph using a minimum mapping unit of 10 ha (20 ha of forest or 20% of the catchment). (**c**) Interpretation based on four large scale photographs using a minimum mapping unit of 4 ha (29 ha of forest or 29% of the catchment)

This means that determining the impact of uncertainty that is related to spatial resolution is extremely difficult. Moreover, using only data of fine spatial resolution does not resolve this problem because such data are merely an additional interpretation of reality and not 'truth.' Moreover, operationally, such data will not exist for an entire area of interest.

18.2.3 Taxonomic Scale and Data Collection

Two final sources of uncertainty are the most subtle, yet potentially have the most profound impacts on the uncertainty of model outputs for any model that employs spatial data — taxonomic scale and subsequent data collection. Variables such as *stems_per_ha* and *soilperm* are based on a spatial taxonomy. A binary taxonomy such as forest and non-forest will lead to less model uncertainty than a taxonomy that contains, say, three, or 10 classes (Fig. 18.4). As the number of classes in a taxonomy increases, the distinctions among different classes decreases, and the spatial organisation of the phenomenon being mapped manifests increasing mixing relative to the taxa. This makes it increasingly difficult to accurately identify each class thereby increasing the uncertainty of model input data and resulting model outputs. The complexity of the landscape also interacts with these factors. That is, if a landscape being mapped is extremely homogeneous, then the complexity of the taxonomical system has little impact on map accuracy and subsequent uncertainty of model outputs.

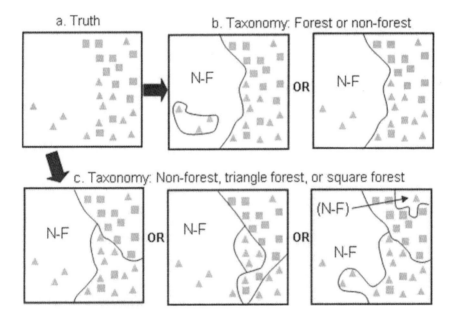

Fig. 18.4. Increased uncertainty as resolution of cartographic taxonomy increases

The issue of taxonomic scale also has profound impacts on the measurement of variables such as *stems_per_ha* and *soilperm*. Such variables are usually determined for a given taxonomical class by taking field-based samples (Fig. 18.5). Whether or not they are representative of the intended class depends on the accuracy of the mapping of those classes. Moreover, the 'representativeness' of the field-based data for a given class depends on the variability of the class relative to the variable being measured. To determine a single soil permeability value for a given class, a number of field-based point samples are taken for each class, and then averaged. Even if a given class is mapped well, if the class is extremely variable (soil type A, Fig. 18.5a), this averaged single value is virtually meaningless for describing hydrological dynamics and producing accurate and/or precise estimates of *outflow*.

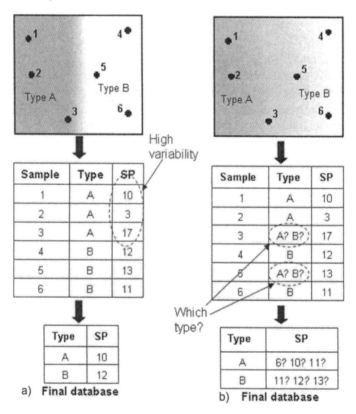

Fig. 18.5. Impact of mapping on data collection; characteristics of soil type A are more variable than those of soil type B. (**a**) Easily mapped soil types with a distinct boundary between the two soil types. (**b**) Difficult to map soil types with an indistinct boundary between the two soil types. SP=soil permeability, numbered dots are ground-based sample points

18.2.4 Summary on Models and Sources of Uncertainty

The five sources of model uncertainty discussed — natural variability, temporal scale, spatial scale, taxonomic scale and data collection — have a two-fold impact on model uncertainty.

The first impact is on model development. If model coefficients are determined empirically, then these five factors will render a model less robust — that is, inapplicable to temporal, spatial and taxonomic scales other than those on which the model is based. Alternatively, if model coefficients are represented as definitional constants, then the definition of 'forest,' for example, used in the model must be the same as the one used to derive the constant. If not, the constant should not be used in the model being developed, and must be determined using empirical data that may be sparse or even unavailable.

Similarly, if a model was developed or calibrated for an area where the maximum rainfall is, for example, 1000 mm/yr, using such a model for an area receiving 2000 mm/yr is inappropriate. Doing so will have unknown impacts on the uncertainty associated with model outputs, and models based on process description will perform better outside their original data range than empirical models. It is nonetheless inadvisable to use any models — process-based or empirical — outside their data range or in violation of underlying assumptions around data and model structure.

The second impact of these factors is in model usage. Models are developed to enable an estimation of some variable (such as *outflow*) for which direct measurements do not exist. When a model is used, therefore, it will produce estimates that are only as accurate and precise as the data used as inputs. Thus even if a given model is somehow 'perfectly' structured and calibrated, data that are subsequently input into the model for predictive purposes will be subject to the five factors mentioned in unknown and difficult to quantify ways — a factor that is exacerbated for data outside the original geographic and numerical range.

18.3 Model Uncertainty and Decision Making

So what does all this mean for those who use models to make decisions? In particular, what does it mean to someone who uses model outputs and wants to consider the associated uncertainty in decision making? As discussed, the reality is that it is extremely difficult for model developers to provide a robust estimate of model reliability. It follows therefore that it is equally difficult for model users to obtain information on the reliability of models.

This does not mean, however, that models should not be used in the decision-making process. In many cases, empirical data do not exist to inform land management decisions that must be made. Hence decision makers must choose between:

- an un-informed guess
- expert opinion and/or local knowledge
- a specially commissioned study
- models.

Depending on the situation, not only are models the most rapid and cost-effective means of producing necessary information, but the information produced is not necessarily less reliable than the information produced by the other methods. Thus the issues surrounding model uncertainty do not mean that alternatives to models are superior. None of the four options mentioned above are without uncertainty, but one or another might provide useful information and be preferable in a given situation, and the alternatives also suffer from an inability to provide useful information about uncertainty.

The risk analysis research community has documented that uncertainty affects the decisions that people make. (See the reading list). At the same time, there are cases in which the uncertainty inherent in models does not affect a decision that is being made. Imagine a situation in which a model states only that 'Option A' will save $1000 and 'Option B' will save $500 — that is, half as much. In this case, in the absence of any additional information, Option A is clearly preferred. Suppose that a decision maker is then told that the uncertainty on the model can be as high as 30%. This would mean that Option A might only provide a benefit of $700 (or as much as $1300) whereas the range on the benefits for Option B would be $350 to $650. Even with a large amount of known certainty, Option A remains superior. The key here is to recognise that if a model suggests that one option is clearly superior to other options, there is probably little need to consider uncertainty.

There may nonetheless be a need for uncertainty information, even if such information is highly subjective in nature. If one has a hypothetical situation for which the consequences are not loss of money but instead something judged to be far more important, decision makers will change the importance of the factors and the information that they use to make decisions. Suppose that a disaster management model estimates that 10 sheep will die in a 25-year flood, 35 in a 50-year flood, and 1000 in a 100-year flood. An economically rational decision maker is likely to try to protect against a 100-year flood, and accept the potential loss of sheep from the

25-year and 50-year floods. However, if the potential loss is human life instead of sheep, policy implications might overwhelm economic rational and result in an effort to protect against 25-year or 50-year floods. In such a situation, uncertainty is also probably irrelevant because the perceived cost — loss of any human life — are the primary consideration.

The essence of the preceding paragraphs is that there are two main factors that affect the importance of uncertainty in decision making — the influence of model outputs on the decision, and the clarity of choice. Figure 18.6 indicates that there is really only one set of circumstances (high model influence and unclear clarity of choice) for which model uncertainty may be a critical component of decision making. In situations where model influence is low, or clarity of choice is clear, uncertainty information is of limited use. However, the prevailing conditions in the majority of decisions are not at the extremes of model influence or clarity of choice, but instead are somewhere in the middle. Thus in the majority of decisions that rely at least partially on model outputs, uncertainty information may have a role to play.

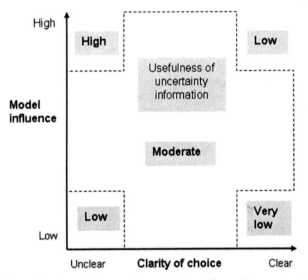

Fig. 18.6. Schematic of usefulness of uncertainty information on decision making as a function of influence of model outputs on the decision of process, and the clarity of choice in the final decision

The availability of uncertainty information will not necessarily make decision making easier. Suppose that one is considering increasing the amount of pasture by 10 ha for a fictitious 100 ha catchment, and there are two choices for where to put it (Fig. 18.7). It is desired to place it in an area that does not eliminate forest, and to place it on an area where soil permeability is high so that animal wastes will be filtered as they reach the watertable. Without consideration of uncertainty, Area 2 is clearly superior — data suggest it has no forest on it, and its soil permeability is higher (10) than for Area 1 (7).

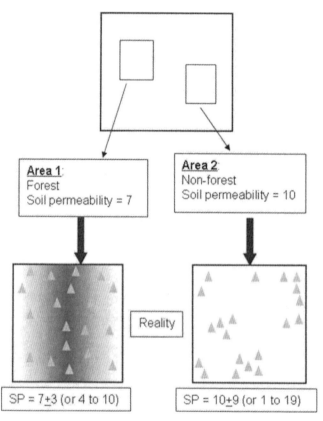

Fig. 18.7. Hypothetical example demonstrating difficulty of using uncertainty information in decision making (see text for explanation; SP=soil permeability)

Suppose now that we have uncertainty information to consider. In doing so, we see that Area 2 was mapped as having no forest because the trees are not concentrated in a single area. In fact, converting Area 2 to pasture will cause a loss of the same number of trees as if Area 1 is converted to pasture. Moreover, the favourable soil permeability value for Area 2 (10) might be as low as 1 or has high as 19, whereas for Area 1 the range is 4 to 10. Confounding this is that Area 2 is mapped as being a single soil type whereas Area 2 is underlain by two soil types that grade into each other gradually — a factor that further complicates the analysis.

The point of this discussion is not that the original decision to convert Area 2 to pasture was wrong, as indeed no evidence has been presented to support that conclusion. Instead, this discussion is presented to demonstrate how the availability of uncertainty information does not necessarily lead to better decisions, or decisions that are easier to make. Given the decision criteria — no loss of soil and high soil permeability — and the real-world characteristics of the two areas being considered (Fig. 18.7), Area 2 might still be preferable given the likelihood that the soil permeability for Area 2 may really be lower than for Area 1. If so, then consideration of uncertainty in this example does not necessarily allow a better decision to be made, nor does it provide more confidence to a decision maker. If anything, the complexity of the uncertainty information makes it more difficult to make a reasoned decision.

This latter point highlights the ultimate point of this chapter. While uncertainty information may be useful to decision makers in many situations, it must be presented in ways that facilitate its communication and understanding by non-experts. A simplistic way of doing this might be to summarise the uncertainty information and present it in tables and graphs. The most useful way of doing this might ultimately be to develop algorithms that optimise a decision based on the uncertainty inherent in both the model components and the ultimate model outputs.

18.4 Conclusion

Uncertainty is inherent in model outputs due to a variety of factors fundamental to models and associated data. The complexity of those factors and the interactions among them make it difficult for model developers to provide model users with estimates of uncertainty information associated with model outputs. This does not invalidate the use of models in the decision-making process, as alternatives to models also have uncertainty. Moreover, there are situations in which the utility of uncertainty information for decision makers is low — such as, when a decision is obvious, or model outputs are not the predominant consideration in decision making. Finally, knowledge of uncertainty is not guaranteed to lead to better decisions, to change a decision, nor to provide assurance that the best possible decision has been made.

Reading List

Andrews C, Hassenzahl D, Johnson B (2004) Accommodating uncertainty in comparative risk. Risk Analysis 24:1323–1335

Borsuk M, Tomassini L (2005) Uncertainty, imprecision, and the precautionary priniciple in climate change assessment. Water Science and Technology 52:213–225

Cameron E, Peloso GF (2005) Risk management and the precautionary principle: a fuzzy logic model. Risk Analysis 25:901–911

Caselton W, Luo W (1992) Decision making with imprecise probabilities: Dempster-Shafer theory and application. Water Resources Research 28:3071–3083

Draper D (1995) Assessment and propagation of model uncertainty. Journal of the Royal Statistical Society Series B-Methodological 57:45–97

Ellsberg D (1961) Risk, ambiguity, and the Savage axioms. Quarterly Journal of Economics 75:643–669

Hoffman E, Hammonds J (1994) Propagation of uncertainty in risk assessments: the need to distinguish between uncertainty due to lack of knowledge and uncertainty due to variability. Risk Analysis 14:707–712

Hope S (2005) Decision making under spatial uncertainty. Unpublished M.Sc. research thesis, University of Melbourne, Australia

Reckhow K (1994) Importance of scientific uncertainty in decision making. Environmental Management 18:161–166

Reichert P, Borsuk M (2005) Does high forecast uncertainty preclude effective decision support? Environmental Modelling and Software 20:991–1001

19 Assessing Water Quality Impacts of Community Defined Land Use Change Scenarios for the Douglas Shire, Far North Queensland

Iris Bohnet[1], Jon Brodie[2] and Rebecca Bartley[3]

[1] CSIRO Sustainable Ecosystems, Tropical Forest Research Centre, Atherton, Queensland, Australia
[2] Australian Centre for Tropical Freshwater Research, James Cook University, Townsville, Queensland, Australia
[3] CSIRO Land and Water, Indooroopilly, Brisbane, Queensland, Australia

Abstract: The Douglas Shire is promoted as the only place in the world where two world heritage areas meet — the Great Barrier Reef (GBR) and the Wet Tropics of Far North Queensland. Over 80% of the shire's land area is World Heritage listed. On the remaining 20%, economic pressures on primary industries coupled with development pressures to subdivide agricultural land for urban and rural residential expansion provide a challenge not only for local people living in the shire, but also for planners and natural resource managers who have a responsibility to implement the Reef Water Quality Protection Plan (Reefplan) (Commonwealth and Queensland Governments 2003). The purpose of the Reefplan is to develop actions, mechanisms and partnerships to halt and reverse the decline in the quality of water flowing in the Great Barrier Reef lagoon.

The research presented in this chapter contributes to this goal by applying a social–ecological framework that links biophysical and social science data and is based on partnerships and local collaboration to achieve sustainability outcomes. This chapter focuses, in accordance with the Reefplan, on water quality and summarises the water quality impacts resulting from six spatially explicit land use change scenarios that were developed in col-

laboration with the local community in the Douglas Shire for the year 2025, based on their aspirations for a sustainable future. The spatially explicit land use change scenarios were used as input into the Sediment River Network Model (SedNet) *to assess and compare the water quality outcomes of the current situation with the community developed scenarios for the future, with the aim of identifying actions to improve water quality.*

The model results show that the changes in land use proposed in all scenarios lead to no or a relatively small reduction in sediment loads exported from the Douglas Shire, indicating that the SedNet *model, which was applied for the whole of catchment, is insensitive to land use changes in the floodplain part of the landscape. Despite the fact that the* SedNet *results did not support community discussion about potential actions to achieve improvements in water quality, the application of the social–ecological framework proved to be an important mechanism to build community capacity for participation in planning for sustainable future landscapes and seascapes.*

19.1 Context and Case Study Location

The coastal catchments adjacent to the Great Barrier Reef — the Great Barrier Reef Catchment Area (GBRCA) — are a region of high economic importance and exceptional environmental value (McDonald and Weston 2004). The Douglas Shire is located in the northern part of the region, and supports the Wet Tropics and the Great Barrier Reef world heritage areas (Williams et al. 2001). The agricultural land, the focus of this study, forms less than 20% of the total shire area and is surrounded by the mountainous Wet Tropics rainforest to the west, and the Great Barrier Reef to the east (see Fig. 19.1).

The need for greater protection of the Great Barrier Reef, a declining local sugar industry and pressures to subdivide agricultural land for urban expansion, provides a challenge not only for local people, but also for planners and natural resource managers who have a responsibility to implement the Reef Water Quality Protection Plan (Reefplan: Commonwealth and Queensland Government 2003). The Reefplan was developed because runoff of land sourced pollutants (suspended sediments, nutrients and pesticides) is believed to be degrading Great Barrier Reef ecosystems (Brodie et al. 2001; Furnas 2003). The goal of the Reefplan therefore is to halt and reverse the decline in water quality entering the reef within 10 years by reducing the load of pollutants from diffuse sources in the water entering the reef, and by rehabilitating and conserving areas of the reef catchment that have a role in removing waterborne pollutants.

Assessing Impacts of Community Defined Land Use Change Scenarios   385

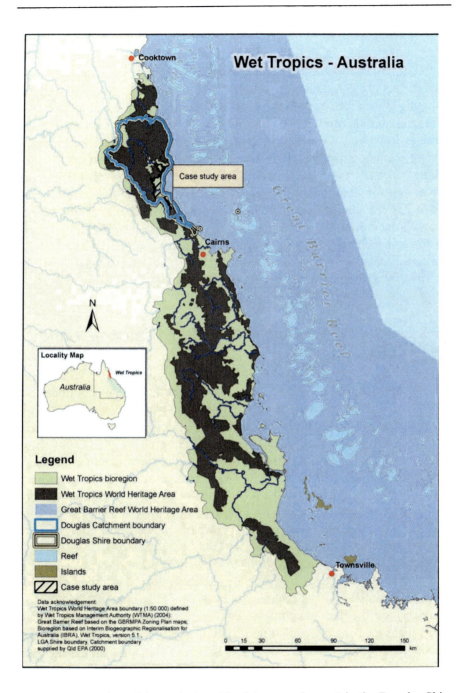

Fig. 19.1. Location of the agricultural land (case study area) in the Douglas Shire within the Wet Tropics bioregion and relative to the Great Barrier Reef and Wet Tropics world heritage areas

The Douglas Shire requires sustainable solutions for its future development that address environmental, social and economic change while enabling remediation of aquatic and terrestrial ecosystem degradation. The goal of this project, therefore, was to develop visions for the future of the Douglas Shire together with the local community that support local industries, communities and ecosystems, while protecting the Great Barrier Reef through improved water quality and hence healthier aquatic ecosystems. One of the key objectives of the project was the translation of broad community visions into spatially explicit land use change scenarios that allow for analysis of water quality impacts on the Great Barrier Reef. Comparing the reef water quality outcomes from the different land use change scenarios with the current situation enables the community, planners and natural resource managers to make more informed and strategic decisions about the future development of the Douglas Shire if reef water quality is imperative for any future development.

19.2 Dialogue over Sustainable Future Landscapes and Seascapes

Dialogue and public participation are essential in the debate about possible future development in the Douglas Shire and elsewhere. Successful implementation of the local planning scheme (Douglas Shire Council 2006), the Far North Queensland natural resource management plan (FNQNRM 2004), and the Reefplan (Commonwealth and Queensland Government 2003) requires the active support and engagement of a wide range of people including industries, community groups, indigenous people, environmentalists, farmers, concerned locals and the general public. Without integrating the diverse points of view of these stakeholders, the implementation of these plans is at risk (Kasemir et al. 2003).

Plans that are consistent with the visions, beliefs, and aspirations of local people will have a greater chance of success than plans imposed without consideration of local opinion, as local people (e.g. land owners, land mangers) are the ones who implement the required changes, proposed in these plans, 'on the ground'. Luz (2000) argues that only with the introduction of a 'social layer' in the superposition of thematic maps, which are generally used in planning processes, can the interests and needs of different landscape users be taken into account as professionally as the mapping of soils, land use and land cover. He also suggests collecting data from local farmers on differences in perception and appreciation of landscape and environment, and perspectives on the future of farming and landscape development, with communication facilitated through workshops, working groups, lectures and site visits.

Communication throughout the planning process is not only important from the point of view of acceptance of the outcomes, but also to enable social learning and encourage a more informed decision-making process about future developments in the Douglas Shire and the GBRCA based on the economic, environmental and social impacts and trade-offs of alternate pathways for change (Barker 2005). The formation of partnerships bringing together scientists, planners, decision makers and local communities presents an opportunity to develop a framework and process for integrating the social dimension, communication and capacity building into landscape planning and analysis for sustainable development in the GBRCA.

19.3 Methodology of an Application of a Social–Ecological Framework for Sustainable Landscape Planning

To create pathways towards a more sustainable future Bohnet and Smith (2007) developed a social–ecological framework for planning future landscapes (Fig. 19.2). Following their framework in broad terms, the following three-stage participatory planning process was chosen to develop a set of land use change scenarios together with the local community based on their aspirations for a sustainable future. In contrast to forecasting or predicting the future, scenarios are vivid stories that are constructed to describe alternative futures that might be very different from the present (Hulse et al. 2004, Nassauer and Corry 2004). The advantage of developing scenarios, in contrasts to forecasts, is the potential to build community capacity and to plan strategically to achieve a desired future outcome that is shared by the community.

19.3.1 Stage I: Community Perceptions and Visions

The first stage involved desktop and field-based assessments to define the character of the Douglas Shire landscape (Swanwick 2002), as well as 19 qualitative semi-structured interviews with a wide range of farmers and land managers in the Douglas Shire (Silverman 2000) and landscape assessments of their properties (Countryside Commission 1993). The goal of the interviews was to discuss issues such as:

- how they use and manage their properties
- why they carry out certain practices and not others

- what their future goals and aspirations are regarding land use and management.

The landscape assessments, which were carried out with interviewees, provided additional information and allowed linking and comparing field data with social data gathered during the interviews. Analysis of interview data and landscape assessments informed the development of two broad community visions for the shire based on the participants' perceptions about the future (Fig. 19.2).

Characteristic photographs from the Douglas Shire landscape were simulated and distinct future paths for landscape development were visualised using the software Adobe Photoshop 7.0. These photorealistic landscape visions were then used as a tool and starting point to communicate to the public possible landscape futures for the Douglas Shire, which promoted public debate about the future (Bohnet 2004). Landscape visualisations were chosen because of their successful application by other landscape researchers and planners to illustrate change in the visual landscape (Al-Kodmany 1999; Schmid 2001). Photorealistic landscape simulations were chosen as a visualisation tool over other tools such as maps, charts, tables, drawings or GIS-based modelled landform surfaces (Jessel and Jacobs 2005; Lange 1999; Palang et al. 2000; Tress and Tress 2003;) because they are easily accessible, attract people's attention and enhance public participation in planning and design.

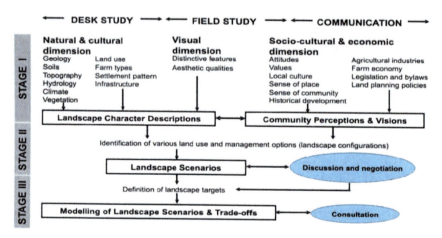

Fig. 19.2. Social–ecological framework for sustainable landscape planning (Source: Bohnet and Smith 2007)

19.3.2 Stage II: Community-driven Landscape Scenarios

Stage I provided the foundation for Stage II in which community workshops were used to further develop the broad landscape visions for 2025 with the wider community and other interest groups, apart from farmers and land mangers. These workshops provided local citizens with the opportunity to discuss their preferred future, what their common and contrasting future priorities were, and how those priorities may translate in the landscape of 2025.

From the workshop discussion, common themes could be identified and spatially explicit land use change scenarios could be derived based on priority themes (Fig. 19.2). Alternate scenarios were drawn from contrasting sets of priority themes. Besides the contrasting priorities themes (i.e. future land use preferences), workshop participants also defined some landscape targets to be achieved under any future scenario. For example, land currently used for agricultural production on steep slopes ($\geq$20%) was proposed to be left to regrow into secondary rainforest. Applying quantitative targets for operational use allowed clear identification of change between the current situation and future scenarios, which is a prerequisite for mapping scenarios. In instances where no quantitative targets were identified by workshop participants it was necessary to translate qualitative information into quantitative or location specific targets. This was not always straightforward, but the information provided by workshop participants indicated at the very least where and what sort of change is favoured by local people in each of the landscape scenarios. For this reason alone, community workshops are of particular importance, as communication and cooperation have been identified as crucial for acceptance and implementation of planning projects (Luz 2000; Luz and Weiland 2001).

19.3.3 Stage III: Modelling of Landscape Scenarios and Assessing Water Quality

Based on the current land use/land cover map developed for this study, the land use change scenarios were mapped in Stage III (Fig. 19.2). The goal was to translate the descriptive visions provided by the local community as accurately as possible into mapped land use change scenarios (Hulse et al. 2002). The resulting scenarios are those that make sense to locals and reflect their aspirations and values — in contrast to scenarios developed by researchers, planners and policy makers. Scenario mapping was done through land allocation modelling, which means assigning one particular land use/land cover to each grid cell or polygon in a digital map according to defined rules. Hierarchical roles (first and second order rules) were derived from the priority themes and landscape targets, defined in the community workshops in Stage II, and mapped in a Geographic Information System (GIS).

Current land use/land cover and that of each future scenario, together with additional information related to management practices (discussed during the community workshops) were used as inputs into *SedNet* to assess the reef water quality impacts of the different scenarios in comparison to the current situation.

SedNet models estimate river sediment loads, by constructing material budgets that account for the main sources and stores of sediment (Wilkinson et al. 2004). *SedNet* uses a simple conceptualisation of transport and deposition processes in streams (Fig. 19.3). Sediment sources, stream loads, and areas of deposition in the system are simulated. The contribution to the river mouth from each subcatchment can be traced back through the system, allowing downstream impacts to be put into a regional perspective.

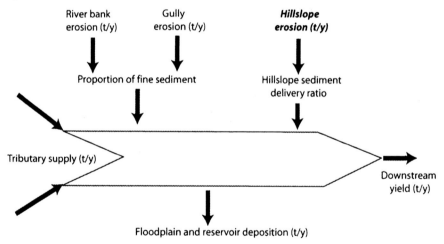

Fig. 19.3. Conceptualisation of the components of the suspended load budget transport and deposition processes in the Sediment River Network Model (*SedNet*) where t/y is tonnes per year (adapted from Wilkinson et al. 2004). Note: only the hillslope erosion component of the budget was evaluated in this study and therefore only suspended (or fine) sediment was evaluated in the final calculations. Bedload is produced from river bank and gully erosion (not hillslope erosion) neither of which was evaluated in these scenarios

The *SedNet* model generates sediment from three erosion processes: hillslope or paddock erosion, gully or drain erosion, and river bank erosion. The methods used to assess each of these sources and to propagate the sediment through the river network are described in detail in Bartley et al. (2004, 2003). It is important to note, however, that in this study the scenarios were only applicable to hillslope or paddock erosion as this is the main process directly impacted by land use/land cover change. It is anticipated that both drain and bank erosion would also be affected by changes in land condition, yet the magnitude of this change is not known and was therefore not addressed in this study. It was believed at the commencement of the study that the principle effect of the land use scenarios on water quality would be on sediment loads rather than specific nutrient or pesticide loads, and hence sediment and the sediment modelling tool *SedNet* were chosen to analyse the effects of the scenarios.

19.4 Results and Discussion

19.4.1 Visions for the Douglas Shire Coastal Landscape

Based on the character of the Douglas Shire landscape, which is defined by a combination of natural and agricultural features (Bohnet 2004), and the qualitative interviews with farmers and land managers in the Douglas Shire, the future visions could be divided into two distinct future pathways:

- a diversified future landscape with continued sugarcane production
- a post-sugarcane landscape.

These visions for the future, described by the interviewees, were presented as landscape simulations based on landscape photographs taken during the landscape assessments (Fig. 19.4a).

In the vision for a diversified future landscape in 2025 with continued sugarcane production, interviewees assumed that declining farm incomes from sugar production and pressures on the local sugar industry led cane farmers to pool their ideas and resources to overcome the crisis. Farm businesses and the cooperative sugar mill diversify in order to produce tropical fruit juices, timber and fibreboards, in addition to sugarcane. In the diversified future landscape sugarcane, tropical fruit, bamboo and rainforest timber is grown (Fig. 19.4b).

Fig. 19.4. (**a**) Douglas Shire coastal landscape in 2003, (**b**) a diversified landscape vision for continued sugarcane production in Douglas Shire 2025, (**c**) a post-sugarcane vision for the Douglas Shire landscape in 2025

The vision for a post-sugarcane landscape is based on the idea, put forward by some interviewees, that development pressures led Douglas Shire council to approve subdivisions on land previously used for sugarcane production forcing the local sugar mill to close down. Some buyers of these blocks choose to carry out some agricultural activities, such as organic farming or other land-based activities with low environmental impact. Subdivisions on hill slopes are only approved under strict development codes and buyers are required to screen their homes with native trees. Sugarcane paddocks are replaced by small-scale cropping, residential development, pastures, regrowth and agroforesty blocks (Fig. 19.4c).

19.4.2 Spatially Explicit Land Use Change Scenarios

Community interest, stimulated by the landscape visions (Fig. 19.4b, 19.4c) enabled the project team to run three community workshops. These landscape visions attracted a diverse range of community members including farmers, traditional owners, long-term residents and newcomers to the area, to attend the workshops. The workshops provided a forum for individuals to present their different viewpoints regarding their aspirations for the future of the area, to discuss their priorities and the potential trade-offs between their priorities, and to learn from other participants. Small workshop groups ensured genuine input from each individual and exposure to different ideas.

The project team facilitated discussion and prompted participants with questions that would allow the information provided by the participants to be translated into spatially explicit land use or management changes. The combined results from all workshops are presented in the following sequence: common themes discussed during the workshops, priorities for the future, and development of spatially explicit land use change scenarios. The combined results from the workshops were then put into *SedNet* to evaluate how the changes impact on end-of-catchment sediment loads (used to represent water quality). Common themes discussed in all workshops were:

- sugar industry survival
- land use diversification options and alternative management practices
- water quality and quantity issues
- protection of the natural environment
- development pressures for urban and rural residential subdivisions.

The results are also summarised in a short report prepared for workshop participants (Bohnet 2004). Discussions about the sugar industry surviving or ceasing in future divided participants into two groups. Therefore, it was decided that the future scenarios developed would need to have two sets of scenarios, similar to the broad landscape visions that stimulated community interest. One set needed to include continued sugarcane production in the future, whereas the other set needed to exclude sugarcane production in the future landscape. The decision to develop two sets of scenarios was made early in each of the workshops in order to progress discussion and participants' thinking about differences between those distinct futures, alternate land uses and priorities.

Three quite different key priorities for the future emerged from the discussions: continued agricultural production, improved water quality, and enhanced biodiversity. These priorities were based on the common themes discussed and on different sets of underlying assumptions and beliefs held by the participants. The core assumptions and beliefs on which people in the Douglas Shire based their priorities for continued agricultural production included:

- agriculture is a means to stop or reduce further rural residential development
- appreciation of local produce that can be associated with the landscape
- opportunity to buy locally grown food
- diversification of agricultural crops reduces farm financial risks, particularly if high value crops are grown in areas 'suitable' for those crops
- move towards more environmentally-friendly/organic farming practices reduces environmental impacts from farming and potential conflicts with the non-farming community
- provision of local employment opportunities.

With regard to the priority of improved water quality, the core assumptions and beliefs of local people included:

- any activity (e.g. agriculture, rural residential development) should not pollute rivers and creeks
- agricultural activities should continue to protect the rural character of the area but with lowest possible impact (e.g. grazing with low stocking rates and minimum fertilizer input)
- obligation to protect the Great Barrier Reef
- necessity to sustain a viable tourism industry.

With regard to enhanced biodiversity as a priority, the core assumptions and beliefs of local people included:

- native vegetation on farms (e.g. remnants) needs to be protected for its intrinsic value
- need to provide habitats for native flora and fauna
- need to improve the coastal environmental to protect Great Barrier Reef (e.g. no sugarcane should be grown in area of frequent flooding)
- move towards more environmentally-friendly/organic farming and management practices will enhance biodiversity on farms and improve water quality
- need to grow crops that are 'suitable' for the area in order to reduce inputs (fertilizers, pesticides) and impact
- necessity to sustain rural character and viable tourism industry.

Discussions provided sufficient detailed information to develop two sets of three spatially explicit future scenarios for 2025 in a geographic information system (GIS).

- Scenario I: continued agricultural production (with continued sugarcane production)

- Scenario II: continued agricultural production (without continued sugarcane production)
- Scenario III: improved water quality (with continued sugarcane production)
- Scenario IV: improved water quality (without continued sugarcane production)
- Scenario V: enhanced biodiversity (with continued sugarcane production).
- Scenario VI: enhanced biodiversity (without continued sugarcane production).

The workshop discussions also provided a wealth of contextual data useful to support the six land use change scenarios. Participants provided examples to underpin their priority and to convince others in the group about their priority considering potential trade-offs (environmental, social and economic) between different priorities. For example, participants who felt that continued agricultural production has to be the main priority suggested a wide range of crops including aquaculture, cocoa, vegetables and agroforesty for agricultural and landscape diversification to achieve a balance between environmental, social and economic outcomes. They argued that these crops would allow for value adding while providing employment opportunities and preserving the character of the area.

The justifications, provided by the participants, to underpin each of the three priorities for the future provided, in essence, the details for development of the spatially explicit land use change scenarios. In addition, quantitative landscape targets such as, all land on slopes steeper than 20% currently used for agricultural production should be left for natural revegetation, were applied to all scenarios, whereas differing width of riparian buffer zones were applied to the improved water quality and enhanced biodiversity scenarios. Land suitability indicators for the crops suggested by the workshop participants were used as criteria to identify the most suitable locations in the landscape for the suggested agricultural crops in the continued agricultural production and enhanced biodiversity scenarios (Wilson 1991).

A summary of the main land use changes and associated GIS-allocation rules is presented in Table 19.1, whereas Fig. 19.5 shows the current land use/land cover map developed for the agricultural land in the Douglas Shire and Fig. 19.6, Fig. 19.7 and Fig. 19.8 present the six land use change scenario maps.

Table 19.1. Summary of the proposed land use changes for the Douglas Shire landscape

2025 Scenario	Main land use change	GIS-allocation rules (replicable criterion) on which land use change is based
Scenario I: Continued agricultural production, with sugarcane	Slopes steeper than 20% and used for agricultural purposes → regrowth Land unsuitable for sugarcane production → cocoa, horticulture, vegetables, sweet corn, agroforestry	All slopes steeper than 20% Land suitability mapping for agricultural crops
Scenario II: Continued agricultural production, without sugarcane	Slopes steeper than 20% and used for agricultural purposes → regrowth All sugarcane → aquaculture, agroforestry, vegetables, sweet corn, horticulture	All slopes steeper than 20% Spatially explicit locations (for some crops) Land suitability mapping for agricultural crops
Scenario III: Improved water quality, with sugarcane	Slopes steeper than 20% and used for agricultural purposes → regrowth Establishment of continuous riparian buffer zones Establishment of wetland buffer zones	All slopes steeper than 20% 50 m riparian buffer zone where native vegetation is missing 100m buffers along conserved coastal wetlands
Scenario IV: Improved water quality, without sugarcane	Slopes steeper than 20% and used for agricultural purposes → regrowth Establishment of continuous riparian buffer zones Establishment of continuous wetland buffer zones All sugarcane → grazing	All slopes steeper than 20% 50 m riparian buffer zone where native vegetation is missing 100 m buffers along conserved coastal wetlands Change sugarcane to grazing
Scenario V: Enhanced biodiversity, with sugarcane	Slopes steeper than 20% and used for agricultural purposes → regrowth Establishment of continuous riparian buffer zones Land unsuitable for sugarcane production → cocoa, agroforestry, coastal wetlands	All slopes steeper than 20% 100 m riparian buffer zone where native vegetation is missing Spatially explicit locations (for coastal wetlands and agricultural crops)
Scenario VI: Enhanced biodiversity, without sugarcane	Slopes steeper than 20% and used for agricultural purposes → regrowth Establishment of continuous riparian buffer zones All sugarcane → horticulture, grazing, agroforestry, coastal wetlands	All slopes steeper than 20% 100 m riparian buffer zone where native vegetation is missing Spatially explicit locations (for coastal wetlands and agroforestry) Land suitability mapping for agricultural crops

Assessing Impacts of Community Defined Land Use Change Scenarios 397

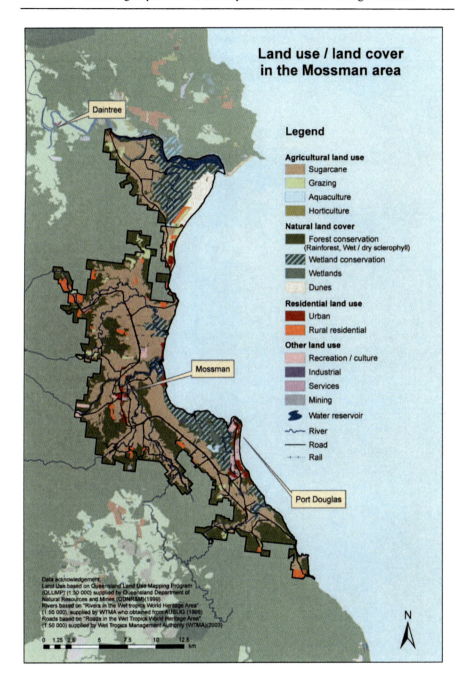

Fig. 19.5. Current (circa 2000) land use/land cover in the Douglas Shire

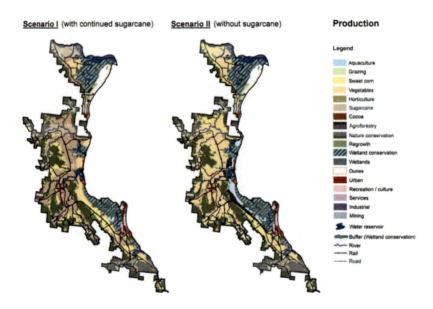

Fig. 19.6. Land use change scenario maps: Scenario I continued agricultural production with continued sugarcane in 2025; Scenario II continued agricultural production without sugarcane production in 2025

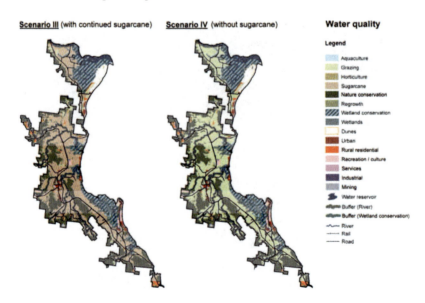

Fig. 19.7. Land use change scenario maps: Scenario III improved water quality with continued sugarcane in 2025; Scenario IV improved water quality without sugarcane production in 2025

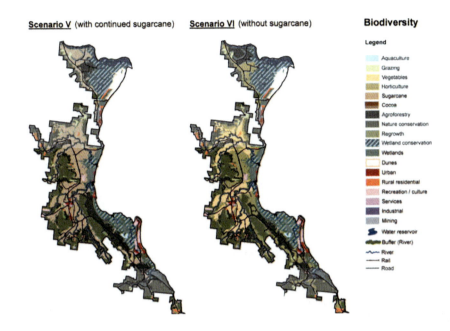

Fig. 19.8. Land use change scenario maps: Scenario V enhanced biodiversity with continued sugarcane in 2025; Scenario VI enhanced biodiversity without sugarcane production in 2025

19.4.3 Inputs into *SedNet* for Water Quality Analysis and Model Results

In addition to changes in land use, land management changes (as suggested by the workshop participants) were also included as inputs into *SedNet*. These management changes for sugarcane included: tillage level (actual for current situation, minimum for all scenarios), headland type (grassed for current situation and all scenarios), and fallow strategy (no fallow for current situation, legume fallow for all scenarios). Management changes for grazing included stocking rate (actual for current situation, reduced for all scenarios). Production systems and management practices were given varying cover factors dependent upon the scenario (Bartley et al. 2004; Merritt 2002). These cover factors influence surface erosion, but not bank or drain erosion in the *SedNet* model. Cover factor grids were created for each scenario for input into the creation of hillslope erosion grids, which formed the final input for the calculation of sediment loads in *SedNet*.

The *SedNet* results indicate that changes in land use and management envisioned in all scenarios lead to no, or a relatively small, reduction in sediment loads exported from the Douglas Shire (Table 19.2). This is not surprising as less than 20% of the area is used for agricultural purposes and most agricultural activities in the shire take place on relatively flat land and coastal floodplains. The original *SedNet* modelling identified that drain erosion (rather than hillslope or paddock erosion) was the most significant erosion process contributing to end-of-catchment sediment loads in the low-lying cane area of the Douglas Shire (Bartley et al. 2004); however, for reasons discussed earlier drain erosion was not evaluated in this study. In addition management practices in the sugar industry in Queensland which address (reduce) soil erosion were implemented widely in the period 1970–2000 (Rayment 2003). These practices include green cane harvesting, trash blanketing and reduced tillage. These practices are believed to have reduced soil erosion in the industry by a factor of approximately ten (Brodie and Mitchell 2005; Rayment 2003).

Thus this study suggests that changes in land use and management from the current situation have little effect in further reducing soil erosion and decreasing suspended sediment delivery to rivers and the coast. There is less than a 3% change in the fine sediment load exported from the catchments for all of the scenarios (Table 19.2). It is anticipated that the error associated with predicting sediment loads from large catchments, particularly with relatively poor data inputs, is at least 10%. Therefore the changes these scenarios are considered to be within the error of model calculations and it is not possible to definitively say that any one scenario will make a difference on end-of-catchment fine sediment loads.

Table 19.2. Results from six land use change scenarios in the Douglas Shire applying the *Sediment River Network Model*

	Sediment supply [Kt/y]				Sediment export [Kt/y]			
Scenario[a]	Drain erosion	Bank erosion	Hillslope erosion	Total supply	Bedload	Suspended	Total exports	% diff[b]
I	8.8	9.98	201.96	220.74	2.24	202.09	220.74	-0.00
II	8.8	9.98	199.37	218.15	2.24	199.75	218.15	-1.16
III	8.8	9.98	201.99	220.77	2.24	202.11	220.76	-0.01
IV	8.8	9.98	196.18	214.96	2.24	196.93	214.95	-2.55
V	8.8	9.98	198.12	216.90	2.24	198.66	216.90	-1.70
VI	8.8	9.98	197.64	216.42	2.24	198.24	216.42	-1.91

[a] I: Production with sugarcane, II: Production without sugarcane, III: Water quality with sugarcane, IV: Water quality without sugarcane, V: Biodiversity with sugarcane, VI: Biodiversity without sugarcane. [b] Percentage difference from current situation.

19.5 Conclusion

Development of future visions with the Douglas Shire community enabled social, economic and environmental issues to be discussed and integrated in six spatially explicit land use change scenarios for 2025. The future scenarios illustrate participants' wide ranging views and preferences related to landscape character and future development of the Douglas Shire landscape. There was broad agreement on the need for environmental, social and economic sustainability; however, contrast among the scenarios highlighted the differences in participants' perceptions and values. Working through the social–ecological framework and formulating landscape visions and spatially explicit land use change scenarios facilitated communication among different social groups, and therefore was an important mechanism for social learning in relation to planning, local development, and the need to think beyond their own property and the shire boundaries.

The small relative difference in sediment delivery to rivers and the coast from the six scenarios reflects the low topography in the Douglas Shire agricultural area. The modelling results did not deliver significant differences between potential future developments as expected by the community, indicating that the *SedNet* model, which was applied to the whole of the catchment, is insensitive to land use changes in the floodplain part of the landscape. This was not realised at the start of the project — the *SedNet* model showed significant differences between scenarios in larger catchments where agriculture is located on steeper slopes (e.g. Bartley et al. 2003).

Despite the fact that the *SedNet* results did not support community discussion about potential actions to achieve improvements in water quality, the application of the social–ecological framework proved to be an important mechanism to develop a set of scenarios and to discuss not only water quality but also biodiversity and economic impacts (Roebeling et al. 2005). Biodiversity values of different land use types have been addressed in a separate study (Metcalfe and Westcott in prep), and suggest that increased connectivity of fragments of remnant native vegetation would be beneficial for a number of floral and faunal groups. Enhanced connectivity could be achieved through:

- revegetation of drainage channels
- land abandonment and secondary forest development (e.g. on steep slopes)
- active revegetation of key areas deliberately for targeted biodiversity benefits or aesthetic reasons

- tree-based agricultural land use options which involve less intensive land management and habitat of greater structural complexity than sugarcane.

However, the magnitude of biodiversity benefits and the particular groups to benefit would vary between land use options and extent (Westcott and Metcalfe pers. comm.[1]). Biodiversity enhancements achieved through re-establishment of riparian buffer zones and coastal wetlands are also likely to improve water quality, however this was not modelled in *SedNet*.

The *SedNet* results also show that it is important to assess the risk from key pollutants (i.e. those of highest environmental risk), before complex scenarios, such as the ones developed in this study, are modelled. In the case of the Douglas Shire it seems now obvious that the key pollutants are nitrate (from fertilizer residues) and possibly a number of herbicide residues (particularly diuron and atrazine), which was not known when this study was carried out. *SedNet*, and its associated nutrient model *ANNEX*, are able to model nutrient transport, which has been done for other parts of the Great Barrier Reef (McKergow et al. 2005); however, neither model can deal with herbicides. The main reason that the *ANNEX* model was not implemented in this study was the lack of nutrient data available for this region with which to populate the model, as well as redevelopment of the *ANNEX* model at the time of this study. However there is a lack of detailed knowledge of the effects of these land uses, and it is accurate to conclude that the land use scenarios analysed in the current study may not have large effects on nutrient and pesticide loads and hence water quality. A number of research needs with respect to water quality modelling have been identified by this study. To ensure adequate calibration of models, the following is required: (a) development of models that can deal with herbicide residue transfer from catchments to marine waters; and (b) more data on the rates and amounts of nutrient movement from different land use types in coastal floodplain environments.

It is important to note that the water quality models currently available for assessing catchment-scale sediment (and nutrient) loads are continually under development. As more monitoring data becomes available, we will be able to improve our understanding of how sediment (and nutrient) loads change with different land use and management configurations, in both time and space, so we can make more reliable assessments. At present, there are a range of different approaches being discussed to help improve this process, including: (a) augmenting the results from annual load mod-

---

[1] DA Westcott and DJ Metcalfe (Research Scientists, CSIRO Sustainable Ecosystems, Atherton, Queensland) 23 November 2007

els (such as *SedNet*) with runoff event-based models that may be more appropriate for identifying subtle changes in land use, and (b) combining water quality targets with land use change targets, as there is an increased acknowledgement that detecting changes in end-of-catchment loads as a result of land use change are difficult to evaluate, particularly for catchments that are not highly modified from their pre-European condition (e.g. Douglas Shire). This continual improvement of water quality models and approaches is important so we can assess the effects of changing from 'traditional' agricultural activities (such as sugarcane cultivation in the Douglas Shire, for which adequate research information exists) to new agricultural mixes, as suggested by the community in the present study, for which little or no research information exists.

Finally, the process followed in the social–ecological framework makes planning for sustainable future landscapes and seascapes socially relevant as the framework links social and biophysical data with local and scientific knowledge to inform landscape visualisation and analysis of a wide range of potential impacts. Analysing community defined land use change scenarios for a multitude of impacts — such as water quality, biodiversity, farm financial performance — allows development of community supported actions to achieve multiple benefits. This makes the framework and process a powerful planning tool that contributes to the development of sustainable landscapes and seascapes and hence meets the goals of the Reefplan.

Acknowledgements

We would like to thank the community members who participated in the project for their time, stimulating discussions and enthusiasm; and the members of the Douglas Shire Joint Venture Partnership for their ongoing support and feedback to the project. We would also like to thank Karl Haug for fieldwork assistance, Jana Kaeppler, Petina Pert and Mick Hartcher for GIS and modelling support, and David Westcott and Dan Metcalfe for valuable scientific discussions on biodiversity. Finally, we would like to thank CSIRO Sustainable Ecosystems for funding this research and Peter Stone and two anonymous reviewers for valuable comments on an earlier draft of this chapter.

References

Al-Kodmany K (1999) Using visualisation techniques for enhancing public participation in planning and design: process, implementation, and evaluation. Landscape Urban Planning 45:37–45

Barker A (2005) Capacity building for sustainability: towards community development in coastal Scotland. Journal of Environmental Management 75:11–19

Bartley R, Henderson A, Prosser IP, Hughes AO, McKergow L, Lu H, Brodie J, Bainbridge Z, Roth CH (2003) Patterns of erosion and sediment and nutrient transport in the Herbert River catchment, Queensland. CSIRO, Canberra, available at http://www.clw.csiro.au/publications/consultancy/2004

Bartley R, Henderson A, Baker G, Bormans M, Wilkinson S (2004) Patterns of erosion and sediment and nutrient transport in the Douglas Shire catchments. CSIRO Land and Water, Atherton, pp 61

Bohnet I (2004) Sustainable landscape planning in the Wet Tropics. Retrieved 25 July 2007, from http://www.cse.csiro.au/research/landscapescenarios/

Bohnet I, Smith DM (2007) Planning future landscapes in the Wet Tropics of Australia: A social-ecological framework. Landscape Urban Planning 80:137–152

Brodie J, Mitchell AW (2005) Nutrients in Australian tropical rivers: changes with agricultural development and implications for receiving environments. Marine and Freshwater Research 56(3):279–302

Brodie JE, Christie C, Devlin M, Haynes D, Morris S, Ramsay M, Waterhouse J, Yorkston H (2001) Catchment management and the Great Barrier Reef. Water Science and Technology 43:203–211

Commonwealth and Queensland Governments (2003) Reef water quality protection plan: for catchments adjacent to the Great Barrier Reef World Heritage Area. Australian Government and Queensland Government

Countryside Commission (1993) Landscape assessment guidance. Countryside Commission, Cheltenham

Douglas Shire Council (2006) Douglas Shire local area planning scheme. Douglas Shire Council, Queensland

FNQNRM (2004) Sustaining the Wet Tropics: a regional plan for natural resource management. Far North Queensland Natural Resource Management Ltd, and Rainforest Corporate Research Centre, Cairns

Furnas M (2003) Catchments and corals, terrestrial runoff to the Great Barrier Reef. Australian Institute of Marine Science, CRC Reef Research Centre, Townsville

Hulse DW, Branscomb A, Payne SG (2004) Envisioning alternatives: using citizen guidance to map future land and water use. Ecological Applications. 14:325–341

Hulse DW, Gregory S, Baker J (2002) Willamette River Basin planning atlas. Oregon State University Press, Corvallis, Oregon

Jessel B, Jacobs J (2005) Land use scenarios development and stakeholder involvement as tools for watershed management within the Havel River Basin. Limnologica 35:220–233

Kasemir B, Jaeger J, Jaeger CC, Gardner MT (2003) Public participation in sustainability science – a handbook. Cambridge, University Press, Cambridge

Lange E (1999) Realität und computergestützte visuele Simulation. Eine empirische Untersuching über den Realitätsgrad virtueller Landschaften am Beispiel des Talraums Bunnen-Schwyz. ORL-Berichte Nr. 106, VDF, Zürich

Luz F (2000) Participatory landscape ecology – A basis for acceptance and implementation. Landscape Urban Planning 50:157–166

Luz F, Weiland U (2001) Wessen Landschaft planen wir? Kommunikation in Landschafts- und Umweltplanung. Naturschutz und Landschaftsplanung 33:69–76

McDonald G, Weston N (2004) Sustaining the Wet Tropics: a regional plan for natural resource management, Vol. 1: Background to the plan. Rainforest Corporate Research Centre and Far North Queensland Natural Resource Management Ltd, Cairns

McKergow LA, Prosser IP, Hughes AO, Brodie J (2005). Regional scale nutrient modelling: exports to the Great Barrier Reef World Heritage Area. Marine Pollution Bulletin 51(1–4):186–199

Merritt WS (2002). Biophysical considerations in integrated catchment management: a modelling system for northern Thailand. Unpublished PhD thesis, Australian National University, Canberra

Metcalfe DJ, Westcott DA (in prep) Distribution of biodiversity and ecosystem processes in a fragmented tropical floodplain

Nassauer JI, Corry RC (2004) Using normative scenarios in landscape ecology. Landscape Ecology 19:343–356

Palang H, Alumäe H, Mander Ü (2000) Holistic aspects in landscape development: a scenario approach. Landscape Urban Planning 50:85–94

Rayment G (2003) Water quality in sugar catchments of Queensland. Water Science and Technology 48:35–47

Roebeling PC, Bohnet IM, Smith DM, Westcott D, Kroon F, Hartcher M, Hodgen M, Vleeshouwer J (2005) Landscapes toolkit for triple-bottom-line assessment of land-use scenarios in Great Barrier Reef catchments. In: Proceedings of MODSIM 2005 Conference, 12–15 December 2007, Melbourne, Australia

Schmid W (2001) The emerging role of visual resource assessment and visualisation in landscape planning in Switzerland. Landscape Urban Planning 54:213–221

Silverman D (2000) Doing qualitative research — a practical handbook. Sage, London

Swanwick C (2002) Landscape character assessment – guidance for England and Scotland. A report by Land Use Consultants for The Countryside Agency and Scottish Natural Heritage, Wetherby, West Yorkshire and Edinburgh

Tress B, Tress G (2003) Scenario visualisation for participatory landscape planning – a study from Denmark. Landscape Urban Planning 64:161–178

Wilkinson SN, Henderson A, Chen Y, Sherman BS (2004) *SedNet*: user guide Version 2.0.0. CSIRO, Canberra

Williams J, Read C, Norton A, Dovers S, Burgman M, Proctor W, Anderson H. (2001) Biodiversity, Australia State of the Environment Report 2001 (Theme Report), CSIRO Publishing on behalf of the Department of the Environment and Heritage, Canberra

Wilson PR (1991) Agricultural land suitability of the wet tropical coast, Mossman and Julatten area. Queensland Department of Primary Industries, Brisbane

20 Analysing Landscape Futures for Dryland Agricultural Areas: a Case Study in the Lower Murray Region of Southern Australia

Brett A Bryan, Neville D Crossman and Darran King

CSIRO, Urrbrae, South Australia, Australia

Abstract: There is an urgent need to reverse the declining environmental condition of rural landscapes across southern Australia. Current approaches focus on natural resources management planning, policy and decision making at the regional level. Regional plans and associated on-ground investment have the potential to have widespread and long-lasting environmental, economic and social impacts. However, rarely are these impacts quantified and clearly understood.

In this chapter we describe part of a large integrated project called the Lower Murray Landscape Futures (LMLF) which aimed to assess the impact of regional plans for the Lower Murray on selected environmental and socioeconomic indicators under alternative future landscape scenarios with input from stakeholders. The dryland component of the LMLF is a large-scale integrated regional planning and landscape futures analysis focussing on issues such as: agricultural production including food, fibre and bioenergy production; soil erosion; loss of terrestrial biodiversity; rising watertables; and the salinisation of the land and waterways. The project was designed to be inclusive and engender collaboration amongst researchers, participation by regional stakeholders, and communication to regional stakeholders and communities.

The intention is to provide useful evidence-based natural resource management planning advice to regional agencies. Landscape futures are plausible spatial arrangements of management actions (vegetation management, ecological restoration, conservation farming, deep-rooted perennials, biomass, and biofuels) that achieve regional natural resource management targets, assessed under six policy options and five climatic and economic scenarios. The triple bottom line impacts of landscape futures under each scenario and policy option were then assessed and visualised. The costs and benefits of landscape futures were compared and the trade-offs assessed to inform regional planning in the Lower Murray.

20.1 Introduction

There is an urgent need to reverse the declining environmental state of rural landscapes across southern Australia (Williams and Saunders 2005). Current approaches focus on natural resource management planning, policy and decision making at the regional level. Regional plans and associated on-ground investment have the potential to have widespread and long-lasting environmental, economic, and social impacts. However, rarely are these impacts quantified and clearly understood. The inherent uncertainty in quantifying future impacts necessitates a scenario analysis and futures thinking approach. This chapter presents an overview of the development and application of such a methodology to the problem of natural resource management planning in the Lower Murray region of southern Australia.

The Lower Murray Landscapes Futures (LMLF) project was conceived in recognition of the need for better informed planning and management in Australia's agricultural regions. The project is an ambitious, tri-state, multi-organisation, and multi-region research collaboration within the lower Murray-Darling Basin, Australia with the following dual aims:

- Aim 1: Assessing the impact of existing natural resource management plans for the Lower Murray on selected environmental and socioeconomic indicators.
- Aim 2: Assessing the impacts of these plans under alternative landscape future scenarios based on the outcomes of the analysis of the existing plans and input from stakeholders.

The LMLF project has two major components — the River Murray corridor and the dryland agriculture components. The focus of the River Murray corridor component is on water quantity, quality and use. The focus of the dryland component is on natural resource management problems in the dryland agricultural areas of the Lower Murray, such as soil erosion, loss of terrestrial biodiversity and rising watertables. Areas of interest in the dryland component include cleared agricultural areas and remnant vegetation, but are exclusive of urban, irrigated, and floodplain areas.

The major objectives of the dryland component are:

Analysing Landscape Futures for Dryland Agricultural Areas   409

- Review all regional plans relevant to the three regional natural resource management agencies participating in this study (the South Australian Murray-Darling Basin Natural Resource Management Board, the Mallee Catchment Management Authority, and the Wimmera Catchment Management Authority) and synthesise a set of quantitative, addressable natural resource management targets for each region.
- Define policy options and landscape futures scenarios.
- Acquire, assemble, model, and integrate a variety of spatial and non-spatial data covering a range of biophysical, ecological, administrative, social and economic aspects of the Lower Murray region under the baseline and landscape futures scenarios.
- Develop models of landscape futures for all policy options and future scenarios using the systematic regional planning framework (Bryan and Crossman in press).
- Analyse, interpret, and visualise alternative landscape futures for the Lower Murray including the environmental, economic and social impact of regional plans and analyse trade-offs under different policy options for the baseline and landscape futures scenarios.

The dryland component of the LMLF project is a large-scale investigation into integrated regional planning and landscape futures analysis. Integrated regional planning and futures analysis is a complex exercise that demands expertise in diverse disciplinary areas. Therefore, the LMLF project was designed to be inclusive and engender collaboration amongst researchers, participation by regional stakeholders, and communication to regional stakeholders and communities. The intention is to provide useful evidence-based natural resource management planning advice to regional agencies. This chapter describes the structure, process and learning achieved from undertaking such an ambitious, integrated analysis of landscape futures with many collaborators and stakeholders.

20.2 Futures Thinking and Scenario Analysis

Scenario analysis and planning has its origin in economic and strategic planning post WWII. Scenario analysis and futures thinking has become an increasingly popular and accepted methodology for planning, especially recently in the field of environmental management and planning (Schwartz 1996; Steinitz 1990). At the global level, a well known application of scenario analysis is the Special Report on Emissions Scenarios (SRES) that underpin the Intergovernmental Panel on Climate Change assessment report (IPCC 2001). Other global scenario modelling of note is the work of the Global Scenario Group (Raskin and Kemp-Benedict 2002) and the Millennium Ecosystem Assessment (Carpenter et al. 2005).

There are many applications of scenario analysis to natural resource management on a regional scale. For example, a comprehensive set of invited papers recently published in Ecological Applications Volume 14 (pp. 311–400) reported on a series of studies into alternative futures analysis for the Willamette River Basin in Oregon, USA (Baker and Landers 2004; Baker et al. 2004). These studies use scenarios to explore alternative futures for many components of natural resource management, including: land and water use (Hulse et al. 2004), agricultural systems (Berger and Bolte 2004), surface water allocation (Dole and Nieme 2004), freshwater biodiversity (Van Sickle et al. 2004), and terrestrial biodiversity (Schumaker et al. 2004). Steinitz et al. (2003) presented a comprehensive analysis of landscape futures for the San Pedro River basin that crosses the United States/Mexico border. Santelman et al. (2004) assessed futures for the agricultural landscapes of Iowa, USA.

As a rule of thumb, future scenarios for environmental applications, including natural resource management, should provide integration across social, economic and environmental elements (Raskin 2005). They should also quantify key variables describing ecosystem and resource condition, span long time horizons of at least several decades, and explore multiple futures that span a broad range of plausible outcomes across the timeframe of projections (Raskin 2005). Hulse et al. (2004) state that futures thinking and scenario analysis contains four steps:

- Defining the assumptions and rules of future alternatives (scenarios).
- Depicting spatially explicit alternatives (scenarios) through land use and allocation models.
- Modelling the effects of land use patterns on natural resources.
- Producing products that characterise the differences between alternatives (scenarios).

However an additional step has also been proposed (Theobald 2005; Wilhere et al. 2007):

- Involve active participation by stakeholders and the public and collaboration among scientists, planners and the public.

The LMLF study follows these five steps in undertaking futures thinking and scenario analyses. We also employ the rules of thumb discussed by Raskin (2005) and examine the impacts of alternative landscape futures across the environmental, economic, and social domains.

20.3 The Lower Murray Landscape Futures study

The Lower Murray study area officially covers the following four regions (Fig. 20.1):

- South Australian Murray Darling Basin Natural Resource Management Board region (hereafter SAMDB)
- Victorian Mallee Catchment Management Authority region (hereafter Mallee)
- Victorian Wimmera Catchment Management Authority region (hereafter Wimmera)
- New South Wales Lower Murray-Darling Catchment Management Authority region, although this region was not included in formal analysis.

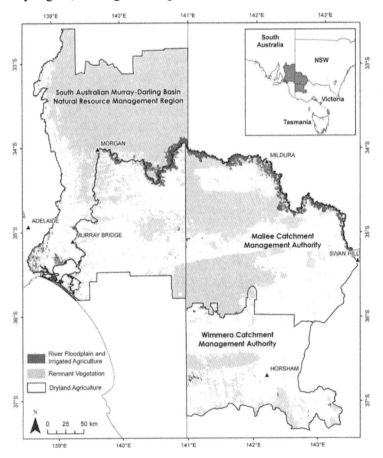

Fig. 20.1. Location and extent of the Lower Murray Landscape Futures dryland component study area

The Lower Murray is a diverse landscape ranging from the moist, hilly uplands of the southern Wimmera and western SAMDB, through large areas of productive agricultural cropping/grazing country, to semi-arid rangelands in the northern parts of the SAMDB. The River Murray also winds its way across the predominantly flat topography, supporting a corridor of high value irrigated agriculture.

The Lower Murray maintains significant biodiversity, land, and water resources. However, the cumulative impact of 90 years of land clearance and productive use has led to the ongoing degradation of these natural resources. Management is required to halt threatening processes and reverse this declining trend in the condition of natural resources in the Lower Murray.

20.3.1 Collaborative and Participatory Approach

Productive collaboration between researchers from several Commonwealth, state government and commercial agencies was essential in analysing a complex problem such as landscape futures over such a large area. The team consisted of a core of geographers and integrated modellers, complemented by collaborators with a range of disciplinary expertise and experience in the Lower Murray. The project team included researchers from: CSIRO Land and Water (CLW): Policy and Economic Research Unit (PERU); Agriculture, Water and Environment (AWE); Australian Research Centre for Water in Society (ARCWIS); CSIRO Sustainable Ecosystems (CSE); Primary Industries Research Victoria (PIRVic); EconSearch Pty Ltd; The University of Adelaide; South Australian Research and Development Institute (SARDI).

Close linkages between researchers and stakeholders was also essential in ensuring a successful research outcome. Participatory approaches were used from the outset of the project to ensure the inclusion of stakeholder input framing the:

- research questions
- definition of policy options and scenarios
- methods and techniques used
- interpretation of the results.

The participatory approach enabled stakeholders to be involved in the processes that shaped the research methods and outcomes. Active co-operation and participation of regional stakeholders in these facets of the LMLF dryland component research helped to:

- clarify the research questions

Analysing Landscape Futures for Dryland Agricultural Areas   413

- increase the relevance of the research
- engender a sense of ownership of the research by regional agencies
- increase the potential for long-term adoption and impact of the research.

The two main mechanisms for encouraging participation by regional stakeholders were through the Project Steering Committee and the Technical Reference Panel. The Project Steering Committee met regularly and the membership consisted of senior people from: CSIRO; University of Adelaide; Primary Industries Research Victoria; South Australian Department of Water, Land and Biodiversity Conservation; Murray-Darling Basin Commission; and the Chief Executive Officers of the SAMDB Natural Resource Management Board, and Mallee and Wimmera Catchment Management Authorities. The role of this group was to provide strategic oversight of the project direction.

The Technical Reference Panel consisted of technically skilled and experienced people from: Primary Industries Research Victoria; South Australian Department of Water, Land and Biodiversity Conservation; SAMDB Natural Resource Management Board; and Mallee and Wimmera Catchment Management Authorities. The role of this group was to provide technical advice on the development of methods, contribute to the development of scenarios, interpretation of results, and broker networks and linkages into other relevant studies and work.

Communication of results was an ongoing focus of the dryland project team. Presentations have been tailored for specific audiences at all levels including international and national conferences, state government workshops, regional agencies, and to on-ground local practitioner groups. Communication of the project is currently geared toward increasing the adoption and impact of the work.

20.3.2 Defining Targets, Scenarios and Policy Options

The regional agencies involved in the LMLF project have developed regional plans and strategies for managing natural resources. The foundation of these plans is a set of targets aimed at achieving regional sustainability. Analysis of regional natural resource management plans identified 15 relevant plans and strategies relevant to the Lower Murray region. Consideration of regional plans and consultation with regional stakeholders revealed four environmental objectives of particular interest in the dryland agricultural areas of the Lower Murray:

- protect and restore ecosystems to enhance terrestrial biodiversity
- reduce deep drainage to prevent rising saline groundwater

414   BA Bryan, ND Crossman and D King

- manage soils to minimise wind erosion
- sequester carbon and reduce carbon emissions to mitigate climate change.

Table 20.1. Summary of quantitative regional targets for the SAMDB, Mallee and Wimmera analysed in the dryland component

Region	Target summary
SAMDB	Manage 50% of remnant vegetation of private land
	Protect 50% of each of six specific threatened communities
	Increase vegetation cover by 1% in the agricultural region
	Planting 25,000 ha of deep rooted perennials
	Managing 40% of high wind erosion risk areas in June each year
Mallee	20% improvement in condition of remnant vegetation across all conservation significance levels
	Increase vegetation cover of Ecological Vegetation Classes to 15% of pre-1750 extent
	20% reduction in groundwater recharge from farming systems
	Reduction in land threatened by salinisation from 10% to 8% of total land surface
	Confine wind eroding land to 3% of land surface in dry years
Wimmera	Protect 750 ha of high quality and 500 ha of low–medium quality remnant vegetation per year (total of 80,000 ha by 2070)
	750 ha of revegetation per year (total of 48,000 ha by 2070) in priority Ecological Vegetation Classes
	750 ha per year revegetation of priority Ecological Vegetation Classes (0 initial Habitat Hectare gain, 75 Habitat Hectare gain annually after 10 years)
	Establish 4,500 ha of tree and native vegetation in local flow systems in highly fractured rocks, 900 ha break-of-slope trees on local flow systems in deeply weathered granites, and 5,000 ha of trees and native vegetation in local and intermediate flow systems in deeply weathered fractured rock
	Establish 4,500 ha of perennial plants in local flow systems in highly fractured rocks, 8,000 ha of perennial plants on local flow systems in deeply weathered granites, and 42,000 ha of perennial plants in local and intermediate flow systems in deeply weathered fractured rock
	5% increase in sustainable land management techniques such as conservation farming

A review of the 15 relevant plans and strategies documented 173 aspirational targets (long-term visionary goals, 50+ year timeframe), 252 resource condition targets (medium-term, 10–20 year timeframe), and 1,252 management action targets (short-term, 1–5 year timeframe). A set of quantitative targets (Table 20.1) were distilled from the regional plans with help from our stakeholders. These targets formed the basis of the analysis of landscape futures because targets are achieved under each landscape future. Six natural resource management actions were identified from regional plans and stakeholder consultation as having potential for addressing the biodiversity, deep drainage, wind erosion and climate change environmental objectives. The six natural resource management actions are explained in Table 20.2.

Table 20.2. Appropriate natural resource management actions for addressing environmental objectives in the Lower Murray

NRM actions	Description
Areas of remnant vegetation only:	
Vegetation management	Including stock exclusion, restoration of local native species, regeneration, weed management
Cleared agricultural areas only:	
Ecological restoration	Restoration of local native species, ecological structure and function
Conservation farming	Incorporation of minimum tillage and stubble retention techniques in crop/fallow rotation
Deep-rooted perennials	Plantation of fodder crops (e.g. lucerne, saltbush) for sheep grazing
Biomass	Oil mallee species grown for processing into activated charcoal, oil, and renewable energy
Biofuels	Wheat/canola rotation (using conservation farming techniques) for ethanol and biodiesel production

The LMLF project used a futures and scenario analysis methodology. In this methodology, landscape futures are alternative arrangements of natural resource management actions (Table 20.2) in the landscape that, together, achieve selected regional targets in the SAMDB, Mallee and Wimmera regions (Table 20.1). In the tradition of futures analysis, landscape futures are generated as plausible, but not necessarily probable, outcomes. Futures were selected that bound the range of possible alternatives and enable the assessment of costs, benefits and trade-offs of policy choices available to regional agencies as a guide for strategic natural resource management decision making. We used a combination of internal and external drivers to define landscape futures. Internal drivers are policy options which are strategic choices that decision makers can make with regard to how they go about achieving regional targets through the location of natural resource management actions in the landscape. External drivers include change in climate and commodity prices.

Six alternative policy options were analysed. Policy options were devised specifically to aid natural resource management decision making through pair-wise comparison. Hence, the impacts of different decision strategies such as cost minimisation (*Cheapest* policy option), single (*Best for Biodiversity* policy option) and multiple (*Best for NRM* policy option) benefit maximisation, cost effectiveness maximisation (*Most Cost Effective* policy option), and sustainability enabling (*Sustainability Ideal* policy option) were assessed relative to each other and relative to having no strategy at all (*Go Anywhere* policy option). Analysis of the impact of alternative policy options enabled us to understand the potential influence of strategic choices (over which the decision maker has full control) on landscape futures. Table 20.3 describes the policy options.

Table 20.3. Description of the alternative policy options assessed in landscape futures analysis

Policy options	Brief description
Go Anywhere	Achieve regional targets by a non-strategic, non-targeted approach of locating NRM actions at random in the landscape
Cheapest	Minimise economic cost of achieving targets including establishment costs and opportunity costs of foregone agricultural production
Best for Biodiversity	Achieve biodiversity targets whilst maximising biodiversity benefits, achieve other targets using the Go Anywhere approach
Best for NRM	Achieve regional targets whilst maximising benefits for the multiple environmental objectives of biodiversity, deep drainage, wind erosion, and climate change
Most Cost Effective	Achieve regional targets whilst minimising economic cost relative to benefits for the multiple environmental objectives of biodiversity, deep drainage, wind erosion, and climate change
Sustainability Ideal	Achieve a new set of regional targets aimed at achieving sustainability whilst minimising economic cost relative to benefits for the multiple environmental objectives of biodiversity, deep drainage, wind erosion, and climate change

Five of these policy options are constrained by the need to meet existing targets, whilst the *Sustainability Ideal* policy option is constrained by a new parsimonious and consistent set of targets based on the best available scientific knowledge and aimed at achieving the long-term aspirational goal of sustainability of regional systems. *Sustainability Ideal* targets include the following:

- At least 15% of each pre-European vegetation community, climate zone, soil type, and bioregion (subregion) must be either under remnant vegetation or targeted for ecological restoration with a minimum size of 1,000 ha.

- At least 30% of each vegetation community, climate zone, soil type, and bioregion (subregion) under remnant vegetation must be targeted for management with a minimum size of 1,000 ha.
- All areas at high risk of deep drainage are addressed by appropriate natural resource management actions.
- All areas at high risk of wind erosion are addressed by appropriate natural resource management actions.

Landscape futures were also analysed under five scenarios which included a baseline scenario, and four alternative future scenarios. Scenarios have a nominal time horizon of 2070, predominantly because of the prevalence of this date in climate change modelling and scenario development used in this study (Bryan et al. 2007a, 2007b). Scenarios are defined by specific levels of climate change (temperature and rainfall) derived from Suppiah et al. (2006) and commodity prices for carbon credits (credits may be acquired through carbon sequestration under ecological restoration), biomass and biofuels. Plausible prices for carbon were defined based on the prices for carbon on the European and New South Wales markets observed over the period of study. Biomass prices were selected for analysis based on a joint analysis of on-farm biomass production and plant-based biomass processing in the South Australian River Murray Corridor (Ward and Trengove 2005). Biofuel prices were selected based on conservative levels of increase in oil prices. The parameters for each scenario are presented in Table 20.4.

The two aims of this project were addressed in an integrated way in this analysis. In addressing Aim 1, the analysis of regional plans, we considered the baseline scenario only (S0). The baseline scenario is based on current (historical mean) climate and commodity prices and as such, there is no market for carbon, biomass, or biofuels. In addressing Aim 2, the analysis of future scenarios (S1–S4, Table 20.4), we considered new markets for carbon trading to encourage ecological restoration, and markets for biomass and biofuel crops to provide incentives for the uptake of these natural resource management actions. Analysis of future scenarios quantifies the influence of external climatic and economic drivers (over which the decision maker has no control) on landscape futures and quantifies the associated impacts.

Landscape futures were generated under the six policy options and five scenarios which, together with the regional targets, drive the distribution of natural resource management actions in the landscape. We assessed the environmental, economic, and social impacts of alternative landscape futures.

Table 20.4. Summary of climate regimes and commodity prices defining each scenario

Scenario	Baseline (S0)	Mild warming/drying (S1)	Moderate warming/drying (S2)	Severe warming/drying (S3)	Mild warming/wetting (S4)
Temperature	Historical mean	1°C ↑	2°C ↑	4°C ↑	1°C ↑
Rainfall	Historical mean	5% ↓	15% ↓	25% ↓	5% ↓
Carbon trading price [$/tonne CO_2^e]	0	15	7	2	15
Biomass price [$/tonne]	0	50	40	30	50
Biofuels price [% of current wheat/canola price]	0	150	130	110	150

↑ = increase, ↓ = decrease

We also examined the costs, benefits and trade-offs of achieving regional targets (Table 20.1) through the various policy options (Table 20.3) and under alternative future scenarios (Table 20.4). The overall structure of the futures analysis is summarised in Fig. 20.2.

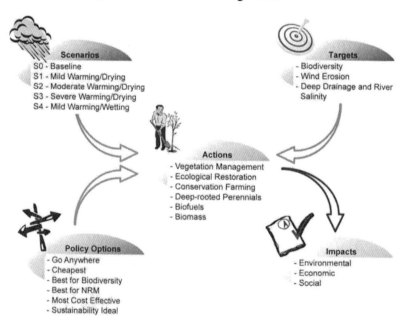

Fig. 20.2. General structure of landscape futures analysis. Note that targets denote formal regional targets

20.3.3 Landscape Futures Analysis

Analysis of landscape futures requires an understanding of the spatio–temporal dynamics of and interrelationships between many complex environmental, economic, and social processes in the landscape. To manage this complexity, the project was designed as a series of modules which were conducted as stand-alone but closely linked pieces of research (Fig. 20.3, Tables 20.5–20.7).

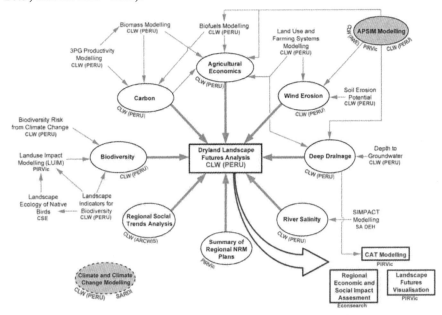

Fig. 20.3. Modular structure of Lower Murray Landscape Futures dryland component

The research modules had different functional roles in the overall project as described in Tables 20.5–20.7. Two external input modules (Table 20.5) provided essential context for landscape futures analysis. Landscape futures analysis has high data demands and requires a comprehensive spatial database describing multiple environmental, economic and social costs and benefits. Several research modules were developed to create a data-rich foundation to provide input into the systematic regional planning module (Fig. 20.3).

Table 20.5. Input modules and key objectives of landscape futures analysis

Input modules	Key objectives of module
External	
Regional social trends analysis	Review regional socioeconomic status and trends using ABS census data and review of previous analyses to inform scenario definition for futures analysis.
Summary of regional NRM plans	Summarise the key NRM issues as identified in the regional plans and distil a set of quantitative targets for the three component regions: SAMDB, Mallee and Wimmera.
Analytical	
Climate and climate change modelling	Model climate surfaces using *BIOCLIM*. Define suitable climate change scenarios and associated estimates of rainfall, precipitation, and atmospheric CO_2 from regional climate models.
APSIM modelling	Model the yield in addition to deep drainage and wind erosion impacts of farming systems in the Lower Murray under the baseline and future scenarios to inform agricultural economics, biofuels, wind erosion and deep drainage analysis.
Landscape indicators for biodiversity	Calculate spatial data layers describing the distribution of landscape indicators for biodiversity for input into the landscape ecology and *LUIM* analyses, and the biodiversity index.
Landscape ecology of native birds	Analyse 'Birds Australia' bird survey data and key relationships between bird diversity and habitat configuration derived to inform *LUIM*.
Land use impact modelling	Develop a Bayesian network for modelling risk to patches of native vegetation to identified threats in the Lower Murray based on workshops with ecologists and other regional experts. Assemble the data required and apply the Bayesian network to calculate risk of remnant vegetation patches.
Biodiversity risk from climate change	Identify using biological survey data plant species sensitive to climate change and create models that map the distribution of habitat suitability for these species. Model the distribution of priority areas for both vegetation management and revegetation to mitigate the effects of climate change on sensitive plant species.
3-PG productivity modelling	Model the carbon productivity associated with ecological restoration of a community comprised of a suite of local native species under the baseline and future scenarios. Model the biomass productivity of eucalypt species under the baseline and future scenarios for input into the biomass economic modelling.
Biomass modelling	Model the spatial distribution of economic returns and carbon benefits from biomass under the baseline and future scenarios.
Biofuels modelling	Calculate the spatial distribution of economic returns and carbon benefits from farming energy crops for biofuel production.
Land use and farming systems modelling	Integrate land use data and map consistently the spatial distribution of land use across the Lower Murray. Characterise the spatial distribution of cropping/grazing rotation frequencies that characterise the farming systems across the region.
Soil erosion potential	Integrate soil database and create a consistent soil erosion potential layer across the Lower Murray region.
Depth to groundwater	Integrate disparate groundwater surface databases across the Lower Murray region with a high resolution DEM to create a depth to groundwater layer.
SIMPACT modelling	Using the *SIMPACT* model, estimate the salinity contribution to the River Murray of deep drainage in dryland agricultural areas.

ABS: Australian Bureau of Statistics. *NRM*: natural resource management. *LUIM*: Land Use Impact Model. *DEM*: digital elevation model.

Thirteen analytical input modules (Table 20.5) involved spatial data acquisition and modelling to provide data layers suitable as input into the six benefit and cost modules (Table 20.6). These benefits and costs modules developed key information about the spatial distribution of the range of environmental and economic benefits and costs of undertaking the six natural resource management actions (Table 20.2) in the landscape under the baseline and future scenarios. The output benefits and costs layers provided the basis for the generation of spatially explicit landscape futures through systematic regional planning (Bryan and Crossman in press) in the core analytical module.

Table 20.6. Benefit and cost modules and key objectives of landscape futures analysis

Benefit and cost module	Key objectives of module
Biodiversity	For areas of remnant vegetation, create an index of biodiversity benefit as priorities for vegetation management based on *LUIM* outputs, and biodiversity risk from climate change. For dryland agricultural areas, create an index of biodiversity benefit as priorities for ecological restoration based on landscape indicators for biodiversity, *LUIM* outputs, and biodiversity risk from climate change.
Carbon	For cleared areas quantify the carbon benefits of ecological restoration, biomass and biofuels, and create a carbon benefits index.
Agricultural economics	Quantify the economic returns from traditional agricultural systems, and the NRM actions of conservation farming systems, grazing deep-rooted perennials, carbon trading, biomass production and biofuels production under the baseline and landscape futures scenarios. Develop an index of net economic returns from changing land use from traditional agriculture to each natural resource management action under the baseline and landscape futures scenarios.
Wind erosion	Develop an index of wind erosion risk under the baseline and climate change scenarios. Quantify the wind erosion mitigation impact of changing land use from traditional agriculture to each natural resource management action.
Deep drainage	Develop an index of deep drainage risk under the baseline and climate change scenarios. Quantify the deep drainage mitigation impact of changing land use from traditional agriculture to each natural resource management action.
River salinity	Identify high priority river salinity mitigation sites in dryland agricultural areas.

Finally, landscape futures were assessed for their environmental, economic, and social impact, visualised, and validated in the output analytical modules (Table 20.7). Primarily, the performance of each landscape future was calculated against a variety of indicators based on the original benefit and cost layers.

Table 20.7. Core analytical module and output analytical modules with key objectives of landscape futures analysis

Analytical module	Key objectives of module
Core	
Dryland landscape futures analysis	Develop systematic regional planning models for analysing the impact of regional NRM plans and landscape futures for the Lower Murray dryland areas. Analyse the impact of regional NRM plans for the baseline scenario (Aim 1) according to six alternative policy options. Analyse the impact of regional NRM plans for four future scenarios (Aim 2) and six policy options.
Output	
Regional economic and social impact assessment	Develop an input-output model to calculate the likely economic and social impacts of natural resource management actions and landscape futures on the regional economies, household income, employment, and population.
CAT modelling	Using the *Catchment Assessment Tool*, build on *APSIM* simulations of deep drainage and expand on the implications of land use change for deep drainage and groundwater in the Mallee and Wimmera regions. Connect a groundwater model to validate and enhance the deep drainage risk layer.
Landscape futures visualisation	Develop an advanced and distributed capability for visualising landscape futures outputs within the *Google Earth* environment to help communicate the results to stakeholders.

In addition, the socioeconomic impact of landscape futures was assessed using indicators such as the contribution to gross regional product, household income, employment and population calculated using regional input-output and demographic models built for the SAMDB, Mallee, and Wimmera regions. The costs and benefits of landscape futures were compared and the trade-offs assessed to inform regional planning in the Lower Murray region.

Systematic regional planning models were developed to quantify the amount and spatial arrangement of the six natural resource management actions in the landscape. Each natural resource management action has spatially heterogeneous benefits for a specific set of environmental objectives (Fig. 20.4). For example, vegetation management addresses biodiversity objectives only whilst ecological restoration addresses all four environmental objectives (biodiversity, wind erosion, deep drainage and climate change). In addition, each has heterogeneous economic costs over the landscape. Hence, quantifying landscape futures involves the complex spatial allocation of natural resource management actions in the Lower Murray such that regional targets are achieved.

Systematic regional planning employs multi-objective programming within a multi-criteria decision analysis framework to spatially allocate actions in the landscape. For each spatial unit i for $i = 1 \ldots n$, the systematic regional planning models selected areas (x_a) of each natural resource management action (a) such that the objective function is minimised subject to target constraints. The outputs of this process are alternative landscape futures.

The regional targets (Table 20.1) form constraints in the multi-objective programming models of landscape futures such that each target is met under each landscape future. Often, regional targets could be operationalised as aggregate constraints on the total area of an action. Thus for example, the following simple constraint was set in implementing the Wimmera CMA target of a 5% increase in conservation farming (*cf*):

$$\sum x_{cf} \geq 0.05G \qquad (20.1)$$

where G is the area of dryland agriculture in the Wimmera CMA.

Other times, regional targets were operationalised as levels of environmental benefit achieved using existing and modelled spatial data (e.g. reduction in land threatened by salinisation from 10% to 8% of total land surface of the Mallee CMA region).

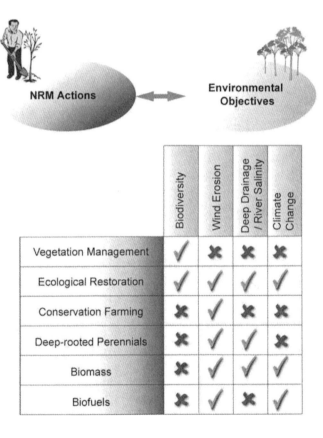

Fig. 20.4. Environmental objectives achieved by natural resource management actions

Largely, the regional plans do not specify precise locations for natural resource management actions in the landscape and as long as the aggregate area of action or amount of environmental benefit is achieved, then the target is met. However, there is a high degree of spatial heterogeneity in the costs and benefits of natural resource management actions in the Lower Murray and the location of actions strongly affects the impacts of achieving regional targets. To understand this effect, landscape futures were generated using five policy options implemented as strategic spatial targeting strategies. Policy options define the objective function in the systematic regional planning models of landscape futures using the layers produced in the benefits and costs module (Table 20.6).

The simplest objective functions which minimise a single factor are the *Cheapest* objective function:

$$\sum_{i=1}^{n}\sum_{a \in A}\left(x_{ia}C_{ia}\right) \tag{20.2}$$

and the *Best for Biodiversity* objective function:

$$\sum_{i=1}^{n}\sum_{a \in A}\left(x_{ia} / B_{ia}\right) \tag{20.3}$$

The *Most Cost Effective* and *Sustainability Ideal* objective functions were the most complex and were structured as weighted multi-attribute utility functions which aimed to minimise the marginal cost of multiple environmental benefits:

$$\sum_{i=1}^{n}\sum_{a \in A}\frac{x_{ia}C_{ia}}{\left(x_{ia}B_{ia} + x_{ia}DD_{ia} + x_{ia}WE_{ia} + x_{ia}CC_{ia}\right)} \tag{20.4}$$

where A is the set of six natural resource management actions, C_{ia} is the net economic cost, B_{ia} is the biodiversity benefit score, DD_{ia} is the deep drainage benefit score, WE_{ia} is the wind erosion benefit score, and CC_{ia} is the climate change benefit score of action a in spatial unit i. Weights were also used to adjust the costs and benefits so that the values are similar in range and influence. Table 20.3 describes in words the objective functions associated with each policy option and Bryan et al. (2007a, 2007b) present the full mathematics. Systematic regional planning models were built for each policy option under each scenario (Table 20.4) using scenario-specific constraint, cost and benefit layers.

20.4 Results

Prior to this analysis, little was known of the impact of regional plans and associated targets. The results of this study present a comprehensive, quantitative analysis of regional plans and landscape futures under six strategic policy options and five scenarios including an assessment of the impacts and trade-offs across multiple environmental, economic and social indicators. We provide below a very broad synopsis of the results, complimented by a specific example for the Wimmera.

Landscape futures vary substantially in the spatial arrangement of natural resource management actions, and in the costs, benefits and level of efficiency. This variation is driven by the choice of strategic policy option in achieving regional targets, by the introduction of new incentives for actions, and by the climatic and economic variability captured in the scenarios.

The most striking outcome of this analysis of landscape futures is that approximately two million hectares of natural resource management actions are required to achieve regional targets in the Lower Murray region. Although some actions (e.g. conservation farming) require minimal change to current land management, other actions require conversion of agriculture to alternative land uses (e.g. trees).

Achieving regional targets is very costly without the introduction of economic incentives for the adoption of natural resource management actions. Cost range from $3.8 million per year for the SAMDB under the *Cheapest* policy option, to $348 million per year for the Mallee under the *Go Anywhere* option. However, the introduction of policy incentives such as carbon trading, and biomass and biofuels industries can greatly reduce the economic impact of achieving regional targets. Landscape futures may have greater economic returns than traditional agriculture when the combination of climatic, economic and policy conditions are right (i.e. milder climate change; high prices for carbon, biomass and biofuels; strategic targeting that considers cost of actions). The maximum impact of this effect is a net positive economic return of up to $173 million per year in S4 for the Wimmera under the *Cheapest* policy option.

The environmental benefits achieved also vary greatly between landscape futures. The magnitude of environmental benefits is driven primarily by the strategic spatial targeting used for locating actions in the landscape and to a lesser extent by climate change. When existing targets are addressed, the *Go Anywhere* and *Best for Biodiversity* policy options achieve the largest total amount of environmental benefits simply due to the magnitude of actions required. The *Best for NRM* policy option addresses the highest environmental benefit sites. The *Most Cost Effective* policy option efficiently trades off benefits for multiple environmental objectives with economic cost. The *Sustainability Ideal* policy option provides high aggregate environmental benefits very cost effectively and achieves long-term sustainability targets.

The outputs of landscape futures analysis are complex and multidimensional. The primary outputs are map-based depictions of natural resource management actions. The impacts of landscape futures were also calculated across a wide variety of indicators. A variety of communication tools were used to convey the outputs to regional stakeholders. An example of some of the map-based and indicator table outputs produced from analysis of the landscape futures of dryland agricultural areas in the Wimmera region are presented in Fig. 20.5 and Table 20.8.

Specifically for the Wimmera:

- Under the baseline scenario (S0), the achievement of existing targets is estimated to have a high net economic cost to the region of between $12 million per year and $290 million per year, depending on policy option. High social costs are also experienced including the loss of between 439 to 2,438 jobs and up 6,024 people to the region.
- Under the future scenarios S1 to S4, the economic impact of achieving regional targets is estimated to range between a net economic cost of $322 million per year to a benefit of $60 million per year. The social impacts of landscape futures on the region ranges from strongly negative effects on employment and population (in the order of thousands of jobs and people) to weak positive effects (hundreds of jobs and people).
- The economic impact of achieving long-term sustainability targets involves substantial economic costs for the Wimmera ranging from a cost of $258 million per year under the baseline scenario to $126 million per year under the future scenario S1.
- Strategic spatial targeting of natural resource management actions can lead to a 154% improvement in efficiency of natural resource management actions.

Analysing Landscape Futures for Dryland Agricultural Areas 427

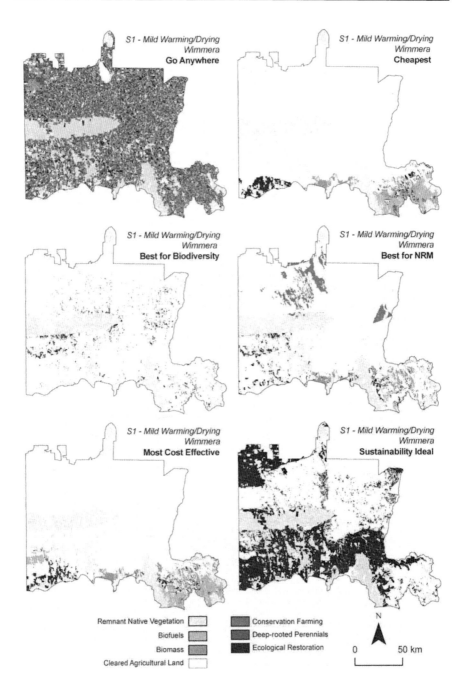

Fig. 20.5. Example of landscape futures in the Wimmera resulting from six policy options under the mild warming/drying scenario (S1)

428 BA Bryan, ND Crossman and D King

Table 20.8. Selected performance indicators of landscape futures by policy option and future scenario for the Wimmera

Scenario	Policy option	Total area of NRM actions required ['000 ha]	Net economic returns [$million]	Total biodiversity benefit score index	High risk deep drainage areas managed ['000 ha]	High risk wind erosion areas managed ['000 ha]	Carbon benefits ['000 cars off the road]	Employment [no. FTE jobs]	Population change [no. of persons]
Baseline (S0)	Go Anywhere	1690	-289.1	1.3	524.0	184	455	-2380	-5800
	Cheapest	270	-11.5	0.5	36.0	6	152	-439	-799
	Best for Biodiversity	1810	-290.7	0.7	568.7	198	139	-2438	-6024
	Best for NRM	270	-17.7	0.7	54.0	31	149	-470	-878
	Most Cost Effective	270	-13.3	0.6	54.1	23	151	-446	-817
	Sustainability Ideal	820	-258.3	5.3	621.7	202	2313	-1794	-3905
Mild warming/ drying (S1)	Go Anywhere	1690	-220.3	1.5	446.4	185	1100	-1037	-1895
	Cheapest	270	60.6	0.6	47.5	7	126	-108	155
	Best for Biodiversity	1810	-234.7	0.8	484.4	198	880	-1238	-2639
	Best for NRM	270	-7.8	0.8	51.0	30	204	-268	-271
	Most Cost Effective	270	56.8	0.7	54.1	10	277	-111	148
	Sustainability Ideal	760	-126.3	5.6	525.5	202	2105	91	2088
Moderate warming/ drying (S2)	Go Anywhere	1690	-229.8	1.5	267.4	358	1060	-1158	-2175
	Cheapest	270	42.1	0.6	34.6	9	108	-159	21
	Best for Biodiversity	1810	-239.3	0.9	290.6	385	853	-1335	-2856
	Best for NRM	270	-15.1	0.8	37.9	50	202	-276	-288
	Most Cost Effective	270	39.1	0.7	51.5	10	276	-120	129
	Sustainability Ideal	730	-160.2	5.2	315.6	397	1952	24	1856
Severe warming/ drying (S3)	Go Anywhere	1690	-190.6	1.8	143.5	752	1011	-1167	-2230
	Cheapest	270	28.2	0.8	5.0	31	88	-222	-150
	Best for Biodiversity	1810	-197.3	1.1	152.9	807	811	-1338	-2893
	Best for NRM	270	-18.8	1.0	25.9	104	197	-283	-310
	Most Cost Effective	270	23.5	0.9	21.7	27	256	-150	47
	Sustainability Ideal	1020	-191.8	9.6	167.8	835	2644	-54	2280
Mild warming/ wetting (S4)	Go Anywhere	1690	-304.0	1.5	572.3	185	1126	-1135	-2074
	Cheapest	270	59.5	0.5	40.9	5	124	-115	143
	Best for Biodiversity	1810	-322.1	0.8	616.3	197	905	-1353	-2863
	Best for NRM	270	-10.2	0.8	54.1	29	205	-279	-292
	Most Cost Effective	270	53.9	0.6	54.2	17	278	-125	119
	Sustainability Ideal	850	-182.9	6.3	675.4	202	2388	128	2391

FTE: Full Time Equivalent.

20.5 Risk, Preference and Strategic Policy Adoption

The results of the analysis of landscape futures can be used to inform and support regional planning and decision making. We assess for the first time the feasibility of achieving existing regional targets through quantifying the costs and benefits associated with achieving regional environmental targets. This is done under a number of alternative policy options and scenarios. A fundamental finding is that a large amount of actions are required to achieve regional targets and the economic cost and flow-on impacts of achieving regional targets are likely to be prohibitively large. However, the introduction of policy innovations such as a carbon trading scheme, and biomass and biofuel industries are likely to reduce the economic impact of achieving regional targets. The results are useful for informing future refinements of regional plans and targets within the ongoing planning cycle.

We are not aware of any regional plans in Australia that have embraced a systematic spatial targeting approach to prioritising investments in natural resource management. However, a major finding of the study is the increased level of cost effectiveness achieved through strategic targeting of natural resource management actions in high benefit and low cost sites in the landscape compared to adopting no strategic targeting in regional planning. This maximises the efficiency of regional investment in natural resource management, whilst at the same time minimising the loss of productive agriculture and the regional economic and social impacts. The results suggest that more than double the environmental benefits can be achieved from a given budget through spatial targeting of actions in the landscape. Given the amount of resources devoted to natural resource management in Australia, the potential for achieving increased environmental benefits is substantial.

These results convincingly support the adoption of intelligent, informed, and strategic spatial targeting of investment in natural resource management actions. However, the objectives of targeting are less clear and are dependent upon the preferences of stakeholders and decision makers. The trade-offs of choosing one strategic direction over another are made explicit in this study but there is no dominant solution. That is, no one strategy performs best on all environmental, economic and social indicators. Hence, there are still trade-offs that need to be made depending on the strategic policy objectives. For example, under the baseline scenario in the Wimmera, the biodiversity benefits of a *Sustainability Ideal* policy are nearly an order of magnitude greater than the *Most Cost Effective* policy, but the trade-off is a much greater decrease in employment, regional population, and net economic returns (Table 20.8).

This leaves regional agencies and stakeholders still with decisions to make. Based on the results of this study, we suggest that decisions about whether to select policy options that favour economic returns over environmental benefits, or different components of environmental benefits (e.g. those options that perform better with respect to biodiversity or carbon, as opposed to deep drainage), should be founded on two factors:

- The preferences and weights that the decision maker attributes to the various objectives, criteria and indicators.
- The performance of each policy option against the more important indicators given uncertainty in future scenarios.

For example, a risk-averse decision maker who wishes to minimise the economic impacts associated with achieving regional targets may be willing to accept the lesser environmental benefits associated with the *Cheapest* policy option as a trade-off for the considerable softening in economic and social cost. Conversely, a more environmentally focussed decision maker may be prepared to take the risk on a high economic cost strategy that produces substantially higher environmental benefits. Further, they may be willing to adopt a set of very expensive long-term goals if there is confidence that the outcome will be more sustainable regional biophysical systems.

20.6 Further Research

There will be a range of different perspectives amongst stakeholders that regional decision makers need to take into account. The *Most Cost Effective* policy option is a good compromise solution because it trades-off economic cost and multiple environmental benefits. From an efficiency perspective, the *Most Cost Effective* policy option is by definition, the most efficient use of regional natural resource management funds. *Most Cost Effective* landscape futures are specific to the scaling and weighting applied to costs and benefit scores. By adjusting the weights of costs relative to benefits in the objective functions we may achieve different landscape futures that are also 'most cost effective'. These alternative *Most Cost Effective* landscape futures are still maximally efficient but their differences reflect a relative change in preference of economic costs or environmental benefits.

Analysis of landscape futures in this study has identified the social, economic and environmental advantages of spatial targeting in achieving regional targets. Ideally we would refine the *Most Cost Effective* landscape futures based upon the actual preferences of stakeholders and decision makers. There are several techniques based on trade-off games that can be used to elicit the relative weights people place on competing objectives such as Analytical Hierarchy Process (Saaty 1980). These techniques could be used in a group decision conference setting to formally quantify the relative importance of economic cost and environmental objectives (in this case enhancing biodiversity, minimising wind erosion and deep drainage, and mitigating climate change). These weights can then be used to generate new *Most Cost Effective* or *Sustainability Ideal* landscape futures using the systematic regional planning engine. This would make explicit the most preferred landscape futures that meet the existing targets or some new set of long-term sustainability targets. This would also further strengthen the participatory approach and provide stakeholders with a greater sense of ownership of the research outcomes.

20.7 Application in Other Regions and Contexts

We have focussed on a quantitative, data-centric approach to the analysis of landscape futures. Conducted within a participatory and collaborative framework, this approach embraces the spatial and process complexity and detail of alternative landscape futures. The landscape futures analysis framework presented here can be adapted to other regions with different natural resource management and planning issues, and with different physical and biotic, economic and social environments. To apply this form of landscape futures analysis to different regions and different contexts, a substantial amount of adaptation is required to understand the environmental, economic, social and spatial nuance of the new region. There will need to be a large data collection and modelling exercise to assemble the multi-dimensional data layers required. Substantial tailoring of existing models and development of new models will also be required.

20.8 Conclusion

The results of this landscape futures analysis can inform regional decision making through making explicit the costs, benefits, and trade-offs of achieving regional environmental objectives under alternate strategic policies. Uncertainty in achieving environmental objectives is explicitly handled through analysis across scenarios that capture the range of future possibilities. However, no single dominant option and obvious choice of strategic direction in regional planning does exist and trade-offs in decision making based on pragmatics and preferences of policymakers are inevitable. Further research should focus on elicitation of decision maker preferences for refinement of strategic policy options that incorporate spatially explicit knowledge on the costs and benefits of natural resource management actions.

Success in undertaking such ambitious aims that consider complex spatial processes over a broad and heterogeneous region is contingent on a trans-disciplinary approach put into effect through a multidisciplinary team of collaborative researchers. Our experiences in the Lower Murray also reinforce the fact that this kind of complex planning and futures analysis works best when regional stakeholders and decision makers enter a cooperative and collaborative arrangement with the researchers and analysts. This collaboration generates a two way flow which enhances regional stakeholders' understanding of the potential of futures analysis and at the same time enhances the research analysts' understanding of the regionally-specific environmental, economic, and social context, problems and priorities.

Acknowledgements

The authors gratefully acknowledge funding from the National Action Plan for Salinity and Water Quality and CSIRO's Water for a Healthy Country Flagship program through the Lower Murray Landscape Futures project. We are also grateful to our project stakeholders including many individuals from: CSIRO; the University of Adelaide; MDBC; Victorian and South Australian State Governments; and the SAMDB, Mallee and Wimmera regional agencies under the banner of the Land Technologies Alliance. Thanks also to John Ward for comments on the manuscript.

References

Baker JP, Hulse DW, Gregory SV, White D, Van Sickle J, Berger PA, Dole D, Schumaker NH (2004) Alternative futures for the Willamette River basin, Oregon. Ecological Applications 14:313–324
Baker JP, Landers DH (2004) Alternative futures analysis for the Willamette River basin, Oregon. Ecological Applications 14:311–312

Berger PA, Bolte JP (2004) Evaluating the impact of policy options on agricultural landscapes: an alternative-futures approach. Ecological Applications 14:342–354

Bryan BA, Crossman ND (in press) Systematic regional planning for multiple objective natural resource management. Journal of Environmental Management, doi:10.1016/j.jenvman.2007.06.003

Bryan BA, Crossman ND, King D, McNeill J, Wang E, Barrett G, Ferris MM, Morrison JB, Pettit C, Freudenberger D, O'Leary G, Fawcett J, Meyer W (2007a) Lower Murray landscape futures dryland component: volume 2 – analysis of regional plans and landscape futures. Land Technologies Alliance and CSIRO Water for a Healthy Country, Canberra

Bryan BA, Crossman ND, King D, McNeill J, Wang E, Barrett G, Ferris MM, Morrison JB, Pettit C, Freudenberger D, O'Leary G, Fawcett J, Meyer W (2007b) Lower Murray landscape futures dryland component: volume 3 – preliminary analysis and modelling. Land Technologies Alliance and CSIRO Water for a Healthy Country, Canberra

Carpenter SR, Pingali PL, Bennet EM, Zurek MB (2005) Ecosystems and human wellbeing: scenarios. Millenium ecosystem assessment report volume 2. Island Press, Washington

Dole D, Niemi E (2004) Future water allocation and in-stream values in the Willamette River Basin: a basin-wide analysis. Ecological Applications 14:355–367

Hulse DW, Branscomb A, Payne SG (2004) Envisioning alternatives: using citizen guidance to map future land and water use. Ecological Applications14:325–341

IPCC (2001) Climate change 2001: mitigation. Third assessment report of the Intergovernmental Panel on Climate Change. Cambridge University Press, Cambridge

Raskin PD (2005) Global scenarios: background review for the millennium ecosystem assessment. Ecosystems 8:133–142

Raskin P, Kemp-Benedict, E (2002) GEO scenario framework. Background paper for UNEP's third Global Environmental Outlook Report. United Nations Environment Program, Nairobi

Saaty TL (1980) The analytic hierarchy process. McGraw-Hill, New York

Santelman MV, White D, Freemark K, Nassauer JI, Eilers JM, Vaché KB, Danielson BJ, Corry RC, Clark ME, Polasky S, Cruse RM, Sifneos J, Rustigian H, Coiner C, Wu J, Debinski D (2004) Assessing alternative futures for agriculture in Iowa, USA. Landscape Ecology 19:357–374

Schumaker NH, Ernst T, White D, Baker J, and Haggerty P (2004) Projecting wildlife responses to alternative future landscapes in Oregon's Willamette Basin. Ecological Applications 14:381–400

Schwartz P (1996) The art of the long view: planning for the future in an uncertain world. Doubleday, New York

Steinitz C (1990) A framework for theory applicable to the education of landscape architects (and other environmental design professionals). Landscape Journal 9:136–143

Steinitz C, Arias H, Bassett S, Flaxman M, Goode T, Maddock T III, Mouat D, Peiser R, Shearer A (2003) Alternative futures for changing landscapes – the Upper San Pedro River Basin in Arizona and Sonora. Island Press, Washington, DC

Suppiah R, Preston B, Whetton PH, McInnes KL, Jones RN, Macadam I, Bathols J, Kirono D (2006) Climate change under enhanced greenhouse conditions on South Australia. An updated report on: Assessment of climate change, impacts and risk management strategies relevant to South Australia. Climate Impacts and Risk Group, CSIRO Marine and Atmospheric Research, Victoria

Theobald D (2005) Landscape patterns of exurban growth in the USA from 1980 to 2020. Ecology and Society 10:32. Available online, http://www.ecologyandsociety.org/vol10/iss1/art32/

Van Sickle J, Baker J, Herlihy A, Bayley P, Gregory S, Haggerty P, Ashkenas L, Li J (2004) Projecting the biological condition of streams under alternative scenarios of human land use. Ecological Applications 14:368–380

Ward JR, Trengove G (2005) Developing re-vegetation strategies by identifying biomass based enterprise opportunities in the mallee areas of South Australia. CSIRO client report for South Australian Department Water, Land, Biodiversity and Conservation

Wilhere GF, Linders MJ, Cosentino BL (2007) Defining alternative futures and projecting their effects on the spatial distribution of wildlife habitats. Landscape and Urban Planning 79:385–400

Williams J, Saunders D (2005) Land use and ecosystems. In: Goldie J, Douglas B, Furnass B (eds) In search of sustainability. CSIRO Publishing, Victoria

21 Applying the *What If?* Planning Support System for Better Understanding Urban Fringe Growth

Christopher Pettit[1], Jessica Keysers[2], Ian Bishop[3] and Richard Klosterman[4]

[1] Landscape Systems Platform, Department of Primary Industries, Carlton, Victoria, Australia
[2] Earth Tech, Melbourne, Victoria, Australia
[3] Centre for Geographic Information Systems and Modelling, The University of Melbourne, Victoria, Australia
[4] Department of Geography and Planning, The University of Akron, Ohio, USA

Abstract: This chapter reports on the application of the collaborative GIS-based What If? *planning support system for managing data and creating exploratory future land use change scenarios. The case study area is Mitchell Shire, a local government municipality located on the urban fringe of the City of Melbourne, Australia. In this research we describe the various modules comprising the* What If? *planning support systems (PSS) and how they have been applied to assist local planners within the shire to better understand land suitability and the probable impacts of projected population growth. The research also aims to increase awareness of the potential benefits of PSS for planning.*

21.1 Introduction

Computers were introduced to urban planning in the 1950s and 1960s (Klosterman 2001). Since then the philosophy of planning has changed considerably from planning *for* communities to planning *with* communities (Forester 1999). The role of computers to support planning practice has in-turn evolved from the application of top–down large-scale urban models to the adoption of participatory spatial planning tools. The development of GIS and its proliferation as a research tool in the 1980s was significant in this evolution (Budic 1994). GIS are recognised as an essential tool for planning, but alone are not sufficient to meet the needs of planners (Klosterman 1999). Planning support systems (PSS) provide a more tailored tool, designed specifically for planning purposes. Through resources like the ESRI Virtual campus course 'Introduction to urban and regional planning using *ArcGIS*' (Pettit and Pullar 2001) planners are provided with a fundamental knowledge of GIS and the necessary skills to apply GIS and PSS in practice.

PSS are a specialised type of spatial decision support system which focuses on assisting planners and decision makers in making better decisions about current and future land uses. PSS offer an evidence-based approach for incorporating socioeconomic and biophysical data to formulate land use change scenarios. There are many tasks PSS can assist with, including: land use/land cover change, comprehensive projections, three-dimensional visualisation, and impact assessment (Klosterman and Pettit 2005). There are also numerous PSS available (Brail and Klosterman 2001; Geertman and Stillwell 2004), which fall into the following modelling categories: large-scale urban, rule-based, state-change, and cellular automata (Klosterman and Pettit 2005).

However, even though there has been a number of PSS developed in the 1990s and 2000s there remain issues in bridging the gap between GIS and PSS (Geertman and Stillwell 2004) and an acknowledged 'bottleneck' in the adoption of PSS tools by planning practitioners. A lack of awareness is a major hurdle in the widespread adoption of PSS in planning practice (Vonk et al. 2005). To disseminate knowledge of the existence of planning support systems, 'real-world example projects ... of PSS application in planning practice will be crucial' Vonk et al. (2005, p. 909). This is precisely what this chapter aims to address in providing a case study of an application of the *What If?* PSS, using real data and planning metrics for the Mitchell Shire, Australia.

21.2 The *What If?* Planning Support System

What If? is a collaborative GIS-based planning support system developed by Klosterman (1999), which is supported by *What If?* Inc. <http://www.whatifinc.biz/>. The PSS is not an academically supported planning tool rather it is a commercial off-the-shelf (COTS) software package. Based on initial feedback from end users of the software, some of the advantages and disadvantages of *What If?* in the context of urban and regional planning in Victoria will be discussed in the ensuing sections of this chapter.

The application of *What If?* version 2.0 — which can be categorised as a rule-based, comprehensive projection, task-oriented PSS — is discussed in this chapter. *What If?* is a stand-alone software package, developed using ESRI's MapObjects embeddable mapping and GIS components <http://www.esri.com/software/mapobjects/index.html>. To date there have been a number of applications of *What If?* (version 1.0) in the United States and elsewhere (Kim 2004; Klosterman et al. 2002, 2006; Li 2003; McClintock and Cutforth 2003; Pettit 2005; Webb 2003). In Australia, *What If?* (version 1.0) has been applied to assist the Shire of Hervey in Queensland to come to terms with the impact of projected population growth contributed to the 'sea-change' phenomena fuelled by tourism and an aging retiring population (Pettit 2005).

The PSS comprises two programs — *Setup* and *What If?*. *What If?* version 2.0 has several significant enhancements from version 1.0. These include:

- the ability to not only project land use, but also population and employment figures
- greater flexibility in the number of land use classes and suitability factors that the PSS can handle, and customisable slider bars to capture planning metrics (ratings and weightings)
- disaggregated reporting on the projections for sub-areas as well as the total study area
- improved cartographic and mapping capability.

What If? requires pre-processing (buffering, union, clip, clean…) of input data layers to create a Unified Analysis Zone (UAZ) file, which is stored in ESRI's shapefile (*.shp) format.

What If? is a PSS that recognises developments that have occurred in urban planning, yet has been developed based on the principle of keeping the models as simple as possible. It is a GIS-based PSS that supports collaborative planning and public participation. *What If?* is a scenario-based, policy oriented PSS that uses GIS data to conduct land suitability analysis, project future land use demand, and allocate these projected demands to the most suitable locations (as described below). It allows users to create future urban growth scenarios and determine the impacts of alternative policy choices on future land use patterns and social trends. *What If?* involves a simple user interface that allows the user to enter data and weightings, generate scenarios and examine results, then go back and alter parameters easily to generate further scenarios. Maps can then be compared within *What If?* which provide a powerful tool for assisting in the collaborative planning process. Figure 21.1 illustrates the *What If?* PSS framework as a planning policy tool.

What If? is made up of three main modules: suitability, demand and allocation. The role each of these modules play in projecting future development patterns is described briefly below.

21.2.1 Suitability Module

The land suitability analysis (LSA) module enables the user to input a number of constraints and opportunities that can be represented geographically. These spatial constraints and opportunity layers are referred to as 'suitability factors'. The LSA approach is based on the simple but effective sieve mapping overlay approach (McHarg 1969). A spatial overlay of all the suitability factors is created and enables users to assign ratings and weightings of importance to specific factors (e.g. distance to railway station) in order to define the overall land suitability for a particular land use (e.g. industrial). *What If?* can produce either maps or reports on the land suitability for a particular land use.

21.2.2 Demand Module

The land use demand module projects the demands for residential, employment-related, preservation, and local land uses. The demand assumptions are driven by the available or calculated projection figures and include projected population and employment trends, future population and employment densities, and desired quantities of open space and recreational land. The demand assumptions are converted into the equivalent land use demands and can be viewed in table format.

What If? can input available demographic projections of population, housing and employment data to derive land use demand. However, if such projections are not available then *What If?* applies a linear extrapolation of existing and/or past trends to project future population, housing and employment figures.

21.2.3 Allocation Module

The *What If?* allocation module projects future land use patterns using both the inputs from the LSA and land use demand modules. The allocation module also takes into consideration public policies (e.g. farm preservation), zoning ordinances, growth controls, infrastructure services (roads, sewers and water) to guide the future allocation of land for particular purposes. The allocation engine incrementally allocates land required for a particular land use on a user-defined time interval. Once the required land for a particular land use is satisfied then the allocation engine proceeds in assigning land for the next prioritised land use until its demand has been met and so on. The allocation engine proceeds in matching land use supply and demand for up to four time step intervals. In the Mitchell Shire case study this has occurred in five-yearly intervals which align with population project years (2006, 2011, 2016 and 2021). However, as *What If?* only supports four time intervals, the final interval has been extended from 2021 to 2031 to allow 30-year land use change scenarios to be developed.

What If? has benefits over other methods of planning as is an integrated tool that allows planning tasks to be performed quickly and easily, and it can be customised to specific database and policy issues (Klosterman 1999). It is designed to identify *what* would happen *if* the underlying assumptions used in a scenario were correct, rather than produce a single 'exact' prediction of the future. Therefore it is intended as a tool to explore likely policy options and facilitate open and ongoing processes of community learning. Feedback from the Mitchell Shire planners suggested that one of the biggest strengths of *What If?* was its ability to analyse, test, see and think about spatial planning policy. However, the complexity or comprehensiveness of such PSS software offer challenges to planning practitioners. Further feedback from Mitchell Shire planners indicated the potential practical applications of *What If?* took quite a bit of time and assistance from tool experts to achieve a sound conceptual understanding of the program.

21.3 Mitchell Shire Application of *What If?*

Mitchell Shire local government authority (LGA) is located 35 to 100 km north of the City of Melbourne, Australia. The shire comprises 2,864 km^2 and is situated along the Hume Highway, connecting Melbourne and Sydney (Fig. 21.2). This area is broadly known as an 'High Amenity Landscape' which is driven by rural lifestyle living (Barr 2005). The natural resource base and rural areas are significant assets to the shire, with wool and beef being the most important types of agriculture (Mitchell Shire Council 2006). Socioeconomic characteristics of the region include: strong population growth; average to low levels of unemployment; and major employment sectors in manufacturing, retail, government administration and defence. The main planning aim for Mitchell Shire is to maintain the urban–rural mix of the area and the integrity and character of the towns (DSE 2006a, 2006b).

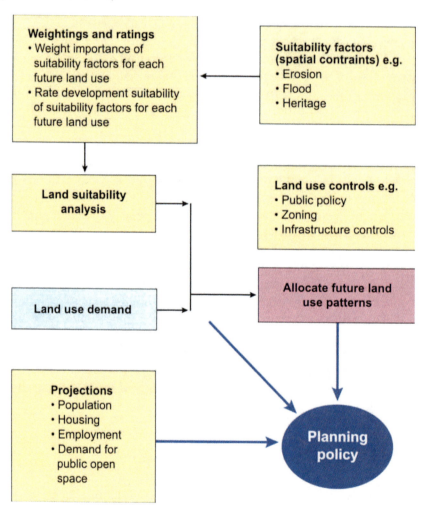

Fig. 21.1. *What If?* planning support system framework for informing planning policy

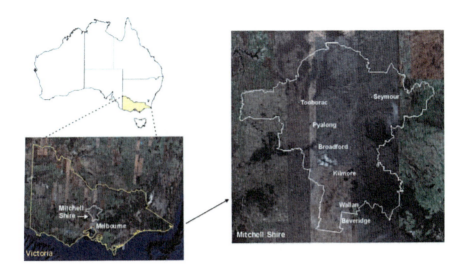

Fig. 21.2. Mitchell Shire locality map

The residential population has grown by 48% over the period 1981–2001 (DSE 2006b) and is currently estimated as 32,600. The location of Mitchell Shire relative to Melbourne strongly influences its growth and development. A large proportion of the shire's workforce commutes to Melbourne for employment, while Puckapunyal Military Base is the largest employer within the shire (DSE 2006b). Continuing population growth will place demands on land, services and infrastructure. Extensive areas around the shire's major towns have been zoned for residential development and the shire wishes to retain tight control of residential expansion into rural areas (DSE 2006a). The Mitchell Shire planners acknowledge the role and relationship of farming land with people, the economy, strategic policy and practice is very important. The planners believe *What If?* is a useful tool to facilitate a greater understanding of what is really happening and how government policy, legislation and controls are impacting the shire.

The *What If?* PSS has been applied to Mitchell Shire to create future land use change scenarios up until 2031, and thus align to the Melbourne 2030 strategic vision (DOI 2002). *What If?* land use change scenarios have been created for 2031 rather than 2030, to fit with Australian Bureau of Statistics (ABS) estimates. The objectives outlined in Mitchell Shire's Local Planning Policies and Municipal Strategic Statement (DSE 2006a, 2006b), as well as those objectives ascertained from discussions with planners from Mitchell Shire, have been the drivers for creating four likely allocation scenarios for 2031:

- *Conservation/Low Growth/Strategic Plan*
- *Conservation/High Growth/Strategic Plan*
- *Conservation/Low Growth/No Strategic Plan*
- *Conservation/High Growth/No Strategic Plan*

The *Conservation* component of these scenarios has been derived from LSA. *Conservation* represents the main planning aim for the Mitchell Shire which is to maintain the urban–rural mix of the area while protecting existing agriculture and productive land from the encroachment of urban growth. This scenario imposes strict restrictions on the supply of suitable land for development in an attempt to preserve the area's rural character. In particular, it incorporates the following assumptions:

- all eleven of the suitability factors are considered when they are appropriate for a particular land use
- development is discouraged from areas of significant land, or where there is vegetation protection
- development (other than agricultural) is severely limited in areas with the highest quality agricultural soils.

The *Low Growth* component of the scenarios relates to land use demand. The *Low Growth* scenario employs low population growth projections, in this case as prepared by the Department of Sustainability and Environment Victoria and assumes that the study area's population will be 48,835 persons in 2031. The *High Growth* option incorporates the comparatively high population growth projections prepared by ID Consulting (2002) and assumes that the study area's population will be 61,502 persons in 2031.

The *Strategic Plan* component of the scenario is part of the allocation module. This component uses the planning zones spatial layer for the shire and aims to determine whether the planning zones existing for Mitchell Shire can handle the projected population growth. The *No Strategic Plan* scenario aims to discern how appropriate or restrictive the planning zones are on the future development of the area and does not restrict 'greenfield' development.

21.3.1 Input Data Layers

A number of spatial and aspatial input data layers have been collated from various government and private sector sources in order to create the four likely allocation scenarios for Mitchell Shire. A number of socioeconomic datasets used to define the demographic projections and land demand scenarios have been used, including:

- current housing unit figures for rural and urban residential areas within Mitchell Shire
- current employment figures by industry sector for Mitchell Shire
- past total population, group quarters and household figures for Mitchell Shire in 1991, 1996 and 2001
- past employment figures by industry sector for Mitchell Shire in 1996 and 2001
- low projected population figures for Mitchell Shire including total population and average household size for the five projection years; 2011, 2016, 2021, 2026 and 2031
- high projected population figures for Mitchell Shire including total population and average household size for the five projection years; 2011, 2016, 2021, 2026 and 2031.
- percentage breakdown density of residential housing types
- future employment density by industry sector
- total hectares of preservation land uses for the five projection years (as above)
- area demand projections for local land uses for the five projection years.

In defining land suitability and allocation requirements, a number of spatial datasets have been integrated into a common geodatabase for Mitchell Shire. Table 21.1 describes the spatial datasets and their source information used to define land use, suitability factors and allocation parameters. Of particular significance to the strategic planning scenarios was the planning zones layer. This layer was switched 'on' when creating the *Strategic plan* based allocation scenarios and switched 'off' when creating the *No strategic plan* allocation scenarios. The *Conservation* scenarios are related to constraint layers determined by the slope, soil, and proximity to infrastructure such as roads and drainage. Roads, rivers and local government area (LGA) boundaries have been used as display layers for visual reference to end users.

Through integration of socioeconomic, environmental and physical factors into the *What If?* spatial planning framework (Fig. 21.2), planners can begin to understand regional dynamics by taking onboard a number of interrelated factors. With regards to the use of PSS such as *What If?*, the perceptions of the Mitchell Shire planners were that the tool has the ability to assist in thinking regionally. This potentially empowers planners to engage in a discourse with other regional practitioners and understand the global context associated with regional issues, contemplate the consequences of local decisions, and have a greater level of regional responsibility associated with such decisions.

Table 21.1. Land suitability and allocation data inputs

Original data layer	Data layer produced for use in UNION file
Mitchell LGA	ABS boundary file
Locality areas	Townships – Corporate Spatial Data Library (DSE)
Statistical local area (SLA)	ABS boundary file
Vacant land	Existing land use – Mitchell Shire Council data
Land use mapping for Victoria	Australian Land Use Mapping classification
Cadastre	Corporate Spatial Data Library (DSE)
10 m contours	Slope derived from Corporate Spatial Data Library (DSE)
Planning zones	Corporate Spatial Data Library (DSE)
Soil	Corporate Spatial Data Library (DPI)
Roads plan 2030[a]	Parcels served by roads in 2030
Drainage	Parcels served by drainage – Mitchell Shire Council data
Floodway overlay and Land Subject to Inundation overlay	State government planning overlay
Erosion Management overlay	State government planning overlay
Salinity Management overlay	State government planning overlay
Rail stations	1 km rail station buffer
Wildfire Management overlay	State government planning overlay
Existing roads	100 m Goulburn Valley Highway/Hume Freeway buffer
Vegetation Protection overlay	State government planning overlay
Significant Landscape overlay	State government planning overlay
Heritage overlay	State government planning overlay

[a] The original data layer was directly used to create the UNION file. *ABS:* Australian Bureau of Statistics.

Figures 21.3 and 21.4 provide examples of two fundamental data inputs used to create land use categories and suitability factors (e.g. suitable soil) respectively. The land use mapping layer has been generated from the standardised Australian Land Use Mapping (ALUM) classification, which has been appended with the more disaggregated urban land use data to generate a comprehensive existing land use map. The soils data comprises mapped parent material lithology combined with landform description. Further interpretation by soil scientists has resulted in a soil suitability layer which rates suitability from very poor to very good. All soil within the Mitchell Shire is considered moderate, poor or very poor. Both of the land use mapping and soils datasets are managed by the Department of Primary Industries (DPI) Victoria. Before this project commenced, Mitchell Shire did not know that this data existed. One of the tangible benefits of this project is an increased awareness of existing natural resource management datasets available. Feedback from the Mitchell Shire planners affirms this by stating one of the strengths in applying a scientific tool such as *What If?* is that it instigates a data- and information-sharing environment which is empowering to local government.

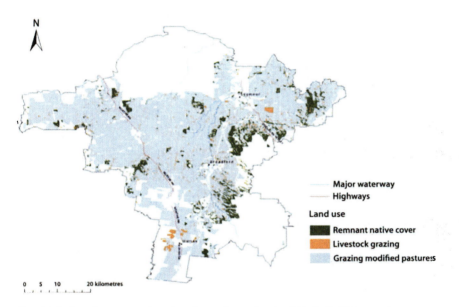

Fig. 21.3. Land use mapping data input layer for the Mitchell Shire study

21.3.2 Land Suitability Analysis

Eleven suitability factors were defined for representing the land suitability for future land use within the Mitchell Shire. Figures 21.5 and 21.6 illustrate the graphic user interfaces (GUIs) for determining user-defined ratings and weightings of importance. These interfaces are a powerful tool for capturing planning metrics. For the first iteration of land suitability scenario generation, the ratings and weightings have been determined by the model experts (the authors). Future work will engage the planners to undertake an iterative process of refining these land suitability variables and their weightings of importance.

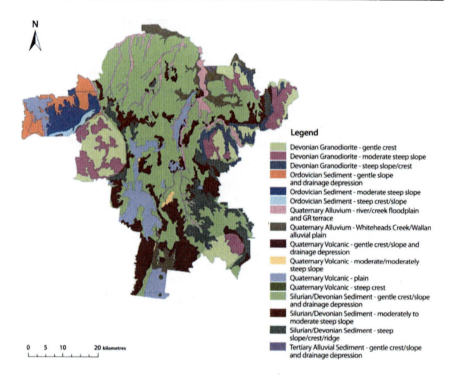

Fig. 21.4. Soil classification data input layer for the Mitchell Shire study

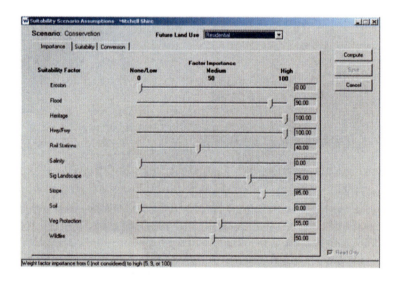

Fig. 21.5. Screen shot showing the *Conservation* scenario–factor importance ratings for residential land use

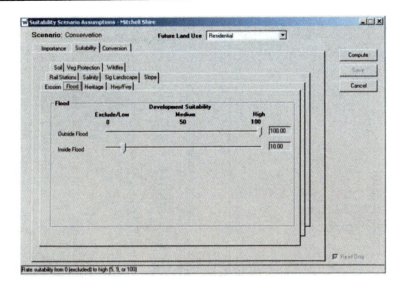

Fig. 21.6. Screen shot showing the *Conservation* scenario–development suitability ratings for residential land use

The first iteration of land suitability scenarios generated for Mitchell Shire resulted in four land use change scenarios. Figure 21.7 illustrates an example of the suitability (ranging from not developable to highly suitable) for future residential land for the low conservation scenarios. The land suitability scenarios are in themselves a valuable planning tool as they enable planners to understand the extent of land suitability for specific land uses such as residential, commercial, industrial, agricultural and conservation.

21.3.3 Demographic Projections and Land Use Demand

There are three sources of population information available to formulate land use demand scenarios for Mitchell Shire, these include:
- past and current population counts from the Australian Bureau of Statistics <http://www.abs.gov.au/>
- DSE population forecasts to 2031 <http://www.dse.vic.gov.au/DSE/dsenres.nsf/>
- ID population forecasts to 2021 <http://www.id.com.au/mitchell/commprofile/>.

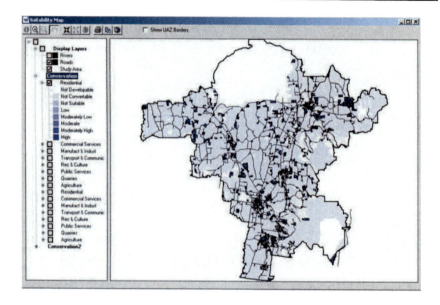

Fig. 21.7. Screen shot of land suitability map for residential land in the Mitchell Shire

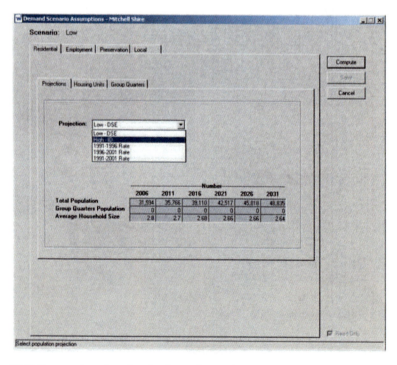

Fig. 21.8. Screen shot showing land use demand assumptions

Employment projection information is currently unavailable for Mitchell Shire and subsequently no employment projection information has been included in the current suite of land use change scenarios for the shire. Ongoing research will develop these employment projections. Figure 21.8 illustrates the land use demand assumptions. The user can select which projection figures are to be used to determined household and employment growth. For the four Mitchell Shire scenarios both the DSE and ID consulting population projections were used. The Mitchell Shire planners were interested in comparing the results of the two population projection forecasts in order to better understand the potential land use conflicts and opportunities which arise from population growth.

21.3.4 Future Land Use Allocation Scenarios 2031

The four scenarios run for Mitchell Shire are a combination of high and low population growth, with and without the existing strategic plan land use control. Figure 21.9 illustrates the future land use allocation scenario for the *Conservation/Low Growth/No Strategic Plan* scenario.

Without taking into account existing planning zones the shire can see where future land use change would likely occur, based on land suitability, with no planning instruments. In this scenario there is some new residential development which would occur outside the existing urban centres, as indicated in Fig. 21.9. *What If?* produced accompanying reports for the changes in land use allocation for each derived scenario. Table 21.2 shows the projected land use change associated with the *Conservation/Low Growth/No Strategic Plan* scenario.

Table 21.2. Projected land use change from 2006 to 2031 under the *Conservation/Low Growth/No Strategic Plan* scenario for the Mitchell Shire. Land area for each land use is given in hectares. Note there has been no change in all other land use categories, as sufficient vacant land classified as Undeveloped exists to satisfy urban and residential land demands

Land use	2006	2011	2016	2021	2026	2031
Rural residential	4,392.12	4,911.14	5,253.49	5,621.61	5,947.50	6,271.95
Undeveloped	19,983.22	19,208.45	18,698.86	18,119.09	17,670.77	17,165.72
Urban residential	1,628.46	1,884.21	2,051.45	2,263.09	2,385.52	2,566.12

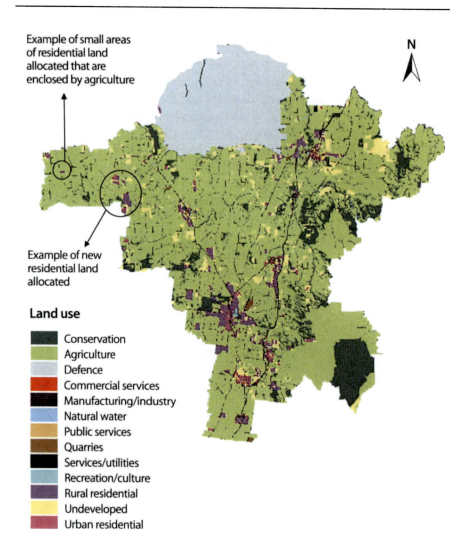

Fig. 21.9. Map of land uses allocated in 2031 for the *Conservation/Low Growth/No Strategic Plan* scenario

Based on the existing data limitation imposed by not having access to any derived employment projections, employment is considered to stay constant it is only urban residential and rural residential land uses which currently absorb any undeveloped land within the shire for urban growth.

Employment projection data would be valuable to derive so that all future land uses (including industrial and commercial land uses) could be allocated in the scenarios, as opposed to being limited to only the range of residential land use types.

21.4 Future Work

A quantitative measure of applying *What If?* in the Mitchell Shire has not yet been determined, nor is it likely to be. Work with the shire is ongoing as the research team continues to refine suitability factors, weightings, ratings, projection figures and further develop and evaluate land use change scenarios. Within the context of the Mitchell Shire case study the research team is also investigating the benefit of producing land use change scenarios in an interactive three-dimensional computer-game environment. Once this is achieved further testing will be undertaken to evaluate if there is added value in communicating land use futures to planners using a virtual world paradigm.

In the general area of PSS research much work is needed to bridge the gap between model expert and planning practitioner. Vonk et al. (2005) have identified there are bottlenecks to adoption of PSS, and that much effort has been placed in developing software tools whilst little has been done in supporting and evaluating their application. As indicated by the Mitchell Shire planners there have been identified barriers in understanding PSS tools such as *What If?* It is therefore identified that further work is required to ensure educational institutions that take responsibility for training planners introduce PSS into their curriculum to ensure that planners are conversant with such tools before entering the work force. Such planners can then be advocates for PSS adoption amongst government and private sector planning agencies.

21.5 Conclusion

This chapter discusses the application of the *What If?* PSS for developing land use change scenarios for a local municipality in the context of Australia. Initial feedback from questionnaires and interviews of planners from Mitchell Shire and managers the State Government of Victoria, indicated the following perceived benefits in applying the *What If?* PSS:

- It provides a useful knowledge-management environment to assist local government in undertaking strategic planning.
- The system enables local government planners to think regionally.
- It empowers local government to critically examine planning policy through an evidence-based science approach
- The lifespan of existing planning strategies can be evaluated against population and ultimately employment projections.

- The system enables a clearer understanding of the impact on farming land from urban growth.

For strategic planning purposes the *What If?* PSS offers an improvement to standard GIS. The PSS reduces the complexity of using map overlays in a GIS as well as reducing the time required to explore a range of land use change scenarios. A strength of PSS tools, such as *What If?*, is that they guide users through a straightforward land use allocation process. The procedure for allocating land is incremental and enables a degree of trade-off to occur between competing land uses, based on user defined weightings and controls. It is a complex process that consumes a lot of time if done manually within a GIS, that is automatically achieved using *What If?*.

PSS such as *What If?* hold a lot of potential for planners, but they will need to consider whether the benefits outweigh the data requirements for running such a data intensive system. *What If?* version 2.0 has been configured to input ESRI Community Tapestry socioeconomic data <http://www.esri.com/data/community_data/community-tapestry/index.html>. However, the automated loading of such data outside the United States is not currently supported. In Australia disaggregated employment projection data are not available at the LGA level. Therefore, the Land Use/Population/Employment option in *What If?* can only be used to project small area land use and population values.

There are many challenges which remain in having PSS such as *What If?* as a common tool on the planning practitioner's desktop computer. Some of these challenges include access to data, training existing staff and educating up-and-coming planners. There are also many perceived benefits in applying PSS in practice which have been discussed in this chapter. However, the real value of applying such PSS tools as *What If?* is in the development of cross-jurisdictional collaborations. This will lead to shared understanding, new learning and robust knowledge management systems to assist Government planners in making better decisions to realise sustainable urban and regional futures.

Acknowledgements

This research was funded from a State Government of Victoria initiative – Our Rural Landscapes <www.dpi.vic.gov.au/orl/> through the Department of Primary Industries. Collaboration with the University of Melbourne was possible through the Department of Geomatics 4th undergraduate project program. Software support was supplied by *What If?* Inc. in the use of

What If? version 2.0. Special thanks to Katie Rizzuto (Planning Strategy Coordinator, Mitchell Shire) for support and endorsement in applying the *What If?* PSS.

References

Barr N (2005) The changing social landscape of rural Victoria, Department of Primary Industries, Melbourne

Brail RK, Klosterman RE (2001) Planning support systems. ESRI Press, Redlands, California

Budic ZD (1994) Effectiveness of Geographic Information Systems in local planning. Journal of the American Planning Association 60(2):244–263

DSE (2006a) Mitchell planning scheme, local planning policies: Victoria's planning scheme's online. Victorian Department of Sustainability and Environment, available at http://www.dse.vic.gov.au/planningschemes/mitchell/home.html

DSE (2006b) Mitchell Planning Scheme, municipal strategic statement: Victoria's planning scheme's online. Victorian Department of Sustainability and Environment, available at http://www.dse.vic.gov.au/planningschemes/mitchell/home.html.

DOI (2002) The State of Victoria, Melbourne 2030: planning for sustainable growth. Victorian Department of Infrastructure, Melbourne

Forester J (1999) The deliberative practitioner: encouraging participatory planning processes. MIT Press, Cambridge

ID Consulting (2002) Mitchell Shire Population Projections 2001–2031. Retrieved 8 May 2006, http://www.id.com.au/mitchell/commprofile/

Geertman S, Stillwell J (2004) Planning support systems: an inventory of current practice. Computers, Environment and Urban Systems 28(4):291–310

Kim J-W (2004) A simulation of the growth of the Seoul Metropolitan's built-up area with a GIS-based PSS model. Journal of the Korean Planners Association 39(7): 69–84

Klosterman RE (1999) The What If? collaborative planning support system. Environment and Planning B 26(3):393–408

Klosterman RE (2001) Planning support systems: a new perspective on computer-aided planning. In: Brail RK, Klosterman RE (eds) Planning support systems: integrating Geographic Information Systems, models and visualisation tools. ESRI Press, California, pp 1–23

Klosterman RE, Pettit CJ (2005) Guest editorial: an update on planning support systems. Environment and Planning B 32(4):477–484

Klosterman RE, Siebert L, Hoque MA, Kim J-W, Parveen A (2002) Using an operational planning support system to evaluate farmland preservation policies'. In: Geertman S, Stillwell J (eds) Planning support systems in practice. Springer, Heidelberg

Klosterman RE, Siebert L, Hoque MA, Kim J-W, Parveen A (2006) What If? evaluation of growth management strategies for a declining region. International Journal of Environmental Technology and Management 6(1/2):79–95

Li Y (2003) Planning support for urban spatial development : a case study of Zhenning county. M. Sc. thesis, International Institute for Geo-Information Science and Earth Observation, Enschede

McClintock T, Cutforth L (2003) Land use impacts for the Black Earth Creek Watershed Modeled with GIS'. ArcNews 25(2): 26–27

McHarg IL (1969) Design with nature. Doubleday – Natural History Press, New York

Mitchell Shire Council (2006) Official website of Mitchell Shire Council. Retrieved 12 August 2006, http://www.mitchellshire.vic.gov.au/

Pettit CJ (2005) Use of a collaborative GIS-based planning support system to assist in formulating a sustainable-development scenario for Hervey Bay, Australia. Environment and Planning B 32(4):523–546

Pettit CJ, Pullar D (2001) Introduction to urban and regional planning using ArcGIS. Environmental Systems Research Institute, available at http://campus.esri.com/

Vonk G, Geertman S, Schot P (2005) Bottlenecks blocking the widespread use of planning support systems. Environment and Planning A 37(5):909–924

Webb BL (2003) An analysis of future land use change in Licking County: an application of an interactive GIS-based planning support system. Denison University, Granville, Ohio

PART 5
LANDSCAPE VISUALISATION

22 Understanding Place and Agreeing Purpose: the Role of Virtual Worlds

Ian D Bishop

CRC for Spatial Information and University of Melbourne, Victoria, Australia

Abstract: Place is a complex concept. We might limit our thinking to a pair of geographic coordinates or we may seek to understand how a place came to be, what processes are happing in it now, or what options exist for its future. To really know a place we should be aware of all these aspects. Some people, based on prior training or life experience, have an ability to 'read' places and immediately understand the geological forces, erosion processes, ecological succession, human activities and climatic constraints which together define character and provide opportunities. Others are not so blessed, and yet agreement on purpose depends upon this prior understanding. Therefore, to reach consensus on land management and policy drivers, we require tools which help all stakeholders to understand what makes a place the way it is and what changes are sustainable. This paper presents recent and ongoing work in environmental visualisation and other tools, such as agent-based modelling, and reviews their demonstrated and potential contribution to understanding and agreement. In particular, work which automates the creation of landscape models, links these to environmental-process simulators for scenario testing, and makes these available as collaborative virtual places is illustrated.

22.1 Introduction

The definition of place is not simple. Much has been written about what constitutes place. Agnew (1987), for example, argues that we can understand place as either:

- location, as a geographic area with attendant political and economic structures
- sense of place, as in the set of sensations and feeling associated with being in a place
- locale, as the setting in which social relationship arise, the neighbourhood or community.

Thus we have the wholly objective view (location), the whole subjective view (sense of place) and the somewhere in between view (locale). Cresswell (2004) argues that in addition we need to consider the meaning of *space* and of *landscape* in order to appreciate the full complexity of place.

Concerning *space*, Tuan (1977) writes:

'What begins as undifferentiated space becomes place as we get to know it better and endow it with value ... the ideas 'space' and 'place' require each other for definition. From the security and stability of place we are aware of the openness, freedom, and threat of *space*, and vice versa. Furthermore, if we think of space as that which allows movement, then place is pause; each pause in movement makes it possible for location to be transformed into place.'

When we consider place in this way it requires people to name it, to map it, to attach meaning to it. We cannot sensibly consider place without this meaning. *Landscape*, on the other hand, is a very visual perspective on place. The landscape is what we can see from a particular location. Topography and surface features come together with our visual abilities. Where a local might see a particular valley or mountain as a place, because of their strong associations, to a visitor it may be just a landscape. As Cresswell (2004) concludes 'we do not live in landscapes — we look at them'. He goes on to argue that place is a way of understanding the world both by the potential to interpret the world as a collection of connected places, and through our tendency to interpret the world from the sometime narrow perspective of 'our place'.

Turning that around and thinking about place in the context of rural landscape management seems to involve learning how to understand someone else's place. This requires knowing not only the names and objective features and processes which occur in that place but also the meanings and attachments of the people for whom it is 'our place'.

In addition to the social significance of place based on human activity, there are physical and biological processes which help to shape place and define its possibilities. Frequently there is, therefore, also a need to understand the geological forces, hydrological processes, ecological succession, and climatic constraints Some people, through their knowledge of the physical or social sciences, have an ability to read the landscape and interpret process underlying what they see. Most of us require assistance to reach such understanding. This paper explores the opportunities to use visualisation and other technologies to communicate and understand place. If a common awareness of place is achieved then people can work together to define and realise purpose.

22.2 Established Options for Understanding Place

People have been communicating place through a variety of media for millennia (Fig. 22.1).

Fig. 22.1. Clockwise: *Ice Landscape* by Hendrick Avercamp (1585–1634); portion of *Sand Dunes* by Robert Frost (1874–1963); *Mont Blanc* by Xaver Imfeld (1853–1909); *Jeffrey Pine, Sentinel Dome* by Ansel Adams (1902–1984)

Landscape paintings are, in many cases, an attempt to communicate much more than just the visual appearance of a view. Aspects of life within the view are also portrayed in order to give a sense of the behaviour and attitude of local people. Landscape poetry was also emerging through the renaissance. Famed for his ability to 'paint' landscapes in words was the American poet Robert Frost.

New technologies provide new media for communication of landscape. Photography has become popular with people in all levels of society for capture of their landscape memories. More ambitious in his use of photography has been Ansel Adams whose moody black and white photography set the standard for interpretive landscape photography. Quite different in character are formal reports and maps which strive for objectivity in reporting the conditions of land. While objective information may be the basis of mapping, there are also aesthetic judgments in the choice of colours or shading and other map elements as pioneered by Xaver Imfeld. This should not change the basic information but may affect the user's interpretation of the base data.

22.3 Emerging Options

A great many new options have emerged with the advent of computer systems capable of rendering landscapes. The literature on the technology is large and growing but there have been few studies of the power of virtual world, or computer simulations, to convey a sense of place. Appleton and Lovett (2003) created realistic landscape images in Visual Nature Studio (VNS) and conducted a survey in which people were asked, 'To what extent do you feel that the style and content of this image allow you to imagine the future landscape that is being considered?' Bishop (2001) explored path and movement choice in virtual landscapes.

These extended earlier validity studies of computer media in aesthetic perceptions (Bishop and Leahy 1989; Vining and Orland 1989; Bergen et al. 1995) but still failed to really explore the extent to which the virtual world can communicate understanding of the real world. Researchers in different contexts have studied ability to navigate, phobic responses and other factors. Wyeld et al. (2007) have used a virtual environment in communicating Australian aboriginal culture and language. Champion et al. (2003) developed an environment and a series of contemporary challenges for understanding Mayan culture. Broadly we know that the virtual world can be a useful surrogate for the real world in certain contexts.

This literature suggests certain key characteristics of virtual environments. A starting list includes:

- faithfulness to reality to enable natural interpretation
- movement and interaction to allow exploration
- access to non-visual information as many variables of importance are non-visual

Understanding Place and Agreeing Purpose: the Role of Virtual Worlds   461

- looking beneath the surface, both physically (at the regolith or deeper) and culturally
- conveying human meanings such as social, cultural events or items of significance
- exposing environmental process enables the results of actions to be seen
- interaction with other people to allow sharing of the experience
- encouraging natural user behaviour in order to support research into real-world decision making
- linking to process and behavioural models to test alternative options.

All this is ideally created at low cost. If the creation of the virtual environments is too expensive it won't happen except in those applications in which money is not an object. These characteristics were the wish list for software developments described below.

22.4 Development Methodology

We are developing visualisation and modelling systems which will help users to understand a real place, as it was, as it is or as it may become, through experience of a virtual place. This includes creation of a realistic visual environment, with representation of the output of environmental process models to show the consequences of human actions, and simulation of human activities, though agents, to show how real people may behave under changing environmental, economic or social conditions.

22.4.1 *SIEVE*

Over the last three years we have developed a suite of software for automating the creation of three-dimensional virtual landscapes, merging these with outputs of environmental process, linking the three-dimensional world in real-time with underlying sources of data, making this virtual world a multi-user collaborative environment and also allowing in-field access through augmented reality (Chen et al. 2006).

Creation of three-dimensional models from two-dimensional data can be a time consuming process with rather repetitive steps: elevation grid to 2.5D surface, image drape, surface object creation, texture capture for surface objects, object placement, and specification of atmospheric conditions. With the right preparation, these procedures can be automated and functional on an organisation server with access to the spatial data infrastructure (SDI). This allows anyone with server access to use a map interface as a means of specifying an area of interest. A single 'export' button sets in motion the procedures to create the corresponding three-dimensional landscape model (Stock et al. in press). The user then downloads the landscape model and loads this in a suitable viewer to experience the virtual environment either as a single user (Fig. 22.2) or as one of several stakeholders in a network-based collaborative virtual environment (CVE). The GIS/server environment in this case is *ArcGIS/ArcServer* (www.esri.com) and the virtual reality interface is called *SIEVE Viewer* and was built on the *Torque Game Engine* (TGE) (www.garagegames.com).

Land management scenarios can be developed in the virtual world, the changes linked back to the GIS where the models determine the consequences of the scenarios. The result is then displayed in the virtual world. For example, Fig. 22.2 shows the terrain as seen in *SIEVE Viewer* cut away, using a clipping plane, and revealing the depth to watertable as generated by a hydrological model (Weeks et al. 2005). The watertable shape is generated from the spatially explicit model output of depth, this is highlighted by the colour — such that shallower dark areas may cause salinity problems, while a stipple pattern on the surface shows the salt level in the groundwater.

In relation to our concept of place it is important to remember Tuan's statement:

> 'if we think of space as that which allows movement, then place is pause; each pause in movement makes it possible for location to be transformed into place'.

When we do fly-throughs in virtual environments we are moving in space, when we stop, review, model and communicate we are beginning to turn the space into place.

a)   b)

Fig. 22.2. A virtual world as created by *SIEVE Builder* and as seen in *SIEVE Viewer:* showing: object detail (**a**), and cut-away to groundwater data (**b**)

22.4.2 Links to Decision Support Systems

The next stages in development of these systems and virtual environments are being designed to broaden the range of applications to include urban environments. Key ongoing developments include extension of the existing object library, intelligent placement of objects (such that trees take their correct growth form and distribution according to biophysical conditions), better navigational and collaboration tools and automated linkage to online land manager support systems such as *eFarmer* (Aurambout et al. 2007). Figure 22.3 illustrates an early test of this linkage as polygons used to define tentative forest plantation options are populated with appropriate species for collaborative review.

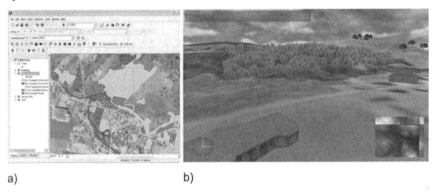

Fig. 22.3. The linkage between candidate decision polygons in the GIS (**a**) and the *SIEVE* expression of chosen land uses (**b**)

22.4.3 Virtual Decision Environment

We have also developed a decision testing environment drawing on visualisation, agent modelling and experimental economics (Spottiswood and Bishop 2005). This virtual decision environment (VDE) simulates a complex mix of economic, environmental and social influences using a combination of autonomous agent modelling, environmental visualisation technology and the procedures emerging in the field of experimental economics. Ecological validity is sustained by the ability of visualisation to provide a natural interpretation of environmental conditions and also observation of the choices of others in the community. Agent modelling provides this social context in the form of decisions made by neighbours and the larger community, calibrated according to observed responses in controlled experimental settings. Finally experimental economics adds real incentives (saliency) to the economic, environmental and social achievements of those immersed in the VDE (Fig. 22.4).

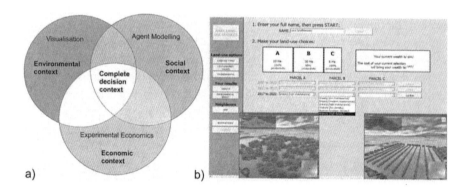

Fig. 22.4. The three technologies/methodologies support the three key aspects of individual decision making. Together they create an integrated virtual decision environment (**a**). The experimental land use selection interface used in conjunction with an immersive display (**b**)

We can replace one agent with a person to test human decision making under a range of external circumstances and draw generalisations which then allow better calibration of the agent models. Policy makers will be able to use the refined agent models to test community response, built from individual decisions, to particular policies. In the current Australian context these policies might relate to climate change, water pricing or land clearing. In this way models (and the supporting visualisation) can reduce uncertainty about public response and hence better fit policy to purpose. There is still much to learn about the extent to which these virtual decisions are a good guide to real decisions in corresponding circumstances. Earlier experiences with the VDE are reviewed by Kennedy and Bishop (Chapter 29).

22.5 Conclusion

Establishment of common purpose depends on a common appreciation of place supplemented by an ability to explore the consequences of decisions from national policy to individual paddock management. The outcomes of policy and management are found not merely in the application of environmental-process models, but also in the ability to model individual decision making, social interactions and consequent emerging regional effects.

This is the role we envisage for a virtual decision environment which encompasses all the features described above.

Virtual environments have the capacity to not only communicate place to those from outside the environment but also to add richness to the understanding of 'our place' among locals. Between individuals, understanding of place will however never be identical and agreement on land purpose will seldom be absolute. An individual land manager must retain autonomy to make decisions which accord with his or her expertise, experience, immediate and longer term needs, willingness to innovate or take risks, availability of capital or labour and so forth. Nevertheless, the premise of ongoing development is that an ability to see the future place, when linked to the environmental-consequences models, forms the basis for development of plans which accord both with the objectives of the individual land manager and agreed regional environmental objectives.

Table 22.1. A subjective assessment of the degree to which we are approaching certain ideal characteristics of virtual environments as they may be applied to place and purpose

Ideal characteristics of virtual environments	Subjective assessment	Subjective grade
Low creation cost	Major advances	A-
Faithfulness to reality	Room to improve	B+
Movement and interaction	Established	A
Access to non-visual information	Very promising	A-
Looking beneath the surface	Needs work	C
Conveying human meanings	Lacking direction	D
Exposing environmental process	Solid beginnings	C
Interaction with other people	Becoming established	B
Encouraging natural behaviour	Developing	B-
Linking to models	Promising signs	C+

A set of characteristics for successful virtual environments were proposed above. Table 22.1 is a very cursory analysis of the degree to which current technology and especially the systems described herein, approach an ideal. Broadly we have come a long way but still have a long way to go. We have shown that:

- virtual worlds need not be expensive to produce
- movement and interaction are quite well established
- virtual worlds are also becoming sufficiently realistic for many purposes.

Where we are still failing is primarily in the area of conveyance of human meaning. While our developments aspire to effective conveyance of place, there is still a lot to be done to move the virtual world beyond being a mere introduction to place and to, some degree, to process. Perhaps when the cyberworld makes better use of the expressive capacities of painting, poetry or photography, then we will really have the tools to effectively communicate and understand place.

Acknowledgements

I am grateful to the CRC for Spatial Information who funded the development of *SIEVE* and to those who have worked on the software: Christian Stock, Alice O'Connor and Alex Chen. Lucy Kennedy (nee Spottiswood) has been part of the development of the VDE concept. Our colleagues in DPI, Chris Pettit in particular, have also shaped my thinking.

References

Agnew J (1987) Place and politics. Allen and Unwin, Boston

Appleton K, Lovett A (2003) GIS-based visualisation of rural landscapes: defining 'sufficient' realism for environmental decision-making. Landscape and Urban Planning 65(3):117–131.

Bergen RD, Ulricht CA, Fridley JL, Ganter MA (1995) The validity of computer-generated graphic images of forest landscape. Journal of Environmental Psychology 15(2):135–146

Bishop ID (2001) Predicting movement choices in virtual environments. Landscape and Urban Planning 56(3–4):97–106

Bishop ID, Leahy PNA (1989) Assessing the visual impact of development proposals: the validity of computer simulations. Landscape Journal 8:92–100

Champion E, Dave B, Bishop I (2003) Interaction, agency and artifacts. digital design: research and practice. In: Proceedings of the 10th International Conference on Computer Aided Architectural Design Futures, Taiwan. Kluwer Academic Publishers, Norwell, Masachusetts, pp 249–258

Chen T, Stock C, Bishop ID, O'Connor A (2006) Prototyping an in-field collaborative environment for landscape decision support by linking GIS with a game engine. Paper presented at Geoinformation, Wuhan, China, 28–29 October 2006

Cresswell T (2004) Place: a short introduction. Blackwell Publishing, Malden

Pettit C, Cartwright WE, Bishop ID, Park G, Ridley A, Kemp O (2007) eFarmer – a web based farm management and catchment planning tool. Paper accepted for International Congress on Modelling and Simulation, 10–13 December 2007, Christchurch, New Zealand

Spottiswood L, Bishop ID (2005) An agent-driven virtual environment for the simulation of land use decision making. Paper presented at International Congress on Modelling and Simulation, 12–15 December 2005, Melbourne, Australia

Stock C, Bishop ID, O'Connor A (in press) SIEVE: collaborative decision-making in an immersive online environment. Cartography and Geographic Information Science

Tuan YF (1977). Space and place: the perspective of experience. University of Minnesota Press, Minneapolis

Vining J, Orland B (1989) the video advantage: a comparison of two environmental representation techniques. Journal of Environmental Management 29:275–283

Weeks A, Beverly C, Christy B, McLean T (2005) Biophysical approach to predict salt and water loads to upland REALM nodes of Victorian catchments. Paper presented at International Congress on Modelling and Simulation. 12–15 December 2005, Melbourne, Australia

Wyeld TG, Carroll J, Gibbons C, Ledwich B, Leavy B, Hills J, Docherty M (2007) Doing cultural heritage using the Torque Game Engine: supporting indigenous storytelling in a 3D virtual environment. International Journal of Architectural Computing 5:418–435

23 Geographic Landscape Visualisation in Planning Adaptation to Climate Change in Victoria, Australia

Ian Mansergh, Alex Lau and Rod Anderson

Department of Sustainability and Environment, East Melbourne, Victoria, Australia

Abstract: Climate change is a global challenge for all scientists the 21st century with a certain amount of 'global warming' already inevitable. The magnitude and risks of climate change are now being more widely appreciated and the need for adaptation, including land use, is becoming a social imperative. Profound changes to ecosystems and biodiversity are predicted, and climate induced migration of biota is envisaged and is already being observed. Although the exact changes remain uncertain, landscapes and biota will be affected at all scales — from the local to sub-continental level. Space for biodiversity to 'self-adapt' is required. Maintaining and restoring ecological connectivity and resilience — biolinks — across landscapes are likely to be crucial aspects of adaptation to climate change. This is particularly so across human modified landscapes (e.g. agricultural landscapes) where the environmental legacy of habitat fragmentation and degradation is already a major global conservation issue. Past and current spatial patterns of human land use and management will also be affected by biophysical drivers and human adaptation.

What will be the function, patterns and processes of future landscapes that we bequeath to future generations under climate change? What they look like will be product of societal choices informed by the community's 'sense of place', that now includes biodiversity conservation. Visualisation tools, particularly when accurately linked to Geographic Information Systems, ecological perspectives and realistic photographic libraries, offer powerful facilities to assist the community, scientists, planners and all major stakeholders to plan for new landscapes that consider climate change, including biolinks. These digitally created landscapes that display physical properties of the real world can provide a useful medium for visualising the results from experimenting with outcomes of different management approaches building in biodiversity conservation. 'Seeing' can augment meaning and 'sense of future place', particularly when linked to a realism derived from ecology. This chapter examines the use and potential of various visualisation tools as part of the emerging debate about biodiversity and adaptation to climate change in south-eastern Australia.

23.1 Introduction

Global warming is the major environmental and economic threat of this century with mass extinctions projected due to a variety of factors including the changing spatial distribution and condition of populations and habitat at sub-continental scales (IPCC 2007; Stern 2006; Thomas et al. 2004). These phenomena will be exacerbated in areas of fragmented and degraded habitats and landscapes (Brereton et al. 1995; Thomas et al. 2004). Biophysical fingerprints of the warming over the 20th century are being documented globally and in Australia (Root et al. 2003; Umina et al. 2005). In Australia this is particularly relevant as it is the driest inhabited continent with the most variable climate and a global mega bio-diverse region. A certain degree of climate warming is now inevitable, thereby increasing the societal need for adaptation to these changes and potential risks (IPCC 2007). Land uses and management (e.g. agriculture, reserve and forestry systems), based on historic allocations and presumption of climate, will also need to adapt to climate change and related consequences. Reconnecting the ecological connectivity of landscapes (biolinks) to allow for species re-colonisation and migration has been called for in various parts of the globe (Brereton et al. 1995; Hilty et al. 2006; Opham and Wascher 2004; Soule et al. 2004). As ecological connectivity has a large visual impact on landscapes, visualisation tools offer an important component to inform the debate about future landscapes. South-eastern Australia now faces a warmer and drier future with irreversible climate changes already in train (Government of Victoria 2006) which means change in both 'natural' biophysical and human-induced landscapes. The capacity to realistically visualise these future landscapes at various scales will assist the emerging societal debate, and consequent planning for more resilient landscapes.

Landscape visualisation has substantial advantages as an investigative and communication medium in designing these new landscapes and enhancing the debate. It allows landscapes to be re-imagined. This is particularly so when combined with the scientific data from ecology. A visual synthesis of vast amounts of environmental historic data is an especially attractive feature as it can rapidly conceptualise the extent of the changes over time and in space. It provides a platform for realistic place-based future scenarios. It can also create meaning to present actions and assist in explaining the manifestations of ecological processes. In this chapter we show a variety of visualisation techniques that have been used and are being developed to provide a better understanding of the past to enhance our capacity to envisage biolink zones in future landscapes. These techniques are important to illuminate the inter-generational equity issues around landscape, biodiversity and natural resource management under changing climate. Visualisation can make a significant contribution to planning these new landscapes and, we argue, is best based within a 'sense of place' and on visual benchmarks of the past linked to ecological concepts and datasets and realistic photo libraries.

We explore these concepts through asking:

- What is the historic context of visualisation in redefining 'sense of place' and landscapes for the future?
- After examining the magnitude of the problem: how can techniques of geographic visualisation help knit a landscape policy narrative from the past to potential futures that include biodiversity conservation (bio-links)?
- How can ecological datasets be utilised and realistic images be incorporated into visualisation of future desired landscapes?

23.2 Context of Visualisation and 'Sense of Place'

Humans imbue landscape with meaning and create and conserve them as manifestations of these meanings which are connected to both the past and projected future (Schama 1998). Colonial societies transformed landscapes from the original into new 'future' productive vistas in their image.

The visual is a powerful, yet often neglected, medium in determining future landscapes. Landscape painting is an early version of visualisation and perceptive artists can reflect and *effect* a changing sense of place. Colonial depictions of 'Europeanised' Australian landscapes, where eucalyptus look like English oak trees, were challenged by the Victorian 'Heidelberg School' in the late 19th century (Clark and Whitelaw 1985).

These images changed the way landscape was perceived, and implicitly what future landscapes could be. Fred McCubbin's iconic *The Pioneers* (1904) hangs in the foyer of the Victorian Parliament depicting the change from pioneering in the Australian bush to the encroachment of civilisation. In the 1930s Arthur Streeton painted *The Vanishing Forests* (1934) — a statement on deforestation of the Dandenong Ranges and the environmental cost of settlement (Smith 1995). Through visualisation, a different landscape 'meaning' was added to colonisation. In the 1940s and 1950s, Sidney Nolan's iconic images of Australia in drought reinforced Australian's appreciation of living *within* the climate and landscapes. Michael Leunig, a nationally recognised Australian artist, articulated these changes in a painting of a mob of native sulphur-crested cockatoos, under which he annotated:

> 'In the tightening uniformity of global culture it is our unique indigenous natural heritage which reminds us of not only the of brilliance and beauty of difference, but also the value of protecting and cherishing what is true to this continent...' (The Age, 22 January 2005, p. 1)

Innovative uses of modern visualisation technology have an important part in re-imaging new biocultural landscapes. To be realised, the technologies must have a realism based on local species, a purposeful view of landscapes based on ecological insights and future climate change.

23.3 Climate Change Predictions and Impacts in South-eastern Australia

Future regional climates are difficult to predict with certainty, but some trends are evident. Increases in CO_2 and changes in the spatial and temporal distribution of climatic variables (temperature, precipitation, etc.) will induce changes to a range of biological assets and ecological processes already severely fragmented and degraded through past land use (NLWRA 2002). Under a high warming scenario, temperatures are projected to increase and by 2070, most of south-eastern Australia (particularly the State of Victoria) will receive 10% to 20% less rainfall than at present (Suppiah et al. 2007). The drought–fire–flood cycle, a persuasive determinant of Australian environments, will accelerate.

Such climate change will also have direct and visual effects on other important environmental drivers in the Australian landscape — stream flows and fire regimes (Hennesey 2005; Jones and Durack 2005; Mansergh and Cheal 2007). These changes will have major effects on the spatial distribution and extent of all life, including resource based industries, such as agriculture, which will have to adapt as the basic biological productivity changes spatially (e.g. water availability). Manifestations of these changes that can be visualised include patterns and processes of land use; structure and function of ecosystems, changing vegetative community structure, the spatial distribution of species and communities, and restoration of connectivity (Lau et al 2006; Mansergh and Cheal 2007).

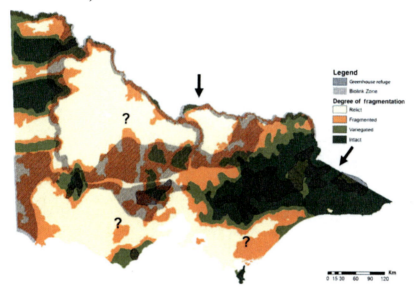

Fig. 23.1. Victorian biolinks and refugia overlaying the degree of fragmentation of native vegetation. Arrows indicate direction of climate-induced movement; question marks indicate potential biolink zones (Bennett et al. 1992; Brereton et al. 1995; fragmentation mapping DSE data, Mansergh and Cheal 2007)

23.3.1 Climate Change and the Need for Ecological Connectivity

Discussing climate change in a recent global review of ecological corridors, Hilty et al. (2006) found it 'hard to imagine any realistic alternative [to large scale linkages] that would be conducive to species persistence'. Opdam and Wascher (2004), in analysing the synergenetic effects of habitat fragmentation and climate change on biodiversity, called for 'bold connectivity zones' and Stern (2006) alluded to them in a economic analysis of adaptation to climate change. Earlier modelled analyses of the impacts of climate change on south-eastern Australian species indicated broad-scale movement of biota, generally to higher latitudes (southwards) and higher altitudes with species vulnerability within the global range of 15% to 37% at risk of extinction. Climatic refugia were also identified. It was concluded that adaptive strategies must include large-scale restoration of ecological connectivity — biolinks (Fig. 23.1) — a new land use to allow space for biota to recolonise and migrate (Brereton et al. 1995; Damschen 2006; Newell et al. 2001; NRMMC 2004; Thomas et al. 2004).

23.3.2 Biolink Zones in South-eastern Australia

The overall objective of biolinks is to optimise the biota's capacity to self-adapt through restoration of habitat heterogeneity, permeability and connectivity across the landscape for multiple species. Traditional 'wildlife corridors' join A to B by a continuous band of more or less homogeneous vegetation generally for a single 'icon' species with harsh boundaries between adjoining land uses (Mansergh et al. 2005). Biolinks may contain such corridors but these zones would feature mosaics and patchworks of native vegetation over much larger areas (100s of kilometres long and 10,000 km^2 in area, Fig. 23.1) where people would live and work.

Within the zones (post-traditional agricultural), the overall density of native vegetation may be greater than 50% with regeneration of long-lived eucalyptus (>150 years) providing a variety of micro-climates and micro-habitats for other biota, for example, tree hollows (Vesk and Mac Nally 2006). The resultant vegetation communities and habitats will not be exact replicates of the past, rather, novel variants responding to the changed climatic conditions. In short, they represent new 'biocultural' landscapes, neither 'conserved' national park nor 'European-like' agriculture. Biolinks would be consciously created and many zones correlate with emerging compatible land-use trajectories (e.g. Amenity Landscapes of Barr 2005). Over time, these zones will look different at all scales. In the course of exploration of biolinks amongst a variety of interested parties (landholders, scientists, policy formulators) it became apparent that visualisation techniques could be used to explain the necessary context (landscape legacy), concepts (ecological connectivity), effects of climate change, and what future landscapes could 'look' like.

23.3.3 Visualisation Tools for Explaining the Context of Biolinks

The potential new land use, biolink zones, have had to be explained to a range of people — both the 'big' picture and implications at smaller visual scales. At a macro statewide level, temporally modelled *Landsat TM* satellite images, including a compilation at pre-settlement (early 19th century) images, allow the visualisation of land use changes since European arrival to present (Fig. 23.2).

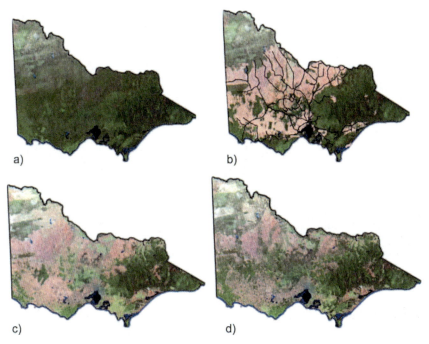

Fig. 23.2. Modelled satellite images of Victoria around (**a**) 1800, (**b**) 1900, (**c**) 2000 (actual) and (**d**) 2100 with revegetation concentrated within biolinks

23.3.4 Visualisation of Environmental Change at a Site over Time

Just as the modelled satellite images provide historical context at the broad scale, early aerial photographs provide a specific place-based visual benchmark in time. The first comprehensive aerial photograph series over the Australian State of Victoria occurred in the 1930s and 1940s, approximately midway between the first wave of large-scale clearing and the present (Mansergh et al. 2006). These images provide a strong basis for comparison to the present and can be overlaid with recent remotely sensed images and data (e.g. LiDAR – Light Detection and Ranging). This also allows the use of three-dimensional geographic landscape models that allow visualisation at the site level. The advantage of such graphic representation is not only its realism but also ability to illustrate the fourth dimension —

time (Fig. 23.3). Decades of past change and *potential futures* can be explored in a relatively short space of time.

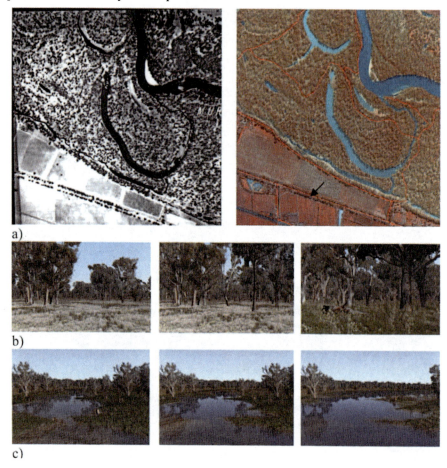

Fig. 23.3. (**a**) Modelling changes in river red gum forest data in 1941 (aerial photo) and 2005 (satellite and LiDAR), note regeneration on sand banks. Arrow indicates roadside site of Fig. 23.4 (**b**) Site in upper landscape showing 1941, present and 2100 views in a virtual landscape as forests mature. (**c**) The virtual landscape can accurately show water levels under any flow regime (see text)

The Murray River is the longest river in Australia, and the Murray–Darling Basin dominates the zone of intensive agricultural production in eastern Australia. The river and surrounding flood dependent red gum forest are a vital east–west biolink at the continental scale and a major ecological transition zone between arid and cool–wet climate zones of Australia (DCE 1992; VEAC 2006; Fig. 23.1). On the basis of spatial information (location and height of all trees), three-dimensional landscape models were built for red gum forest sites, going backwards to 1941 and forwards in time to what it could become if managed as a biolink (Fig. 23.3). Different growth stages of the forest look different and provide different ecological attributes.

Apart from illustrating change in the forest, such realistic three-dimensional models can be used to communicate the ecological effects of changes of climate:

- seasonal flooding on plant condition and distribution
- changes in composition and distribution of animal habitat (e.g. tree hollows, Fig. 23.3).

Realistic geographic visualisation also allows evaluation of the effects of potential policies critical to climate change adaptation and biolinks (e.g. C-sequestration in native vegetation or plantations, management of roadside vegetation). Linkages to ecological datasets allows attributes (e.g. habitat connectivity, species persistence, carbon sequestration) to be quantified. Policy consequences, including sense of place, can be accurately 'seen' and evaluated in virtual reality (Fig. 23.4).

Fig. 23.4. Roadsides contain remnant trees that are critical to biolinks and 'sense of place'. Red gum and box trees on roadside site (indicated by arrow in Fig. 23.3a), showing 2005 photograph (**a**); in virtual geographic landscape (**b**); with native vegetation protected and poplar plantation removed (**c**); and, native vegetation removed (**d**)

Although highly degraded at present, riparian vegetation provides crucial habitat refuges and a network of arteries within and feeding into biolink landscapes (Mac Nally et al. 2000; SRWSC 1983). By taking a real Victorian landscape, presently denuded of native vegetation, the riparian zone can be progressively 'revegetated' in virtual reality and can be viewed from various perspectives, scales and time periods (Fig. 23.5). The three-dimensional view is recognisable to stakeholders, including landholders and neighbours who can evaluate which future alternative view conforms to their sense of place. The 'realism' of this view is dependent upon the visualisation using recognised species and vegetation — a realistic library of images. As these early examples demonstrate, visualisation offers an exciting medium for exploration of biodiversity and adaptation to climate change. To fully realise the potential of visualisation we now examine linking further developments through geographic databases and ecological research.

Fig. 23.5. Real Victorian landscape (DEM) in virtual reality, allowing visualisation of revegetated riparian corridor. (**a**) Landscape with a background of mosaic on topographically diverse hills. (**b**) A three-dimensional landscape model allows detailed scale

23.4 Realism behind Visualisation Technology

Landscape visualisation has been used in simply explaining complex ecological issues vital to the establishment of biolinks, for example, vegetation condition (Lau et al. 2006; Mansergh et al. 2006) and the legacy of habitat fragmentation (Fig. 23.2). There is a need to further develop the technology and link it to the evolving sciences of ecological connectivity and catchment management overall. The GIS-modelling environment is an important component of this research to which the visualisation may be linked. We see visualisation becoming an integral component of research into landscape change. New visualisation techniques potentially provide a method to 'see' landscape implications of scenarios (GIS-derived) and policy responses, and rapidly modified these scenarios with new information.

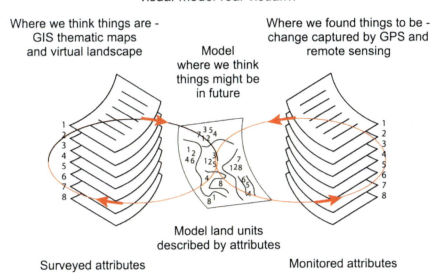

Fig. 23.6. Scheme represents the flow of information essential for planning and implementation of biolinks. The layers on the left represent spatially the current state of land attributes. The attributes become a subject of forward modelling allowing for exploration of different planning scenarios — the middle layer. Layers on the right represent spatially the state of same land attributes after a period of time. The feedback loop is shown by arrows. To be effective in planning and implementation of biolinks, geographic landscape visualisation needs to be a part of information loop with stake holders' feedback

There are two relevant elements to this realism:

- the realism of the underlying ecological analysis and future climatic change (which requires the visualisation medium to be connected to GIS modelling and analysis of real biophysical data)
- the realism embedded in the end product (which requires an extensive three-dimensional library of regional relevant images of species and communities).

As new specific knowledge of ecological connectivity becomes available, this provides a basis for modelling with GIS to provide alternative spatial configurations from the theoretically optimal to those with lesser functionality. To be effective in biolink planning and establishment, geographic landscape visualisation needs to be part of this information monitoring loop (Fig. 23.6). The loop links landscape surveys, predictive scenario modelling and monitoring surveys in a flow of information that is best represented, due to its complexity, by visual means (Asner et al. 2007). Emerging sophisticated catchment modelling (Eigenraam et al. 2005) allows quantification of many ecosystem services, including biodiversity, water and carbon sequestration. A cross-disciplinary approach to capture landscape complexity in digital models allows a multitude of scenarios each of which will look and perform differently under climate change. These developments are progressing, however, the realisation of portraying biolink landscapes, native vegetation and indeed a 'local sense of place' is severally inhibited by the lack of a three-dimensional electronic library of native Australian flora.

23.5 Realism at the Front End

Planning for ecological connectivity and habitat diversity (e.g. age classes) will manifest itself differently in different landscapes and have different visual impact over time. Most aspects are theoretically capable of rapid and realistic three-dimensional visualisation. However, the lack of a full range of visual objects and modelling structures necessary to drive the creation of images hinders the process of landscape visualisation, specifically for biolinks where local species are imperative.

Trees (long-term ecological assets) have different visual impacts at different stages in their life cycle — seedling, sapling and young tree to maturity — with the ecologically important tree hollows developing after 100 to 120 years (Manning et al. 2006; Fig. 23.3 and Fig. 23.8). Under climate change, the understorey and ground cover may evolve to take different forms, or the modelling may show that components of the canopy itself may change. Visualisation is potentially able to show the visual and ecological effects of fire and recovery of habitats, consequences of management regimes, risk factors (weed invasion) and stochastic events (flood and fire). If these are to be 'seen' at site-scale the images assist landholders' and the community's knowledge of the patterns and processes of future local environments.

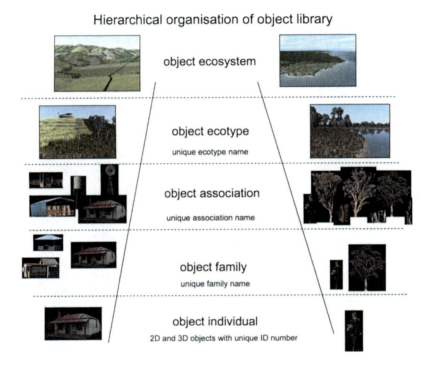

Fig. 23.7. An example of a hierarchical organisation of two- and three-dimensional objects. Well-designed storage of landscape objects facilitates an efficient retrieval and allocation of appropriate objects in a virtual space. Organising individual uniquely numbered objects into hierarchically structured groups that mimic natural patterns of the object's occurrence in real landscapes ensures populating of virtual landscape becomes more reliable, allowing for better quality control. A well-designed database of objects is an essential part of geographic landscape visualisation

Currently most environmental planning and management projects consider the use of landscape visualisation as an extraneous addition. To fully benefit from seeing the past, present and possible future of a landscape under climatic changes, it is critical to develop the potential of visualisation so that it can be embedded in the framework and output of projects. This, in turn, can improve stakeholder involvement in planning and implementation feedback. Realistic visualisation of spatial habitat heterogeneity requires a complex range of objects and entities that change over time. Thus it is also essential to invest in incrementally building a hierarchical object library (Fig. 23.7), so that the construction of a realistic visual landscape model is possible. A hierarchical structure will ensure a much faster retrieval of groups of objects that represent complex entities such as vegetation communities (Ecological Vegetation Classes — EVC — can contain more than 100 species). Also it is practical to maintain two-dimensional and three-dimensional objects of the same landscape element in varying file types and sizes to aid in rendering time (Fig. 23.8).

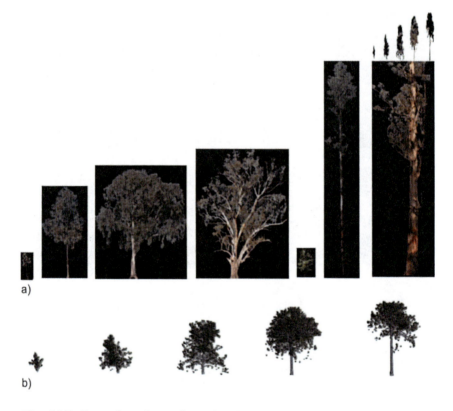

Fig. 23.8. Examples of two-dimensional *Eucalyptus* spp. objects (**a**) and three-dimensional volumetric objects needed for Australian plant library (**b**) showing trees in different serial stages. Both two- and three-dimensional types representing different ages of plants need to be prepared and stored (three-dimensional volumetric tree model produced by *Bionatics*)

In the development of biolinks, geographic visualisation has the potential to play a powerful augmenting role to spatial analysis. It will help to close communication gaps and promote debate between scientists, policy makers and land managers all of whom will need to be active participants in the planned biolinks.

23.6 Future Directions

Geographic visualisation, as evidenced in Al Gore's movie *An Inconvenient Truth,* is a powerful tool for explaining impacts of climate change. A crucial next step is to provide the capacity to fully use visualisations in the place-based adaptation debate. Although uncertainty is inherent in climate change modelling and related impacts, the broad trends on biodiversity and land use are potentially profound. Adaptation to some amount of inevitable global warming will become increasing more importance. As an adaptive response, biolinked landscapes represent a major new land use at the continental scale (Fig. 23.1). Building new biocultural landscapes will require social debate. What will our future landscapes look like under climate change, and what can we do to adapt and conserve the elements we value?

New forms of visualisation offer exciting possibilities to inform this debate. The scientific imperatives of restoring ecological connectivity with biolinks need to be within the domain of the social landscape or 'sense of future place'. A major gap in visualisation being able to fulfil this role is the lack of an accessible, effective three-dimensional image library of the diversity of Australian biota, particularly plant species. Realistic future visualisation of local–regional landscapes, using only foreign images creates, at best, false impressions and, at worst, cartoon-like landscapes.

The last twenty years has seen rapid evolution of the main geographic tool — GIS. A noticeable change in GIS developments is the rise of three-dimensional ability (Asner et al. 2007) from gimmicky addition to a more sophisticated volumetric tool, allowing viewing of maps and images to drape on digital terrain models. Additionally, the process of three-dimensional developments is aided by popular demand for more realistic representation of landscape features, largely thanks to the rapid development of *Google Earth* and the game industry. Using ecological research, geographic landscape visualisation is an excellent tool for illustrating patterns of changes in condition and composition of flora and restoration of vegetation over time (Parkes et al. 2002; Fig. 23.3). Looking to the future of landscape visualisation it is possible to imagine individual virtual landscape objects with programmed physical and behavioural properties. These would have an important 'educative' factor in adaptation to climate change. A virtual three-dimensional tree, herb or grass could have information about its size, seasonal appearance, occurrence in the floristic associations and sensitivity to climate parameters. The rules of allocation of such objects during the construction of a virtual landscape could have better quality checks in terms of correct location and association. More importantly, spatial allocation of objects could be a stochastic modelling exercise in a real-time display of a virtual landscape.

However, well chosen narrative remains the most vital component of landscape visualisation. This is often overlooked when technological pursuits dominate visualisation efforts. In the development of planned biolinks, geographic visualisation has the potential to play a powerful augmenting role to spatial analysis. It will help to close communication gaps and promote debate between scientists, policy makers and land managers all of whom will need to be active participants in the planned biolinks.

23.7 Conclusion

Inevitable climate change will initiate major land use change and have large effects on biodiversity. Re-evaluation of what we want future land-scapes *to be* and *to provide* is an important societal question. Adaptation to climate change requires us to re-envisage landscapes. Early examples of visualisation have shown it to be effective in communicating the problem of biodiversity under climate change and potential solutions. Biolinks is a new land use — a risk management strategy for biodiversity that requires new forms of landscapes to be developed. Scientific research (e.g. spatial manifestation of 'ecological connectivity') will continue to evolve (and be informed by climate science) with results increasingly derived from mod-elling within a GIS framework. Digital landscape visualisation is a power-ful communication medium to synthesise and add meaning to these results. These qualities can be further enhanced through progress in the digital simulation of landscape objects. The digitally-created landscape — which displays physical properties of the real-world related to ecological meaning — has the potential to become a powerful medium for experimenting with different management approaches and land use policy development, as well as informing public debate. Such technologies will be increasingly needed for appreciating land use issues that are the result of the impacts of a changing climate.

Acknowledgements

We are grateful to assistance of Fiona Ferwerda (Fig. 23.1), Nevil Amos (Fig. 23.2), Fiona MacKenzie, Gordana Marin and many others in DSE that provided encouragement and criticisms.

References

Asner GP, Knapp DE, Kennedy-Bowdoin T, Jones MO, Martin RE, Boardman J, Field CB (2007) Carnegie Airborne Observatory: in-flight fusion of hyperspectral imaging and waveform light detection and ranging (wLiDAR) for three-dimensional studies of ecosystems. Journal of Applied Remote Sensing, Vol. 1, 013536 (13 September 2007)

Barr N (2005) The changing social landscape of rural Victoria. Department of Primary Industries, Melbourne

Brereton R, Bennett S, Mansergh I (1995) Enhanced greenhouse climate change and its potential effect on selected fauna of south-eastern Australia: a trend analysis. Biological Conservation 72:339–354

Clark J, Whitelaw B (1985) Golden summers, Heidelberg and beyond. Interntional Cultural Corporation of Australia, Sydney

Damschen E, Haddad N, Orrock J, Tewksbury J, Levey D (2006) Corridors increase plant species richness at large scales. Science 313:1284–1286

DCE (1992) Draft flora and fauna guarantee strategy. Department of Conservation and Environment, Melbourne

DSE (2006) Climate Change in Victoria: a summary. Department of Sustainability and Environment, Melbourne

Eigenraam M, Stoneham G, Beverly C, Todd J (2005) Emerging environmental markets: a catchment modelling framework to meet new information requirements. In: Proceedings of the OECD Workshop on Agriculture and Water Sustainability, Markets and Policies. 14–18 November 2005, Adelaide

Hennessy K, Lucas C, Bathols J, Nicholls N, Suppiah R, Ricketts J (2005) Climate change impacts on fire weather in south-eastern Australia. CSIRO and Bureau of Meteorology, Melbourne

Hilty JA, Lidicker WZ, Merenlender AM (2006) Corridor ecology: the science and practice of linking landscapes for biodiversity conservation. Island Press, Washington

IPCC (2007) The physical science basis: summary for policymakers. International Panel on Climate Change, United Nations Environment Program, Paris

Jones R, Durack P (2005) Estimating the impacts of climate change on Victoria's runoff using hydrological sensitivity model. CSIRO and Department of Sustainability and Environment, Melbourne

Lau JA, Amos N, Parkes D, Mansergh I (2006) Imagining the future: visualising sustainable landscapes – 'a picture is worth a thousand words'. International Landcare Conference, 8–11 October 2006, Melbourne

Leunig M (2005) The Age, 22 January 2005, p 1, available at www.theage.com.au

Mac Nally R, Soderquist TR, Tzaros C (2000) The conservation value of mesic gullies in dry forest landscapes: avian assemblages in the box-ironbark ecosystem of southern Australia. Biological Conservation 93:293–302

Manning A, Fischer J, Lindenmayer D (2006) Scattered trees are keystone structures – implications for conservation. Biological Conservation 132:311–321

Mansergh I, Cheal D (2007) A contribution to protected area planning and management for eastern Australian temperate forests and woodland ecosystems under climate change – a landscape approach. In: Protected areas: buffering nature against climate change: Proceedings of a WWF-Australia and IUCN World Commission on protected areas symposium, 18 June 2007, Canberra, pp 58–72

Mansergh I, Anderson H, Amos N (2006) Victoria's living natural capital – decline and replenishment: 1880–2050 (Part 1 & 2). Victorian Naturalist: 123:4–28, 288–322

Mansergh I, Cheal D, Amos N (2005) Biolinks: the Journey. In: the great greenhouse gamble – A NSW Nature Conservation Council Conference, 15–16 September 2005, Powerhouse Museum, Sydney. Available online at www.nccnsw.org.au

Newell G, Griffioen P, Cheal D (2001) The potential effect of 'greenhouse' climate warming scenarios upon selected Victorian plant and vegetation communities. Arthur Rylah Institute for Environmental Research, Melbourne

NLWRA (2002) Australians and natural resource management 2002. National Land and Water Resources Audit, Canberra

NRMMC (2004) National biodiversity and climate change action plan 2004–2007, National Resource Management Ministerial Council. Department of Environment and Heritage, Canberra

Opham P, Wascher D (2004) Climate change meets habitat fragmentation: linking landscape and biogeographical scale levels in research and conservation Biological Conservation 117:285–297

Parkes D, Newell G, Cheal D (2003) Assessing the quality of native vegetation: the 'habitat hectares' approach. Ecological Management and Restoration 4(S):29–38

Root T, Price J, Hall K, Schneider S, Rosenzweig C, Pounds J (2003) Fingerprints of global warming on wild animals and plants. Nature 421:57–60

Schama S (1998) Landscape and memory. Vintage Publications, New York

Smith G (1995) Arthur Streeton, 1867–1943. National Gallery of Victoria, Melbourne

Soule M, Mackey B, Recher H, Williams J, Woinarski J, Driscoll D, Dennison W, Jones M (2004) The role of connectivity in Australian conservation. Pacific Conservation Biology 10:266–279

SRWSC (1983) The state of the rivers: Victoria, Australia. State Rivers and Water Supply Commission, Melbourne

Stern N (2006) The Stern Review: the economics of climate change. Cambridge University Press, UK

Thomas C, Cameron A, Green R, Bakkenes M, Beaumont L, Collingham Y, Erasmus B, de Siqueira M, Gralnger A, Hannah L, Hughes L, Huntley B, van Jaasvel A, Midgley G, Miles L, Ortega-Huerta M, Peterson A, Phillips O, Williams S (2004) Extinction risk from climate change. Nature 427:145–148

Umina P, Weeks A, Kearney M, McKechnie S, Hoffmann A (2005) A rapid shift in a classic clinal pattern in Drosophila reflecting climate change. Science 308:691–693

VEAC (2006) River red gum investigations: descriptive report. Victorian Environmental Assessment Council, Melbourne

Vesk P, Mac Nally R (2006) The clock is ticking – revegetation and habitat for birds and arboreal mammals in rural landscapes of southern Australia. Agriculture. Ecosystems and Environment 112:356–366

24 Visualising Alternative Futures

William Cartwright

School of Mathematical and Geospatial Science, RMIT University Melbourne, Victoria, Australia

Abstract: In Australia, many rural towns have experienced a loss of population in conjunction with the loss of buildings. These buildings can be considered to be key elements for appreciating the heritage of the development of rural living spaces. Demolishing buildings of note has assisted in the demolition of the ability for the current generation of Australians to appreciate the life style in rural towns. This problem has been compounded in peri-rural areas, where urban growth has exerted pressure in the demand for increased space. This can lead to heritage buildings and unique streetscapes being sacrificed to the demands of urban growth. Therefore there is a need to provide tools for better understanding the development pressures on rural and peri-urban areas, and tools as a means to appreciate these now disappeared environments. The use of virtual models of Australian towns can help people to visualise what exists and the pressures to develop. If these towns and environments have already been transformed, the use of virtual models allows people to visualise 'how they might be' if the town had remained intact. This is seen as a potential tool for understanding what we have and the heritage we have lost — showing the current generation what these environments once were, or to demonstrate past errors to decision makers.

Many of these buildings have disappeared without a trace. Buildings removed in the haste to modernise or expand were not always properly recorded (if recorded at all) neither were the buildings measured or photographed for posterity. Therefore, how might visualisations be built from a combination of information about buildings that remain and those that do not? A project that captured images of existing buildings and sourced images, photographs, drawings, paintings and other data from archives and exhibitions, and then developed three-dimensional visualisations that merged all of these inputs was used to build a virtual environment for use to appreciate what built environments are, might be (if developments took place), and could be (if significant buildings were not demolished or removed).

This chapter describes three projects: one that built a virtual environment for the peri-urban township of Barwon Heads, Victoria, Australia (built for presenting planning development proposals to the residents of the township); an exploratory project in the City of Melbourne, Australia; and a project dealing with proposed urban landscapes in Sydney Road, Brunswick, Australia. As the Barwon Heads virtual environment presented the existing situation only, there was no way for users of the system to fully appreciate what would be lost if 'breakout' development was allowed to occur that would forever destroy the unique characteristics that make up the fabric of this Australian township. The City of Melbourne project provided a composite three-dimensional model by combining imagery of existing heritage buildings with the 'missing' buildings of the city — those that were removed in the heady development days of the 1960s. The chapter considers whether the methods developed for the City of Melbourne project could be used as a model for developing better virtual environments for peri-urban and rural applications.

24.1 Introduction

The move to three-dimensions in order to experience worlds better, or differently, has been commented upon by Batty (1995). He noted that to study cities in any manner it will be necessary to use diverse methods of computation, varying from the straightforward browsing of digital data to much more sophisticated methods of simulating futures. He said, 'We will, in fact, make a distinction between real cities as viewed using computers and abstract cities as simulated on computers' (Batty 1995, p. 4).

Batty's view regarding developing virtual cities is focused on building virtual cities, but his comments are relevant to the development of visualisations of rural landscapes. These are based on existing geographies and use existing geographical frameworks upon which to build 'other' geographies — alternative geographies.

A number of research linked projects which use Virtual Reality Modelling Language (VRML) for developing non-immersive Virtual Reality (VR) products than can be used by the general public and professional planners to explore what alternative futures might look like were developed at RMIT. These were the:

- Barwon Heads peri-urban development visualisation tool
- Central Business District of Melbourne *What the City Might Be?* prototype
- Sydney Road prototype.

The aim of these projects was to explore the use of non-immersive three-dimensional virtual worlds to ascertain their effectiveness in communicating information about actual environments, those that might have been if buildings had remained intact, and possible future scenarios.

24.2 The Barwon Heads Peri-urban Development Visualisation Tool

Building and evaluating previous models of inner urban areas using VRML provided much information about how to develop non-immersive virtual reality products for web delivery. Evaluations carried out with prototypes with community stakeholders illustrated the usefulness of such web-delivered tools (Cartwright et al. 2005). This project was built on the foundations of previous products and had to be:

- built using open standards software VRML
- web-delivered
- usable on relatively modest computers with a web browser with a VRML plug-in.

The study area of Barwon Heads (Victoria) was chosen, as it is a typical peri-urban township that is under pressure by developers to expand. It is located on a site surrounded by the Barwon River, the ocean and wet areas, as can be seen by the aerial photograph in Fig. 24.1. It can be seen as an island township, almost surrounded by water. The Planning Scheme, managed by the City of Greater Geelong, limits township growth to ensure the unique attributes of the town, which attracted residents to it in the first place, would not be spoilt by uncontrolled expansion. However, as is typical with coastal townships, there is much pressure by developers to expand. In 2005 the Lechte Corporation proposed to the City of Greater Geelong a plan to develop the Sea Bank Estate at the edge of the town. This was a 150-lot subdivision of the last large parcel of undeveloped land (Pipers Mills Oakley 2005). The site is shown by the rectangle in Fig. 24.1.

Fig. 24.1. Aerial photograph of Barwon Heads township

As there was much debate about the project in the town of Barwon Heads, it was decided to build a prototype product that could be used to support community consultation. One problem that community members have when considering planning applications that affect them is the lack of access to sophisticated tools. They rely mainly, if not solely, on paper maps and associated products, like aerial photographs and architectural drawings.

VRML was used as a development tool due to existing in-house expertise in using the scripting language to develop virtual models of parts of cities. VRML provides open extensible formats to be used and the 'built' worlds could be constructed in web browsers that included a VRML plug-in. VRML is an extensible interpreted language and it became an industry-standard scene description language. It is used for three-dimensional scenes, or worlds, on the Internet. To produce three-dimensional content two-dimensional components are defined or drawn, and the viewpoint specified. Once defined, the drawing package renders the three-dimensional image onto the screen. VRML code defines objects as frameworks that are rendered. This makes file sizes very small. The appearance of rendered surfaces can also be modified using different textures. By using the computer's fast processing speeds, and specifying multiple, sequential viewpoints 'walk-throughs' or 'fly-throughs' can be produced.

It was decided to model the 'strip' of houses that were located directly opposite the proposed development site (Fig. 24.2), as well as the development site itself (Fig. 24.3). All houses in the strip were photographed and VRML models built of each (Fig. 24.4). All vegetation and street furniture was photographed and representative models built of every item. The idea was to provide a tool that would enable residents to locate their own property and then to 'see' what the completed development 'across the road' would be like.

Fig. 24.2. Looking north along Golflinks Road in the Barwon Heads VRML world. The area to be subdivided is at the left of the image

Fig. 24.3. Looking north in the development area in the Barwon Heads VRML world (Golflinks Road is at the right)

From the evaluation of a model developed previously of Sydney Road, Brunswick (Cartwright et al. 2005), it was known that community users of these packages want detailed visualisations of buildings. Therefore much effort was made to build faithful representations of houses in the model. Figure 24.4 shows one of the houses in the model.

Fig. 24.4. Model of a house on Golflinks Road, Barwon Heads (looking southwest)

The model was completed and plans for evaluation formalised. As can be seen from the figures above, the development area was populated with existing buildings and a communications tower. The next step was to complete the model by adding numerous completed project houses, like those shown in Fig. 24.5.

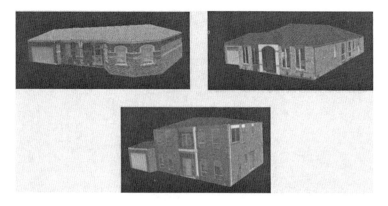

Fig. 24.5. Models of project houses

However, the plan for development received almost 'immediate' council approval. Therefore the tool was not utilised in the decision-making process. However, it did provide generous information about how the processes for developing community collaboration support tools may be used for exploring peri-urban areas.

Constructing the model did show that a useful tool could be provided relatively quickly. However, to provide an enhanced picture of the unique characteristics of such peri-urban townships it might be useful to make heritage buildings more prominent and to add missing buildings, so as to illustrate how these townships had already been visually degraded by the removal of older buildings. A virtual environment had already been built to explore the potential of combining missing buildings with existing ones (Cartwright 2006). This seems to be a technique that could be applied to build more appropriate representations of rural and peri-urban towns. This project is described in the following sections.

24.3 The Central Business District of Melbourne *What the City Might Be?* Prototype

A prototype model was built of part of the Central Business District in Melbourne (Australia) to test whether virtual environments could depict an alternative scenario of a city that would exist if significant buildings still remained. A model was built to fuse together existing buildings of significance with missing buildings — those removed hastily in the building boom that occurred in Australia in the 1950s, 1960s and early 1970s. The aim of producing the model was to develop a simple, accessible demonstration prototype that could be used to demonstrate the potential that three-dimensional simulations provide for better appreciating what the city might have been if all significant buildings remained intact. The prototype was used as a test bed to develop techniques for sourcing imagery (current and historical), capturing and processing images of standing buildings, and actually building the model using VRML and creating a usable web-delivered product.

24.3.1 Marvellous Melbourne

This section provides information about 'Marvellous Melbourne', when the city was built upon the prosperity of the Gold Rushes in the post-1860s (Davidson 1978, 2004). Melbourne was established on the banks of the River Yarra in 1835 by settlers from Tasmania — formally the Port Phillip

Association — and led by John Batman. Soon after, in 1837, it was laid out in a formal grid pattern by Robert Hoddle (City of Melbourne 2006).

By 1860 the city had reached its final form (City of Melbourne 2006) and during the 1880s the population was almost 500,000 — rivalling the size of many European cities at that time (see Statham 1989 regarding Melbourne's development). It had ornate office buildings up to 12 storeys high that were compared to those of New York, London and Chicago (Museum of Victoria 2004). The city boomed from 1880 to 1890, fired by the massive amounts of gold being discovered in Victoria (Melbourne being its capital city) and general business optimism of the time. Streetscapes like that shown in Fig. 24.6 displayed this wealth, with grand buildings lining most streets. The city was alive with commerce and frenzied building, and re-building, activities.

This activity came to an abrupt halt in 1891, and by 1892 many banks and building societies closed — by mid-1892, 21 financial companies were in suspension (City of Melbourne 2006). However, the grand city buildings had been established as a record of what wealth and pure optimism had achieved.

Fig. 24.6. Collins Street looking east from Elizabeth Street in Melbourne during the late 1880s. Photographer: Charles Rudd (Source: State Library of Victoria)

24.3.2 Melbourne and the Removal of Significant Buildings

In 1956 Melbourne hosted the Olympic Games. Post-games the redevelopment of the city began — during the 1950s, 1960s and the early 1970s many landmark buildings, most irreplaceable, were removed by wreckers. The boom demolition era was between 1962 and 1971 (Annearer 2005). Examples of grand buildings demolished during this period included the:

- Equitable Life Assurance Society building, on the corner of Elizabeth and Collins streets (Fig. 24.7a)
- Federal Hotel (Fig. 24.7b)
- Colonial Mutual Life building (Fig. 24.7c).

These three buildings are just a few examples of the architectural richness of Melbourne that has disappeared.

Fig. 24.7. Examples of grand buildings demolished: (**a**) The Equitable Life Assurance building demolished in 1959 (Source: *The Age*); (**b**) The Federal Hotel (Photographer: Wolfgang Sievers; Source: National Library of Australia); (**c**) Demolition of the Colonial Mutual Life building, 1960 (Source: Museum of Victoria)

Whilst widespread demolition no longer occurs, there still exists the potential for government and developer alike to covet sites upon which historic buildings sit. As recently as 2004 some of Melbourne's historical buildings were under threat (Rundle 2004), hence the interest in developing a model that could be used for better appreciating the existing, and lost, buildings and streetscapes.

24.3.3 Prototype World

The prototype model covered two city blocks: Block 1 bounded by Swanston Street, Flinders Street, Russell Street and Flinders Lane; and Block 2 bounded by Swanston Street, Flinders Lane, Russell Street and Collins Street. These blocks were chosen as they are typical, insofar as they contained a number of significant buildings that had not been removed in the haste for redevelopment and a number of 'holes' that needed to be filled with sourced historical photographs. Typical of the buildings that escaped the wrecker's hammer in the study area is the Regent Theatre and the Temperance and General Building. The site chosen was significantly large enough to provide a challenging area to model, but not so big that it would be unworkable. That is, the process of taking many images, cleaning them (removing objects like trees, cars and people that are captured in the image that, if not removed, would appear as a two-dimensional image in the three-dimensional world, thus degrading the perception of the Virtual World) and then stitching them together to form one image should not take too much time to complete. The aim was to develop research and production techniques that could be extended to cover the whole of Melbourne's Central Business District at a later stage.

The first step was to undertake a survey of the area to record where significant buildings still stand. Two maps were produced, one showing the situation in 1960 and the other for late 2005. Next, the buildings of note still standing in the study area were surveyed to ascertain position, use and building height. Each building façade was photographed for use in 'stitching' the images onto the sides of VRML primitive shapes. The next stage was to develop the three-dimensional model with VRML.

Fig. 24.8. The Temperance and General Building composite image

Some of the images that would be included as textures in the VRML world were stitched from numerous individual images. As the buildings are multi-storey, the images needed to be further manipulated to remove the scale distortions throughout the image. For example, the Temperance and General Building comprised a composite of many images. The composite image is shown above (Fig. 24.8) and how it appeared in the final world can be seen in Fig. 24.9.

Fig. 24.9. Detail of the Temperance and General building as seen in the VRML model

As the images sourced from historical archives were black and white, the final process was to convert colour images to grey tone images. Whilst this conversion causes the VRML world to lose some impact, this process ensured that a uniform model resulted. Existing buildings could perhaps be shown in colour, and the removed buildings illustrated in black and white. This combination will need to be evaluated for its potential to provide more useful information.

The major change to the study area is at the north-west corner (corner of Swanston and Collins streets). This area, now mainly open space was once clothed in buildings — the Queen Victoria Buildings, City Club Hotel and Cathedral Hotel were demolished in the 1960s. The Queen Victoria Buildings were demolished for the development of the City Square (Davidson 2004). Also lost were the lanes and associated alleyways which once separated these buildings — Queens Walk, Quirk Alley, Argus Alley and Regent Place. These ceased to exist after building demolition and subsequent land parcel consolidation enabled the construction of the first City Square, which was later removed and redeveloped.

A number of the buildings in the study area were gutted and rebuilt with the façade of the old building left on the outside of a modern interior. A few buildings have been demolished and replaced with modern designs. These sites were the focus for uncovering historical photographs to fill the gaps in the model's streetscapes. The oldest buildings have not been demolished, but the façades have been changed so often over the years they are now unrecognisable as being from the 1850s, except for their lack of height which remains at two storeys. Similarly, most of the street level shop fronts have been modernised even if the building façades have not. St Paul's Cathedral did not have spires until 1926 and was not as imposing as it is today. The model was built with the spires included.

The 'missing' building sites, such as the Queen Victoria Buildings (on the corner of Collins and Swanston Streets, inside the box in the photograph in Fig. 24.10) were mapped and building images retrieved from various sources. These included the State Library of Victoria, the Museum of Victoria, the National Library of Australia, the City of Melbourne, *The Age* newspaper archives and a number of web sites that focus on Melbourne.

Fig. 24.10. Swanston Street looking south from Little Collins Street. Photographer: Kerr Brothers Photographic Studio 1915. (Source: State Library of Victoria)

Fig. 24.11. The model, looking towards the south-west from the north-east corner of the study area

All existing and missing buildings in the study area were subsequently inserted into the model. The completed world is shown in Fig. 24.11. Many of the source images were low resolution, but were used, as no alternatives were available. Below are sample images of the Queen Victoria Buildings — only 642 x 423 pixels in Fig. 24.12a and 125 x 235 pixels in Fig. 24.12b.

Fig. 24.12. Queen Victoria Buildings: (**a**) image is 624 x 423 pixels, (**b**) image is 125 x 235 pixels (Source: Walking Melbourne)

Looking at the Queen Victoria Buildings specifically, to be usable a number of images had to be manipulated and joined to produce a composite image. Each 'component' of the image had to be manipulated to remove as much distortion as possible. The VRML model texture dictated that 'front-on' photography was used, however as these images did not exist a 'best fit' path was followed. Whilst unsatisfactory with respect to image quality and integrity, the non-existence of alternative images made this the only path of operation. Figure 24.13 shows how this composite image appears in the VRML model. Further research needs to be done to ascertain whether this image is acceptable as part of the complete model (Fig. 24.11), or if further processing is necessary. (The Queen Victoria Buildings are at the bottom right of the model depicted in Fig. 24.11.)

Fig. 24.13. Composite image of Queen Victoria Buildings as it appears in the VRML world

24.3.4 Initial Impressions

The model works as a tool to illustrate how the study area would look if all significant buildings remained. It works effectively using the combination of *Microsoft's Internet Explorer* browser and the *The BitNet Management* VRML browser plug-in. While seeming odd initially, the black and white model allows the city buildings to be adequately visualised. Buildings still standing are easily recognised and their 'rebuilt' neighbours provide information that was hitherto unavailable in a composite model. As can be seen in Fig. 24.11, some of the building images are still to be sourced. When images are found they will be processed and subsequently added to the model. The final verdict about whether the model works needs to be made after the product is evaluated. This evaluation will ask users two questions: Stage 1 of the evaluation process will ask 'Is it usable?', and Stage 2 will ask 'Is it useful?'

The initial evaluation, Stage 1, will be employed to ascertain whether there were any difficulties in its use. This will be undertaken using a qualitative evaluation with an expert group of users to test the tool itself. When evaluating the models developed for previous applications the Reeves and Harmon (1993), the 'User Interface Rating Tool for Interactive Multimedia' was used to make an initial evaluation of the product (Cartwright et al. 2005). Stage 2 of the evaluation will address the usefulness of the product. Here, usability issues will be addressed and used as the basis for formulating qualitative and quantitative evaluation questions.

24.4 Visualising Proposed Landscapes: Sydney Road, Brunswick

Plans to enhance the gateway to Sydney Road were proposed to the Sydney Road Brunswick Association (SRBA) and the City of Moreland by the company Village Well (www.villagewell.org). Their jointly-developed concept for the shopping precinct was: 'Sydney Road is an eclectic destination that offers a diverse and authentic experience of multi-cultural Melbourne.' Village Well developed a number of concepts for signs and banners, and developed a number of scenarios for community comment by placing proposed signage into static photographs.

In discussion with local residents, a number of comments were made about how difficult it was to visualise Sydney Road after it had been 'treated' with the signage and banners developed by Village Well. Typical of these comments was: 'I can't see you need one really. I just can't imagine it. If you showed me a design, it might look quite good but I can't imagine it.' Therefore it was decided to 'import' Village Well's imagery into the Sydney Road section of the existing model and to evaluate its usefulness for discussing potential street treatments.

For this project, 'branding' imagery developed by the SRBA's consultant, Village Well, was imported into the model. This allowed the 'look and feel' of the branding to be visualised by SRBA members prior to accepting the concept. The branding imagery was placed on banners, as shop window decals and advertising on the sides of trams placed in Sydney Road.

Figure 24.14 shows banners above Sydney Road and on light poles inserted in the VRML world. Figure 24.15 shows detail of one of the banners on a light pole and the decals on shopfront windows.

Fig. 24.14. Banner placed in the VRML world, Sydney Road looking north from Park Street

Fig. 24.15. Branding banner placed on light pole and on shopfront windows in the VRML world, east side of Sydney Road just north of Park Street

A video of the model in use was provided to Village Well for presentations to community groups. The usefulness of the banners and other signage were not evaluated.

24.5 Usefulness of the Prototypes

The *What the City Might Be?* prototype was developed to provide a test-bed for developing and evaluating techniques that could be applied to the building of landscapes that contained numerous buildings. Initial impressions were that the model works using a combination of high resolution and low resolution images. The need to seek all images at high resolution was deemed not to be necessary when presenting an area that might contain many missing buildings.

The Sydney Road prototype was developed to learn how proposed developments would impact on an existing streetscape. It was used for community briefing sessions.

For rural and peri-urban applications the techniques could be used to build virtual environments that show not just how a township is now, but also how it might be if many significant buildings remained, and how it might be if certain developments took place. For building built/historic visualisations, access to complete or high resolution imagery cannot always be guaranteed, so low resolution images, scanned paintings or sketches could be used. For future scenarios all that is required is concept imagery that can be imported into the VRML world. Such visualisations can have the potential to not just show what is in the township now, but also the scenarios of a lost heritage and future prospects. A composite tool (showing present, past or future possibilities) can provide a more emotive picture of rural and peri-urban townships which can provide a better picture of what has already been lost and what still might be lost if inappropriate development or re-development occurs.

24.6 Conclusion

This chapter has described the concepts and development techniques used to produce the Barwon Heads peri-urban development visualisation tool. It described the background behind the Central Business District of Melbourne *What the City Might Be?* and the Sydney Road prototypes. It also addressed whether the techniques used to develop the Sydney Road product might be applied to future products like the Barwon Heads visualisation tool.

An evaluation program is continuing to test the usefulness of models that combine existing and missing buildings for providing a tool to support community input into the decision-making process. The development and testing of the Central Business District of Melbourne *What the City Might*

Be? prototype is continuing and other rural and peri-urban townships are being investigated for their potential to be used to further test the techniques developed. The lessons learned developing urban applications can be applied to rural and peri-urban applications. The use of VRML models have the potential to empower community members by provisioning them with tools to better understand alternative futures that development might impose. They also provide the means to better appreciate what has been lost and what townships might look like if they had remained intact.

Acknowledgements

The Barwon Heads and Sydney Roads projects were supported by a VRII (Virtual Research Institutes Initiative) grant from RMIT.

The author acknowledges the production work done on the project by Dane McGreevy, Scott Furey, Carol Farr, Joanna Skorkowska, Frank Seebach, Florian Ploetz, Ercan Kesbir and Tobias Röseneder.

References

Annearer R (2005) A city lost and found – Whelan the Wrecker's Melbourne. Black Inc, Melbourne

Batty M (1995) The computable city. In: Proceedings of the m-squared Conference, University College London. Retrieved 13 February 2001, http://www.geog.ucl.ac.uk/casa/melbourne.html

Cartwright WE (2006) Using 3D models for visualising 'the city as it might be.' ISPRS Technical Commission II Symposium, Vienna: International Society for Photogrammetry and Remote Sensing and Spatial Information Sciences, pp 115–120

Cartwright WE, Pettit C, Nelson A, Berry M (2005) Towards an understanding of how the 'geographical dirtiness' (complexity) of a virtual environment changes user perceptions of a space. In: Zerger, A. and Argent, R.M. (eds) MODSIM 2005 International Congress on Modelling and Simulation. Modelling and Simulation Society of Australia and New Zealand, 12–15 December 2005, Melbourne

City of Melbourne (2006) History and heritage. Retrieved 20 December 2006, http://www.melbourne.vic.gov.au/info.cfm?top=52&pg=702

Davidson G (1978) The rise and fall of marvellous Melbourne. Melbourne University Press, Melbourne

Davidson G (2004) The rise and fall of marvellous Melbourne (revised edn). Monash University Press, Melbourne

Museum of Victoria (2004) Marvellous Melbourne. Retrieved 20 December 2006, http://www.museum.vic.gov.au/marvellous/1880s/index.asp

Pipers Mills Oakley (2005) Briefings autumn 2005. Retrieved 11 April 2007, http://www.millsoakley.com.au/cms/download.asp?moduleID=2070&filenum =1.

Reeves TC, Harmon SW (1993) User interface rating tool for interactive multimedia. Retrieved 19 March 2002, http://mime1.marc.gatech.edu/MM_Tools/ UIRF.html20.

Rundle (2004) Who will save Melbourne from the wrecker's ball? *The Age* 15 March 2004, retrieved 20 December 2006, http://www.theage.com.au/articles/ 2004/03/14/1079199092582.html?from=storyrhs

Statham P (1989) The origins of Australia's capital cities. Cambridge University Press, New York

Photographic sources

Fig. 24.6
Photographer: Charles Rudd. Source: State Library of Victoria, retrieved from http://www.slv.vic.gov.au/pictoria/a/1/6/doc/a16743.shtml

Fig. 24.7
The Equitable Life Assurance Society building. Source: *The Age,* retrieved from http://www.theage.com.au/news/books/wreck-ruin-and-glory/2005/07/31/1122748527204.html.

Federal Hotel. Photographer: Wolfgang Sievers. Source: National Library of Australia, retrieved from http://nla.gov.au/nla.pic-vn3305975-v.jpg

Demolition of the Colonial Mutual Life Building. Source: Museum of Victoria, retrieved from http://www.museum.vic.gov.au/colonial/demolition.asp

Fig. 24.10
Source: State Library of Victoria, retrieved from http://www.slv.vic.gov.au/ pictures/0/0/0/doc/pi000467.shtml

Fig. 24.12
Queen Victoria Buildings. Source: Walking Melbourne, retrieved from http:// www.walkingmelbourne.com/building_profile.php?ID=604

25 Virtual Globes: the Next GIS?

Jean-Philippe Aurambout[1], Christopher Pettit[1] and Hayden Lewis[2]

[1] Department of Primary Industries, Parkville Centre, Victoria, Australia
[2] Department of Primary Industries, Tatura Centre, Victoria, Australia

Abstract: Improvement in the processing power and the graphic memory of desktop computers, coupled with increased Internet accessibility and connectivity has supported the prolific adoption of digital globe technologies by broad sectors of society. The recent release of freely downloadable digital globes, such as Google Earth *and* NASA World Wind, *has sparked an enormous public interest and increased people's awareness of spatial sciences. The ease of use of digital globes and their capacity to display spatial information make them a powerful tool to communicate and make data accessible to a range of users including decision makers, researchers and the general public. As a result digital globes present an enormous potential for the communication of scientific information to a wider audience. The ubiquitous nature of digital globe technologies provides significant opportunity for the science community to communicate information and share the results of often complex models with people who traditionally could not operate or access spatial technologies such as GIS, remote sensing and visualisation products. This paper presents an overview of a range of different digital globes currently available and their underlying structures and features. Through a case study approach we illustrate the strength and weaknesses of five major digital globes (*Google Earth, NASA World Wind, ESRI ArcGIS Explorer, Skyline Globe *and* Dapple Earth Explorer) *and evaluate their potential applicability in the fields of agriculture science, natural resource management and spatial planning.*

25.1 Introduction

Virtual globes, also known as virtual hyperglobes, digital globes, and earth browsers are scale-bound structured models of celestial bodies presented in virtual space in their undistorted three-dimensional wholeness (Riedl 2007). They allow the visualisation of digital images on a three dimensional virtual globe structure. According to Riedl (2007), virtual globes incorporate features and functionality that provide significant advantage over traditional spatial data mapping interfaces, as:

- the earth imagery displayed on a globe structure is free of distortion
- data displayed in virtual globes can be viewed at any scale and from any angle
- virtual globes provide a large degree of interactivity, allowing the user to move to different locations and visualise different type of spatial data.

This technology is a major improvement in the area of spatial data visualisation and may lead towards the development of former US Vice President Gore's concept of a Digital Earth as:

> 'a multi-resolution, three-dimensional representation of the planet, into which we can embed vast quantities of geo-referenced data' (Gore 1998).

The recent release of freely downloadable virtual globes providing access to online imagery repositories has sparked enormous public interest and increased people's awareness of geographical information sciences. The ease of use of virtual globes and their capacity to display spatial information offers a strong potential to communicate spatial data and it is believed that virtual globes could lead to a democratisation of GIS technology (Butler 2006).

By providing underlying high quality satellite and aerial imagery, virtual globes provide an excellent tool to publish spatial information, aid the teaching of earth sciences (Lisle 2006) and communicate and make data and landscape models more accessible to decision makers, the broader research community and the general public.

This chapter reports on research undertaken to assess the potential and capacity of five freely downloadable virtual globes to display and share GIS data with stakeholders in local and regional agriculture and natural resource management (ranging from the farmer to the policy maker). This research critically examines the multitude of virtual globe technology applications currently available, and in particular focuses on five applications (judged as the most suitable) and assesses their respective potential to be used to communicate data relevant to natural resource management.

25.2 Methodology

An initial Internet search to identify existing applications of virtual globe technology (performed between January and April 2007) yielded a list of 41 software products (Table 25.1). Each of these software products was investigated and virtual globes judged to present the most potential to communicate, share and create spatial information were selected for in-depth testing.

The pre-selection process consisted of accessing online applications or downloading them onto a test computer and then evaluating for a limited range of functionality including: navigation, pan, zoom and data import. These pre-selection criteria assured that data could be uploaded onto the virtual globes and visualised from different angles, elevations and locations. Virtual globes that did not provide basic and easy to use navigation functionality or which were restricted to certain locations or did not accept data inputs from the user were rejected outright from further evaluation. This process yielded a list of five virtual globe applications: *Google Earth, NASA World Wind, ArcGIS Explorer, SkylineGlobe* and *Dapple Earth Explorer* (Table 25.1).

Google Earth is a virtual globe software product launched in 2005 by Google™ that maps the earth by superimposing satellite imagery, aerial photography and GIS data over a three-dimensional globe. This software is based on the Keyhole Markup Language (KML), an XML-based language initially developed by Keyhole Inc. (acquired by Google in 2004) for describing three-dimensional geospatial data and its display in application programs (Wikipedia 2007). The tested version was *Google Earth* 4.02737 (downloaded from http://earth.google.com/).

NASA World Wind is a virtual globe software product initially developed by NASA Learning Technologies and released in 2004 as an open source program. *NASA World Wind* provides three-dimensional interactive globes of the Earth, the Moon, Mars, Venus and Jupiter (including its four moons). It is based on Microsoft.NET technology and makes use of Extensible Markup Language (XML) and Web Map Services (WMS). The tested version was *NASA World Wind* 1.400 (downloaded from http://worldwind.arc.nasa.gov/download.html).

512   J-P Aurambout, C Pettit and H Lewis

Table 25.1. Selection of virtual globe applications with high potential

Applications of virtual globe technology[a]	Applications satisfying pre-selection criteria
Nintendo Wii Weather	*Google Earth*
Microsoft Live Local 3D	*NASA World Wind*
Dapple Earth Explorer	*ArcGIS Explorer*
Wayfinder Earth	*SkylineGlobe*
ESRI ArcGIS Explorer	*Dapple Earth Explorer*
Volvo Ocean Race Virtual Spectator	
Erdas Imagine Virtual GIS	
Google Earth	
Global-i	
Punt	
Ping 3map	
EarthSLOT	
osgPlanet	
ESRI ArcGlobe	
NASA World Wind	
Table	
Eingana	
Celestia	
Skyline Software TerraSuite	
GeoFusion GeoPlayer	
SRI Terravision	
Lunar Software Earthbrowser	
Hipparchus	
Mark Pesce's WebEarth	
Windows Live Local	
GeoVirtual GeoShow3D	
Viewtec TerrainView	
Gaia	
Virtual Spectator	
PYXIS	
Talent Cruiser	
Virtual Terrain Project	
Earthsim	
GeoVirtual	
GRIFINOR	

[a]http://www.nanocarta.com/wiki/index.php?title=Virtual_globes

ArcGIS Explorer is virtual globe software product developed by ESRI designed to access online GIS content from *ArcGIS Server*, *ArcIMS®*, *ArcWeb Services*, and *Web Map Service* (WMS) and overlay them on a three-dimensional globe. It was launched in 2005 as a beta version. The tested version was *ArcGIS Explorer* 9.2380 (downloaded from http://www.esri.com/software/arcgis/explorer/index.html).

SkylineGlobe is a three-dimensional web portal developed by Skyline Software Systems and is designed to be a 'turn-key' solution for businesses to add their targeted content and tools in a virtual globe environment. *SkylineGlobe* was launched in 2006 and makes use of the *TerraExplorer* plug-in to display its three-dimensional maps in a web browser. We tested *SkylineGlobe* by downloading *TerraExplorer* 5.0.2 from http://www.skylineglobe.com/SkylineGlobe/WebClient/PresentationLayer/Home/Index.aspx?.

Dapple Earth Explorer is a virtual globe derived from the *NASA World Wind* open source project. It is based on the *Geosoft* open source software and is designed to visualise, present and share geoscientific data in a three-dimensional environment. Launched in 2006, it is primarily targeted towards professional earth scientists. The tested version was *Dapple Earth Explorer* 1.020.0 (downloaded from http://dapple.geosoft.com/).

The general features and capacities of each virtual globe product were evaluated using a set of forty criteria (Table 25.2) which encompassed:

- hardware requirements
- type and quality of data provided
- capacity to import GIS features
- capacity to allow data manipulation
- capacity to share data
- degree of openness and customisation potential
- overall performance and stability.

While only free versions of virtual globes were tested in this study, professional and enterprise solution versions of *SkylineGlobe* and *Google Earth* are available and they may provide more advanced functionalities than those described here.

The capacity of each virtual globe product to import and display GIS data was evaluated through the application of a standard Geographical Information System (GIS) dataset. This dataset (see Fig. 25.1) was composed of three ESRI shapefiles: a point shapefile (representing building location), a line shapefile (representing the road transportation network), and a polygon shapefile (representing land uses), as well as a georeferenced TIF raster image. We chose to test these applications using shapefiles rather than other GIS data formats, as it is one of the most generically used proprietary GIS vector formats.

Table 25.2. In-depth evaluation criteria

Category	Detailed criteria
Hardware requirements	Platform (Windows, MAC, Linux)
	Price and accessibility: availability of pro versions and availability of support
	Computer requirements
	Online or application on hard drive
	Internet speed requirements
Type and quality of data provided	Data type available
	Quality of underlying data
	Number of layer databases available
	Match with the underlying imagery data (boundaries, lines features)
	Recognise address information and give road directions
Capacity to import GIS features	Import modules and editing modules
	Recognition of KML and quality of import and number of files imported
	Import of shapefile
	Import of image and grid file
	Ease of import
	Capacity to support three-dimensional objects with texture and underground elements
	Capacity to interact with existing GIS software
	Possibility to import GPS coordinates
	Recognition of the spatial reference of the data exported
Capacity to allow data manipulation	Possibility to create data within the software: polygon, lines, place markers
	Capacity to move data layers in the display
	Allow font and colour editing
	Gives a legend
	Capacity to have labels on roads and rivers
	Capacity to export data straight from GIS and keep legends and other elements
	Flexibility of legends and colour
	Flexibility to have clickable links, images and other links
	Perform data analysis or run data models
	Possibility to display time series
Capacity to share data	Share data with client (KML or other file format) export
	Capacity to share data with client (they can download it and use it)
	Possibility to import from virtual globe to GIS
	Ease of export
Degree of openness and customisation potential	Availability of extensions to simplify exports and imports
	Degree of openness and possible customisation and ease of customisation and script writing.
	Estimation of knowledge level necessary to use the tool for natural resource management
	Compatibility with other software (VSN, objects, xml etc.)
	Simplicity of export

Fig. 25.1. Illustration of the test GIS dataset composed of a point shapefile, line shapefile, a polygon shapefile and a TIF image (aerial imagery). This test data is localised in the Bet Bet catchment, part of the Victorian North Central Catchment Management Authority region. The colour scheme displayed in this ArcGIS screen shot were tentatively replicated in each virtual globe tested

All virtual globes were installed and evaluated on a HP XW8200 workstation, running Windows XP Pro, with 2 GB RAM, Dual Xeon 2.80 GHz processors and an nVIDIA Quadro FX 1400 128 MB video card. The computer was connected to a high speed Internet connection with a band width access up to 10.0 MB/s.

25.3 Results

25.3.1 Hardware

All considered applications required high speed broadband Internet connections ($\geq$ 768 KB/s) as all retrieved digital imagery and elevation data from remote data servers.

Google Earth was the only virtual globe application available across the Linux, Macintosh and Windows operating systems, while the other products could be run in the Windows environment only. However, this specificity may be temporary as *NASA World Wind* and *SkylineGlobe* are currently developing versions compatible with Linux and Macintosh.

SkylineGlobe being a plug-in running within an Internet browser had the lowest computing requirements. The other virtual globe applications, among which *Google Earth* was the most resource hungry, were running from the user's computer and their performance was therefore strongly influenced by the computer processor speed, video card and available RAM. Results are given in Table 25.3 (see Appendix).

25.3.2 Background Data

All five virtual globe products were based on a structure of tiled satellite and aerial imagery draped over a digital elevation model of the earth. The resolution of imagery available for the Australian territory was the lowest in *SkylineGlobe* (in which high resolution imagery is limited to the USA) and in *ArcGIS Explorer*, while *Dapple Earth Explorer* and *NASA World Wind*, through the use of *Microsoft VirtualEarth* imagery data server, achieved better resolutions (Table 25.4). However, the imagery accessible through these four virtual globes was relatively poor over less populated areas where high resolution imagery is not freely available. *Google Earth*, which uses it own proprietary imagery server, provided the highest resolution imagery and the largest coverage for Australia.

All virtual globes except *Dapple Earth Explorer* provided layers displaying place names and transportation networks that could be overlaid on top of the imagery. *ArcGIS Explorer* provided a worldwide basic coarse resolution transportation network, while *SkylineGlobe* provided detailed information for the USA only. *NASA World Wind*, through the use of Windows VirtualEarth data server, provided a good resolution transportation network, while *Google Earth* provided the most detailed and best coverage for transportation features in Australia. The availability of place names and transportation networks was associated with query functionalities that allowed place finding in *NASA World Wind* and exact address match in *Google Earth* and *SkylineGlobe* (only in the USA).

The capacity to display three-dimensional objects was available in *Google Earth*, *NASA World Wind* and *SkylineGlobe* (which was the only virtual globe to display animated three-dimensional objects). However, this functionality is only currently being used in *Google Earth* and *SkylineGlobe* which provide overlays of three-dimensional buildings on top of imagery for selected cities of the world. *Google Earth* is the only virtual globe that allows the user to import three-dimensional objects (created in *Google Sketchup* or downloaded from the Google 3D warehouse) to any chosen location.

NASA World Wind and *Dapple Earth Explorer* were the only virtual globes providing access to non-imagery data such as instant weather, (*NASA World Wind*) and scientific earth data by accessing multiple external data servers such as Geosoft DAP servers, NASA servers, and USGS data servers. Results are given in Table 25.4 (see Appendix).

25.3.3 GIS Data Import

ArcGIS Explorer, SkylineGlobe and *NASA World Wind* provided shapefile import modules designed to allow the overlay of ESRI shapefile features within the three-dimensional environment of virtual globes. Despite being unable to directly import shapefile features, *Google Earth* could also display GIS features once these were converted into the KML format. On the other hand, the import or display of shapefile features appeared impossible in *Dapple Earth Explorer*. While diverging in their capacity to import shapefiles, all considered virtual globes advertised the capacity to import imagery through the use of raster layer import tools (Table 25.5 in Appendix).

The testing of the different data import modules using our sample GIS dataset (Fig. 25.1) showed that:

- Point and line shapefile features could successfully by imported in *ArcGIS Explorer* (Fig. 25.2) and *SkylineGlobe*. However, the import of polygon features appeared to be incomplete and did not allow the display of multiple colours within the same feature.
- The shapefile import module in *NASA World Wind* was not functional, but point and line shapefiles could nevertheless be imported through non-trivial user manipulation (requiring the editing of XML files and the copy of files to specified folders).
- Shapefile import required preliminary data reprojections into geographic coordinates to be imported in *NASA World Wind* and *SkylineGlobe*, but not for *ArcGIS Explorer* which accepted projected data.
- All shapefile formats, in any projection type, could easily be exported into KML through free (export to KML) or proprietary extensions (Arc2Earth) to *ArcGIS* or *ArcMap*. Features converted to KML retained predefined formatting (colour scheme, line width and icons), allowing datasets brought into *Google Earth* to look almost identical to their original counterparts in GIS software (Fig. 25.4)
- All virtual globes tested could import custom imagery (Fig. 25.2–25.4), provided their spatial extend (coordinates of the corners) were manually specified.

Fig. 25.2. Screen shot of *ArcGIS Explorer*: point, line features and imagery were imported using an inbuilt import module while the polygon feature was imported through a KML file

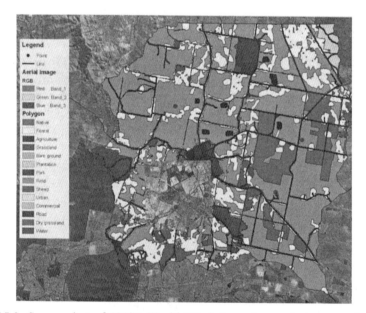

Fig. 25.3. Screen shot of *NASA World Wind*: the polygon feature was imported through a KML file and the screen overlay, point and imagery were added through an edited XML file

All virtual globes except *Dapple Earth Explorer* could import GIS features in the KML format. However, complex combinations of GIS features within KMZ files were fully supported in *Google Earth* only. In *AcrGIS Explorer* and *NASA World Wind*, the use of KML file could palliate dysfunctions in polygon shapefile import modules and allow the display of polygons with multiple colours (Fig. 25.2, Fig. 25.3). However KML polygon features were not recognised by *SkylineGlobe*. Results are given in Table 25.5 (see Appendix).

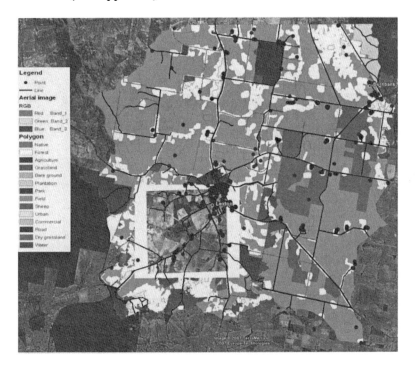

Fig. 25.4. Screen shot of *Google Earth*: all shapefile features and the screen overlay were imported through a KML file, the image overlay was imported with the inbuilt image module

25.3.4 Display and Data Manipulation

GIS data creation as well as editing of point, line and polygon features was possible in *Google Earth*, *ArcGIS Explorer* and *SkylineGlobe*. Data editing of point, line and polygon features was possible in *NASA World Wind*, but data creation was limited to point features (through downloadable extension IGE). *Dapple Earth Explorer* did not allow any creation or editing of GIS features (shapefile or raster).

The symbology used in GIS data layer is often difficult to interpret. The use of a legend with appropriate cartographic symbology is paramount to the proper communication of spatial data. *ArcGIS Explorer* was the only virtual globe providing an inbuilt tool to directly create a legend from the GIS data displayed (Fig. 25.2); however this tool did not work for data imported from KML files. The lack of a legend tool could be partially alleviated in *Google Earth* (Fig. 25.4) and *NASA World Wind* (Fig. 25.3) through the use of pre-created legend images that could be inserted in the view via image overlay functions manually imbedded in KML (see Code 1) or XML (see Code 2) files.

Code 1. Example KML script to create a screen overlay. This code resizes and overlays an image over the virtual globe dataset

```xml
<?xml version="1.0" encoding="UTF-8"?>
<kml xmlns="http://earth.google.com/kml/2.0">
<ScreenOverlay>
<name>Absolute Positioning: Top left</name>
<Icon>
<href>E:\projects\Place_and_purpose\Virtual_globe_data_
test\Viewer_comparison\legend2.png</href>
</Icon>
<overlayXY x="0" y="0.6" xunits="fraction" yunits="fraction"/>
<screenXY x="0" y="0.6" xunits="fraction" yunits="fraction"/>
<size x="0.12" y="0.5" xunits="fraction" yunits="fraction"/>
</ScreenOverlay>
</kml>
```

Code 2. Example XML script to create a screen overlay. This code resizes, names and overlays an image over the virtual globe dataset

```
<?xml version="1.0" encoding="UTF-8" ?>

-<LayerSet Name="Bet_Bet" ShowOnlyOneLayer="false" ShowAt-
Startup="false"    xmlns:xsi="http://www.w3.org/2001/XMLSchema-
instance"
xsi:noNamespaceSchemaLocation="C:\WorldWind\bin\Debug\Confi
guration\WorldXMLDescriptors\EarthLayers\LayerSet.xsd">

-<ScreenOverlay ShowAtStartup="true">

<Name>Legend Bet Bet</Name>

<ImageUri>C:\Program        Files\NASA\World        Wind
1.4\Config\Earth\legend2.png</ImageUri>

<StartX>250</StartX>

<StartY>200</StartY>

<Width>161</Width>

<Height>648</Height>

<Opacity>255</Opacity>

<ShowHeader>false</ShowHeader>

</ScreenOverlay>
```

The tested virtual globes also presented functionalities that could enhance the value of GIS data:

- All virtual globes except *Dapple Earth Explorer* supported the imbedding of a clickable hyperlink, image display and html functionality with point features. This feature could be particularly useful to associate advanced descriptions or metadata to spatial data.
- The display of data as time series was possible in *NASA World Wind*, *SkylineGlobe* and *Google Earth*. However, *Google Earth* was the only software product to provide tutorials on how to effectively use this functionality in KML.
- All virtual globes except *Dapple Earth Explorer* allowed the creation of fly through allowing the display of GIS features in a pre-defined order and angle.

Results are given in Table 25.6 (see Appendix).

25.3.5 Data Sharing

The export and sharing of data into virtual globes was possible in *Google Earth* (through data conversion into a .kml file), *ArcGIS Explorer* (through data conversion into an .nmf file), and *SkylineGlobe* (through data conversion into a .fly file). Data sharing in *NASA World Wind* was possible but required the creation of add-ons that are complex to develop and install on users' computers. *Dapple Earth Explorer* offered the possibility to share views with other users, but did not allow the sharing of GIS data.

The transfer of GIS data created in virtual globes into standard GIS software was also possible in *SkylineGlobe* through the export to a shapefile module (available in the professional version) and through third party extensions converting *Google Earth* KML files to shapefiles.

25.3.6 Openness and Customisation

Google Earth, *ArcGIS Explorer* and *SkylineGlobe* presented fixed interfaces that could not be freely modified or customised. *SkylineGlobe* did offer the possibility of advanced customisation of the user interface and tools available, but only in its professional version.

The open source nature of *NASA World Wind* and *Dapple Earth Explorer* allowed for advanced customisation of the world viewer interface, the tools available, and the software and databases with which it can interact. However, while this open structure provided flexibility, the full potential of these virtual globes could only be realised through advanced programming in the C# language, which requires expert Information Communication and Technology (ICT) expertise.

25.3.7 Performance

The ease of usage and intuitiveness of user interfaces varied across software. Navigation tools in *Google Earth*, *SkylineGlobe* and *ArcGIS Explorer* were more effective than those in *NASA World Wind* and *Dapple Earth Explorer*, which lacked navigation toolbars.

The capacity of each virtual globe to consistently display data also varied. *Google Earth* and *ArcGIS Explorer* could handle large files (>100 MB) and multiple datasets without crashing. *NASA World Wind* and *SkylineGlobe*, although less suited for large datasets, were also very stable environments. *Dapple Earth Explorer* was prone to crashes, particularly when retrieving scientific data from different servers.

25.4 Discussion

All of the virtual globes investigated, by providing access to contextual three-dimensional imagery and relief data as well as enhanced navigation and search functionalities, have the capacity to greatly enhance the accessibility and visualisation quality of GIS data. However, despite continuing improvements in spatial information importer modules, the transfer of vector and imagery data to virtual globes still remains imperfect and requires significant coding from the user to reach acceptable cartographic display standards (displaying data with a clear and readable symbology and providing a complete legend identifying each data feature displayed).

Of the five virtual globes investigated in this study, *Google Earth* presented the most accurate and best performing virtual globe to display and share GIS data. This was mainly due to the high quality and large coverage of the imagery it provided, and the availability of external computer programs to convert GIS data and imagery to KML. *SkylineGlobe*, although limited by low quality imagery data (over Australia) and improper display of polygon features, also presented very useful functionalities (capacity to be displayed in any web browser and direct import of GIS features in particular) that conferred it some advantages over *Google Earth*.

NASA World Wind required advanced computer programming skills to customise it and is thus difficult to use as a development platform by nonspecialist users. However, the *NASA World Wind* API is completely open for development, unlike *Google Earth* and therefore potentially offers the open source community greater flexibility in developing custom-built functionality.

The results from our research indicate that although *Dapple Earth Explorer* appeared unsuited for the creation and import of GIS data, *Google Earth*, *NASA World Wind*, ESRI *ArcGIS Explorer* and *SkylineGlobe* all presented a strong potential to be used to display GIS data and make it accessible to decision makers, broad research communities and to the general public. With the recent proliferation of virtual globe technologies for displaying and navigating spatial information, only time will tell if they develop into the next GIS.

We would like to point out that this study only considered the use of freeware versions of virtual globes and that the proprietary versions (*Google Earth Pro* and *SkylineGlobe Pro* in particular) may present enhanced functionalities (GIS data import modules, support for a wide variety of data format, better software stability), alleviating some of the shortcomings present in the free versions.

25.4.1 Applications

Butler (2006) reports that, 'easy-to-use virtual globes will facilitate the communication of spatial information between stakeholders and government agencies'. To test the applicability of virtual globe technology in the fields of agriculture sciences, natural resource management and land use planning, we developed a set of five case studies aimed at communicating data to clients. These applications were developed in the Keyhole Markup Language (KML), using proprietary export tools (the free 'Export to KML' *ESRI ArcMap* extension) and manual editing, and were displayed in *Google Earth*. *Google Earth* was preferentially chosen over other virtual globes for its high resolution imagery of rural Australia, providing a suitable context for the data displayed, and the availability of tools to export GIS features with pre-formatted symbology directly into KML.

The first application (Fig. 25.5) aimed at presenting soil landforms in the Corangamite Catchment Management Authority region in Victoria. Soil samples were extracted along a gas pipeline crossing the catchment from east to west, providing a unique insight into Victorian soils.

Fig. 25.5. Screen shot of a KML presenting soil landforms visualised in *Google Earth*

For each sampled locations, the following information was recorded:
- the coordinates of each sample point
- the soil type present at the site

- a detailed description of the soil structure
- a photograph of the soil profile.

The use of virtual globe technology allowed the synthesis of these GIS data, descriptive text and images within a spatially dynamic and interactive interface. The developed application allowed the:

- draping of geomorphic units on top of aerial imagery
- visualisation of pictures of the soil profiles for each sampled location
- creation of hyperlinks allowing the user to access detailed descriptions of soil types for each sampled location.

The second application focused on visualising spatio–temporal data of the movement of fish through the Gippsland Lakes system in south-east Victoria. A number of black bream have been tagged and monitored using acoustic receivers. The tracking of fish enables scientists to identify valuable fish habitats and to determine how fish are likely to respond to changes in the environment. Through the use of the time stamp feature available in *Google Earth*, temporal animations of individual fish movements could be visualised along with associate attribute data including tag identification number and depth below sea level (See Fig. 25.6).

Fig. 25.6. Spatial–temporal visualisation of Fishtrack data for the Gippsland Lakes

The third application displays the results from the Analytical Hierarchy Process (AHP) predicting change in land use suitability for wheat and barley in north-western Victoria (Mallee, Wimmera and North Central Catchment Management Authority regions) between the year 2000 and 2050 (Hossain et al. 2006). Through the use of the time stamp technology a temporal animation of changes in land suitability allows to identify regions where crop production will be affected by climate change (Fig. 25.7).

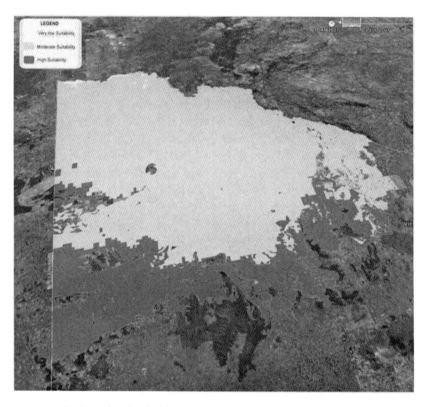

Fig. 25.7. Display of soil suitability for barley in north-western Victoria

The fourth application focuses on visualising the outputs from the *eFarmer* farm management tool (Fig. 25.8). *eFarmer* is a tool that enables farmers to create their own farm management plans and for this information to be stored in a common data repository and shared with catchment management authorities. The use of the virtual globe technology enables the overlay of spatial information depicting current and proposed land management actions from farmers, and subsequent actions (such a timber production or planting native seeds) to be represented in three-dimensions.

In this application the capacity of *Google Earth* to display three-dimensional objects was used to place tree objects (created in *Google SketchUp*) in the landscape.

Fig. 25.8. Display of three-dimensional ecological vegetation classes in *eFarmer*

The fifth application focused on presenting data from the Waterwatch Australia project in the Goulburn Broken catchment. Waterwatch was established in 1993 as a national community water quality monitoring network, and encourages Australians to become involved in the protection and management of their waterways and catchments. The Waterwatch program is structured around an online database containing records of stream quality (based on biological and habitat assessments plus physical and chemical water tests). The developed application (Fig. 25.9) made use of virtual globe point hyperlink technology to link the Waterwatch online database with:

- a GIS layer of rivers in the Goulburn Broken catchment
- coordinates of Waterwatch sites
- upstream and downstream photos of each site
- a written description of each site.

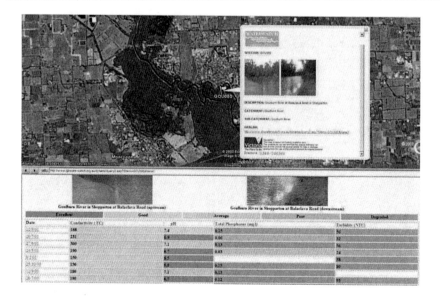

Fig. 25.9. Screen shot of Waterwatch *Google Earth* application

The use of this application represents an improvement over the current Waterwatch two-dimensional map interface in that it allows locations to be visualised in a three-dimensional environment (displaying hills and valleys) and therefore improving the understanding of the surrounding environment at the Waterwatch sites.

The development of these applications showed that virtual globes have the potential to greatly enhance the presentation and visualisation of spatial data. By providing worldwide elevation and digital imagery data, virtual globes provide a landscape context and can improve the capacity of users to 'relate' to GIS data. We used the capacity of virtual globes to create clickable links and associate text, images or web hyperlinks with various spatial data features, thereby greatly increasing the amount of information that could be shared and communicated with stakeholders. The use of the time stamp feature also showed that virtual globes could display temporal changes in spatial data and help visualise dynamic changes in landscapes. Finally we showed that three-dimensional trees and other vegetation objects could be displayed within the virtual globe's environment and further increase the realistic feeling of visualisations of agricultural and ecological spatial data. Although our applications were only developed for *Google Earth*, most of features utilised were available in some of the other virtual globes evaluated (*Google Earth* being the only virtual globe supporting all features) and similar visualisations could have been produced using a combination of *SkylineGlobe*, *NASA World Wind* and *ArcGIS Explorer*.

Our research showed that virtual globes could greatly enhance the way GIS data can be visualised and presented, however, these applications were not tested for their usability and the perceptions of users to these spatial data interfaces will be evaluated in the next phase of the research.

25.5 Conclusion

In this research we have investigated the strength and weaknesses of five major virtual globe environments and their capacity to be used to present GIS data in the context of agricultural sciences and natural resource management. We have shown that despite recent developments in conversion tools, the link between GIS data and virtual globes requires further work. Improvements in the support of various GIS data formats (polygon features and raster in particular) are paramount to allow data export to virtual globes. Major improvements in the symbology that can be used to display imported data will also be necessary to match the cartographic quality of current GIS software. We believe the development of user friendly tools (removing the need for code writing by the user) as well as exporting exact maps and legend features directly from GIS packages into virtual globes could lead to their widespread use to display GIS data.

Through a series of case studies we have exemplified that virtual globe technology can readily be adapted to enhance the communication and accessibility of spatial information. We believe that future improvement in the compatibility between GIS and virtual globe technologies will change the way in which spatial data can be used to inform natural resource management and land use planning.

References

Butler D (2006) The web-wide word. Nature 439:776–778
Gore AA (1998) The digital earth: understanding our planet in the 21st century. California Science Center, Los Angeles
Hossain H, Sposito V, Evans C (2006) Sustainable land resource assessment in regional and urban systems. Applied GIS 2:21–24
Lisle RJ (2006) Google Earth: a new geological resource. Geology Today 22:29–32
Riedl A (2007) Digital globes. In: Cartwright W, Peterson MP, Gartner G (eds), Multimedia cartography. Springer, Heidelberg, pp 255–266
Wikipedia (2007) www.wikipedia.org. Retrieved 19 March 2007

Appendix: Hardware results

Table 25.3. Hardware results summary

Software (Platform)	Price	Computer requirements	Internet speed requirements	Application type
Google Earth (PC, MAC, Linux)	Free *Google Earth Pro*: $400: *Google Earth Plus*: $20 *Google Earth Enterprise*: $100,000	CPU: 2.4 GHz, RAM: 512 MB Hard disk: 2 GB free space Graphics Card: 3D-capable with 32 MB of VRAM	768 KB/sec	Installed on hard drive and locally caching data
NASA World Wind (PC)	Free	CPU: 1 GHz RAM: 256 MB Hard disk: 2 GB of disk space Graphic card: 3D-capable	Broadband (unspecified)	Installed on hard drive and locally caching data
ArcGIS Explorer (PC)	Free	CPU: 1.5 GHz RAM: 512 MB Hard disk: 52 MB free space Graphic card: 24-bit capable and 32 MB memory	Broadband (unspecified)	Installed on hard drive and locally caching data
SkylineGlobe (PC)	Free *SkylineGlobe Pro, Business Solution, Integrator Solution, Local Government Solution and Enterprise Solution* (unknown cost)	CPU: none RAM: 256 MB RAM Hard disk: none Graphic card: 32 MB of memory	Broadband (unspecified)	Online application, no data cashed
Dapple Earth Explorer (PC)	Free	CPU: 2.4 GHz RAM: 512 MB Hard Disk: 2 GB free space Graphics Card: 3D-capable with 64 MB of memory	768 KB/sec	Installed on hard drive and locally caching data.

Virtual Globes: the Next GIS? 531

Table 25.4. Background data result summary

Software	Imagery	Imagery resolution over Australia	Transportation network and place finder in Australia	Support of 3D objects	Availability of other data for Australia
Google Earth	Satellite and aerial imagery	High over urban and rural areas	Detailed transportation layer and address matching place finder	Yes	None
NASA World Wind	Satellite and aerial imagery	High over urban but low over rural areas	Transportation network and place finder (no address matching)	Yes	Yes, global clouds, scientific visualisation, Landsat7, Geocover data.
ArcGIS Explorer	Satellite and aerial imagery	Low urban and rural areas	Basic transportation layer and place finder (no address matching)	No	None
SkylineGlobe	Satellite and aerial imagery	Low urban and rural areas	None (only available in the USA)	Yes	None
Dapple Earth Explorer	Satellite and aerial imagery	High over urban and low over rural areas	Transportation network but no place finder	No	Yes, Geosoft data

532   J-P Aurambout, C Pettit and H Lewis

Table 25.5. GIS data import results summary

Software	Shapefile import module	Image import module	Recognition of KML data	Import of GPS coordinates	Data format supported
Google Earth	No	Yes	Yes, full compatibility. Complex KMZ supported	No (possible in *Google Earth Plus*)	Raster image format (jpg, bmp, tga, png, gif, tif)
NASA World Wind	Yes but not functional	Yes	Yes partial compatibility Complex KMZ not supported	Yes (GPS tracker plug-in)	ESRI shape-file Raster image format (jpg, png)
ArcGIS Explorer	Yes	Yes	Yes partial compatibility, Complex KMZ	No	ESRI shape-file Raster image format (img, bmp, jpg, gif, tif, ArcInfo, DTED, ERDAS, ECW) Geodatabase
SkylineGlobe	Yes	Yes	Yes partial compatibility. Polygon colours not supported	Yes	ESRI shape-file
Dapple Earth Explorer	No	Yes	No	No	Raster image format (tif)

Table 25.6. Display and data manipulation results summary

Software	Creation of GIS data	Creation of a legend	Support clickable hyperlinks	Creation of fly-through
Google Earth	Yes	No (possible through screen overlay)	Yes	Yes
NASA World Wind	Yes (point features only)	No (possible through screen overlay)	Yes	Yes
ArcGIS Explorer	Yes	Yes	Yes	Yes
SkylineGlobe	Yes	No	Yes	Yes
Dapple Earth Explorer	No	No	No	No

26 A Virtual Knowledge World for Natural Resource Management

Christopher Pettit and Yingxin Wu

Department of Primary Industries, Parkville Centre, Victoria, Australia

Abstract: There is a wealth of natural resource management information available to land managers, farmers, researchers and the public via a plethora of websites and portals. However, how engaging and user-friendly are such online resources? This chapter explores the development of an online virtual world for presenting natural resource information to diverse audiences.

The Natural Resource Management (NRM) Virtual Knowledge World has been created using the Virtual Reality Modelling Language (VRML). Visitors to this virtual world can explore the Bet Bet sub-catchment in North Central Victoria and learn about such natural phenomena as biodiversity, climate and soil health. Visitors can explore simulated landscape processes such as flooding, erosion and salinisation. Point and click functionality enables the visitor to download objects from a three-dimensional object library which includes a number of built structures, trees, shrubs and grass species. By navigating further into the virtual world the visitor can enter a natural resource management virtual shop and access a range of information products pertaining to landscape models, tools, frameworks, workbooks and reports. The next step in this research is to evaluate the pedagogical value of such virtual worlds for enhancing the understanding of natural resource phenomena across a diverse range of user groups from policy makers to school students.

26.1 Introduction

To ensure that policy makers can make more informed and timely decisions about the sustainable use of natural resources there is need to research the potential of enhanced communication tools. Landscape visualisation tools such as computer game engines (e.g. *Torque* and *Unreal*) and virtual globes (e.g. *Google Earth* and *NASA World Wind*) can be used to accelerate the understanding of complex landscape models to end users, spanning from communities to policy makers. Significant research has been undertaken in developing landscape visualisation products of 'real' geographies (Bishop and Lange 2005; Cartwright, et al. 2007a). A lesser developed body of research has been undertaken in visualising 'fictitious' geographies — for example, Dodge and Kitchin (2001) mapped the geographies of cyberspace, while Cartwright et al. (2007b) created information landscapes. In this research we focus on the development of an interconnected 'real' and 'fictional' virtual natural resource management world.

There is a wealth of natural resource management data, information and knowledge available to catchment managers, farmers, researchers and the public via a number of Internet sites and portals. Examples of such portals include:

- Victorian Resources Online (VRO)
  <http://www.dpi.vic.gov.au/dpi/vro/vrosite.nsf/pages/vrohome>
- Australia Agricultural Resource Online (AANRO)
  <http://www.aanro.net/>
- Australian Natural Resources Data Library (ANRDL)
  <http://adl.brs.gov.au/anrdl/php/>
- Atlas of Canada <http://atlas.nrcan.gc.ca/site/english/index.html>
- Consultative Group for International Agriculture Research (CGIAR)
  Consortium for Spatial Information (CGIAR-CSI)
  <http://csi.cgiar.org/index.asp>.

These sites contain a wealth of natural resource management data and information, which is made accessible via standard web pages, hyperlinked documents, data download centres and interactive map servers. What these sites do not deploy is an interactive three-dimensional environment enabling users to serendipitously navigate and access available information products. In our research we have attempted to develop a more innovative approach in presenting natural resource management data and information products using a three-dimensional virtual-world environment.

In this chapter we discuss the development of a virtual reality front-end to a natural resource management information portal known as the *Natural*

Resource Management (NRM) Virtual Knowledge World. The *NRM Virtual Knowledge World* has been developed using Virtual Reality Modelling Language (VRML) and is based on the *game player* metaphor. The game player metaphor is one of nine metaphors proposed by Cartwright (1999) to complement the existing map metaphor used for representing and accessing geographical information. Through applying gaming skills users can navigate through an information landscape and interact with NRM artefacts including three-dimensional objects, sound files, spatial datasets, documents and databases.

26.2 Virtual Worlds

Virtual worlds are an abstraction of reality developed using three-dimensional and sometimes four-dimensional (time) constructs. Virtual worlds can be constructed as fictitious geographies or real geographies. Examples of virtual world modelling on fictitious geographies include:

- *Activeworlds* <http://www.activeworlds.com/>
- *Cybertown* <http://www.cybertown.com/main_ieframes.html>
- *Second Life* <http://www.secondlife.com/>.

Each of these virtual worlds have citizens who interact with one another within an online virtual environment, populated by thousands or millions of three-dimensional objects including avatars, building, trees and the like. 'Fictitious' geographies can comprise of both fictional and non-fictional content. For example, avatars in *Second Life* represent real people within a fictitious landscape. Since 1995 when the first of these virtual worlds was established (*AlphaWorld*) millions of people worldwide have taken up a presence in one or more of these virtual worlds.

To date the use of virtual world technologies for constructing real geographies are predominantly focused on the development of virtual models of buildings and cities. For example, Hudson-Smith and Evans (2004) have constructed a three-dimensional virtual London model, Pettit et al. (2006) have constructed a three-dimensional neighbourhood area around the Jewell Train Station in Melbourne, and Barton and Parolin (2004) have constructed a three-dimensional virtual model of a public housing estate in Sydney. However, there has been limited interest in applying virtual world technologies for constructing rural and natural resource focused applications. Examples include the Coffey et al. (2007) coastal seabed models, and the Arrowsmith et al. (2005) virtual Grampian's field trip model.

A number of the virtual worlds referred to above (Arrowsmith et al. 2005; Coffey et al. 2007; Pettit et al. 2006) have been created using VRML, an open standard format for representing three-dimensional interactive vector graphics. VRML files, known as world files (*.wrl), can integrate three-dimensional graphics with multimedia files such as animations, sounds and lighting, to represent real-world objects and landscapes. VRML worlds can be displayed via web browsers with appropriate viewer plug-ins. VRML can be distributed to a wide range of users from decision makers to the public, via CD-ROM or via hosting an online virtual world sites. VRML was first specified in November 1994 and has been designed particularly for use on the World Wide Web. VRML has been used to create virtual worlds such as *Cybertown* (VRML2.0 released in 1997) and has been applied widely for educational and research purposes. Due to its open format and the ability to easily share three-dimensional models, VRML has been used to create the *NRM Virtual Knowledge World*.

26.3 *NRM Virtual Knowledge World*

Virtual world technologies offer a potentially innovative approach for making accessible a wide range of natural resource management information products to both decision makers and communities. In this chapter we discuss the development of version 2.0 of a *NRM Virtual Knowledge World*. Version 1 was originally presented to the Victoria Catchment Management Council in December 2005 and at the Land Water Australia workshop in April 2006. Feedback on content and usability of the product resulted in the development of version 2.0.

To date the prototype *Virtual Knowledge World* has been developed for NRM data, information and knowledge products in Australia. The modelling structure for the virtual world is illustrated in Fig. 26.1. Visitors (also referred to as users) to the virtual world enter via the *NRM Knowledge Management Hub* represented as a map of Australia with hotlinks to the *Bet Bet Virtual Landscape* and *Victorian Virtual Knowledge Arcade* (Fig. 26.2).

A Heads Up Display (HUD) located at the bottom right corner of the *NRM Knowledge Management Hub* gives 'where you are' information by pointing out the current location on a minimised map of the virtual world. The incorporation of the HUD is considered an important navigation tool as it minimises the chance of the user getting lost in the virtual world. The hub was created by importing and re-sampling the Geoscience Australia 9″ digital elevation model (DEM) to a 20 km grid cell resolution and draping over a satellite image of Australia. Ultimately the knowledge hub will contain links to additional Australian virtual landscape products as they are developed.

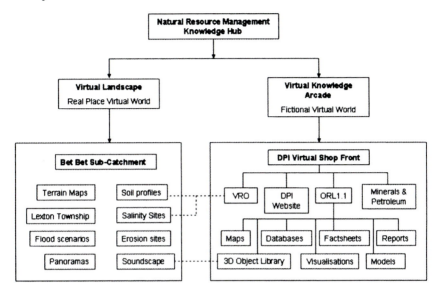

Fig. 26.1. Modelling structure of the *NRM Virtual Knowledge World*

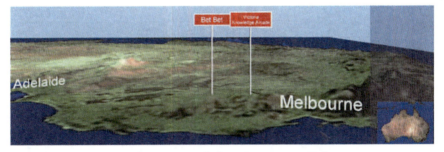

Fig. 26.2. Screen shot of *NRM Knowledge Management Hub*

26.4 *Bet Bet Virtual Landscape*

The Bet Bet sub-catchment provides a suitable 'real world' landscape to visualise as there are a number of agricultural industries in the area including grazing, cropping and production forestry. There has been some significant landscape degradation in the form of salinisation and subsequently there has been extensive data acquisition and environmental modelling work undertaken in the area. From a natural resource management and an environmental education perspective there are significant natural and human induced landscape processes occurring and therefore it makes an interesting rural landscape to simulate.

The Bet Bet Creek catchment is a relatively small catchment of approximately 9600 hectares lying in the south-west corner of the Loddon River catchment (Fig. 26.3). The major town centre in the area is the township of Lexton. The Murray Darling Basin, in which the Bet Bet Creek catchment lies, is identified as the major source of salt (>40,000 tonnes per annum) entering the Boort irrigation area from the Loddon dryland catchment areas (Connor et al. 2004). Recharge in the Bet Bet area contributes more salt per volume of drainage to local rivers than any other sub-catchment in the region (Connor et al. 2004). With salinity and water quality being two of the most important issues in the area, significant work has been undertaken in understanding vegetation, geomorphic, soil, land capability and geohydrological conditions (Day 1985; Lawrence 2006; Westbrooke and Hives 1991).

Similarly to the *NRM Knowledge Management Hub*, the Bet Bet sub-catchment virtual landscape has a HUD plus an additional menu panel where the user can further explore the existing landscape mosaic and some of the current and potential natural and human induced impacts including erosion, salinity and flooding.

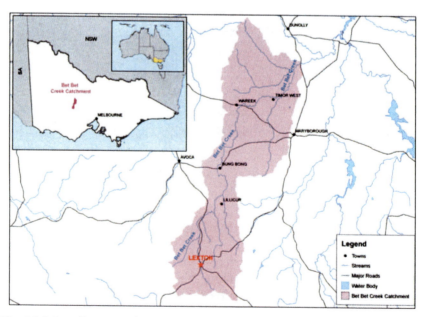

Fig. 26.3. Locality map of the Bet Bet sub-catchment

To make the landscape more realistic a high resolution aerial photo has been used to drape over the terrain map generated from a 1:25,000 scale DEM. The landscape has been populated by a number of trees and buildings geo-referenced to their appropriate locations according to the aerial photo (Fig. 26.4). Through the button menu on the HUD, the aerial photo can be replaced by a road and river base map, or a thematic height classified DEM image, to provide a better context of the landscape.

The *Bet Bet Virtual Landscape* is composed of two areas of interest. The first area of interest is the township of Lexton, comprising a 2.3 km by 2.8 km map extent. The second area of interest is located 4 km north-west of the township, with a 1.4 km by 2.5 km map extent which captures several bore holes and soil pit sites. In total, the *Bet Bet Virtual Landscape* contains over 360 tree objects and 60 built structures comprising rural homesteads, water tanks, tin sheds and commercial establishments.

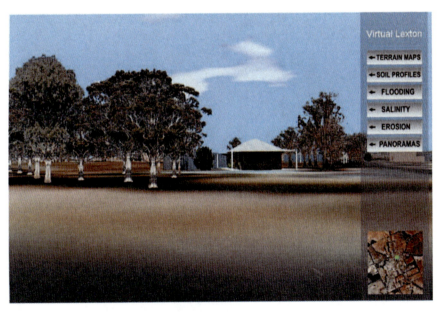

Fig. 26.4. Screen shot of *Bet Bet Virtual Landscape*

Through this virtual landscape users can learn more about the existing landscape processes by exploring bore holes and soil pit sites. Users can click on abstract icons within the virtual landscape to activate hyperlinks to additional natural resource information repositories such as VRO. Within the *Bet Bet Virtual Landscape* there are three bore sites used to measure and model salinity level, which also provide links to related groundwater reports. When a user clicks on a spherical information marker indicating a bore hole site, the related groundwater report is accessed from VRO (Fig.

26.5). There are also three soil pit sites within the virtual landscape. These are clustered at the end of Yalong Road, about 6 km north-west of the township of Lexton. When a user clicks on the related spherical information marker, a soil profile is extruded above the virtual landscape. By clicking on the soil profile the user is taken to further information on the soil profile morphology residing on VRO (Fig. 26.6). VRO <http://www.dpi.vic.gov.au/dpi/vro/vrosite.nsf/pages/vrohome> is a natural resource information portal comprising over 9000 pages of information delivered via standard html web pages. The cross-referencing of material from the virtual landscape to VRO enables the user to drill in to a desired depth of knowledge, accessing the most definitive local and expert information available.

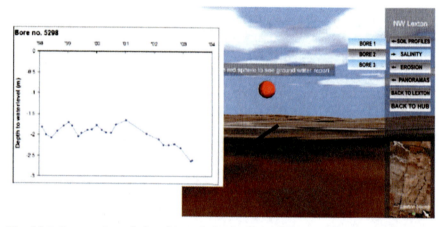

Fig. 26.5. Screen shot of virtual bore hole site hyperlinked to groundwater report

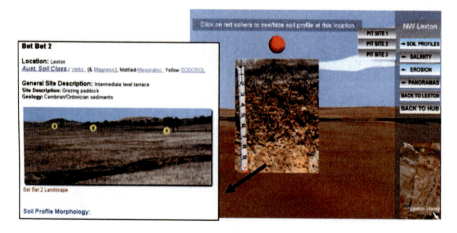

Fig. 26.6. Screen shot of virtual soil profile hyperlinked to soil morphology information

Landscape processes such as erosion and flooding have been captured within the *Bet Bet Virtual Landscape*. There are two sites where a series of photos have been captured depicting gully erosion within a paddock along Yalong Road. These photos have been arranged in a slideshow format (Fig. 26.7).

Fig. 26.7. Screen shot of gully erosion site in the *Bet Bet Virtual Landscape*

The virtual landscape also includes scenarios for '1-in-50' and '1-in-100' year floods. As there are no recorded flood events for the Bet Bet River these scenarios are fictitious examples of how flood information can be displayed and communicated to users (Fig. 26.8). For research in visualising real flood scenario data refer to Basic and Nuantawee (2004) where they have developed a VRML/PHP prototype for improving flood risk communication to the public.

Fig. 26.8. Screen shot of a '1-in-100' year flood event in Lexton

A number of cubic panoramas have also been created within the *Bet Bet Virtual Landscape*. *Stitcher Express* 2.0 software has been used to generate three cubic panoramic views to enhance the virtual tour experience of the user. Each panorama comprises approximately 60 photos which have been stitched together to enable the user to:

- further investigate the Lexton township (Fig. 26.9)
- view gully erosion occurring within a farm paddock
- view the 300 million year old landscape mosaic comprising the Bet Bet sub-catchment from the vantage point on top of hill (Fig. 26.9).

Using the navigation functionality available through *Quicktime* and other panorama viewing packages, the user can zoom in and zoom out of a panorama to further explore the scene.

Environmental acoustics have been incorporated into the *Bet Bet Virtual Landscape*. The sonograms have been acquired as part of a project to test soundscapes for measuring biological activity within farm and natural habitats (Lavis et al. 2006). In total there are eight sonogram files embedded within the virtual landscape. Within the VRML code the sonograms have been placed at the location of a tree or animal. Emitters have been set up at to span a 10 m radius so when the user is within 10 m or less the sonogram will be audible.

Fig. 26.9. Screen shots of extracts from panoramas within the *Bet Bet Virtual Landscape*: Lexton township (**a**) and farm hilltop (**b**)

By including sound as an element of the virtual landscape, it is believed a better understanding of landscape context can be achieved by the user and ultimately assist in improved navigation through the recognition of sound cues. Previous research by Pettit et al. (2003) has examined the integration of spatial, visual and acoustic information. However further work is required to evaluate the tangible benefits of acoustics for improving the understanding of context and navigability of virtual landscapes.

Embedded within the *Bet Bet Virtual Landscape* is the concept of engineered serendipity as a means for discovering additional layers of information (Cartwright 2004). This is best exemplified through the ability to point and click on an object in the virtual landscape to find out more about the object. For example, a user may want to know more about a particular tree species. By clicking on this tree object further information is made available through a pop-up window that provides details on the type of tree including its botanical name and the size the file which comprises this tree (Fig. 26.10). The file size information is provided so that users can ascertain whether or not they wish to proceed in downloading the individual tree object.

Instead of providing a standard search or query web interface users are encouraged to explore the virtual landscape and subsequently discover new pieces of information. By further clicking on this tree object a third tier of information is revealed — the three-dimensional object library. The three-dimensional object library comprises three categories of objects: flora, fauna and built structures. There currently reside over 140 objects available in a number of image file formats including: VRML, X3D, 3DS and GIF (Fig. 26.11). Users can point and click individual objects to open and view, or download to use in the creation of their own virtual landscapes

(Fig. 26.11). The objects contained with the library are those used to create the *Bet Bet Virtual Landscape*.

Fig. 26.10. Screen shot showing the ability to discover additional layers of information through the use of pop-ups

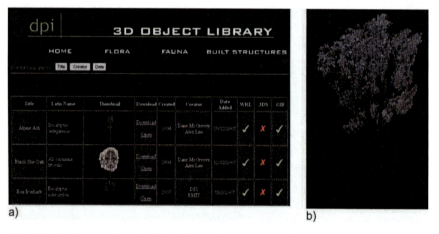

Fig. 26.11. Screen shot of object library (**a**) and an individual tree object (**b**)

26.5 *Victorian Virtual NRM Knowledge Arcade*

Like the *Bet Bet Virtual Landscape*, the *Victorian Virtual NRM Knowledge Arcade* is accessed via the *Knowledge Management Hub* (Fig. 26.1). The

Victorian Virtual NRM Knowledge Arcade was first coined by Christine Forster (Chair of the Victorian Catchment Management Council) in 2005. The *Virtual Knowledge Arcade* (Fig. 26.12) is populated with a number of virtual knowledge shop fronts (19 individual shops) comprising all Catchment Management Authorities, Water Authorities and relevant State Government Departments which comprise the Victorian Catchment Management Framework.

Fig. 26.12. Screen shot of *Victorian Virtual NRM Knowledge Arcade* comprising 19 shop fronts

The Department of Primary Industries (DPI) Virtual Shopfront is populated with a range of data, information and knowledge products, particularly the project outputs from the Our Rural Landscape (ORL) 1.1 New Dimensions for Agricultural Landscapes. Users are able to enter the virtual shop and browse a number of data, information and knowledge products including maps, databases, fact sheets, reports, visualisation products and three-dimensional objects. Figure 26.13 illustrates the virtual workspace where the user can click on a drawer and access a model database which provides information on more than 60 modelling tools developed specifically to assist catchment managers and planners in understanding landscape processes. Models which have been discussed in other chapters: *Land Use Impact Model* (*LUIM*), *Catchment Analysis Tool* (*CAT*), and *Slope Land use Evaluation Urbanisation, Transportation and Hillshading* (*SLEUTH*), are captured in this virtual workspace.

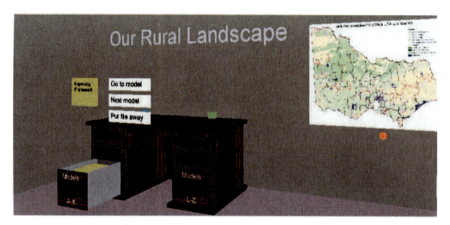

Fig. 26.13. Screen shot of the DPI Virtual Shopfront showing landscape model database and virtual map wall

Embedded within the *NRM Knowledge Arcade* is the concept of nested virtual worlds. This means users can explore worlds within worlds. This is exemplified through the virtual book case located within the DPI shop. The virtual book case contains a number of research products produced from the ORL project. The ORL project was a four year Victorian State Government Initiative focused on developing innovative technologies for the sustainable development of Victoria's food and agriculture sector.

There are more than 27 ORL research reports, which have been categorised via a shelf labelling system based on natural resource management themes such as soil health, climate change and biodiversity. Within the virtual book case is the knowledge management book shelf where users can open and download visualisation products such as movies and digital globe Keyhole Markup Language (KML) virtual tours (Fig. 26.14). Users can begin to view virtual worlds within virtual worlds — this provides a different way to articulate and categorise landscape process, form and function. The 'nested virtual worlds' concept is another example of engineered serendipity (Cartwright 2004) encouraging greater flexibility in searching and exploring information on the World Wide Web.

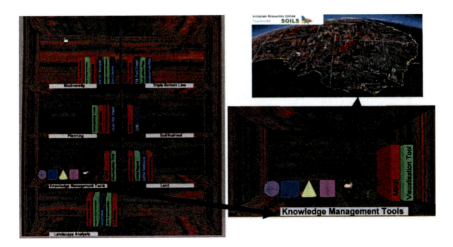

Fig. 26.14. Screen shot showing 'nested virtual worlds' concepts: a *Google Earth* virtual tour is embedded within the virtual book case

26.6 Future Work

The *NRM Virtual Knowledge World* presented in this chapter has yet to be made publicly accessible via the Internet. Users can currently view video clips representations of the *Virtual Knowledge Arcade* and the three-dimensional object library via the VRO Geographical Visualisation Portal <http://www.dpi.vic.gov.au/dpi/vro/vrosite.nsf/pages/geovis>. Future work will see the *NRM Virtual Knowledge World* accessible via the VRO Geographical Visualisation Portal which will make the virtual information landscape available to communities and the general public. Additional cross-referencing of material embedded within VRO with the virtual landscape via hyperlinks is also an important next step.

As previously discussed in this chapter, the *NRM Virtual Knowledge World* has been constructed using the open source VRML programming language. Future work would see the translation of the VRML code to X3D, which uses the eXtensible Markup Language (XML) to represent the geometry and behavioural capabilities of VRML. A tool such as X3D-Edit could be used to manually convert each VRML tag and attribute to a corresponding X3D format. The translation of the VRML code to X3D is considered an important next step so that it is compatible with other extensible geographical standards such as Geographical Markup Language (GML) and Keyhole Markup Language (KML).

Additional research is required to evaluate the usability and usefulness of both fictitious and real virtual worlds such as those comprising the *NRM Virtual Knowledge World*. Such an evaluation should be undertaken in the context of current Web 2.0 thinking around second generation web-based communities and collaboration between users. The value of the virtual world paradigm for collaborating and making accessible data, information and knowledge is becoming increasingly relevant with the advent of a range of new collaborative web delivery platforms, including Youtube <http://au.youtube.com/>, Flickr <http://flickr.com/>, and Myspace <http://www.myspace.com/>. The evaluation will result in a better understanding of the value of virtual worlds for complementing and enhancing existing Web 1.0 hosted services.

26.7 Conclusion

With the recent proliferation of the Internet as a mainstream technology, and as faster Internet connectivity becomes realised, it is envisaged that virtual world technology will gain increasing importance for both e-science and e-business. Our research provides an example of a prototype virtual world constructed with the goal of increasing the understanding of landscape processes, and the data and modelling tools available to catchment managers and planners for making more sustainable land use decisions.

It is envisaged that the NRM virtual world presented in this chapter will serve to support e-science knowledge management information technology infrastructure. This e-science infrastructure will enable greater access to NRM information by the community and decision makers, supporting knowledge brokerage (Ewing 2006) across NRM themes (e.g. land health, biodiversity and climate change) and ultimately facilitate more efficient catchment knowledge exchange.

There exist a number of technologies for creating virtual worlds, such as: VRML, X3D, gaming engines (*Torque* and *Unreal*), and digital globes (*Google Earth* and *NASA World Wind*). These virtual world technologies offer exciting possibilities on how information can be made available to decision makers, researchers and the general public. This will facilitate the rise of geo-collaboratories such as the Human Environmental Research Observatory (MacEachren et al. 2006). Virtual world interfaces to such geo-collaboratories will provide an important technology infrastructure to facilitate better information accessibility and enhanced collaboration to support us in addressing critical global problems such as climate change.

Acknowledgements

This project has been funded via the Our Rural Landscape DPI Initiative. Professor William Cartwright and Tim Germanchis from the School of Mathematical and Geospatial Sciences, RMIT University are acknowledged for their guidance and assistance in building parts of the *NRM Virtual Knowledge World*. Dane McGreevy is also acknowledged in building version 1.0 of the *NRM Knowledge Management Hub* and *Knowledge Arcade*.

References

Arrowsmith C, Counihan A, McGreevy D (2005) Development of a multi-scaled virtual field trip for the teaching and learning of geospatial science. International Journal of Education and Development using Information and Communication Technology 1(3):42–56

Barton JB, Parolin B (2004) A spatial decision support system for the management of public housing: case studies and approach to interactive visualisation. In: Timmermans LJP, Timmermans H (eds) Recent advances in design and decision support systems in architecture and urban planning. Kluwer Academic Publishers, Netherlands, pp 69–84

Basic F, Nuantawee M (2004) Generating a VRML world from database contents: illustrated by application to flood risk communication. Journal of Spatial Science 33(1):37–47

Bishop ID, Lange E (eds) (2005) Visualisation in landscape and environmental planning: technology and applications. Taylor & Francis, London

Cartwright W (1999) Extending the map metaphor using web delivered multimedia. International Journal of Geographical Information Science 13(4):335–353

Cartwright W (2004) Engineered serendipity: thoughts on the design of conglomerate containing geoviz tools and geographical new media artefacts. Transactions in GIS 8(1):1–12

Cartwright W, Peterson MP, Gartner G (eds) (2007a) Multimedia cartography. Springer-Verlag, Berlin

Cartwright W, Williams B, Pettit C (2007b) Realising the literate traveller. Transactions in GIS 11:9–27

Coffey J, Beard DJ, Ryan DA (2007) Visualising coastal seabed characteristics: using VRML models to present three dimensional spatial data via the web. Journal of Spatial Science 52(1):133–144

Connor J, Ward J, Thomson D, Clifton C (2004) Design and implementation of a land holder agreement for recharge credit trade in the upper Bet Bet Creek catchment, Victoria. CSIRO Land and Water Report S/04/1845, Canberra

Day C (1985) A study of the geomorphic, soil and geohydrological conditions of the Timor West/Black Ranges area. Department of Conservation, Forests and Lands, Victoria

Dodge M, Kitchin R (2001) Mapping cyberspace. Routledge, London

Ewing S (2006) The truth is out there! Knowledge sharing in Victoria's catchment. International Landcare Conference, 8–11 October 2006, Melbourne, Australia

Hudson-Smith A, Evans S (2003) Virtual cities: from CAD to 3-D GIS. In: Longley P, Batty M (eds) Advanced spatial analysis. Redlands, California, ESRI Press, California, pp 41–60

Lavis T, Williams DG, Gage SH, Parker AJ, Avery AL (2006). Trialling the use of innovative audio techniques as an indicator of biodiversity. International Landcare Conference, 8–11 October 2006, Melbourne, Australia

Lawrence G (2006) Multi-scale mapping for better salinity management. Focus on Salt 39:8–10

MacEachren AM, Pike W, Chaoqing WY, Brewer I, Gahegan M, Weaver SD, Yarnal B (2006) Building a geocollaboratory: supporting Human–Environment Regional Observatory (HERO) collaborative science activities. Computers, Environment and Urban Systems 30:201–225

Pettit C (2006) Geographical visualisation a participatory planning support tool for imagining landscape futures. Applied GIS 2(3):22.1–22.17

Pettit C, More G, Cartwright W, Burry M (2003) Synthesising spatial, visual and acoustic city information for better understanding and navigation, GeoCart Conference, 12–14 February 2003, Taupo, New Zealand

Pettit C, Cartwright W, Berry M (2006) Geographical visualisation a participatory planning support tool for imagining landscape futures. Applied GIS 2(3):22.1–22.17

Westbrooke M, Hives N (1991) The vegetation and pastures of the Lexton Landcare area. University of Ballarat, Victoria

27 Computer Games for Interacting with a Rural Landscape

Brian Quinn and William Cartwright

School of Mathematical and Geospatial Science, RMIT University Melbourne, Victoria, Australia

Abstract: Computer games and, increasingly, mobile games, are becoming useful viewers of three-dimensional landscapes. A recognisable landscape can be made in computer games with some (but not a great deal of) expertise using the game editor. Modelling buildings, creatures, vegetation and people takes considerable time, skill and money but certain editable computer games and their editing community provide many models and scripts, which provide control functions at no cost. Editing games is termed 'modding'. The scenes created in the games can be recorded as a movie, used in presentations or delivered in other applications, such as mobile games using mobile phones.

Computer games, when applied to learning, can provide the means to deliver educational programs that teach by game play. In the context of visualising landscapes as part of an educational package, computer games' software packages provide exceptional (when cost is considered) platforms for the development of teaching packages that involve landscape simulation.

This chapter provides information about a research and development initiative that is building and evaluating the usefulness and effectiveness of commercial off-the-shelf computer games for building and delivering educational packages. The prototype package will be applied to training volunteer fire fighters in an Australian context. This chapter chiefly concerns the making of visualisations for the Bushfire Rescue Game. *There is a section on disasters, and some suggestions on how decision making and networking can be assisted during an emergency by the use of training games. This is followed by ideas from cognitive science on navigation and decision making. Conversation Theory then links engagement between people and machine productions like visualisations to produce new ideas. Visualisations are then discussed in the context of decision making in time-critical situations and some terms to describe types of visualisations are elaborated. There is an introduction to some terms used in describing aspects of games. Aspects of the editing of selected games are then described and illustrated. Finally future directions for research are summarised together with a conclusion.*

27.1 Introduction

The motivation for the *Bushfire Rescue Game* was the personal experiences of the first listed author as a volunteer fire fighter. Large areas of eastern Australia experienced disastrous bushfires on 16 February 1983 (Ash Wednesday fires). Seven lives were lost, 29,500 ha of land destroyed and 628 buildings were burned between East Trentham and Mt Macedon in central Victoria (DPI 2006). The Newham CFA truck's water pump failed close to midnight on Ash Wednesday, as the high temperatures vapourised the petrol in the fuel line. Without water the truck was engulfed by burning embers. The fire truck Captain slowly rolled up all the hoses and the truck proceeded back to the watering spot where a mechanic got the pump going again and returned to the fire front. The Captain's demonstration of cool headedness is a long lasting memory.

In an account of a fatal incident during this disaster, Packham and Pierrehumbert (1990) state:

'five of these thirteen victims died in the immediate vicinity of their homes....With the benefit of hindsight it is evident that successful survival strategies require an understanding of the importance of the wind change. In fact, of the 47 deaths on Ash Wednesday, 46 died from injuries sustained immediately after the wind change'.

In the *Bushfire Rescue Game* the key event is a wind change. The *Bushfire Rescue Game* is a response to the loss of life that in some cases was avoidable because it was known that there was going to be a wind change. The game will be a pilot mobile location-based game for an iPAQ pocket computer <www.hp.com.au/ipaq/> with a mobile phone and a geographical positioning system (GPS) linked together by a Bluetooth wireless network. The game will utilise recordings of visualisations created in the computer games that are described later in this chapter. The *Bushfire Rescue Game* is being made using the game editor Mscape <www.mscapers.com>. The game's purpose is to train people for time sensitive decision making in emergency and disaster situations.

In a paper on the integration of game-based systems and incident command technologies for emergencies, McGrath and McGrath (2005) state:

'Network-centric emergency response is emerging as a new paradigm, driven in large part by technologies derived from the military sector...Game engines are particularly well suited for multiplayer, real time representations of reality, and game-based emergency response simulations integrated with information systems can provide a powerful method for evaluating human system integration.'

The development of the *Bushfire Rescue Game* will help answer some key questions:

- Which theories of game and game design, psychology, spatial cognition, navigation, decision making and learning are most suitable for developing design principles for a time-sensitive decision-making game using visualisations on computer and mobile devices?
- What are the design principles for visualisations that best support decision making under stressful conditions?

The *Bushfire Rescue Game* involves time-critical decision making, a category of spatial decision making. Also, before addressing the computer game per se, two foundation theoretical aspects need to be reviewed — cognitive science and Conversation Theory.

Andrienko et al. (2007) divide spatial decision making into three types: site selection, temporal- and location-based planning, and time-critical decision making in emergencies. Regarding the delivery of information for time-critical decision making Andrienko et al. (2007) state:

'Enabling the use of rapidly changing, heterogeneous information for time-critical decisions is a fundamental challenge that integration of geovisual analytics with computational spatial decision-making tools needs to address.'

It is proposed that computer and mobile games are a good way of combining aspects of a time-critical spatial problem, such as organising an evacuation or rescue. The *Bushfire Rescue Game* is about assisting school buses that are in danger of being trapped by a bushfire. The game employs several devices accessing heterogeneous sources of information.

Before we consider games we will look at cognitive science and what it reveals about navigation and decision making. This leads to some terms we can use to define game views or visualisations of scenes. These visualisations are used in a game that trains for decision making — the *Bushfire Rescue Game*.

27.2 Cognitive Science

How do we perceive a scene, make a decision and develop successful courses of action? What happens in the brain and will that information assist us in designing mobile games with better delivery of visual and other information for decision making?

O'Keefe (1976) discovered place cells in the hippocampus of the rat brain. Place cells in the hippocampus record short-term memories of where we have been, and produce the egocentric map (Hermer and Spelke 1994, 1996). The egocentric map is achieved by recording the direction of each straight lined segment (or vector) that we walk. Each egocentric vector's angle to the nearest perceived boundary of the area is recorded. The boundary might be the walls of the room or the distant bounding hills of a valley. The egocentric maps in the hippocampus are recorded by the parahippocampus into long-term memory with the addition of iconic objects to create the allocentric map. These memories of objects or entities are linked in the mind to associated physical and affective memories (Hartley et al. 2006). Place cells in rats, and presumably humans, also record slope being traversed, and thus vertical movement is mapped (Jeffery et al. 2006). Thus the egocentric and allocentric maps are in three-dimensions.

After studying rats learning to navigate mazes, Tolman (1948) concluded that they form a mental model or representation of the environment. This model he termed the cognitive map. O'Keefe and Nadel (1978) in their book *The Hippocampus as Cognitive Map* identified the hippocampus as the location of the cognitive map, and described how it might work:

> 'Briefly, a cognitive map would consist of two major systems, a place system and a misplace system. The first is a memory system which contains information about places in the organism's environment, their spatial relations, and the existence of specific objects in specific places. The second, misplace, system signals changes in a particular place, involving either the presence of a new object or the absence of an old one.'

Thus the cognitive map is not merely about place but about the relationships of objects and their properties including their temporal existence, that is, when they appear, change or disappear. Although the allocentric map is about place and the entities therein, the cognitive map is also about events. The cognitive map attends to entities and their context — location, affective and physical — and has understanding of cause and effect relationships.

Games about decision making should be designed such that they take into account the natural ways we understand the locational, affective and physical context of entities in a temporal event such as a disaster. Each person possesses their own cognitive map which changes as it integrates new experiences and knowledge from books, the Internet and other people. The cognitive maps of two or more people can be combined by conversation, sharing visualisations and other types of interaction.

27.3 Conversation Theory

A literature review by Naismith et al. (2004) indicates that Conversation Theory (Pask 1976) with refinements (Laurillard 2002) is a useful theoretical basis for the design and analysis of mobile games. Conversation Theory says that learning occurs when a teacher discusses a theme or ideas with a learner. The two-way discussion or dialectic produces learning, as stated by Naismith et al. (2004, p. 15):

> '...Conversation Theory (Pask 1976)...describes learning in terms of conversations between different systems of knowledge. Pask was careful not to make any distinction between people and interactive systems such as computers, with the great advantage that the theory can be applied equally to human teachers and learners, or to technology-based teaching or learning support systems.'

A human being's system of knowledge, based on the cognitive map and a computer's system of knowledge can thus be mutually intelligible.

Pask (1992) suggests that virtual reality should be employed to make the 'otherwise incomprehensible clear'. This was written in 1992 and perhaps he might now advocate the employment of computer games to visualise and learn about the 'otherwise incomprehensible'.

In an emergency situation (as well as in games about decision making in disasters), the participants need to have 'conversations' so that new ideas can emerge. These conversations include interactions with computers and visualisations. The conversations lead to the merging of cognitive maps, thus to learning and better informed decision making.

27.4 Visualisations

How can we design, make and deliver a visualisation so that it produces a new idea in the cognitive map of the viewer? In the case of decision making in time-critical situations, how do we design and deliver the visualisation? How can we make visualisations that provide readily accessible information about entities and their contexts, so that the mind can better understand causes and effects and see problems and solutions? Is the answer to the last question, that it might be possible in a three-dimensional computer game?

After testing the utility of simulations on subjects and looking at affective responses using animations of urban scenes, Bishop and Rohrmann (2002) state:

> 'Main results are that simulations were perceived as valid and acceptable, that appraisals differ according to lighting and time of day conditions, and that provision of sound enhances the perceived quality of presentations'.

This suggests that the affective dimension should be attended to, adding sounds, lighting appropriate to time of day and weather conditions. This is achieved readily in the computer game editors described later in this chapter.

27.4.1 Viewing Simulations

A pilot flying an aircraft during take off and landing requires time critical information to perform the task. Augmented visualisation has been long used and is vital to safe air travel and effective air combat. The cockpit instrumentation in modern planes employs visualisations showing various views from the aircraft, often augmented with overlaid icons of other planes and the recommended taxiing and flight path. The types of views provided by the instrumentation give an indication of what kinds of displays have been found to be beneficial in a highly time-critical and stressful situation.

Wickens (2002) divides displays for aircraft pilots into three types. The egocentric view is the forward perspective view from the cockpit. The three-dimensional exocentric viewpoint shows the game view, behind, slightly above and at a fixed distance — termed 'tethered' — from the aircraft being flown. A two-dimensional coplanar, or double display, shows a map view of the plane and contour map of terrain above, together with a side view of the plane and terrain below. In all three displays, the required forward path of the aircraft is marked together with a small icon for the aircraft. Foyle et al. (2005) based on Wickens (2002), terms these egocentric display, exocentric (perspective) display and exocentric (two-dimensional plan view) display respectively.

Here the exocentric (perspective) display will be termed allocentric perspective and the egocentric view becomes egocentric perspective view. Avatar or vehicle control is much easier in allocentric perspective views, as one can see to the sides. We also introduce the term 'egocentric plan view' in which the map rotates with the player's direction of travel, the player's icon is central to the map and only a local part of the whole game space map is shown. By contrast, in allocentric plan view the map does not move, the player's icon moves about the map and the map shows the complete game space. Allocentric plan view is better for both detecting other players, and strategic and group decision making. Egocentric plan view is better for individual navigating and tactical decision making. One does not need to 'rotate the map in the mind', the computer is doing the rotation for you thus relieving some of the cognitive load and freeing up mental space for time-critical decision making.

27.4.2 Mobile and Computer Games

Increasingly the use of computer and mobile games has been recognised as a useful environment for delivering cartographic and other spatially visualised information:

> 'Contemporary cartography also strives for what computer games are trying to achieve, namely a shift in thinking that stresses that better tools can be produced by combining traditional representation methods with simulation' (Cartwright 2006).

The use of visualisations using games machines is acknowledged by Germanchis et al. (2004):

> '...rationale behind their use is that games offer an intuitive organisation of spatial objects that replicates or reflects the real world'.

These ideas suggest that computer games are an appropriate method of representing geospatial data and thus for problem solving purposes.

Games are a complex human activity ranging from sport and hunting (principally a diversion) to economic activity and warfare (vitally important). Computer game theory is an eclectic mix of fields that describe and explain the forms and functions of computer games. It includes mathematical game theory (Smith 2005), together with literary, narrative- and media-based analysis of games (Ryan 2001). In addition, computer usability type studies have led to 'interactivity' as an explanatory field (Mortensen 2002). Computer games have varying proportions of narrative or interactivity which assists in determining their game genre (Kücklich 2001). These can be first person shooter (FPS), role playing game (RPG) and massively multiplayer online role playing game (MMORPG).

Computer games often involve a non player character (NPC). In order for the NPC to be believable and a worthy opponent, it must have some decision-making intelligence. This is termed 'game artificial intelligence'. Much of game artificial intelligence is distinct from academic artificial intelligence in that it uses cheats such as preset paths in a scene. The NPC does not really have to work out a route to the target itself <http://www.answers.com/topic/game-artificial-intelligence>.

This is unlike the decision-making process of successful military leaders such as Napoleon, as described by Von Clausewitz (1968). He writes that they examine the battlefield and after a moment's refection a 'coup d'oeil' or flash of insight occurs. The leader sees the problem and solution in one moment and then determines appropriate courses of action from experience (Duggan 2005). This battle process is a 'conversation' with the enemy. It results in courses of action that determine who wins and who loses the game of battle. This decision-making process is also used by humans in games, whether fighting a bushfire or during conflicts with neighbours over resources.

27.5 Game Development

Entities in a computer game, such as NPCs, have game artificial intelligence. This gives them the ability to behave in realistic ways. In most games they either fight against, or are allied with, the player. The game artificial intelligence controls whether they will ally with you or not, what other functions they can undertake (such as guarding a bridge or the need to seek a fresh supply of energy if they run low). The NPCs use path finding artificial intelligence. In the games described it is a simple system using invisible paths already laid down in the scenes. These paths are constructed in the game editor.

A script provides a set of rules guiding the behaviour of an entity so that it reacts appropriately to events in the game. All the games discussed have access to scripting editors. In *Unreal Tournament 2004* the script editor can be opened by right clicking on an object of interest, then selecting the script editor from the choices available. The language resembles Java or a simplified version of C++. Scripting tutorials are available. There are game proprietary script editors such as UnrealScript for *Unreal Tournament 2004* and LindenScript for *Second Life*. Lua is a free scripting language (from www.lua.org) used by many games.

The games can be used as a viewer of user-created local terrain with an overlain map. This can be made relatively easily by someone with some game editing knowledge. Scenes of the terrain can be used to make videos that can be displayed on a website, shown with a multimedia projector to an audience, or displayed on a pocket computer or mobile phone. The games can also be used as a simulation where the object is to learn about problems in an area and help a community decide ways it can rally resources to remedy the situation. Games may be single player or multiplayer.

The computer games examined are in two categories: commercial off-the-shelf (COTS) games, and MMORPGs (massively multiplayer online role playing game). COTS games are generally purchased in a retail store and MMORPGs are found on the web and usually require a subscription. The COTS games outlined are:

- *Trainz* from Auran <www.auran.com>
- *Unreal Tournament 2004* (UT2004) from Epic Games <www.epicgames.com>
- *Farcry* from Ubisoft <www.ubisoft.com>.

The MMORPG example is *Second Life* from Linden Labs <www.lindenlab.com>.

27.5.1 *Trainz*

Trainz is made by Auran, an Australian company. There is a large community of *Trainz* enthusiasts. Tutorials on editing the game are provided by websites <such as http://www.virtualrailroader.com/DEMTrainz.pdf>. The scene in Fig. 27.1 shows the *Trainz* editor with terrain at Hanging Rock, Victoria, with bushfire smoke animations and emergency services in action. The helicopter is flying along a preset aerial path that can be seen strung between the tops of the dashed vertical lines. A fire truck follows the thicker line on the ground. Editor functions are reached via tabs at the right hand side.

Trainz has a simple game artificial intelligence. Vehicles follow precise tracks. Points can change a train's direction of travel. The game's artificial intelligence controlled how trains respond to signals, run to a timetable and halt at specified stations. Goods-trains deliver and pick up goods-wagons in preset routines. This could be used in a simulated emergency that involves a train crash with a subsequent emission of a toxic gas or radioactive cloud.

Figure 27.2 is an allocentric perspective view of a helicopter. Fire trucks, trees and a bushfire have been added to a container port model obtained from the Auran Download Centre. The smoke, water, container cranes and vehicles are animated. Cities with traffic can also be produced. *Trainz* is a useful program for constructing rural and urban scenes including moving vehicles; however, it is single player game. *Trainz* is an Australian made game with a large range of Australian rural and small town type buildings that can be used to simulate an Australian rural or peri-urban scene with fire and other emergency services in action.

Fig. 27.1. *Trainz* editor showing pylons that carry the invisible tracks for the helicopter

Fig. 27.2. *Trainz* screen shot showing fire at a shipping container port

27.5.2 *Farcry*

Farcry is a three-dimensional game with an editor. The terrain can be overlain with a user produced map or texture. Vegetation, water features and buildings are rendered attractively.

The egocentric perspective view in Fig. 27.3 shows Hanging Rock in the Macedon Ranges. There is excellent detail in the grass and the rocks, randomly rotated and varied in scale, look somewhat like the Hanging Rock terrain. In an egocentric perspective view of the reservoir in Fig. 27.4, animated water is pouring over the dam wall. The terrain has been overlain with contours and geology. The colours in the geology are somewhat washed out, and water features are difficult to produce.

Farcry's artificial intelligence, which determines the behaviour of creatures or bots (also called NPC), can be changed using the script editor. *Farcry* was found to be useful, especially for creating views that established the setting of the Macedon Ranges in the *Bushfire Rescue Game*.

Fig. 27.3. *Farcry* screen shot showing view from Hanging Rock looking towards Mt Macedon

Fig. 27.4. *Farcry* screen shot showing Mt Macedon Reservoir with contours and geology

27.5.3 *Unreal Tournament 2004*

The images in this section are the results of experiments with *Unreal Tournament 2004* to see what kinds of animations, images and maps can be displayed within the game. Timers can be added to objects and particle emitters such that a bushfire can spread on a set schedule.

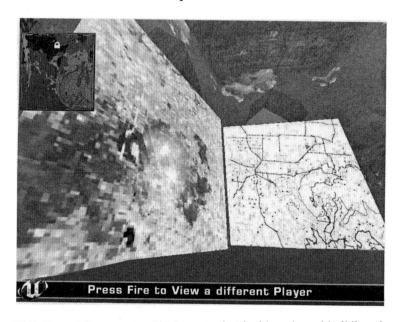

Fig. 27.5. *Unreal Tournament 2004* screen shot inside a domed building showing interactive maps

The allocentric plan view of the game space is in the top left of the scene in Fig. 27.5. It shows the geology of the Mt Macedon district, as well as player and iconic object positions. This map stays in the top left corner and is part of the 'heads up display'. The satellite image of the Macedon Ranges area (on the board to the left of the scene) has an emitter overlying an invisible teleport trigger. An emitter is a particle forming device that can be set to look like flames, smoke and water fountains. The player walks or runs through the emitter on the map and is teleported to the map position indicated by the base of the emitter's position on the satellite image. Teleporting can be very disorienting. The roadmap on the right hand board has an animated arrow that partly follows a road route. Through the archway, the game terrain can be seen with an overlay of geology and contours. This room is designed to be the start place for a static game where a briefing about a bushfire rescue occurs.

In *Unreal Tournament 2004* the game artificial intelligence behaviours are manifested by actors and game objects. These sets of behaviours are called Classes. One is the Monster Class. It contains the controls for a set of behaviours that play animations depending on the type of monster or creature involved and the situation encountered. Together, the game and scripting editors allow these behaviours and animations to be changed. These scripts can be adapted for use with models of animals such as the kangaroo (Fig. 27.6) or a sheep.

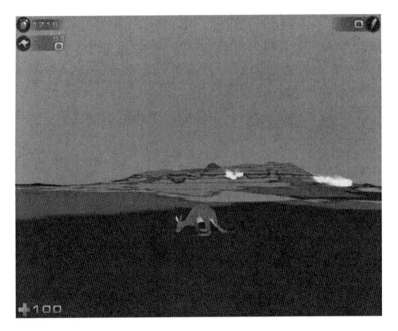

Fig. 27.6. *Unreal Tournament 2004* screen shot view of terrain with geology and scripted kangaroo

Unreal Tournament 2004 can be used to create scenes. Maps can be overlain on the terrain to provide information about a problem and promote conversation amongst participants to find a solution. These maps can be created in a GIS and display information such as fuel load in the forested areas. The white smoke clouds are bushfires controlled with timing triggers.

27.5.4 *Second Life*

MMORPGs (massively multi-playing online role playing games) like *Second Life* are multi-user games found on the Internet. *Second Life* is free to download, and players can buy buildings as well as many other pre-built models and scripts. Scripts can be written and edited in LindenScript. Editing of objects is accomplished in the game and does not require the player to exit and open a separate game editor. Height maps for an area can be imported and will then display a 'real terrain'. However ownership of an Island (called a sim) is required to access this function. This process is greatly assisted by the freely available Bailiwick application <http://www.spinmass.com/bailiwick.html>. Unfortunately maps cannot be laid over the terrain. In the Figs. 27.7 and 27.8 below, the Macedon Ranges are in reality about 10 km to the side. In the game scenes the avatars are therefore about forty times too big. This is a significant difficulty.

Fig. 27.7. *Second Life* screen shot showing terrain with links to websites shown

Second Life is minimised when a player right clicks on the board (in Fig. 27.7 this is to the right of the avatar). The same area is then opened in *Google Earth* <www.earth.google.com>. The board furthest to the right will open to a *Wikimapia* of the same area <www.wikimapia.org>. There are bushfires burning on the side of Hanging Rock.

In Fig. 27.8 the GMap-Track website <www.gmap-track.com> is open at the bottom left. The map shows the location a mobile phone with a GPS and GMap-Track application. It will also show a group of people involved in a location game. It could also show the location of emergency vehicles.

Second Life uses simple textures on the terrain; however, overlaying maps is not directly possible. The allocentric perspective view in Fig. 27.8 shows, to the front of the avatar, a model of the terrain overlain with the geology and contours, and in a second semi-transparent overlay, the road network. This model is termed a 'sculpty' in *Second Life*. The model was made using Terrain Sculptor, programmed by Patrick Rutledge (download free at http://spinmass.blogspot.com/). The ability to show block diagrams of terrain overlain with maps provides a way around the lack of an overlay capability for the terrain in a *Second Life* island (or sim). These are useful ways to introduce and debrief a mobile game, as three-dimensional models with overlays can be displayed, and maps and videos can be triggered at equivalent locations in the virtual terrain (as in the mobile game).

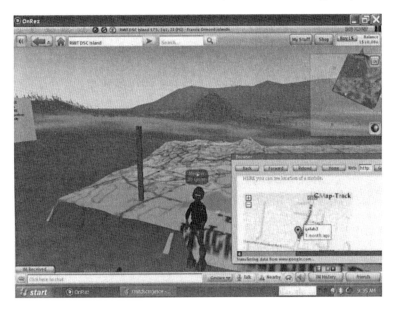

Fig. 27.8. Screen shot of *Second Life* showing OnRez viewer

This outline of several games gives an idea as to how they can be used in training, and how by making videos of scenes in the game editor they can then be displayed in another game or by projection to an audience.

27.6 The *Bushfire Rescue Game*

The *Bushfire Rescue* mobile game will use the Mscape application. Mscape uses Flash format movies and can access the web (via Bluetooth) as well as play music and recorded voice and sounds. It can also display pictures. Figure 27.9 shows the editor with the first author's farm as the base map. It shows a preliminary experiment for the development of the *Bushfire Rescue Game*. The picture at left is a movie of Hanging Rock made in the *Farcry* game. The player entering a circle triggers an event. In this case the small icon representing a player is in the circle at the bottom right and has triggered the Hanging Rock movie.

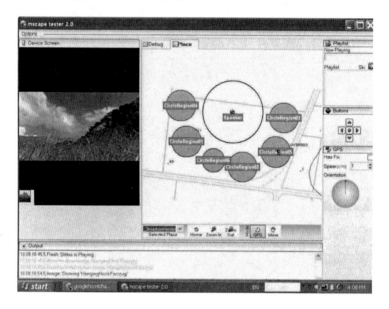

Fig. 27.9. The Mscape editor

Fig. 27.10. Mscape on the iPAQ pocket computer

A game can be seen running in Fig. 27.10. In this version the trigger circles can be seen. In the *Bushfire Rescue Game* these would be hidden. The Hanging Rock movie occupies the whole screen when it is running, replacing the map. The visualisations will be made using the computer game editors described later.

The mobile *Bushfire Rescue Game* will use visualisations of scenes delivered in Flash format recorded from computer games. They will be based on terrain of the Macedon Ranges. Some games have a better library of models for a scene involving fire trucks and airplanes, whereas others have a better library of vegetation. Vegetation can be important means of adding realism, especially smaller types like grass (Appleton and Lovett 2003). Figure 27.3 shows an example of such foreground vegetation in the *Farcry* game.

In addition to the mobile version of the *Bushfire Rescue Game*, a desktop computer game version will be made. Geographical positioning system (GPS) logs from the mobile game can be overlain on the terrain in the desktop version. Visualisations can be triggered in the computer game's terrain as in the mobile game. This allows for the debriefing of mobile game participants. It may also be used as an introduction to the game and as the basis of a stand alone game. As outlined earlier, sharing, combining and altering cognitive maps through conversation promotes understanding and better decision making, and conversations can be with a person or visualisation.

The mobile and computer game are to be designed with this in mind. However questions still to be addressed include:

- Do the games appropriately enable conversations with people and technologies to create understanding?
- Do the games allow this understanding to demonstrably inform decision making and the development of courses of action?
- Do the debriefing and introduction games appropriately assist the intended learning?

The games will be shown to members of fire fighting organisations for critical comments on their validity for training. This may assist in answering the further questions of:

- Do the games assist in learning about wind changes and their critical importance in turning an ordinary bushfire into a disaster?
- Are games like the *Bushfire Rescue Game* seen as appropriate for training by fire fighting organisations?

27.7 Conclusion

Three COTS games and a MMORG have been investigated and found to be suitable in various ways for displaying spatial and temporal information. *Trainz* is useful for simple animated scenes where free three-dimensional travel is not required. *Farcry* is a very useful game with realistic vegetation and outdoor scenery and a relatively easy to use editor. *Unreal Tournament 2004* is also a very useful game with, after much practice, a relatively user friendly editor.

The web links from *Second Life* to videos at Youtube <http://www.youtube.com> or maps and scenes at *Google Earth* and Wikimapia is a very useful feature, however it opens the websites outside of the game. Some people find this disturbing. The OnRez viewer for *Second Life* though, enables viewing of websites from within the game. The mobile *Bushfire Rescue Game*, using a mix of devices and Mscape, also allows for links to websites from within the game, in this case triggered by location. In both cases the transition is relatively seamless, minimising the feeling of disruption.

The ability to link seamlessly within games to websites on the Internet is important. In this chapter this capability is displayed by the mobile game *Bushfire Rescue* and its mix of devices, and by *Second Life* (particularly the OnRez viewer version). The capacity to better reveal the context of entities in a game, will lead to a better understanding of causes and effects. This 'conversation' enables the production of a cognitive map that sees the problem and the solution and helps develop and manage courses of action.

References

Andrienko G, Andrienko N, Jankowski P, Keim D, Kraak MJ, MacEachren A, Wrobel S (2007) Geovisual analytics for spatial decision support: Setting the research agenda. International Journal of Geographical Information Science 21(8):1365–8816. Retrieved 30 November 2007, http://www.informaworld. com.ezproxy.lib.rmit.edu.au/10.1080/13658810701349011

Appleton K, Lovett A (2003) GIS-based visualisation of rural landscapes: defining 'sufficient' realism for environmental decision making. Landscape and Urban Planning 65:117–131

Bishop ID, Rohrmann B (2003) Subjective responses to simulated and real environments: a comparison. Landscape and Urban Planning 65:261–277

Cartwright WE (2006) Exploring games and gameplay as a means of accessing and using geographical information. Human IT 8(3):28–67

DPI (2006) Bushfires report. Department of Primary Industries Victoria. Retrieve-ded July 2006, http://www.dpi.vic.gov.au/DSE/nrenfoe.nsf/LinkView/FAAF0 80E6756F7904A25679300155B2B7157D5E68CDC2002CA256DAB0027EC A3

Duggan W (2005) Coup d'oeil: strategic intuition in army planning. Retrieved 20 August 2007, http://www.StrategicStudiesInstitute.army.mil/

Foyle DC, Andre AD, Hooey BL (2005) Situation awareness in an augmented re-ality cockpit: design, viewpoints and cognitive glue. In: Proceedings of Hu-man-Computer Interaction International 2005. Las Vegas, Nevada, 22–27 July 2005. Retrieved 7 December 2007, http://human-factors.arc.nasa.gov/ihi/hcsl/pubs/Foyle_HCI2005.pdf

Germanchis T, Pettit C, Cartwright W (2004) Building a three dimensional geo-spatial virtual environment on games technology. Journal of Spatial Science 49(1):89–96

Hartley T, Bird CM, Chan D, Cipolotti L, Husain M, Vargha-Khadem F, Burgess N (2006) The hippocampus is required for short term topographical memory in humans. Hippocampus 17(1):34–48

Hermer L, Spelke E (1994). A geometric process for spatial reorientation in young children. Nature 370:57–59

Hermer L, Spelke E (1996) Modularity and development: the case of spatial reori-entation. Cognition 61:195–232

Kücklich J (2001) Literary theory and computer games. In: COSIGN 2001. Pro-ceedings of the 1st Conference on Computational Semiotics and New Media. Amsterdam, 8–9 September 2001. Retrieved 12 December 2006, http://www.cosignconference.org/cosign2001/papers/Kucklich.pdf

Jeffery KJ, Anand R, Anderson MI (2006) A role for terrain slope in orienting hippocampal place fields. Experimental Brain Research 169(2):218–25

Laurillard D (2002) Rethinking university teaching: a framework for the effective use of educational technology (2nd edn). Routledge Falmer, London

McGrath D, McGrath SP (2005) Simulation and network-centric emergency re-sponse. In: Proceedings of Interservice/Industry Training, Simulation and Education Conference, November 2005, p 1. Retrieved 12 July 2006, http://www.ists.dartmouth.edu/library/163.pdf

Mortensen T (2002) Playing with players: potential methodologies for MUDs. In: Aarseth E (ed) Game studies (issue 2). Retrieved December 2006, http://www.gamestudies.org/0102/mortensen

Naismith L, Lonsdale P, Vavoula G, Sharples M (2004) Literature review in mo-bile technologies and learning. NESTA Futurelab Series, Report 11:15. Re-trieved 5 December 2007, http://www.futurelab.org.uk/resources/documents/lit_reviews/Mobile_Review.pdf

O'Keefe J (1976) Place units in the hippocampus of the freely moving rat. Ex-perimental Neurology 51:78–109

O'Keefe J, Nadel L (1978) The hippocampus as a cognitive map. Oxford Univer-sity Press. Retrieved 7 December 2007, http://www.cognitivemap.net/

Pask AGS (1976) Conversation theory: applications in education and epistemol-ogy. Elsevier, Amsterdam

Pask AGS (1992) Interactions of Actors (IA), theory and some applications, incomplete manuscript. Retrieved 12 January 2008, http://www.cybsoc.org/PasksIAT.PDF

Packham DR, Pierrehumbert C (1990). Bushfires in Australia: a problem of the weather. Bulletin of the World Meteorological Organisation 39:21

Ryan M (2001) Narrative as virtual reality, immersion and interactivity in literature and electronic media. The Johns Hopkins University Press, Baltimore and London

Smith J (2005) The problem of other players in game collaboration as collective action. Paper presented at Digital Games Research Association 2005, Simon Fraser University, Burnaby, British Columbia, Canada. Retrieved 10 July 2006, http://www.gamesconference.org/digra2005/viewabstract.php?id=48

Tolman EC (1948) Cognitive maps in rats and men. Psychology Review 55:198–208

Von Clausewitz C (1968) On war. Penguin, London

Wickens D (2002) Situation awareness and workload in aviation. Current Directions in Psychological Science 11(4):128–133

28 Automated Generation of Enhanced Virtual Environments for Collaborative Decision Making Via a Live Link to GIS

Tao Chen[1], Christian Stock[2], Ian Bishop[1] and Christopher Pettit[3]

[1] Department of Geomatics, University of Melbourne, Victoria, Australia
[2] Cooperative Research Centre for Spatial Information, Carlton, Victoria, Australia
[3] Department of Primary Industries, Parkville Centre, Victoria, Australia

Abstract: Decision making in environmental planning and sustainable development relies on a comprehensive understanding of current scenarios and envisioning of possible future changes. With rapidly increasing computing power and graphic display capabilities, virtual environments have proven to be a powerful medium for visualising spatial information and sharing experiences about exploring a common virtual study area with other participants. The integration of virtual-environment technology and decision-making practices will allows for enhanced understanding of the real environment and its associated dynamic natural processes, hence providing an opportunity for greater involvement in community decision making. This chapter will illustrate our approach to rapidly generating an enhanced and dynamic virtual environment in real time through a live linkage as part of our visualisation package SIEVE (Spatial Information Exploration and Visualisation Environment). The live link (SIEVE Direct) provides a network link between a game engine and a geographic information system (GIS) in order to help communities envision landscape scenario changes and evaluate the decisions made in conjunction with the selected scenarios. The performance of this solution has been examined in a particular case study area of the Toolibin catchment in Western Australia.

28.1 Introduction

Visualisation is increasingly used in infrastructure planning and landscape simulation. Infrastructural entities (factories, roads, wind turbines and bridges, etc.) were rendered in three-dimensions using computer aided design. The purpose of early simulation work in industry was to impress the public with the scope of the investment and the wonderful visual representation development would make possible. Although the entities themselves were drawn, their context was often neglected. The first examples of computer-based landscape visualisation came into being in the 1970s. The early standard was set by wholly computer drawn images with standing arrows representing trees for forest applications (Myklestad and Wagar 1977).

Since the 1979 'Our National Landscape' conference on applied techniques for analysis and management of the visual resource in Nevada, USA, there have been a number of endeavours on constructing virtual landscape environments as a visual management tool for landscapes and environmentally sustainable development, such as AMAP (De Reffye et al. 1988), IMAGIS (Perrin et al. 2001) and Smart Forest (Orland 1997; Orland et al. 1994). These developments, however, did not allow people to be immersed into the virtual environment nor did they support multi-user participation in the same context. There was no anticipation that this technology could be used to interactively communicate with public communities. With the introduction of quality rendering algorithms and specialised graphic processors in personal computers, interactive simulation of changing environments began to prevail along with rapid evolution of the computer games industry in the past five to ten years. However, a trade-off between realism and interactivity remained in visualisation of landscapes. This gap had not been bridged until recent products were developed (e.g. Paar and Rekittke 2005).

Recent examples include developments that focus on the potential of virtual environments to be used interactively by members of the general public to support participation in decision making for natural resource issues (e.g. Kwartler and Bernard 2001; Lovett et al. 2002; Stock and Bishop 2005). The interaction developed between the participants and computer simulated world can take place either locally through a community workshop or remotely through networked virtual environments. However, the use of growing spatial data infrastructure (SDI) in such a scenario is still restricted as accessing the digital SDI data remains cumbersome and linking spatial analysis capabilities to three-dimensional models and simulating these models is very time consuming and resource intensive.

An interactive dynamic visualisation includes various complex representations of natural process models and a flexible transition between different stakeholder's view points for exploration in three-dimensions. It can provide a better presentation over the environmental model changes, and thus a better understanding of the real world, than a static representation or an animation clip which is identical for all users (Rheingans 2002). For example, farmers are concerned about how their crops will be affected by the increasing salinity over a period of time while ecologists may be interested to examine the dieback of tree species and the loss of habitat created by the same salinity. GIS and remote sensing packages are currently the most commonly used solutions for addressing such environmental and agricultural issues. However:

> 'general purpose GIS were originally designed for the classical two-dimension application areas, and hence often do not reflect the state of the art landscape visualisation. Moreover, if a GIS implements its own visualisation subsystem, it will be difficult or even impossible to integrate cutting-edge visualisation technology' (Döllner 2000, p.1).

A general purpose GIS traditionally using two-dimensional geographic data for analysis and display can no longer satisfy the complex real time dynamic visualisation demands and cannot be deployed as a generic platform for virtual environments. Therefore, independent visualisation toolkits have to be used for developing the visualisation of virtual environments (Döllner 2000).

Nowadays, computer game technology offers an option for improving the seamless and rapid integration of spatial data, functionalities and fluid realistic and unrestrictive visualisation for generating virtual environments. Today's sophisticated game software fuses several highly integrated packages called engines, each with a specific functionality. A terrain engine creates and renders the ground, while a physics engine handles the motion behaviour of characters and objects, such as realistic simulation of gravitational forces and kinetics (Herwig et al. 2005). A graphics user interface creation toolset and in-game editors enrich the experience of a user-friendly interaction. A server–client networking structure and customised multiple user interface with a rich set of avatar models supports multiple participants to share and explore the common world with customised view points and the ability to communicate with each other in a chat box.

The research presented in this chapter draws on a game engine-based *SIEVE* (Spatial Information Exploration and Visualisation Environment) and a real-time bidirectional coupling solution with the GIS called *SIEVE Direct*. The objectives are:

- To design a process for rapid generation of a dynamic virtual environment and visualisation of the associated landscape features and complex scenario changes.
- To populate the virtual environment with advanced analytical and display functions for better interaction with and understanding of the environment, hence helping public communities and policy makers to make better decisions.

28.2 Background

SIEVE was designed to enhance the understanding of landscape scenario outputs and associated impacts by allowing multiple users to explore and discuss data collaboratively, hence leading to making better decisions with regard to landscape planning and environmental management. This virtual environment simulation package was developed from the commercial low-cost *Torque Game Engine* (TGE) (see O'Connor et al. 2005 for more details). Virtual environments of user specified areas can be easily generated through an integrated linkage of GIS software to the game engine. The TGE platform was chosen as it is suitable for landscape simulation by providing:

- efficient scene rendering
- support of a range of three-dimensional format conversion plug-ins from various independent vendors in the marketplace
- interactive manipulation of surface objects
- in-built provision for multiple user access through networks (this is particularly important).

One of its critical components is called *SIEVE Builder* which converts SDI-based geographic information in the selected area such as a DEM, a ground texture image and optional point features linked to three-dimensional objects to a mission file (.mis) and a terrain model (.ter). A component called *SIEVE Viewer* loads up the mission and terrain files and renders the resulting three-dimensional virtual world.

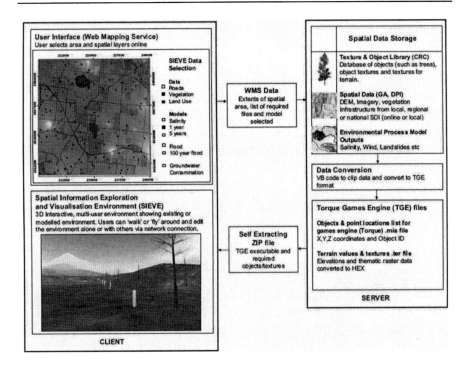

Fig. 28.1. *SIEVE* components (O'Connor et al. 2005, p. 3)

However, this is a one-way process (Fig. 28.1) from two-dimensional spatial data to a virtual environment. A geographic feature in geodatabase is represented as a three-dimensional landscape object in the scene graph of *SIEVE*. Similar to an XML (eXtensible Markup Language) compliant format, a generic mission file contains a description of the environmental nature (e.g. sky, sun and atmospheric features) as well as the properties of the three-dimensional landscape objects in the area. The complexity of different types of three-dimensional objects supported by TGE varies. A simple DTS format three-dimensional object only has the properties of: object identification, model-file name and store location, coordinates (x, y, z) in the *Torque* world system, size and rotation — while a FoliageReplicator object includes up to 30 more attributes such as foliage count, minimum and maximum size, view distance, cull resolution, fading-in and -out distances, light reflectance and so on.

It takes *SIEVE Viewer* on average ten minutes to load up a mission file for the environment containing approximately 10,000 individual three-dimensional DTS objects. In the previous development, if the geodatabase was altered (e.g. appending or deletion of geographic features) a new mission file had to be created by the *SIEVE Builder* from scratch to keep the virtual environment up-to-date to the changes made. To make the scenario editing and updating easier and faster, users can now immediately see the suggested changes made in the geodatabase and evaluate the resulting impacts on the environment at run-time manner instead of repeating the whole exporting process (Chen et al. 2006). This makes possible dynamic simulation on how a virtual environment will behave in response to the updated geodatabase in real-time. Moreover, object manipulation (e.g. appending, editing or deletion of three-dimensional landscape objects) in the virtual environment can be transferred back via a return path to update the geodatabase.

28.3 Methodology

SIEVE Direct, an extension to the existing *SIEVE* system, serves as a means of live linkage between GIS and a virtual environment built at runtime. It has been developed largely based on the Client/Server (C/S) structure that enables data transfer between GIS software and *SIEVE* over a local area network (LAN) or the Internet. ESRI *Desktop ArcGIS* was chosen as it provides a built-in Integrated Development Environment (IDE) for developers to program with powerful ESRI *ArcObjects* to tailor their various customisations.

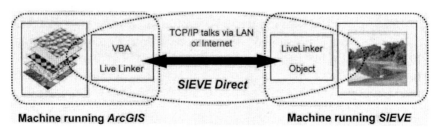

Fig. 28.2. Overview of *SIEVE Direct* architecture

SIEVE Direct consists of two main software components (Fig. 28.2). GIS-to-*SIEVE Live Linker*, a VBA (Visual Basic for Application) program with the embedded Microsoft Winsock ActiveX Control and Windows Socket Application Programming Interface (API), as well as some necessary libraries supported by ArcObjects in *ArcGIS*, serve as a common graphic user interface (GUI) designed to send data to *SIEVE* for creating and updating the virtual environment and to receive data from *SIEVE* for updating the geodatabase. On the *SIEVE* side we customised the TGE source code in C++ by creating a class called LiveLink and adding it to the

TGE's network layer that handles receiving and delivering data packets through TCP/IP (Transmission Control Protocol/Internet Protocol) via the network. The procedure called 'Process' in a LiveLink object performs validating, interpreting and parsing the in-coming data packets to separate character strings and feeds them in sequence into the appropriate type of three-dimensional model for processing. During the initialisation of *SIEVE*'s game libraries, a network socket is automatically bound to a fixed port number on the IP address of the computer and *SIEVE* starts to listen to requests for connection and data transfer as soon as it is completely launched. On the other side, a socket is created and bound to a fixed port on the IP address of the computer where *ArcGIS* is running during the initialisation of the GIS-to-*SIEVE Live Linker* program, and the program is ready to send requests for connection to the listening port and afterwards for data transfer with *SIEVE*. Through TCP/IP, therefore, both applications are able to communicate and exchange data via a connection established by two sockets bound on each computer IP address and port number through the network.

TGE was inherently designed from the foundations up to offer robust client/server networked simulations. The design for the networking model was driven by performance over the Internet. It has successfully managed to deal with three fundamental issues of network simulation programming: limited bandwidth, packet loss and latency (covered in TGE 2007 Chapter 6). This enables a robust and sound synchronisation between server and client. An instance of *SIEVE* can be set up as a dedicated server, a client, or both a client and a server. If the scenario is a client and a server, it still behaves as a client connected to a server — instead of using the network, however, a NetConnection object on the network layer allows for a short-circuit link to another NetConnection object in the same application instance. This is particularly useful for using *SIEVE Direct* on a local machine. In a distributed decision-making environment, if an instance of *SIEVE* is set up as dedicated server and it is used to communicate with a remote GIS, instances of *SIEVE* as client are able to share the common virtual space and immediately reflect the environmental model changes on landscape as driven by the GIS users, or GIS-based modelling, via the network.

SIEVE Direct relies on an application message request/response mechanism over TCP/IP. A message, mostly known as a data packet, is designed in ASCII format which is easy to read and interpret. It consists of two components: a unique data header and data items in sequence that make a character string. Data header and items are delimited by space. For instance, when a message is received by *SIEVE*, a LiveLink object will validate and handle the received message. Depending on the header information, the delimited data items will be parsed and fed into the pre-defined procedure for process. To perform the function of position synchronisation from two-dimensional map to three-dimensional world, for example, the data string is designed as:

```
pos <transformed coordinate value along x-axis in SIEVE world>
<transformed coordinate value along y-axis in SIEVE world >
```

in which the x and y values are calculated through multiplying the actual x and y values in Universal Traverse Mercator (UTM) system by the ratio of Torque world size to the actual area size. Upon connection with *SIEVE* the *ArcGIS* user can simply click the mouse on anywhere within the area of interest on the two-dimensional map, and *SIEVE* will acquire the delivered message, parse and feed it into the procedure that re-sets the avatar's position in the virtual environment in accordance with that of the avatar on the two-dimensional map. Meanwhile, movement of the avatar in *SIEVE* can also be continuously reflected and symbolised by that of a pre-set icon on the two-dimensional map.

In addition to position synchronisation between two-dimensional map and the three-dimensional virtual environment, there are a wide range of functionalities implemented in *SIEVE Direct* including:

- rapid automated revegetation simulation in the virtual environment based on Ecological Vegetation Class (EVC) types from the geodatabase by a FoliageReplicator object in *SIEVE* (Fig. 28.3)
- editing of vegetation cover types and other parameters such as density, plant instance size, growing speed and visibility distance
- using *ArcGIS* to create and edit selected features in association with their represented three-dimensional models and the impact on the virtual environment of the suggested changes will be reflected in *SIEVE* (e.g. variation of underground watertable as a result of the revegetation practice, and death simulation of vegetation caused by flooding). The visual effects can be immediately examined
- using *ArcGIS* to query and zoom in to any selected feature that has been rendered in *SIEVE*
- transferring edits on properties of the selected landscape feature in the virtual world back to geodatabase for updating.

Being an extension to *SIEVE*, the *Direct* facility was designed to run on the basis of a start-up mission and a terrain model that make a virtual environment. Both of them can be generated by *SIEVE Builder* as the first step.

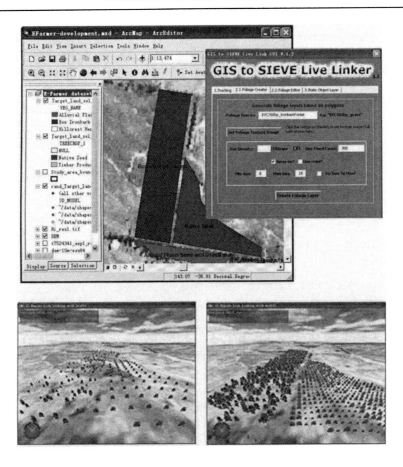

Fig. 28.3. *ArcGIS* users can control the vegetation cover changes in *SIEVE*

Such a preliminary virtual environment can either be a barren landscape or contain the initial surface features exported by *SIEVE Builder*. Various landscape features represented by points and lines in a geodatabase, such as power plants, cottages, trees, rocks, roads and fences, are in the TGE-specific DTS or DIF (Digital Interface Format) file formats that can be directly exported by 3DMax from the common three-dimensional file formats (e.g. 3DS and OBJ). For rapid rendering a large volume of objects of the same type, such as a cluster of grass, a dozen hectares of eucalyptus trees or a large flock of sheep that is normally stored in a geodatabase as polygon, it is preferred to use FoliageReplicator or ShapeReplicator class in *SIEVE*. They share very similar properties and are able to construct up to tens of millions of objects with a similar appearance in just a few seconds. *SIEVE Direct* facility enables users to specify properties — such as object appearance file, count of object instances, object size range, minimum and maximum allowed terrain slopes and sway magnitude — and render a FoliageReplicator or ShapeReplicator object on the fly. A FoliageReplicator object uses a two-dimensional PNG (Portable Network Graphics) image file for appearance and instances of the object always rotate to face the viewer in simulation. In contrast, a ShapeReplicator object uses a three-dimensional DTS file for object appearance. Although ShapeReplicator class provides a better visual effect in simulation than FoliageReplicator, FoliageReplicator consumes less computing resource memory than its counterpart and it generally suffices in a massive landscape simulation.

28.4 Case Study and Discussion

The functionality of *SIEVE Direct* can be extended on demand according to various simulation requirements for various landscape scenarios. In addition to offering significant visual representations to policy makers to help make better decisions, this may benefit a range of professional users including foresters, rural and urban planners, and landscape architects. The benefits of using a virtual environment that live-links to GIS to enhance perception of and better interaction with our rural landscape for the agricultural communities have been explored and evaluated using the Toolibin catchment area in Western Australia (Fig. 28.4).

Fig. 28.4. Aerial view of the Toolibin catchment area and Toolibin Lakes (picture adopted from George 1998)

The Toolibin catchment is located approximately 250 km south-east of Perth at the headwaters of the Blackwood/Arthur River drainage system in the central wheat belt of Western Australia. It has an area of approximately 47,000 ha and an average annual rainfall of 400 mm. Like much of the region it has been extensively cleared (90%) for agriculture over the past ninety years. The catchment now faces serious problems with secondary salinity (approximately 6% of the catchment has been severely salt-affected) and it is predicted that at equilibrium, as much as 25% to 30% could become saline if the water balance is not altered and groundwater levels continue to rise (George 1998). An imbalance between incoming rainfall and the water use by current farming systems has brought about substantial rises in watertables and widespread salt encroachment. The cumulative effects of dryland cropping and grazing in a Mediterranean climate are also taking their toll, and today waterlogging, erosion and soil structure decline are posing serious threats to agricultural productivity (Hearn 1988).

Lake Toolibin (32°55'S, 117°36'E), situated at the southern end of the catchment remains one of the few relatively fresh inland wetlands in the south-west of Western Australia and is an important breeding habitat for native waterfowl. Its deterioration in recent years has resulted in an inter-departmental investigation aimed at saving the lake. In order to reduce the salinity level and the negative impact on the ecological system led by the salinity issue in the area, it is required that massive revegetation be integrated into the farming systems. These trees are intended to help increase water use and help control salinity, provide soil and livestock shelter, and also may provide commercial returns.

Using *SIEVE*, we have developed a three-dimensional dynamic virtual environment that links to the back-end GIS for the study area in the Toolibin catchment. The entire study area is approximately 24,050 m by 24,050 m and the minimum elevation is 294 m with a height range of 94 m.

Major tree species in each of nine vegetation classes number from three to seven, with four dominant species in the study area: acuminate jam, York gum, salmon gum and mallet (Fig. 28.5). The objective of this project is to help agricultural professionals and rural communities to visualise surface vegetation changes over time, envision desired future landscape scenarios after a simulated revegetation practice, and improve understanding of groundwater level changes resulting from the revegetation. As a result, better decisions regarding revegetation and land use in the catchment can be expected.

Fig. 28.5. Aerial view for revegetation at Toolibin Lake in Western Australia (source: Dept. of Conservation and Land Management, Western Australia, 1995–96) and four typical tree species in the Toolibin catchment

The entire simulation system is based on *SIEVE Viewer* and ESRI's *Desktop ArcGIS* running on separate computers, whose IP addresses are known, and connected via the network. The GIS contains the DEM, aerial texture imagery for the study area, and the initial environmental models and process for the current scenario, and those for the five sequential scenarios. Each simulated scenario spans ten years, and contains a certain magnitude of vegetation (approximately 2,000 trees of various types, see Fig 28.6) that are, as a whole, represented by a single polygon layer in *ArcGIS*, as well as an underground watertable model. Given the fact that each polygon in a particular layer may embrace multiple vegetation species, within that layer, the ecological class name or bio-types, the identification numbers (IDs) of vegetation species contained and general plant density are stored as a record in the GIS attribute table. In association with the species ID, the name or alias, the store location of the three-dimensional model file, and the minimum and maximum heights for the particular species are stored in a separate text file for the species properties in the machine running GIS. The actual three-dimensional model files for various vegetation species are stored in the machine running *SIEVE*.

Fig. 28.6. Screen shot of the virtual environment of the Toolibin catchment simulated by *SIEVE*

The simulation starts by an initial scenario that represents current condition with approximately 15,500 landscape features including trees, shrubs, grasses, large rocks, cottages and an underground watertable. Once a *SIEVE Direct* connection is established between *ArcGIS* and *SIEVE*, users can simply click a button on the toolbar of *ArcGIS* to launch the revegetation program. Starting with the first scenario through to the fifth one respectively, *SIEVE Direct* automatically goes through every polygon feature in the layer, reads the attribute table with the ecological class, species IDs contained and plant density, and links the ID to the corresponding species property file. The coordinates in the Universal Traverse Mercator project (UTM) of each vertex of the polygon are transformed to those in the Torque World system for positioning the plant cluster. Species count in that region can be calculated by the region area and plant density. Once all necessary properties for the polygon are prepared, the *SIEVE Direct* link organises and wraps it in a message string with an attached header and sends the entire message string to *SIEVE* over the network. Upon receiving the message string, *SIEVE* validates the header and parses the rest of the message string to separate parameters to construct a FoliageReplicator or ShapeReplicator object on the fly (depending on the input model file). Along with the automated vegetation placement, ID number, model file and proposed elevation of the watertable for each of the five scenarios are also sent via *SIEVE Direct* to change the underground watertable level and appearance upon completion of the revegetation for that scenario. Figures 28.7–28.10 show four simulated virtual scenarios of the potential revegetation process for the Toolibin catchment area in Western Australia.

In addition, users are able to pause and restore the procedure of revegetation at any time through the *SIEVE Direct* link and explore and make comments on the virtual environment being changed during the course of revegetation. In this way, users are able to visualise the model output and dynamic scenario of revegetation simulation and increase of watertable level, and envision land cover changes over a certain period of time.

Properties of all rendered objects in *SIEVE* are stored in the Content Tree that displays when users toggle the Editor Mode. Every object is assigned a unique identity during its construction on the fly, which is a combination of the identities of the input layer, feature and species from GIS, so that *Arc*GIS users can easily fly to the object of their interest in *SIEVE* by simply highlighting the feature in GIS and sending the request via the *SIEVE Direct* link.

Automated Generation of Enhanced Virtual Environments   585

Fig. 28.7. Screen shot of simulated virtual scenario: current condition

Fig. 28.8. Screen shot of simulated virtual scenario for year 2020

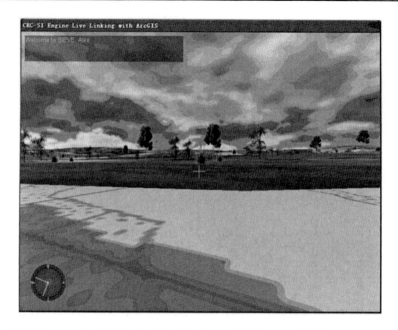

Fig. 28.9. Screen shot of simulated virtual scenario for year 2030

Fig. 28.10. Screen shot of simulated virtual scenario for year 2040

28.5 Conclusion and Outlook

Virtual environment technology, as a powerful medium for visualisation of geographic information (MacEachren et al. 1999), allows for enhanced interaction with, and understanding of, landscape function, structure and behaviour as well as its associated dynamic natural processes. Furthermore, it enables an integration of the outcomes of social, economic and environmental (triple bottom line) analysis through the visual sense to convey meaning to a broad range of participants including rural and regional planners, managers, communities, stakeholders and policy makers (Pettit 2006).

We have developed a landscape simulation methodology called *SIEVE Direct* that allows for rapid automated generation of high performance immersive virtual environments enhanced by the user specified landscape alternatives to help people learn about landscape systems and examine the likely consequences of practice changes. This is achievable through the use of some underpinning e-science technologies. In this research, we use TCP/IP message exchange to keep both geodatabase and three-dimensional virtual representation staying in synchronisation at run-time. The methodology presented in this chapter can be easily adapted to other study areas given sufficient data, appropriate model calibration and certain customised development. In future research, we will incorporate the functions of *SIEVE Exporter* and *SIEVE Direct* for quicker processing. We will also explore the possibility of developing a smart virtual landscape through *SIEVE Direct*. Such a virtual landscape will intelligently respond to the environmental variables, such as levels of soil erosion, salinity and moisture applied by a geodatabase, and it will automatically change land or vegetation covers accordingly.

Acknowledgements

The Cooperative Research Centre for Spatial Information (CRCSI) who funded the *SIEVE* development, Department of Primary Industries (DPI), Victorian Government and the University of Melbourne are acknowledged for their contribution to this chapter.

References

Chen T, Stock C, Bishop ID, O'Connor A (2006) Prototyping an in-field collaborative environment for landscape decision support by linking GIS with a game engine. Paper presented at the 14th International Conference on Geoinformatics, 28–29 October 2006, Wuhan, China

De Reffye P, Edelin C, Francon J, Jaeger M, Puech C (1988) Plant models faithful to botanical structure and development. Computer Graphics 22:151–158

Döllner JKH (2000) An object-oriented approach for integrating 3D visualisation systems and GIS. Computers and Geosciences (26):67–76

George RJ, Beasley R, Gordon I, Heislers D, Speed R, Brodie R, McConnell C, Woodegate P (1998) Evaluation of airborne geophysics for catchment management for agriculture, fisheries and forestry – Australia and the National Dryland Salinity Program. The National Airborne Geophysics Project, Land and Water Australia, Canberra

Hearn SJ (1988) Soil conservation and management strategies for the Toolibin catchment. Resource Management Technical Report No.75. Department of Agriculture and Food, Western Australia

Herwig A, Kretzler E, Paar P (2005) using games software for interactive landscape visualisation. In: Bishop ID, Lange E (eds.) Visualisation in landscape and environmental planning: technology and applications. Taylor & Francis, London, pp 62–67

Kwartler M, Bernard RN (2001) CommunityViz: an integrated planning support system. In: Brail RK, Klosterman RE (eds) Planning support systems: integrating geographic information systems and visualisation tools. ESRI Press, Redlands, California, pp 285–308

Lovett A, Kennaway J, Sünnenberg G, Cobb D, Dolman P, O'Riordan T, Arnold D (2002) Visualising sustainable agricultural landscapes. In: Fisher P, Unwin D (eds) Virtual reality in geography. Taylor & Francis, London, pp 102–130

MacEachren AM, Edsall R, Haug D, Baxter R, Otto G, Masters R, Fuhrmann S, Qian L (1999) Virtual environments for geographic visualisation: potential and challenges. In: Proceedings of the 8th International Conference on Information and Knowledge Management, Association for Computing Machinery, 2–6 November 1999, Kansas City, Missouri

Myklestad E, Wagar JA (1977) PREVIEW: computer assistance for visual management of forested landscapes. Landscape Planning 4:313–331

O'Connor A, Bishop ID, Stock C (2005) 3D visualisation of spatial information and environmental process model outputs for collaborative data exploration. In: Proceedings of the 9th International Conference on Information Visualisation (IV'05), 6–8 July 2005, Greenwich

Orland B (1997) Smart Forest-II: forest visual modeling for forest pest management and planning. Retrieved 4 August 2006, from http://www.imlab.psu.edu/smartforest

Orland B, Radja P, Su W (1994) Smart Forest: an interactive forest data modelling and visualisation tool. In: Proceedings of the Fifth Forest Service Remote

Sensing Applications Conference, 11 April 1994, Salt Lake City, Utah, pp 283–292

Paar P, Rekittke J (2005) Lenné3D – walk through visualisation of planned landscapes. In: Bishop ID, Lange E (eds) Visualisation in landscape and environmental planning: technology and applications. Taylor & Francis, London, pp 152–162

Perrin L, Beauvais N, Puppo M (2001) Procedural landscape modeling with geographic information: the IMAGIS approach. Landscape and Urban Planning 54:33–47

Pettit C (2006) Visualisation techniques for envisioning landscape futures. ISPRS II/5 Workshop, 8 July 2006, Institute for Geography and Regional Research, University of Vienna

Rheingans P (2002) Are we there yet? Exploring with dynamic visualisation. Institute of Electrical and Electronic Engineers Computer Graphics and Applications 22(1):6–10

Stock C, Bishop ID (2005) Helping rural communities envision their future. In: Bishop ID, Lange E (eds) Visualisation in landscape and environmental planning: technology and applications. Taylor & Francis, London, pp 145–151

TGE (2007) Torque Game Engine online user documentation and technical forums. Retrieved 20 August 2007, from http://www.garagegames.com

29 Land Use Decision Making in a Virtual Environment

Lucy Kennedy and Ian D Bishop

Department of Geomatics, The University of Melbourne, Victoria, Australia

Abstract: This chapter outlines a study that investigates the use of a virtual decision-making environment as a tool for better understanding individual land use choice behaviour. This chapter describes a preliminary study into the extent to which peoples' land use decisions are affected by varying the visual and social context provided in the virtual environment. Early results indicate that individuals' land use choices do vary under different configurations of the virtual environment, although not always as anticipated. Under more complex configurations of the virtual environment, users tend to deviate from what would be predicted based on their stated value priorities. Several users found that the three-dimensional visualisation component of the virtual environment was useful for conveying information, but it was not shown to affect their behaviour. Adding social context to the virtual environment, in the form of an agent-based model simulating the behaviour of neighbours, resulted in minor changes to users' behaviour, particularly when the agents were programmed to make environmentally-friendly land use choices. Following further development and testing, it is anticipated that the virtual land use decision-making environment could be a useful tool for gaining a better understanding of individual decision making. This information could then be used as the basis of improved models of individual land use choice behaviour, and the development of more effective land use policy in response to significant changes in decision contexts such as climate change.

The chapter begins with a brief overview of some key theories about people's decision-making behaviour, namely rational decision making and the role of values and attitudes in determining people's behaviour. This is followed by a description of the components of the virtual environment, including an agent-based model, three-dimensional visualisation, and a user interface built in Microsoft Access. *The series of experiments conducted as a preliminary investigation in to the use of the virtual environment are then described, and the results discussed. Overall, some general trends were found, but the limited statistical significance of the results indicates that people's decision-making processes within the virtual environment are more complex than originally anticipated, and more testing will be required to fully understand this.*

29.1 Introduction

The development of a virtual land use decision-making environment as a tool for capturing and investigated individual behaviour offers several advantages over other data collection methods. Stated preference studies (asking people to rate their preferences for different features, or what they would do in different scenarios) are simple to conduct, but the answers provided may not match how respondents would behave in the real-world. This limitation is overcome by revealed preference methods, which involve investigating people's behaviour in the real-world; however this approach also has drawbacks, namely a lack of control over the experimental conditions, and difficulty in inferring individual behaviour from large-scale trends (Swait and Adomowicz 2001).

Using features from the field of experimental economics, a virtual land use decision-making environment has been developed for this study. The virtual environment endeavours to bridge the gap between stated and revealed preference techniques by allowing for investigation of individual land use decision making in a simulated rural environment — with control over the social, economic and environmental variables that may influence land use choice behaviour.

This chapter describes a study conducted to explore the use of the virtual land use decision-making environment as a tool for understanding individual land use choice behaviour. The study investigated the following research questions:

- Does knowledge of the choices made by neighbouring farmers in the region (social context) influence people's land use choice behaviour?
- Do variations in the mode of visual representation of information (visual context) influence people's land use choice behaviour?
- To what extent are people's economic and environmental attitudes correlated with their land use choice behaviour?

29.2 Rational Decision Making

There are a number of theories that have been proposed to explain people's land use choice behaviour. While their foundations are theoretically sound, their application to complex real-world scenarios is limited.

The Theory of Rational Choice plays a significant role in economics; the notion that a 'rational' decision maker will choose, from a set of alternatives, the one that will be of maximum benefit to them. The alternative the individual chooses is optimal, according to the individual's utility function, which is based on the individual's unique set of preferences, motivations and goals (Simon 1997).

There are a number of assumptions underlying this theory. Firstly, the theory assumes that each individual has a utility function, which is given prior to calculations and based on their preferences and motivations. The individual attempts to maximise this utility function when making a decision, thereby allowing them to make consistent choices among all possible alternatives. The theory also relies on the notion that the individual is aware of all possible options, and has unlimited computational abilities and foresight to know the consequences of each.

In the real-world, these assumptions are rarely met. For example, the preferences and motivations held by an individual — their utility function — are not necessarily known explicitly even by the individual. Furthermore, in all but the simplest real-world scenarios, the consequences of the full set of decision alternatives cannot be fully calculated. Even if it were possible, the time and money required to do so would be prohibitive in most situations. When combined with the impact of external factors, such as social networks, which can also have an influence, the decision maker will invariably make inconsistent choices.

29.2.1 Values, Attitudes and Behaviours

Values can be considered ideals or standards used by people to guide and evaluate their lives and the world around them (Rokeach 1973). While values are generally thought of as an abstract concept, attitudes are the expression of values in a specific context or situation. Values and attitudes are often inherently linked with behaviour. The attention given to values and attitudes by social psychologists actually stems from the presumption that they are the precursors to social behaviour (Rokeach 1973). As Bardi and Schwartz (2003, p. 1207) highlight:

'unless there is a clear link between values and behaviour, there is little point in efforts to establish and change values in daily conducts, such as in education and the mass media'.

In terms of developing effective land use policy, understanding the link between people's values, attitudes and their behaviour has significant implications. Programs designed to increase pro-environmental attitudes, for example, are of little benefit if the shift does not lead to the subsequent adoption of more environmentally-friendly behaviours (Gardner and Stern 2002).

Ajzen's (1991) Theory of Planned Behaviour provides a formal theory to describe how an individual's behaviour is driven by their attitudes and beliefs. The theory states that:

'...a person's intentions to perform a particular behaviour are the immediate antecedents to the actual behaviour. These intentions themselves result from both one's attitudes and the subjective, value-based assessment of the norms of one's society or group' (Bell et al. 1978, p. 33).

There is some question over the extent to which the Theory of Planned Behaviour holds true in the real-world. De Oliver (1999), for example, found that there was very little correlation between people's stated attitudes towards water usage, their self-reported water usage, and the amount of water they actually used, as measured from their water meters. Although this also demonstrates the potential problems in self-reporting techniques often used to collect data, it highlights that the link between people's attitudes and behaviour is not always clear.

29.3 Methodology

For policy makers, the value of any behavioural theory depends in its ability to describe or predict behaviour in real-world situations. While the decision-making theories outlined above offer some insight into simple decision-making processes, their application in complex real-world situations is limited. Understanding real-world decision making requires studying real-world behaviour to discover if, or under what conditions, the theories hold true. The virtual environment used in this study has been developed to enable individual land use choice behaviour to be observed, and the influence of different variables to be investigated, in a controlled environment. Participants in the study took part in a series of six choice experiments, making land use choices for three parcels of land over 25 years. In each of the six experiments, the information provided to them, or the level of visual and social context, was varied. The economic and environmental consequences of the land use choices made by the subjects were then compared to their value priorities, as measured by a short version of the Schwartz value survey (European Social Survey 2007; Lindeman and Verkasalo 2005).

The virtual environment simulates the social, economic and environment aspects of the real-world rural environment in which land use choices are made. Details of this are outlined below.

29.3.1 Social: Agent-based Modelling

Agent-based models (ABM) involve modelling individual or micro-scale decisions that lead to changes and patterns emerging at the macro-scale. The ABM used in the virtual environment simulates the land use choices being made by other farmers in the region. The model is based on the *FEARLUS* model developed at the Macaulay Land Use Research Institute (Gotts et al. 2003). The model has been converted to a vector format (rather than a raster format) to make it more suitable for use in a particular context in which all agents can be considered as neighbours.

In the virtual environment the ABM provides social context for the user. In the real-world a farmer may be influenced by their neighbours either through simply observing their farming behaviour, or through more formal interaction such as Landcare groups. This type of information would provide the farmer with some understanding of the extent to which their neighbours are employing environmentally-sustainable practices. They are also likely to have at least some idea about the financial status of other farmers in the region, through informal indicators such as the car they drive, the house they live in and the school they send their children to. In the virtual environment the environmental and economic consequences of the neighbours' decisions are conveyed to the user via a simple thematic map. The map is updated at every time step in the simulation to reflect the changing land use types chosen by each neighbour.

29.3.2 Environmental: Three-dimensional Visualisation

The user can see their land parcels, and those of neighbouring farmers, in a three-dimensional representation of the landscape. The visualisation is updated, in real time, each time the land use decisions are made by the user and their neighbours. The three-dimensional visualisation of the landscape is generated using *SIEVE* (*Spatial Information Exploration and Visualisation Environment*), a virtual environment package developed by Cooperative Research Centre for Spatial Information, Australia. *SIEVE* is a virtual environment built on the *Torque Game Engine* platform. The package allows for two-dimensional data stored in a GIS to be viewed and explored in three-dimensions. An exporter tool, supplied with the software and installed in *ArcMap*, converts two-dimensional spatial information in addition to image files stored in an image library, into a three-dimensional virtual world. A live link between the GIS and *SIEVE* can be created using another extension installed in *ArcMap* (see Chen et al. 2007 for further details). For this study, this live link enables land use changes made by the user, and their neighbours', to be updated in the GIS (the user does not see this step), then viewed in three-dimensions through *SIEVE*. The user can then navigate their way around the environment and observe the changes occurring in the landscape.

The visualisation provides visual context for the user, enabling them to see the environment they are working in, in a format that is familiar (Fig. 29.1). This may allow the user to more easily interpret the changes occurring in the landscape. It is anticipated that the use of *SIEVE* to generate a three-dimensional representation of the features and terrain of an actual landscape will be particularly useful for future studies using the virtual environment. If participants from a rural community were viewing their own properties in the virtual environment, their decisions are likely to more accurately reflect the real decisions they would make.

Fig. 29.1. Three-dimensional visualisation generated in *SIEVE* showing (**a**) high maintenance grazing, (**b**) low maintenance walnuts

29.3.3 Economic: Experimental Economics

The virtual environment uses aspects of experimental economics methods. Experimental economics involves creating choice experiments in a controlled laboratory environment. The virtual environment described in this chapter adds realism to a standard laboratory environment through the use of the visualisation and ABM.

In the virtual environment the user has a range of land use types to choose from. Each land use type has economic and environmental consequences. Information on these consequences is contained in a simple *Access*-based database that also shows the user how their wealth and environmental impact is changing over time (Fig. 29.2). As in the real-world, the user can look at as much or as little of the information available to them as they think is necessary to arrive at a land use choice.

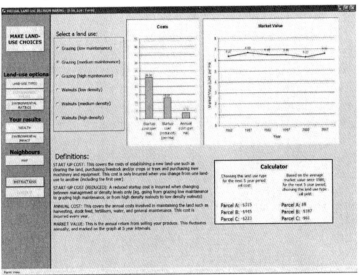

Fig. 29.2. Screen shot of the economic information provided to the user: graphs showing start-up and annual costs and returns for medium maintenance grazing

29.3.4 Experiment Design

The virtual environment described above was used to measure participants' land use choice behaviour. Each participant completed six experiments, with varying visual and social context provided in each. This provided data that could be used to investigate the influence that the three-dimensional visualisation and ABM components of the virtual environment had on the land use choices made by participants. Participants also completed a questionnaire to measure their value priorities, which were then compared with their behaviour under the different experimental conditions.

Participants

Twenty-nine people participated in the study. Subjects included: students from the Department of Land and Food Resources (11), the Department of Architecture (5) or the Department of Geomatics (6); as well as University of Melbourne academics, or professionals with experience with geographic information systems (7). Consequently, there was significant variation in subjects' knowledge and experience of both farming and spatial concepts such as mapping and three-dimensional visualisation. The subjects' were asked to rate their experience and knowledge of these between zero (no experience) and five (expert). Subjects were chosen from these diverse backgrounds to allow for investigation of whether an individual's background influences their use of the virtual environment (for example, Do Geomatics students rely more on the two-dimensional and three-dimensional visualisation? Do Land and Food Resources students focus more on the land use types and their economic and environmental consequences?), and their subsequent land use choice behaviour.

Scenario Development

In each experiment participants were asked to pretend to be a farmer and manage three parcels of land for 25 years. At the end of each five year time span, participants were required to make land use choices for their parcels of land from two land use types, grazing or walnut plantation, each of which had three management levels (i.e. grazing low maintenance, grazing medium maintenance, grazing high maintenance, walnuts low density, walnuts medium density or walnuts high density).

Each of the six alternatives had varying economic and environmental consequences, the details of which were available to the participant in the form of graphs and tables. The choice of land use types required users to trade-off the environmental and economic consequences of the six land use type options depending on their personal goals in the game. Participants were paid performance based incentives to encourage them to make more efficient decisions. Efficiency was not based on either economic or environmental performance alone but on achievement on both scales relative to possible outcomes.

Experimental Modes

Participants took part in six experiments, each involving a different configuration of the virtual environment. In each experimental mode the visual and social context provided to the user was altered to allow for a better understanding of the influence of each of these aspects on individual land use choice behaviour. The economic and environmental consequences of each land use type remained the same in each experimental mode, so that the impact of the social and visual context could be isolated. The six experimental modes are described below:

- Mode 1: Participants make land use choices in isolation. The economic and environmental consequences of each land use type option are described, but there is no visualisation and no neighbours.
- Mode 2: Participants see their neighbours' decisions on a two-dimensional GIS map, along with their economic and environmental consequences. Neighbours are programmed to tend towards making environmentally-friendly land use choices.
- Mode 3: Participants see their own, but not their neighbours', land parcels in the three-dimensional visualisation.
- Mode 4: Participants see their own, and their neighbours', land parcels in the three-dimensional visualisation. Neighbours are programmed to tend towards making environmentally-friendly land use choices.
- Mode 5: Participants see their own, and their neighbours', land parcels in the three-dimensional visualisation. Neighbours are programmed to tend towards making economically-driven land use choices.
- Mode 6: Participants see their own, and their neighbours', land parcels in the three-dimensional visualisation. Neighbours are programmed to make random land use choices.

Measuring Value Priorities

A short version of Schwartz's value survey was used to measure participants' value priorities (European Social Survey 2007; Lindeman and Verkasalo 2007). Schwartz's Value Theory is based on the premise that that there is a universal set of values held by all people, but individuals vary in priority they place on each of the values in the set (Rokeach 1973). From this basis, Schwartz (1992) developed a set of relations between 10 motivationally distinct value types; self-direction, stimulation, hedonism, achievement, power, security, conformity, tradition, benevolence and universalism. He further broadened the structure to four higher-order value types, which form two axes along which an individual's value priorities can be described; self-transcendence versus self-enhancement and conservation versus openness to change (Schwartz 1992).

The survey used in this study measures an individual's self-transcendence score (their willingness to put the needs of others over their own), and a conservation score (the extent to which the individual conforms to the standards or norms of society as opposed to following their own ideas and beliefs). The conservation score will be referred to in the remainder of this chapter as the 'conformance' score to avoid confusion with conservation in terms of environmentally-sustainable behaviour.

29.4 Environmental and Economic Efficiency: Results and Discussion

Users' economic and environmental results at the completion of each experiment were converted into a value between zero and one, to denote the proportion of the maximum available score that the user achieved. This provided a measure of how efficient, both economically and environmentally, the user was in the experiment. These two efficiency scores were combined to produce an overall efficiency score for each experimental mode, for each user. The economic and environmental efficiency scores for each user in the six experimental modes are shown in Fig. 29.3.

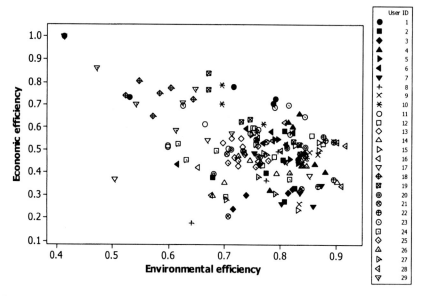

Fig. 29.3. Economic efficiency versus environmental efficiency per user

The graph demonstrates that among the 29 participants there is a wide range of land use choice behaviour in the virtual environment. However, while there is some variation, users generally tend to make land use choices that result in similar outcomes. User 18, for example, consistently made economically efficient land use choices in each of the six experimental modes. User 22, on the other hand, has tended to make more environmentally-efficient land use choices. It appears that although the configuration of the virtual environment does make some difference to subjects' land use choice behaviour, each individual has an overriding goal or preference for the desired economic and environmental consequences of their behaviour. The figure also shows that the highest economic efficiency could only be gained by sacrificing environmental outcomes. Overall efficiency can be interpreted as distance from the origin which means that many solutions were inferior (i.e. closer to the origin) through scoring less than possible on both economic and environmental outcomes.

29.4.1 Complexity

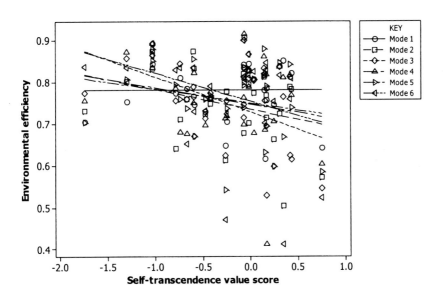

Fig. 29.4. Environmental efficiency versus self-transcendence value score

Figure 29.4 indicates that the complexity of the virtual environment has influenced subjects' behaviour. According to the Theory of Planned Behaviour (Ajzen 1991), an individual's behaviour is the result of both their attitudes towards the behaviour as well as the norms of society. Their more specific attitudes are formed based on their broader value priorities. From this premise, it was expected that a person with a high self-transcendence value score would aim to achieve a high environmental, rather than economic, efficiency. As Fig. 29.4 indicates, however, this was only found to be the case in the simplest configuration of the virtual environment

(Mode 1). In Mode 1, where there is no visual or social context provided, there is a slightly positive correlation between users' self-transcendence score and their environmental rating. Individuals with a higher self-transcendence score (indicating they are motivated by the welfare of others rather than by their own needs) made more environmentally-based land use choice decisions. In all other experimental modes, when users are provided with a three-dimensional visualisation of the landscape, and/or information about their neighbours' behaviour, the opposite trend occurs. This comparison indicates that when there is less complexity in the virtual environment, as in Mode 1, subjects' rely more on their value priorities to guide their decisions. When faced with increased complexity, as in modes 2 to 6, users' behaviour becomes less aligned to their stated value priorities. It appears that while the Theory of Planned Behaviour holds true in simple scenarios (e.g. Milfont and Gouveia 2006; Schultz and Zelzny 1998), when complexity is added it fails to adequately explain the resulting behaviour. Refinement of the theory is required to fully explain individuals' behaviour in this type of scenario. The fact that complexity in the virtual environment results in people behaving in ways that are inconsistent with their value priorities also has consequences for policy makers. Programs designed to change people's values may have limited impact on their behaviour, due to the complexity of the environment in which farmers operate.

29.4.2 Visualisation

Figure 29.5 and Fig. 29.6 demonstrate that adding the three-dimensional visualisation to the virtual environment induces little difference in the environmental and economic efficiency of subjects' land use choices whether the ABM is included (Fig. 29.5: Mode 2 versus Mode 4) or not (Fig. 29.6: Mode 1 versus Mode 3). This is not to say that the visualisation does not assist in conveying information to the users. As one participant noted after completing experiment 3 '…seeing the visualisation of my parcels on their own brought to light the lack of organisation in my farm. I had walnuts on both sides with grazing in the middle.' Similarly, another user commented that '…it was very useful having the visualisation of my choices as I was able to use it as a quick reference when investigating the enviro[nmental] and cost effects of my choices'. Users may have utilised the visualisation of the landscape to make their land use choices, even though its influence is not clear.

Land Use Decision Making in a Virtual Environment   603

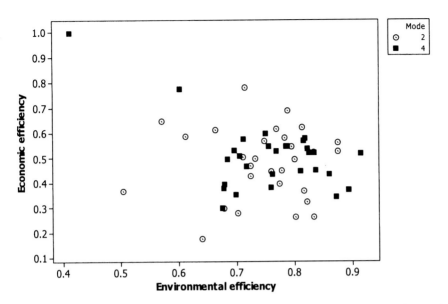

Fig. 29.5. Environmental efficiency versus economic efficiency in experimental Mode 2 (no three-dimensional visualisation, pro-environmental neighbours) and Mode 4 (three-dimensional visualisation, pro-environmental neighbours)

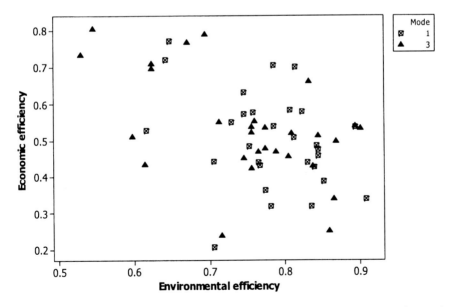

Fig. 29.6. Environmental efficiency versus economic efficiency in experimental Mode 1 (no three-dimensional visualisation, no neighbours) and Mode 3 (three-dimensional visualisation, no neighbours)

29.4.3 Social context (ABM)

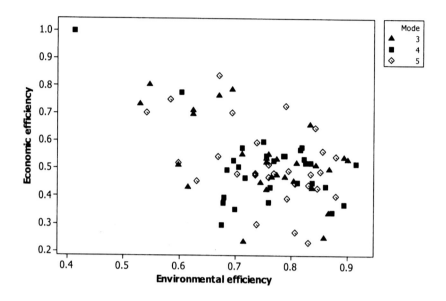

Fig. 29.7. Environmental efficiency versus economic efficiency in experimental Mode 3 (three-dimensional visualisation, no neighbours), Mode 4 (three-dimenional visualisation, pro-environmental neighbours), and Mode 5 (three-dimensional visualisation, economically-driven neighbours)

Figure 29.7 shows that when users are surrounded by environmentally-driven neighbours (Mode 4), compared to no neighbours (Mode 3), the results are more tightly clustered towards the environmentally-efficient end of the scale. When surrounded by economically driven neighbours (Mode 5), compared to no neighbours (Mode 3) however, there is no clear difference in the spread of results. This may indicate that people generally try to make money regardless of what their neighbours are doing, but they require peer guidance to induce them to behave in an environmentally-friendly manner.

While some inferences can be made based on the graph above, t-tests on the difference in efficiency between the three modes do not produce statistically significant results. The seemingly weak impact of the ABM on users' behaviour may be due to users' not building a sense of trust in their neighbours, and hence not following their behaviour. In the future, the virtual environment will be expanded to enable multiple players to play at once. If users can discuss options with some neighbours, and are unaware that other neighbours are computer-driven agents, they may feel they can trust the decisions of surrounding farmers, and possibly follow their lead more closely.

29.4.4 Value Priorities

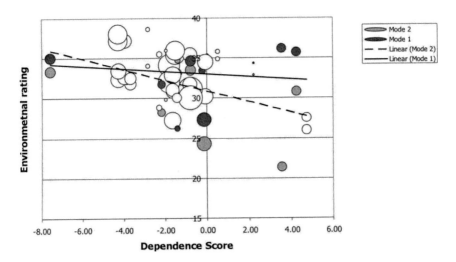

Fig. 29.8. Results for Mode 1 (no three-dimensional visualisation, no agents) and Mode 2 (no three-dimensional visualisation, pro-environmental agents). Shaded circles indicate positive self-transcendence score, open circles indicate negative self-transcendence score

It was anticipated that participants with a high conformance score (indicating that they prefer to follow other people) would follow the behaviour of their neighbours more closely, and subsequently have a higher environmental rating score in Mode 2. Figure 29.8 shows, however, that the opposite occurred. When users with higher conformance scores were surrounded by environmentally-friendly neighbours (Mode 2), they had lower environmental scores than when they had no neighbours at all (Mode 1). This was potentially due to users with a negative conformance score having a positive self-transcendence score, meaning that the pro-environmental behaviour of the neighbours was in line with their values anyway. However, as shown in Fig. 29.8 this was not the case, with the majority of users with a negative conformance score also having a negative self-transcendence score (shown as open circles on the graph).

29.4.5 Experience

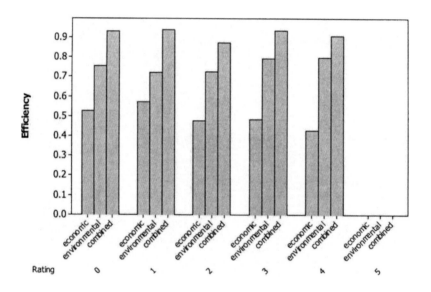

Fig. 29.9. Average economic, environmental and combined efficiency per rating for farming experience (0=no experience, 5=expert)

The histogram in Fig. 29.9 shows the average environmental, economic and combined efficiency scores according to the users' self-rated experience with, or knowledge of, farming. There is a slight decrease in economic performance as farming experience increases which is partly off-set by an increase in environmental performance. The slight trend shown is not statistically significant though, which indicates that using a random sample of participants, as opposed to a group of experienced farmers for example, should be sufficient for future studies using the virtual environment as a tool for investigating land use choice behaviour.

29.5 Conclusion

The study outlined in this chapter is an initial investigation into the use of a virtual environment as a tool for measuring land use choice behaviour. The diversity of the initial results suggested that tracking behaviour in hypothetical situations is more complex than was envisaged in the original experimental design. Nonetheless, we feel that the approach described here is the appropriate starting point for further assessment and development. Results which defy expectations raise a wide range of interesting questions. With further refinement the virtual environment could provide a useful laboratory environment in which economic, social and climate conditions can be varied, and individual responses to these changes can be measured.

The results of this study indicate that current models that describe the relationship between individuals' attitudes, values and behaviour — such as the Theory of Planned Behaviour — have limited application in this complex decision space. The creation of new theories to describe the link between people's values and their behaviour would be invaluable for rural land use policy makers. Improved knowledge of how people's values relate to their behaviour, for example, will allow policy makers to see whether attempts to change people's behaviour through shifting their values will lead to subsequent changes to their land use choice behaviour.

The use of the virtual environment to gain a better understanding of individual land use choice behaviour will provide data for the creation of more realistic agent-based models. These can then be used to better understand the impact of the accumulated individual decisions at a regional scale. By understanding and modelling individual land use behaviour, policy makers will be able to better understand likely consequences of different policies under different conditions with greater confidence in the outcomes, resulting in better informed and more effective policy decisions.

References

Ajzen I (1991) The theory of planned behaviour. Organisational Behavior and Human Decision Processes 50:179–211

Bardi A, Schwartz S (2003) Values and behaviour: strength and structure of relations. Personality and Psychology Bulletin 29(10):1207–1220

Bell P, Greene T, Fisher J, Baum A (1978) Environmental psychology. Harcourt College Publishers, Philadelphia

Chen AT, Stock C, Bishop ID, Pettit C (2007) Automated generation of enhanced virtual environment for collaborative decision making via a live link to GIS. Paper presented at Place and Purpose Conference, 30–31 May 2007, Bendigo, Australia

De Oliver M. (1999) Attitudes and inaction: a case study of the manifest demographics of urban water conservation. Environment and Behaviour 31:372

European Social Survey (2007) Retrieved 15 March 2007 http://www.europeansocialsurvey.org

Gardner GT, Stern PC (2002) Environmental problems and human behaviour. Custom Publishing, Boston

Gotts NM, Polhill JG, Law ANR (2003) Aspiration levels in a land use simulation. Cybernetics and Systems 34(8):663–683

Lindeman M, Verkasalo M (2005) Measuring values with the short Schwartz's value survey. Journal of Personality Assessment 85(2):170–178

Milfont TL, Gouveia VV (2006) Time perspective and values: an exploratory study of their relations to environmental attitudes. Journal of Environmental Psychology 26:72–82

Rokeach M (1973) The nature of human values. Free Press, New York

Schultz PW, Zelzny LC (1998) Values and pro-environmental behaviour: a five country survey. Journal of Cross-Cultural Psychology 29(4):540–559

Schwartz SH (1992) Universals in the content and structure of values. Advances in Experimental Social Psychology 25:1–65

Simon HA (1997) Models of bounded rationality: empirically grounded economic reason (vol 3). The MIT Press, London

Swait J, Adamowicz A (2001) Choice environment, market complexity, and consumer behaviour: a theoretical and empirical approach for incorporating decision complexity into models of consumer choice. Organisational Behaviour and Human Decision Processes 86(2):141–167

Index

2

2CSalt, 60, 61, 65

A

Ackoff's framework, 25
adaptive management, 31, 159–177
agent-based model, 463, 464, 595, 607
agricultural restructuring, 308, 310, 322
AMAP, 572
Analytical Hierarchy Process, 75, 84, 347, 526
ANNEX, 402
ANUClim, 56, 57, 108
APSIM, 420
ArcMap, 596
artificial intelligence, 557, 559, 561, 563
ASSESS, 76
Australian Greenhouse Office, 43
Australian Land Use Mapping, 113, 444

B

Barwon Heads visualisation tool, 491, 505
Base Flow Separation, 60
Bayesian Belief Network, 90, 181, 348
BC2C, 59, 61
Bet Bet Virtual Landscape, 537, 539, 541, 542
biodiversity, 121, 122, 128, 132, 135, 168–177, 183, 184, 197, 236–248, 279, 284, 410, 421, 471, 473, 478, 480, 483, 484
biolink, 470–483
BioMetric, 143–146, 151, 152
Bionatics, 482
bioregion, 283
Broadhectare Study, 100
Bushfire Rescue Game, 552, 553, 565, 567

C

Catchment Analysis Tool (CAT), 49–59, 69, 108, 109, 114, 422, 545
Catchment Management Framework (CMF), 108, 109
cellular automata, 353–356
Choice Theory, 329
CLASS, 54
climate change, 279, 414, 417, 420–425, 431, 478–484, 548
Climate Change, 60
cognitive map, 554, 555, 567, 568
community, 254–266, 269, 270
computer game
    Farcry, 559, 561
    Second Life, 559, 563, 565
    Trainz, 559
    Unreal Tournament 2004, 558, 559, 562, 563
Conversation Theory, 553, 555
Corporate Spatial Data Library (CSDL), 56, 108

D

Dapple Earth Explorer, 511–523

610 Landscape Analysis and Visualisation

Decision Aiding System (DAS), 328, 329, 333, 347
Decision Support System (DSS), 463
Decision Theory, 165, 166, 177
decision-making process, 22, 26, 33, 77, 79, 109, 159, 160, 368, 378, 381, 387, 505, 558, 594
DEFINITE, 75
deltatron, 355–357
digital elevation model (DEM), 55–57, 108, 146, 420, 536, 539, 574, 582
Driving Forces–Pressure–State–Impact–Response model (DPSIR), 279, 281, 299

E

E2, 54
eco–civic region, 266–269
ecological connectivity, 470–484
EcoTender, 108
emergency response simulation, 553, 555, 559
environmental indicator, 277, 280, 281
evaluation, 29–32, 42, 503
evidence-based policy, 29, 31, 36, 44
Extensible Markup Language (XML), 511, 517–520, 547, 575

F

FEARLUS, 595
FEFLOW, 54
fly-through, 462, 492, 532
future landscape, 470, 471, 474, 483
futures thinking, 408–410

G

gene flow, 212–219, 232
General System Theory, 280
generalised additive model (GAM), 146, 149, 150
genetic marker, 213, 216, 222

geodatabase, 575–579
groundwater flow system (GFS), 59, 65
geographic information system (GIS), 21, 74, 75, 98, 141–146, 150, 235, 237, 248, 436, 437, 479, 483, 484, 563, 573, 576, 582, 596
  ArcGIS, 75, 145, 462, 576–578, 584
  ArcGIS Explorer, 511–523, 528
  ArcInfo, 75, 76
  ArcView, 76
  GIWIN, 76
  IDRISI, 75, 76
  ILWIS, 76
  MapInfo, 76
geographic positioning system (GPS), 552, 567
global warming, 470, 483
Google, 41
Google Earth, 422, 483, 511, 515–519, 522–524, 534
governance, 102, 103
graphic user interface (GUI), 75, 445, 573, 576
GRASP, 146, 151, 152
Greenhouse in Agriculture, 38, 42, 43

H

habitat, 124–130, 133, 183, 185, 186, 188, 198–200, 237, 239, 243, 470, 473, 474, 477–481
habitat suitability, 124, 183, 185, 191, 194, 197
heads up display (HUD), 536, 538, 562
HERO, 76
human–environment system, 79, 91

I

IMAGIS, 572

indicator, 240, 245–247, 278–288, 294, 295, 296, 299–301, 306–310, 312, 315, 321–323, 408, 420–429

innovation diffusion theories, 115

Integrating Farming Systems into Landscapes, 35, 36, 38

International Panel on Climate Change (IPCC), 34, 409

IWM, 76

K

Keyhole Markup Language (KML), 511, 514, 517–524, 533, 546

knowledge hierarchy, 19, 20

L

land suitability analysis (LSA), 438, 442

land use

change, 66, 184, 198, 353–358, 362–365, 395, 401, 403, 439, 475, 484

change scenario, 436, 441, 447, 448, 451, 452

decision making, 592

Land Use Impact Model (LUIM), 114, 420, 545

land value, 295–297, 307, 310, 313, 317–320

Landsat TM, 106, 146, 149–152

landscape

connectivity, 237, 245–248

decision framework, 25

ecology, 237–240

futures analysis, 409, 416–422, 426, 431

genetics, 213, 214, 224

model, 475, 478, 482

mosaic, 538, 542

planning, 387, 388

process, 50, 68, 368

reconstruction, 183, 184, 200

restoration, 237, 239, 247

simulation, 572, 574, 580, 587

landscape analysis, 300

Coastal Landscape, 293–299

Semi-arid Landscape, 285–292, 299

Landscape Preferencing, 60

live link, 79, 571, 576

Land Management Advice System (LMAS), 76

Lower Murray Landscape Futures (LMLF), 408–412

M

MEACROSS, 76

migration, 306, 312, 316, 318

Millennium Ecosystem Assessment framework (MA), 279–285, 299, 300

mobile game, 553, 555, 557, 565–568

model

calibration, 53, 61, 65

uncertainty, 369, 371, 375, 377–379

validation, 68

MODFLOW, 53, 59, 60, 61

MODIS, 106

molecular

ecology, 213–215, 224

population biology, 212–226, 231

MULINO-DSS, 76

Multi-Criteria Analysis Shell for Spatial Decision Support (MCAS-S), 74, 79–83, 87, 89, 91

multi-way analysis, 80, 88, 89

two-way analysis, 80, 87, 88

N

NASA World Wind, 511, 512, 516–519, 521–523, 528, 534

navigation, 553

nested virtual worlds, 546, 547

nesting of regions, 259, 264, 265, 269, 270

next user, 29, 30, 36, 38

O

Organisational Innovation Process, 97, 100–105
Our Rural Landscape (ORL), 27, 545, 546

P

pair-wise comparison, 75, 84, 90
perception
  community, 386, 388, 401
PERFECT, 52
perspective, 121–124, 127–129, 133, 134
  allocentric, 556, 559, 564
  egocentric, 556, 561
place, 256, 259, 260, 269, 457–459, 462, 464, 466
Place and Purpose conference, 19
planning support system (PSS), 436, 437, 438, 441, 451, 452
Platform for Environmental Modelling Support (PEMS), 99–103, 107–115
  crop monitoring and forecasting, 105
  land use modelling, 105, 111, 113
  market-based approach, 105, 108
  wildfire planning, 105, 109–112
policy–science interface, 46
post-productivist transition, 305, 322
Preference Prediction, 331–335, 338, 342, 345, 347
  Santa Barbara case study, 334, 335, 347
Pressure–State–Response model (PSR), 279, 280, 299
Principal Components Analysis, 306, 311
priority-setting process, 77
process-based model, 26, 53
PROMETHEE, 76
Prospect–Refuge Theory, 123

R

Rational Choice Theory, 330, 593
Reef Water Quality Protection Plan (Reefplan), 384, 386, 403
regional investment strategy, 91
regional target, 414
remote sensing, 141, 142, 146, 148, 150–154
resource governance, 254–259, 262, 267–270
revegetation, 64–67, 73, 77, 82, 85, 91, 168, 170, 171, 183–186, 189–192, 194, 196–200, 475, 582
risk analysis, 378

S

Salinity Investment Framework 3, 42
Sawlogs for Salinity, 42
scenario, 20, 186, 198, 204, 226, 239, 244–248, 301, 327, 329–331, 334, 353, 354, 357, 389, 580–582, 592, 598, 602
  analysis, 52, 54, 63, 66, 67, 198, 237, 247, 408, 409, 415
  evaluation, 335
  land use change, 386–389, 393, 395, 398, 401
  merit rating, 332, 333, 336–343, 345, 346, 358
science–policy interface, 23, 32–35, 42–45
script editor, 558–561
Sediment River Network Model (SedNet), 390–393, 399–402
SHE, 53
SIEVE, 462–466, 573, 575, 577–579, 583, 595
  Builder, 574, 578
  Direct, 576, 577, 587
  Viewer, 574
sieve mapping, 100
SILO, 57
SimLab, 76

SIMLAND, 76
SIMPACT, 420
simulated virtual scenario, 585
SkylineGlobe, 511–513, 516–519, 521–523
Slope Land use Elevation Urbanisation Transportation and Hillshading (SLEUTH), 114, 353–355, 358–364, 545
Smart Forest, 572
social landscape, 306, 315–319, 321–323
    Amenity Farming Landscape, 318, 319
    High Amenity Landscape, 319, 439
    Intensive Agriculture Landscape, 319
    Production Landscape, 316
    Transitional Landscape, 317, 318
social learning, 387, 401
social–ecological framework, 387, 388, 401, 403
socioecological process, 301
Soil Health, 42, 43
soundscape, 542
space, 458, 462
spatial
    model, 19, 21, 151
    scale, 26, 34, 258, 374
Spatial Data Infrastructure (SDI), 98–101, 115, 572, 574
Spatial Decision Support System (SDSS), 328, 329, 347
SPOT5, 146, 149
Steinitz, 24
sustainability, 82, 86
SWAT, 54
Sydney Road prototype, 491, 503, 505
systems-based model, 368

T

temporal scale, 372
Terrain Analysis, 60

Theory of Planned Behaviour, 330, 331, 594, 601, 602, 607
Theory of Reasoned Action, 330
Theory of Social Marketing, 40
three-dimensional
    object library, 482, 482, 543, 544, 547
Torque Game Engine (TGE), 462, 534, 574, 577, 596
trade-off, 49, 73, 75, 77, 90, 387, 393, 395, 409, 415, 418, 425, 429, 431, 572
triple bottom line (TBL), 110, 114, 278, 587

U

uncertainty, 368–370, 372–375, 377–381
Unreal Game Engine, 534
urban and regional planning, 437

V

values and behaviour, 594, 607
Value Theory, 599
vegetation
    condition, 140–143, 149–152, 164, 168, 170–176
    mapping, 130
Victorian Greenhouse Office, 43
Victorian Resources Online (VRO), 37, 38, 301, 534, 540
virtual
    environment, 460–463, 465, 495, 581, 596
    globe, 510, 511
    landscape, 461, 476, 481, 483, 538–543, 547
    model, 492
virtual globe application
    crop production, 527
    eFarmer, 526–528
    Fishtrack, 526
    soil landform visualisation, 524
    Waterwatch, 527
Virtual Knowledge Arcade, 547

Virtual Knowledge World, 536
Virtual Reality Modelling Language
  (VRML), 490–495, 498, 499,
  502, 504–506, 535, 536, 541–
  543, 547
virtual world, 491, 535
  Activeworlds, 535
  Cybertown, 535
  Second Life, 535
visual globe application
  eFarmer, 527
visualisation, 459, 461–464, 479,
  483, 552, 555, 596
  augmented, 556
  building, 494
  computer game, 552, 567
  geographic, 471, 483, 547, 587
  historic, 505
  human perspective, 129
  landscape, 19, 29, 89, 122, 129,
    184, 197, 301, 388, 403, 458,
    471, 479, 480–484, 572, 573,
    595, 602

peri-urban, 491, 505, 560
spatial data, 510, 526, 528
three-dimensional, 436, 595–598
virtual environment, 573, 574

W

walk-through, 462, 492
wellbeing, 295
  community, 21, 30, 289, 299
  human, 280, 281, 285, 298, 300
What If?, 435, 437, 439, 441, 443,
  451, 452
What the City Might Be? prototype,
  491, 495, 505, 506
woodland birds, 183, 195
workshop, 40, 41, 87, 331, 338, 342,
  348, 389, 392–395, 399

Z

Zhang, 59, 61